Student Solutions Manual

Topics in Contemporary Mathematics

TENTH EDITION

Ignacio Bello
University of South Florida

Anton Kaul
Cal Poly, San Luis Obispo

Jack R. Britton
University of South Florida

Prepared by

Ann K. Ostberg
AIB College of Business/Grace University

CENGAGE Learning

Australia • Brazil • Mexico • Singapore • United Kingdom • United States

CENGAGE Learning

ISBN-13: 978-1-285-42074-5
ISBN-10: 1-285-42074-8

Cengage Learning
200 First Stamford Place, 4th Floor
Stamford, CT 06902
USA

Printed in the United States of America
1 2 3 4 5 6 7 17 16 15 14 13

Table of Contents

Chapter 1 Problem Solving

Section 1.1 Inductive Reasoning and Deductive Reasoning

Study Tips

Problem solving is at the heart of mathematics. So, do you know how to solve problems? This section provides a procedure to solve problems. Knowing that you have a procedure to follow will help in conquering your fear and/or disability in solving word problems.

Know the problem-solving procedure:
Think: **RSTUV**
Note: In short, **RSTUV** stands for
1. *Read the problem…*
2. *Select the unkown…*
3. *Think of a plan…*
4. *Use the techniques you are studying…*
5. *Verify your answer.*

Word of Advice
You may find it helpful to write these steps on a notecard for easy reference.

To understand inductive reasoning:
Think: repeated observations
Note: The conjecture (general conclusion) may or may not be true.

To understand deductive reasoning:
Think: general principles
Note: The specific conclusion will be true.

To solve application problems:
Think: RSTUV

Word of Advice
Be sure to read through the examples carefully. Your ability to problem solve can only be developed through practice. So, work as many problems as you can and you will find your ability to solve problems increase.

1.1 Exercises

1. **Step 1.** Understand the problem.
   Step 2. Devise a plan.
   Step 3. Carry out the plan.
   Step 4. Look back.

3. What does the problem ask for? (What is the unknown? After all, if you don't know the question, how can you find the answer?)

5. Look under the "Monthly Fee" column and look for a fee of $15.00. AT&T has the $15.00 per month plan.

7. If you send 900 text messages a month, the most economical plan would be Sprint (1000 messages). This plan would cost $10.00 per month.

9. For the average teenager sending 3339 messages per month, the best choice would be to have an unlimited plan. AT&T and Sprint both charge $20.00 per month, so either company would be the most economical.

11. To get the 2nd term (2), you added 1 to the 1st term.
    To get the 3rd term (4), you added 2 to the 2nd term.
    To get the 4th term (7), you added 3 to the 3rd term.
    To get the 5th term (11), you might add 4 to the 4th term.
    In general, the pattern to get the $(n + 1)^{st}$ term is: Add n to the nth term
    The 6th and 7th terms would be
    $11 + 5 = 16$ and $16 + 6 = 22$.

13. Note that the odd numbered terms are always 1's and the even numbered terms are multiples of 5. Thus, the 7th and 9th terms are 1's and the eighth term is the next multiple of 5 after 15, that is 20. Hence, the next three terms are 1, 20 and 1.

15. Going clockwise, the shaded region is moved 1 place, 2 places, 3 places and so on. The next three moves will move the shaded region 4, 5 and 6 places. The answer is shown.

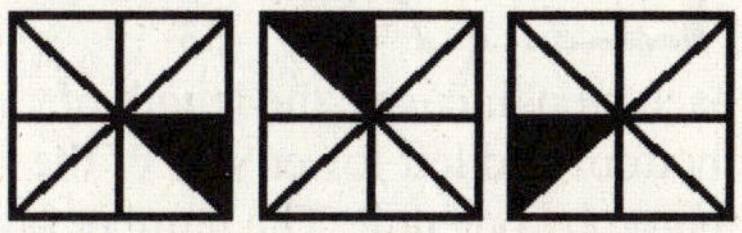

17. The numbers in the denominator are obtained by doubling. Thus, each term is half the preceding term. The next three terms are $\frac{1}{16}, \frac{1}{32}$, and $\frac{1}{64}$.

19. The odd numbered terms are 1, 2, 3 and we hypothesize 4, 5, 6, ... and the even

numbered terms are 5, 6 and we hypothesize 7, 8, 9, ... The next three terms are 7, 4, 8 as shown.
1, 5, 2, 6, 3, 7, 4, 8

21. a.

15

b. The rows are constructed by adding one more dot than on the preceding row. The next triangular numbers after 10 are $10 + \mathbf{5} = \underline{15}$ $15 + \mathbf{6} = \underline{21}$ and $21 + \mathbf{7} = \underline{28}$.

c. Following the pattern after the 7th triangular number which is 28, the 10th triangular number is: $28 + 8 + 9 + \mathbf{10} = \underline{55}$. (Problem 23 develops the formula for triangular numbers. You might want to give it a try.)

23. a. $1 + 2 + 3 + 4 + 5 + 6 + 7 + \mathbf{8} = \underline{36}$
b. 36
c. $1 + 2 + 3 + ... + \mathbf{12} = \underline{78}$
d. 78
e. $1 + 2 + 3 + ... + (n-1) + n = \dfrac{n(n+1)}{2}$
f. $\dfrac{100 \cdot 101}{2} = 50 \cdot 101 = 5050$

25. a.

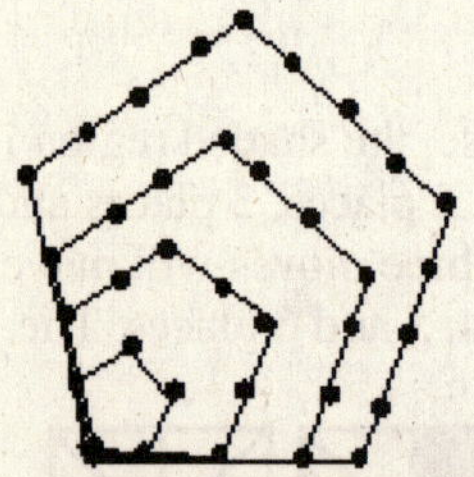

35

b. At each step, increase the length of the bottom and left lower side of the pentagon by one unit. The number of dots on each side is increased by one unit. (This is adding recursively: if $p(n)$ = the number of the nth pentagonal number, then $p(n+1) = p(n) + 3n + 1$.)

c. The 6th pentagonal number is 51.

27. Here is a summary of the information shown in the figure:

Sides	4	5	6	7
Diagonals	1	2	3	4

The number of diagonals is three less than the number of sides. Thus, $10 - 3 = 7$ diagonals can be drawn from one vertex of a decagon. (Geometrically , you connect the originating vertex with each of the others (yielding $n - 1$ connections) of which two are sides; leaving $n - 3$, which are diagonals.)

29. a.

Start	5	10	20	100
	12	17	27	107
	36	51	81	321
	30	45	75	315
	10	15	25	105
End	5	5	5	5

The final result is always 5.

b.

Start	n
	$n + 7$
	$3(n + 7) = 3n + 21$
	$3n + 21 - 6 = 3n + 15$
	$\dfrac{3n+15}{3} = n + 5$
End	$n + 5 - n = 5$

31. a.

Start	5	10	20	100
	10	15	25	105
	40	60	100	420
	20	30	50	210
End	10	10	10	10

The final result is always 10.

b.

Start	n
	$n + 5$
	$4(n + 5) = 4n + 20$
	$\dfrac{4n+20}{2} = 2n + 10$
End	$2n + 10 - 2n = 10$

33. a. It is always 4.
b. If you pick any number and follow the instructions you eventually get to a number less than or equal to 10. For any of these numbers the pattern leads to the number 4.

35. a. $(1+2+3+4)^2 = 1^3 + 2^3 + 3^3 + 4^3$
$(1+2+3+4+5)^2$
$= 1^3 + 2^3 + 3^3 + 4^3 + 5^3$
$(1+2+3+4+5+6)^2$
$= 1^3 + 2^3 + 3^3 + 4^3 + 5^3 + 6^3$
b. The square of the sum of the first n counting numbers equals the sum of the cubes of these numbers.

37. The number of units of length of the pendulum is always the square of the number of seconds in the time of the swing.

39. a. For each inch increase in foot length, the shoe size increases by 3: 12, 15, 18
b. We can see from the table that each unit increase in size corresponds to a $\frac{1}{3}$ of an inch increase in length. Thus, a 2 unit increase in size (from 6 to 8) corresponds to a $\frac{2}{3}$ in. increase in length, from 9 to $9\frac{2}{3}$ in.

41. Answers may vary.

43. Answers may vary.

45. The reasoning is going from general (average toilet uses 3 gallons per flush) to specific (your toilet uses 3 gallons per flush), so this is deductive reasoning.

47. The reasoning is going from general (save when computer is in "sleep mode") to specific (your computer is in "sleep mode"), so this is deductive reasoning.

Section 1.2 Estimation: A Problem-Solving Tool

Study Tips

One of the most powerful tools in solving problems is estimation. Estimation saves time and helps us understand if the answer we are getting is a reasonable one. (Remember the fifth step in the **RSTUV** procedure? Verify!)

To round numbers:
Think: if 5 or more, round up
Note: Be sure to identify the place to be rounded, then look to the digit to the right.

To estimate:
Think: rounding
Note: Estimates can vary based on what is required in the problem. Check with your instructor for clarification, if you are not sure.

Word of Advice
There are mathematical demands in adult life and estimation will be a useful tool as you read your electric meter or estimate your taxes. Pay close attention to the examples given in the text.

1.2 Exercises

1. 416.38 rounded to the nearest 100 is 400. \$30.28 rounded to the nearest dollar is \$30. Thus, a reasonable estimate of the value of the investor's stock is $400 \cdot \$30 =$ \$12,000.

3. \$7.80 → \$8.00
\$2.29 → \$2.00
\$3.75 → \$4.00
\$1.85 → \$2.00
\$2.90 → \$3.00
Estimate: \$19.00

5. $6 \cdot 150 = 900$ gallons is a good estimate.

7. a. $\dfrac{4256}{14{,}053} \approx 0.303$ (to 3 decimal places)
b. $\dfrac{4300}{14{,}100} \approx 0.305$ (to 3 decimal places)

9. $\dfrac{9 \cdot 14}{140} = \dfrac{9}{10} = 0.900$

11. It takes 4 to 8 hors d'oeuvres per person, so for 100 persons it takes 100 times as much, that is, 400 to 800 hors d'oeuvres.

13. It takes $\frac{1}{3}$ lb of boneless meat or fish per person, so for 100 persons it takes 100 times as much or $100 \cdot \dfrac{1}{3} = 33\dfrac{1}{3}$ lb. To the nearest pound, 33 lb are needed.

15. It takes $\frac{1}{4}$ lb of raw pasta per person, so for 100 persons it takes 100 times as much or $100 \cdot \frac{1}{4} = 25$ lb of pasta.

17. a. The reading is: 5 1 8 2
b. $5182 - 5102 = 80$ KWH
c. $\$0.08 \cdot 80 = \$6.40$
d. $30 \cdot \$6.40 = \$192.00$

19. a. The reading is: 7 0 0 1
b. $7001 - 6951 = 50$ KWH
c. $\$0.08 \cdot 50 = \$4.00$
d. $30 \cdot \$4 = \$120.00$

21. $\$4681.25 + 0.25(\$60{,}000 - \$34{,}000)$
$= \$4681.25 + 0.25(\$26{,}000)$
$= \$4681.25 + \$6500.00$
$= \$11{,}181.25$

23. a. $H = 2.89 \cdot 15 + 27.81$
$= 43.35 + 27.81 = 71.16$ in.
b. Rounded to the nearest whole number,
$H = 3h + 28$
When $h = 15$, $H = 3 \cdot 15 + 28 = 73$
The difference is
$73 - 71.16 = 1.84$ in.

25. $\text{BMI} = \frac{705W}{H^2} = \frac{705 \cdot 150}{68^2} \approx 22.87$ (to the nearest hundredth). Since 22.87 is greater than 18 and less than 24, the person's BMI is normal.

27. $W = \frac{G^2 \cdot L}{330} = \frac{70^2 \cdot 66}{330} = 980$ pounds
Measured in 100 pound units, this is 9.8 or about 10 to make it safe. The horse needs 0.6 gal of water, 1 lb of hay and $\frac{1}{2}$ lb of grain for each 100 lb of body weight. Thus, the horse needs:
$0.6 \cdot 10 = 6$ gal of water
$1 \cdot 10 = 10$ lb of hay
$\frac{1}{2} \cdot 10 = 5$ lb of grain

29. $C = 596 + 0.0019V + 21.7A$
$= 596 + 0.0019 \cdot 500{,}300 + 21.7 \cdot 5$
$= 596 + 950.57 + 108.50$
$\approx \$1655$

31. a. For the first 2 years, we have 24 human years. For the next 3 years, we have $3 \cdot 4 = 12$ human years. Thus, for $2 + 3 = 5$ years, we have $24 + 12 = 36$ human years.
b. For the first 2 years, we have 24 human years. For the next 8 years, we have $8 \cdot 4 = 32$ human years. Thus, for $2 + 3 = 5$ years, we have $24 + 32 = 56$ human years.

33. The distance between the intersection of 90 and 128 to the intersection of 90 and 495 is about one inch on the map or about 15 actual miles.

35. The distance between the intersection of 90 and 290 to the intersection of 90 and 86 is about one inch on the map or about 15 actual miles.

37. The distance between the intersection of 90 and 32 to the intersection of 90 and 91 is about 1.5 in on the map or $1.5 \cdot 15 = 22.5$ miles.

39. The distance in problem 34 is 22.5 miles. If the car makes 20 miles per gallon, it would need $\frac{22.5}{20}$ gallons that would cost
$\$4.00 \cdot \frac{22.5}{20} = \$4.50.$

41. The distance in problem 36 is $1\frac{1}{4}$ in. on the map or 18.75 miles. If the car makes 20 miles per gallon, it would need $\frac{18.75}{20}$ gallons that would cost
$\$4.00 \cdot \frac{18.75}{20} = \$3.75.$

43. A gray whale consumes $\frac{268{,}000}{30} \approx$ 8933.33 lb of amphipods per day.

45. A killer whale eats $\frac{14{,}000}{30} \approx 466.67$ pounds of herring per day.

47. For a 150 lb male: Multiply the body weight 150 by 10 which is 1500; Add twice the body weight or 300; The BMR is $1500 + 300 = 1800$.

49. Answers may vary.

51. $23\text{ft} \cdot 5.5\text{ mi} \cdot \dfrac{5280\text{ft}}{1\text{ mi}} \cdot 2 = 1{,}335{,}840\text{ft}^2$

Section 1.3 Graph Interpretation: A Problem-Solving Tool

Study Tips

Developing your problem-solving skills continues with learning how to interpret a graph. This section will discuss three types of graphs:

Circle graph (pie chart)
Bar graph
Line graph

The focus will be on interpreting the graphs. You will not have to construct the graphs, they will be provided for you. Your task is simply to interpret them. Take time to carefully read the examples.

1.3 Exercises

1. a. The largest category of materials is the one that covers the *most area*, that is, Paper. You can reach the same conclusion by noting that Paper is the category with the highest percent (41%).
   b. The smallest category of materials is the one that covers the *least area,* that is, Plastic (7%).
   c. Since 1000 tons of materials are sent to the landfill each day, the percent of each category times 1000 will be the number of tons of that category going to the landfill each day. Paper (41%) will have $0.41 \cdot 1000 = 410$ tons. Plastic (7%) will have $0.07 \cdot 1000 = 70$ tons.

3. a. The cheese produced the most is Cheddar (36%).
   b. The cheese produced the least is Swiss (2.8%).
   c. The second most popular cheese is Mozzarella (30.6%).

5. a. Water is used the most for bathing (30%).
   b. 30% of $500 = 0.30 \cdot 500 = 150$ gal used for bathing.
   c. The dishwasher uses 3% of the water and the toilet leak uses 5%, thus the toilet leak uses more water.
   d. The dishwasher uses 3% of the water, which represents 5 gallons. The faucet uses 12% of the water, which is 4 times as much, that is, the faucet uses $4 \cdot 5 = 20$ gal of water.

7. a. Paper (40%) is the most prevalent item in average trash.
   b. Yard trimmings (18%) is the second most prevalent.
   c. It would contain 40% of $50 = 0.40 \cdot 50$ or 20 lb of paper; it would also contain 18% of $50 = 0.18 \cdot 50$ or 9 lb of yard trimmings

9. a. Oil (33%) produces the most energy.
   b. Nuclear (5%) produces the least energy.
   c. Natural gas (18%) produces the least energy.

11. a. "No drinks per day" means "None" and the bar representing "None" is about 114 units long (actually, it is 114.4). Thus, the systolic blood pressure for young adults consuming no drinks per day is 114.4.
    b. The bar representing < 1/day is a little less than the bar of part (a), so the approximate answer is 114.
    c. The bar corresponding to 1–< 2/day is about 111.2 long. (Answers may vary.)
    d. The lowest blood pressure corresponds to the category 2–< 3/day (110 systolic blood pressure).
    e. The longest bar (highest blood pressure) corresponds to > 3/day (almost 120 systolic blood pressure).

13. a. The number of fatalities with a negative blood alcohol level was 39.
    b. Blood alcohol levels (BAL) .10–.19, .20–.29, and .30+ define a legally drunk person. There are $13 + 14 + 2 = 29$ people who were legally drunk.
    c. The most prevalent BAL for those legally

drunk was .20– .29. Fourteen (14) people had that BAL.

15. a. The longest bar represents the age group with most of the fatalities. This is the 20–29 age group with 40 fatalities.
b. The shortest bar represents the age group with the least fatalities. This is the 13–15 age group with 1 fatality.
c. The number of fatalities involving people who are less than 50 years old is $5 + 1 + 31 + 40 + 22 + 27 = 126$. For more than 50 years old: $28 + 9 + 19 + 17 + 2 = 75$. More fatalities involve people less than 50 years old.
d. The age group 90+ had only two fatalities. Answers may vary. The number in that age group is small compared to the other groups.

17. a. 85% of the 3000 adults owned a cell phone, which is $0.85 \cdot 3000 = 2550$ adults.
b. 59% of the 3000 adults owned a desktop computer, which is $0.59 \cdot 3000 = 1770$ adults.
c. 52% of the 3000 adults owned a cell phone, which is $0.52 \cdot 3000 = 1560$ adults.
d. The difference between the number of people that owned a laptop and those that owned a desktop computer is $1770 - 1560 = 210$ people.

19. a. The Chocolate Fudge ice cream is about 260 calories ($\frac{1}{2}$ cup) or 520 per cup.
The Chocolate Fudge yogurt is about 180 calories ($\frac{1}{2}$ cup) or 360 per cup.
The difference (per cup) is $520 - 360 = 160$ calories.
b. 2 cups of the Chocolate Fudge ice cream has $2 \cdot 520 = 1040$ calories.
2 cups of the Chocolate Fudge yogurt has $2 \cdot 360 = 720$ calories.
The difference is $1040 - 720 = 320$ calories. (Or simply double the answer in part a.)

21. a. The longest bar represents the most popular item sold, Cuban toast.
b. The second longest bar represents the second most popular item, Cheese toast.
c. $60 + 20 + 10 + 15 = 105$ breakfasts.
Since each breakfast uses $\frac{1}{4}$ of a loaf of bread, we need $\frac{1}{4} \cdot 105 = 26\frac{1}{4}$ loaves, or 27 loaves, of Cuban bread.

23. a. The horizontal bar corresponding to the password **123456** represents approximately 3100 persons (go down to the horizontal axis to estimate).
b. The horizontal bar corresponding to the password **consumer** represents approximately 250 persons.
c. The horizontal bar corresponding to the password **lifehack** represents approximately 700 persons.

25. a. From 1960, move up to the Total population line (red), from that point move to the left to estimate about 175,000,000.
b. From 1970, move up to the Total population line (red), from that point move to the left to estimate about 200,000,000.
c. From 1990, move up to the Total population line (red), from that point move to the left to estimate about 250,000,000.
d. From 2010, move up to the Total population line (red), from that point move to the left to estimate about 300,000,000.
e. From 2020, move up to the Total population line (red), from that point move to the left to estimate about 325,000,000.

27. a. From 1°C, move to the right to the line (red), from that point move down to read the year 2040.
b. From 2.5°C, move to the right to the line (red), from that point move down to read the year 2100.

29. a. Go to 9 in human years, move up to the line (red), then to the left to 60 dog years.
b. Go to 65 in dog years, move to the right to the line (red), then down to about 10 human years.
c. Go to 21 in dog years, move to the right to the line (red), then down to about 2 human years.

31. a. Go to 4 months badger age, move up to the line, then to the left to about 3 kg.
    b. Go to 8 months badger age, move up to the line, then to the left to about 8 kg.
    c. When the line becomes nearly horizontal, it represents when badgers stop growing. This is at about 9 months.

33. Go to 6 on the horizontal axis and up until you meet the lower curve (red), from that point move to the left to the vertical axis to estimate the balance to be $570.

35. Go to 48 on the horizontal axis and up until you meet the top curve (blue) , from that point move to the left to the vertical axis to estimate the balance to be $300.

37. Answers will vary.

39. Answers will vary.

41. a. Look for the years where the blue line (placebo group) is below the red line (medicine group) on the graph titled Heart Attack. This occurs during years 1–7.
    b. Look for the years where the blue line (placebo group) is below the red line (medicine group) on the graph titled Stroke. This occurs during years 2–7.
    c. Look for the years where the blue line (placebo group) is below the red line (medicine group) on the graph titled Blood Clots. This occurs during years 1–7.
    d. Look for the years where the blue line (placebo group) is below the red line (medicine group) on the graph titled Breast Cancer. This occurs during years 4–7.
    e. Look for the graph which has the greatest distance vertically between the blue line (placebo group) and the red line (medicine group). This occurs at year 7 of either the breast cancer or the stroke group. The numerical difference is about 0.005.
    f. Look for the graph where the red line (medicine group) is above the blue line (placebo group) for the most years. The condition where the medicine group fared better than the placebo group was breast cancer. This occurred during years 0–4.

Chapter 1 Practice Test

Study Tips

Attitude is extremely important when perfoming mathematics. Keep a positive attitude as you prepare and when you take your exam.

It is important to prepare properly. Find out what type of problems will be on your actual text, and how long you will have to take it. Take the practice test that is in your text using the time limit set by your instructor. Find your weaknesses and remedy them before your actual test. Remember that the best way to study for a mathematics test is to practice, practice, practice! We will have more test taking tips later. Here are the solutions to the practice test.

1. **R**ead the problem
   Select the unknown
   Think of a plan
   Use the techniques you are studying to carry out the plan
   Verify the answer

2. Inductive reasoning is the process of arriving at a general conclusion on the basis of repeated observations of specific examples.

3. Look at the difference between successive terms as shown

```
        1    2    7    19    41    76
Diff      1    5    12    22    35
Diff         4    7    10    13
Diff            3    3    3
```

The third differences are constant (3), so the next number can be constructed by addition. Add the last diagonal from bottom to top. We obtain the next number in the pattern,
$3 + 13 + 35 + 76 = 127.$

```
     1    2    7    19    41    76    127
Diff 1    5    12    22    35    51
Diff    4    7    10    13    16
Diff       3    3    3    3
```

Now, we can use the 127 to continue the last three rows as shown. The next term now is

3 + 16 + 51 + 127 = 197. If you do this one more time, you will find the next term to be 289. Once you get four terms you can show that all the following terms can be obtained from the formula:

$$a_{n+1} = 3 + 3a_n - 3a_{n-1} + a_{n-2}$$

Thus, the next three terms after 76 are 127, 197, 289.

4. a. Select a number: n
   Multiply by 4: $4n$
   Add 6 to the product: $4n + 6$
   Divide the sum by 2:
   $$\frac{4n+6}{2} = 2n + 3$$
   Subtract 3 from the quotient:
   $$2n + 3 - 3 = 2n$$
   b. Using 1, the final result is 2.
   Using 10, the final result is 20.
   Using 100, the final result is 200.
   c. The conjecture is that the final result is twice the original number.

5. a. 319.<u>2</u>6 → 319.3
   Since the 6 after the 2 is greater than 5, add one to the 2.
   b. <u>3</u>19.26 → 300
   Since the 1 after the 3 is less than 5, leave the 3 alone; add 0's.

6. a. 6064 KWH
   b. 6064 – 6002 = 62 KWH
   c. $\$0.10 \cdot 62 = \$6.20$
   d. $30 \cdot \$6.20 = \$186.00$

7. a. Female: $H = 28.6 + 2.5(15) \approx 66$ in. (rounded from 66.1)
   b. Male: $H = 32.2 + 2.4(15) \approx 68$ in. (rounded from 68.2)

8. a. Public Safety
   b. 1/2
   c. Sales tax

9. a. About \$50 million
   b. About \$52 million
   c. About \$116 million
   d. About \$75 million
   e. About (\$75 – \$50) million = \$25 million

10. a. About 6.91%
    b. About 6.76%
    c. About 6.91 – 6.76 = 0.15%
    d. They seem to be decreasing.

Chapter 2 Sets

Section 2.1 Sets: A Problem-Solving Tool

Study Tips

While your instructor will most likely want you to get the right answer, it is important to understand what you are doing. In this section, sets will be used as another problem-solving tool. In mathematics, the word "set" is an undefined term, that is, a term used without specific mathematical definition and so we are limited to describing the properties of a set.

Well-defined set: stay away from subjective words like "good," "bad," and "beautiful.

Set membership: this is describing the set by the following methods. (See table on page 48.)

a. *Verbal or written description*: writing the set in words
b. *Roster method*: listing the elements separated by commas between braces
c. *Set-builder notation*: using a property to define the set

Caution
The empty set is denoted by $\varnothing$ or { }, but not {$\varnothing$}.

You might want to remember the following:

1. *Equality of sets*: same elements but do not have to be in the same order
2. *Subset*: a set is a subset of another if every element in the set is also in the other set, or if there is no element in the set that is not an element in the other set
3. *Proper subset*: all the subsets of a set except the subset that is equal to the set
4. *Universal set*: all the elements under discussion
5. *Number of subsets of a set*: determine the number of elements in the set (n) and then calculate 2^n, which will be the number of subsets
6. *Empty set*: { } or $\varnothing$; the empty set is a subset of any set.

2.1 Exercises

1. People do not agree on the meaning of "grouchy," so this description does not define a set.

3. A set

5. A set

7. People do not agree on what is "good," so this description does not define a set.

9. a. *Incorrect.* The letter D is not an element of A.
   b. *Correct.* Desi is an element of A.
   c. *Incorrect.* Jane is an element of A, not the other way around.
   d. *Correct.* The letter D is not an element of A.
   e. *Incorrect.* Jane is an element of A

11. x is an element of the set X. Fill the blank with $\in$.

13. A is not an element of the set X. Fill the blank with $\notin$.

15. 2 is an element of the set $\{2,4,6,8\}$. The statement is True.

17. 0 is not an element of the set **N** (the set of counting numbers). The statement is False.

19. 5 is an element of the set **N** (the set of counting numbers) and is odd. The statement is True.

21. The set consisting of the first and last letters of the English alphabet.

23. The set consisting of the names of the first biblical man and woman.

25. The set of counting numbers from 1 to 7. Note that the numbers do not have to be in any specific order.

27. The set of odd counting numbers from 1 to 51.

29. The set of hours slept per day by 15- to 19-year-old women.

31. {Dioxin, Xylene} is the set of compounds that were found in everybody's tissue.

33. {1, 2, 3, 4, 5, 6, 7}

35. {0, 1, 2, 3, 4, 5, 6, 7}

37. The word "between" is to be taken literally. The 3 and the 8 are not elements of this set. The answer is {4, 5, 6, 7}.

39. Identify the set of generations that greater than 50% go online wirelessly. These generations are Millennial, Gen X, and Younger Boomers. The set is {*M*, *X*, *Y*}.

41. Identify the set of generations where the members are younger than 65. These generations are Millennial, Gen X, Younger Boomers, and Older Boomers. The set is {*M*, *X*, *Y*, *O*}.

43. The ranking numbers of the cars made by Volkswagen corresponds to {4, 6}.

45. The ranking numbers of the cars with manual transmissions corresponds to {4, 6, 9, 10}.

47. The ranking number of the car that has the highest 150 MPG corresponds to {1}.

49. The ranking number of the car that makes more than 50 highway MPG corresponds to {1}.

51. The set A = {4, 8, 12, 16, ...} and the set B = {2, 4, 6, 8, ...}, so these sets are not equal ($A \neq B$).

53. The set A is the empty set and the set B is the empty set. Both A and B are empty, so these sets are equal ($A = B$).

55. a. Every element of *A* is an element of *B*, and every element of *B* is an element of *A*. Thus, $A = B$.
    b. Because 0 is an element of *C* but not of *A*, $A \neq C$ is correct.
    c. Because 0 is an element of *C* but not of *B*, $B \neq C$ is correct.

57. ∅, {a}, {b}, {a, b}; the first three of these are proper subsets of the given set.

59. ∅, {1}, {2}, {3}, {4}, {1, 2}, {1, 3}, {1, 4}, {2, 3}, {2, 4}, {3, 4}, {1, 2, 3}, {1, 2, 4}, {1, 3, 4}, {2, 3, 4}, {1, 2, 3, 4}; all but {1, 2, 3, 4} are proper subsets of the given set.

61. ∅, {1}, {2}, {1, 2}; the first three of these are proper subsets of the given set.

63. Since there are 4 elements in this set, there are 2^4 or 16 subsets.

65. There are 10 elements in A, so A has 2^{10} or 1024 subsets.

67. Note that $32 = 2^5$, so there are 5 elements.

69. Since $64 = 2^6$, there are 6 elements.

71. Yes. Since ∅ has no elements, there is no element of ∅ that is not in ∅. Furthermore, every set is a subset of itself.

73. $B \subseteq A$ because every counting number that is divisible by 4 is also divisible by 2. (A is not a subset of B because a number can be divisible by 2 and not by 4. For instance, 6 is divisible by 2, but not by 4.)

75. a. There are 5 toppings, so you have 5 choices.
    b. There are 10 subsets with 2 elements in each. So you have 10 choices. (Assume no double toppings. Try writing these out.)
    c. You can choose which two toppings you don't want. So the answer is 10 as in (b). (Assume that a topping cannot be used more than once on a single pizza.)

77. Since $256 = 2^8$, you would need 8 different condiments.

79. Answers will vary.

81. Answers will vary.
    a. The empty set ∅ does not contain any elements, so it is true that $\emptyset \notin \emptyset$.
    b. The set {∅} contains the element ∅, so it is true that $\emptyset \in \{\emptyset\}$.
    c. The set {0} contains the element 0, so it is true that $\emptyset \neq \{0\}$.
    d. { } is another way to express the empty set ∅.

83. a. If $g \in S$, then Gepetto shaves himself, which contradicts the statement that Gepetto shaves all those men and only those men of the village who do not shave themselves. Hence, $g \notin S$.
    b. If $g \in D$, then Gepetto does not shave himself, and so by the same statement, he does shave himself. Thus, there is again a contradiction and $g \notin D$.

85. The word "*non-self-descriptive*" cannot be classified in either way without having a contradiction. If it is an element of *S*, then it is a self-descriptive word, which contradicts the definition, "*non-self-descriptive is a non-self-descriptive word.*" On the other hand if non-self-descriptive is a non-self-descriptive word, then it is an element of *S*, which is again a contradiction.

Section 2.2 Set Operations

Study Tips

With the goal of being able to solve application problems, you have to learn how to find the following with sets:

Intersection of sets *A* and *B*:
Think: all the elements that are in *A* and also in *B*

Union of two sets:
Think: the elements that can be in *A* or in *B* or in both

Difference of sets *A* and *B*, (*A* – *B*):
Think: the elements have to be in *A* but not in *B*

Complement of a set:
Think: all the elements that are in the universal set *U* and not in *A*, that is, the difference of *U* and *A*, $U - A$

2.2 Exercises

1. a. $A \cap B$, the set of all elements in both *A* and *B*, is {1, 3, 4}.
   b. $A \cap C$, the set of all elements in both *A* and *C*, is {1}.
   c. $B \cap C$, the set of all elements in both *B* and *C*, is {1, 6}.

3. a. $A \cap (B \cup C)$, the set of all elements in both *A* and $B \cup C$, is {1, 3, 4}.
   b. $A \cup (B \cap C)$, the set of all elements in *A* or in $B \cap C$, is {1, 2, 3, 4, 5, 6}.

5. $A \cup (B \cup C)$, the set of all elements in *A* or in $B \cup C$, is {1, 2, 3, 4, 5, 6, 7}.

7. $A \cap (B \cap C)$, the set of all elements in *A* and in $B \cap C$, is {1}.

9. a. $A \cap B$, the set of all elements in both *A* and *B*, is {c}.
   b. $A \cap C$, the set of all elements in both *A* and *C*, is ∅. Note: the set {*a*, *b*} is an element of *A*, but *a* and *b*, separately, are not elements of *A*.

11. a. *Correct.* The set {*b*} is a subset of the set $A \cap B$.
    b. *Incorrect.* The set {b} is not an element of the set $A \cap B$.

13. a. *Correct.* The set {a, b, c} is a subset of the set $A \cup B$.
    b. *Correct.* The set {a, b, c} is an element of the set *A*, so it is an element of the set $A \cup B$.

15. a. A', the set of elements in *U* but not in *A*, is {*b*, *d*, *f*}.
    b. B', the set of elements in *U* but not in *B*, is {*a*, *c*}.

17. a. $(A \cup B)'$, the set of elements in *U* but not in $A \cup B$, is ∅. Note that $A \cup B$ includes all the elements in *U*.
    b. $A' \cup B'$, the set of elements in A' or in B' is {*a*, *b*, *c*, *d*, *f*}.

19. a. $(A \cap B) \cup C'$, the set of elements in $A \cap B$ or in C', is {*c*, *e*}. Note that $A \cap B$ is {*e*}.
    b. $C \cup (A \cap B)'$, the set of elements in *C* or not in $A \cap B$ is {*a*, *b*, *c*, *d*, *f*}.

21. a. $A' \cap B$, the set of elements not in A but in B, is $\{b, d, f\}$. Note that A' is $\{b, d, f\}$.
    b. $A \cap B'$, the set of elements in A but not in B, is $\{a, c\}$. Note that B' is $\{a, c\}$.

23. a. The elements in C' are c and e and those in $(A \cap B)'$ are $a, b, c, d,$ and f. Thus, $C' \cup (A \cap B)' = \{a, b, c, d, e, f\}$.
    b. $C' = \{c, e\}$, $A \cup B = U$, so $(A \cup B)' = \varnothing$. Thus, $C' \cup (A \cup B)' = \{c, e\}$.

25. a. This is the set U with the elements in A taken out, that is, $\{b, d, f\}$.
    b. This is the set U with the elements in B taken out, that is, $\{a, c\}$.

27. a. This is the set of elements not in B. The answer is $\{2, 3\}$.
    b. This is the set U with the elements in B taken out. The answer is the same as in part a, $\{2, 3\}$.

29. $\varnothing'$, the set of all elements in U that are not in the empty set, is U.

31. The set of all elements that are in both A and the empty set is $\varnothing$.

33. This is the set of all elements that are in both A and U, that is, A.

35. This is the set of elements that are both in A and not in A, that is, $\varnothing$.

37. This is a double negative; the elements that are not in "not A" are, of course, in A. Thus, the answer is A.

39. To include A, you must have the elements 1, 2, 3. Then, to include B, you must have the element 4, and to include C, you must have the element 5. Thus, the smallest set that can be used for U is $\{1, 2, 3, 4, 5\}$.

41. The set of traits that are only mentioned once are {Beauty, Consideration, Kindliness, Friendliness, Helpfulness, Loyalty}.

43. This is the set of traits that are in both M_w and M_m, so the answer is {Intelligence, Cheerfulness, Congeniality}.

45. This is the set of traits that are in all four of the sets, {Intelligence, Cheerfulness}.

47. The traits only mentioned once are {Is aware of others, Follows up on action}.

49. S_1' would be all elements in the universal set that are not in set S_1. The set S_1' is {Follows up on action}.

51. Refer to the diagram. The set of all five categories is {p, n, a, t, l}.

53. C' is a set of all elements in the set of all five categories {p, n, a, t, l} that are not in set C {n, t, l}. Thus, C' is {p, a}.

55. The set of categories H that account for the highest portion of your score would be payment history (35%). Set H = {p}.

57. H' is the set of elements in {p, n, a, t, l} that are not in H. Thus, H' is {n, a, t, l}.

59. $(H \cup C)'$, the set of elements not in either H or in C, is {a}.

61. $(H \cap C)'$ is the set of elements in {p, n, a, t, l} that are not in $H \cap C$. Thus, $(H \cap C)'$ is {p, n, a, t, l}.

63. Compare both lists to find that the set of names that appear only once is {Blogspot.com, Baidu.com, Qq.com, MSN.com, MySpace.com, Blogger.com}.

65. M', the set of all names on the second list that are not on the first list, is {MSN.com, MySpace.com, Blogger.com}.

67. Look for 2.4 as the first number in the second column, Eng Size/Cylinders. The set with a 2.4-liter engine is {Toyota}.

69. Look for the highest mpg as the first number in the third column, MPG/City/Hwy. The set with the highest fuel economy in the city is {Toyota (Prius)}.

71. Look for the highest mpg as the second number in the third column, MPG/City/Hwy. The set with the highest fuel economy in the city is {Toyota (Prius)}.

73. Look for the smallest first number in the second column, Eng Size/Cylinders. The set with the smallest engine is {Toyota (Prius)}.

75. Look for the highest number in the fourth column, Annual Fuel Cost. The set with the highest annual fuel cost is {BMW, Lexus}.

77. Answers will vary. *A* and *B* have no elements in common. (Set examples may vary: *A* may be a set of cats and *B* a set of dogs.)

79. Answers may vary. All the elements of *A* are elements of *B*, and all the elements of *B* are elements of *A* ($A = B$). (Set examples may vary: the same set defined two different ways.)

81. a. This is the set of characteristics that are in both columns of the table: {Long tongue, Skin-covered horns, Native to Africa}
    b. The same answer as in part a.
    c. This is the set of characteristics that occur in either column of the table: {Tall, Short, Long neck, Short Neck, Long tongue, Skin-coverd horns, Native to Africa}
    d. G' = {Short, Short neck}
    e. O' = {Tall, Long neck}

82. $M \cap D$, the set of males who are 35 years or older, is the first item in the 35 and older row: 3,293,000 elements.

83. $F \cap D$, the set of females who are 35 years or older, is the second item in the 35 and older row: 685,000 elements.

85. Look for the smallest number in either the second or third columns. This is the set of 12-17 year old females: $F \cap A$.

87. There are no persons that are both male and female. This set is empty.

89. $F \cap B$, the set of females with a bachelor's degree, is the second item in the Bachelor's degree row: $60,293.

91. $A \cap F$, the set of females with an associate degree, is the second item in the Associate degree row: $39,935.

93. Average earnings of males with a High School degree; $43,493.

Section 2.3 Venn Diagrams

Study Tips

It has been said that a picture is worth a thousand words. A *Venn diagram* is a picture of the relationship between sets. Each set is represented as sets of points inside a circle.

Procedures for creating a Venn diagram with set A and set B.

Intersection of two sets: draw vertical lines inside of *A* and horizontal lines inside of *B* and where the lines *intersect* is the intersection. If there is no intersection the sets are called *disjoint*.

Union of two sets: draw vertical lines inside of *A* and continue drawing vertical lines inside *B*. Wherever there are vertical lines represents the union of the sets.

Complement of set *A*: represent the universal set *U* by a rectangle outside the circles *A* and *B*. The complement of *A* consists of all points in *U* *not* in *A*.

Important use of Venn diagrams: These pictures can be used to verify the *equality* of certain sets.

Caution

Sometimes students get confused by the placing of the name of the set on the diagram. Both diagrams on the next page show a set *A* inside a universal set *U*. It does not matter if the set *A* is labeled outside or inside the circle. In both cases, we have a set *A* inside a universal set *U*.

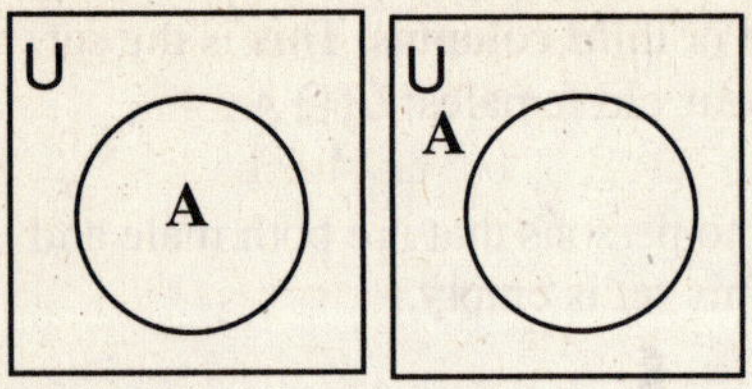

2.3 Exercises

1. **Step 1**: Draw a diagram with two circles and label the regions *w*, *x*, *y*, *z*, as shown.

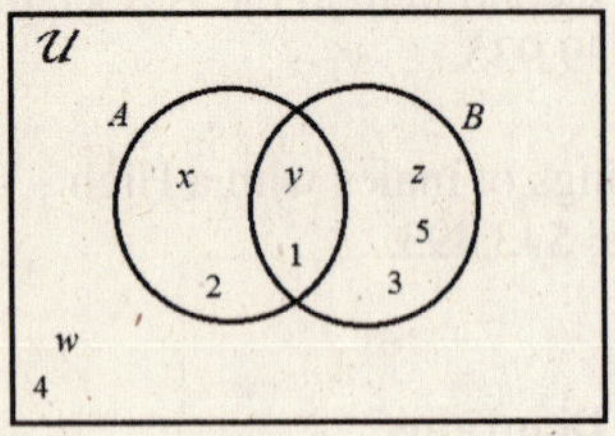

Step 2: Select the element that is in both *A* and *B*, the number 1. Write 1 in region *y*.
Step 3: Select the element that is in *A* but not in *B*, the number 2. Write 2 in region *x*.
Step 4: Select the elements that are in *B* but not in *A*, the numbers 3 and 5. Write 3 and 5 in region *z*.
Step 5: Select the element that is not in *A* or *B*, the number 4, and write 4 in region *w*. This completes the diagram.

3. Draw a Venn diagram labeled as in the figure. Then look at the regions corresponding to the various sets. *A*: regions 1, 3; B' (regions outside of *B*) regions 1, 4; $A \cap B'$ (regions in both *A* and B'): region 1. Shade region 1 for the desired diagram.

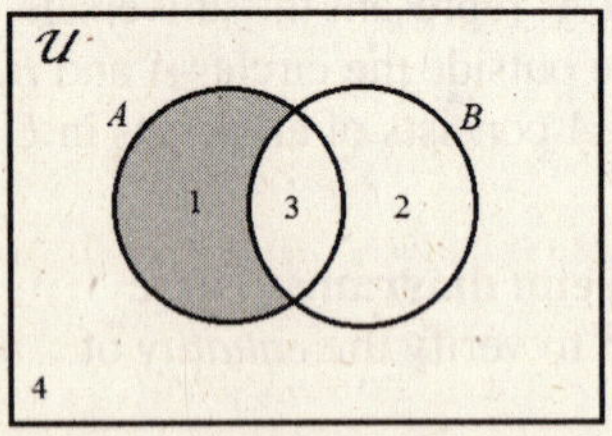

5. Draw a Venn diagram labeled as in the figure.Then find the regions corresponding to the various sets. $A \cup B$ (regions in *A* or *B*): regions 1, 2, 3; $A \cap B$ (regions in both *A* and *B*): region 3; $(A \cup B)-(A \cap B)$: Take the region in $A \cap B$ away from the regions in $A \cup B$, leaving regions 1, 2. Shade regions 1, 2 for the desired diagram.

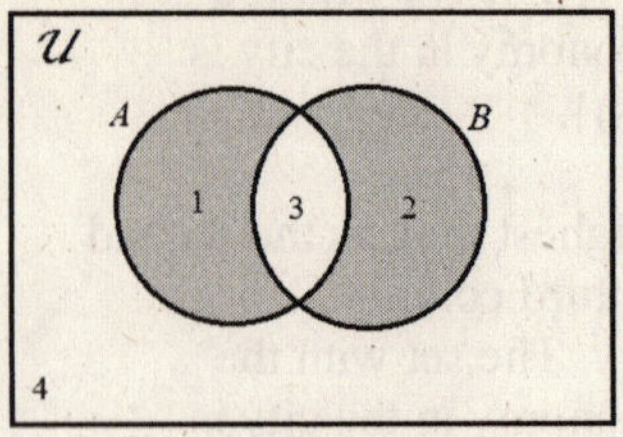

7. Draw a Venn diagram labeled as in the figure. The region corresponding to A' is the entire area outside of circle *A*, regions 2, 4. The region corresponding to B' is the entire area outside of circle *B*, regions 1, 4. Thus, the region corresponding to $A' \cap B'$ is the region outside both circles, region 4.

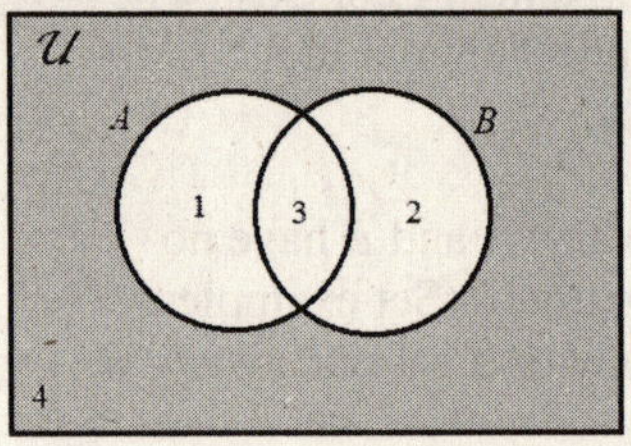

Refer to the following diagram for solutions to excersises 9 – 15.

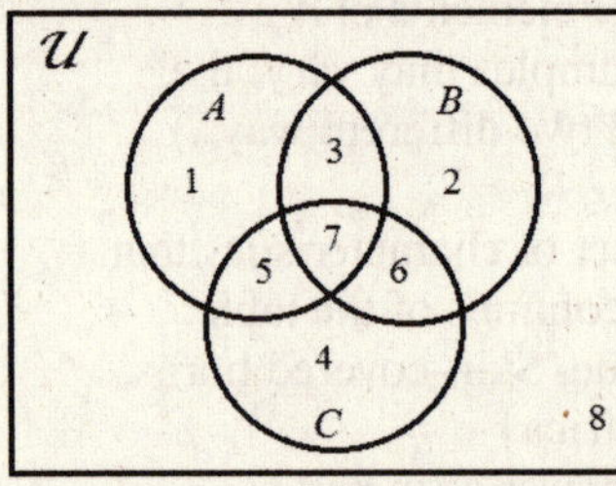

9. *A*: regions 1, 3, 5, 7; $B \cup C$ (regions in *B* or *C*): regions 2, 3, 4, 5, 6, 7; $A-(B \cup C)$: Take the regions common to *A* and $B \cup C$ away from the regions in *A*, leaving the answer: region 1.

11. $A \cap B \cap C$ (regions common to *A*, *B*, *C*): region 7; $A \cap B$ (regions in both *A* and *B*): regions 3, 7; $(A \cap B \cap C)-(A \cap B)$: Take the regions common to $A \cap B \cap C$ and $A \cap B$ away from those in $A \cap B \cap C$. This leaves no regions, so we get the empty set, $\varnothing$.

13. $A \cup B'$ (regions in *A* or outside of *B*): regions 1, 3, 4, 5, 7, 8; *C*: regions 4, 5, 6, 7; $(A \cup B') \cap C$ (regions in both $A \cup B'$ and *C*): regions 4, 5, 7.

15. $A \cap B'$ (regions in *A* and not in *B*): regions 1, 5. $(A \cap B') \cup C$ (regions in either $A \cap B'$ or *C*): regions 1, 4, 5, 6, and 7.

17. This consists of the region that is outside all three of *A*, *B*, *C*: region 8.

19. This includes all the elements that are outside of *A* or outside of *B*. Thus, everything except the region common to the two circles is shaded.

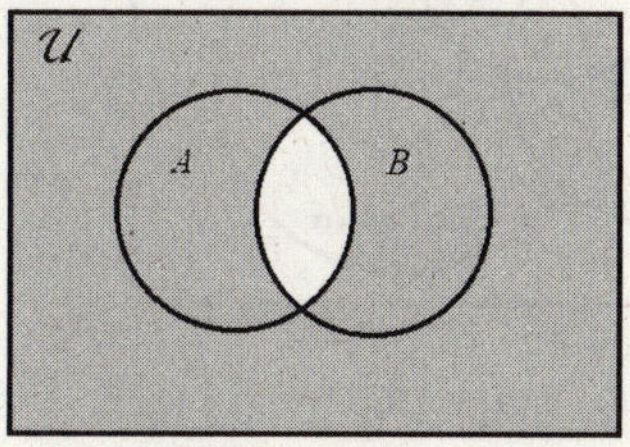

21. This includes all the elements that are in *A* and also in *B*, but not in *C*. Thus, the region that is common to *A* and *B* but is outside of *C* is shaded.

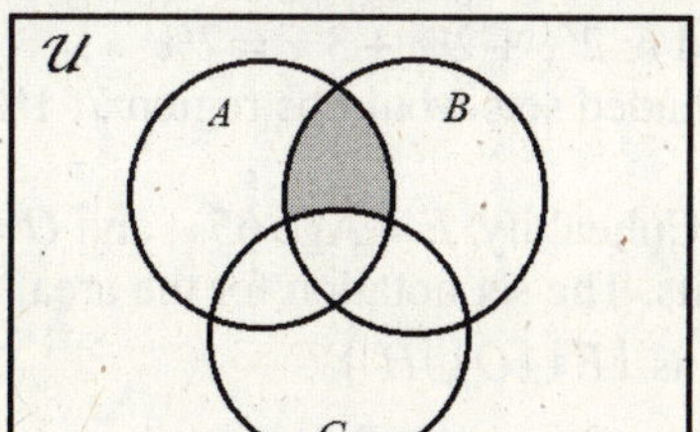

23. This includes all the elements of *B* that are not elements of *A*. Thus, the region in *B* that is outside of *A* is shaded.

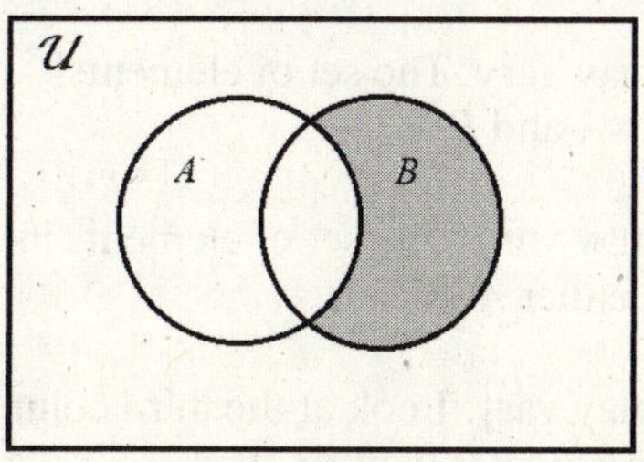

25. This includes all the elements that are in both *B* and *C*, but not in *A*. Thus, the region common to *B* and *C*, but outside of *A* is shaded.

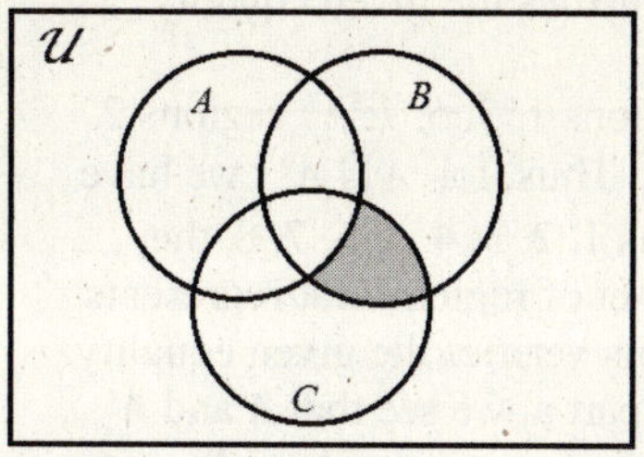

27. Because the intersection of *A* and *B* is given equal to *B*, all of *B* is contained in *A*, as shown in the diagram.

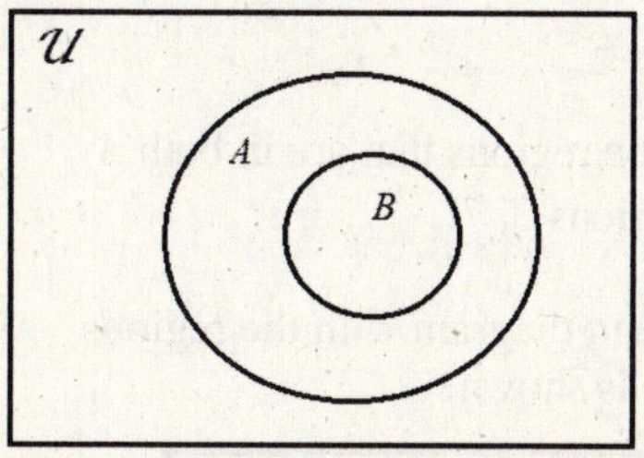

29. Note that $A \cap B$ contains all the elements common to *A* and *B*, so that $A \cap (A \cap B)$ is the same as $A \cap B$. Because $A \cap (A \cap B) = A$, all the elements of *A* must be in *B*. Thus, the circle *A* was drawn inside the circle *B*.

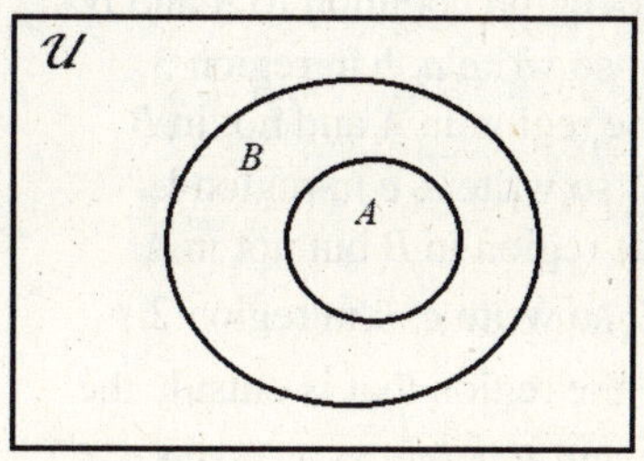

31. a. *A*: regions 1, 3, 5, 7; $B \cup C$: regions 2, 3, 4, 5, 6, 7. Thus, for $A \cup (B \cup C)$, we have regions 1, 2, 3, 4, 5, 6, 7. Similarly, we have the following correspondences. $A \cup B$: regions 1, 2, 3, 5, 6, 7; *C*: regions 4, 5, 6, 7. So for $(A \cup B) \cup C$, we have regions 1, 2, 3, 4, 5, 6, 7. This verifies the given equality.

b. *A*: regions 1, 3, 5, 7; $B \cap C$: regions 6, 7. Thus, for $A \cap (B \cap C)$, we have region 7. Also, $A \cap B$: regions 3, 7; C: regions 4, 5, 6, 7. Thus, for $(A \cap B) \cap C$, we have region 7. This verifies the given equality.

33. a. A: regions 1, 3, 5, 7: A': regions 2, 4, 6, 8. Thus, for $A \cup A'$, we have regions 1, 2, 3, 4, 5, 6, 7, 8, the same set of regions that represents U. This verifies the given equality.
    b. From part a, we see that A and A' have no elements in common. Therefore, $A \cap A' = \varnothing$.
    c. $A - B$: regions 1, 5; $A \cap B'$: regions 1, 5. This verifies the equality.

35. $A \cap B$ (The regions that are in both A and B): regions 3, 7.

37. Draw a Venn diagram with the regions numbered as shown.

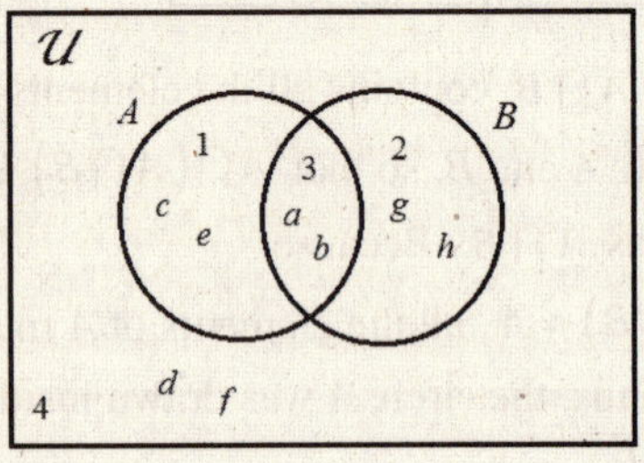

$A \cap B$: the region common to A and B is region 3, so write a, b in region 3.
$A \cap B'$: the region in A and not in B is region 1, so write c, e in region 1.
$A' \cap B$: the region in B but not in A is region 2, so write g, h in region 2.
$(A \cup B)'$: the region that is outside the union of A and B is region 4. Write d, f in region 4. Now, you can read off each of the required sets:
    a. $A = \{a, b, c, e\}$, $B = \{a, b, g, h\}$, $U = \{a, b, c, d, e, f, g, h\}$
    b. $A \cup B = \{a, b, c, e, g, h\}$
    c. $(A \cap B)' = \{c, d, e, f, g, h\}$

39.

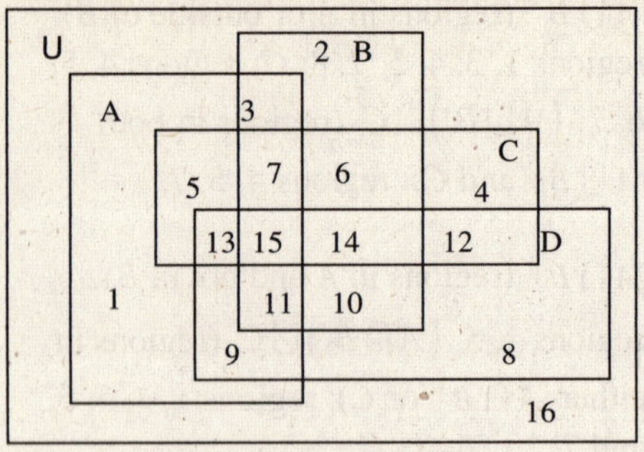

Refer to the following diagram from Example 6, with regions numbered for clarity, for solutions to exercises 41 – 43.

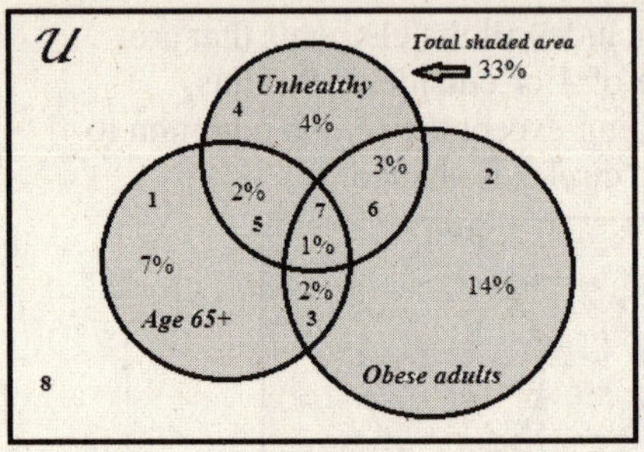

41. a. Obese but not unhealthy would be regions 2 and 3: 14% + 2% = 16%
    b. Obese, not unhealthy, and under 65 would be region 2: 14%

43. a. In two shaded sets only would be regions 3, 5, and 6: 2% + 2% + 3% = 7%
    b. In all shaded sets would be region 7: 1%

45. Let H' = Unhealthy, E = Age 65+, and O = Obese adults. The set notation for the area not shaded is $(E \cup O \cup H')'$.

47. The number one Web site in set E and in set R is Facebook.

49. Answers will vary. 51. Answers will vary.

53. Answers may vary. The set of elements common to A and B.

55. Answers may vary. The set of elements in U and not in either A or C.

57. Answers may vary. Look at the third column for numbers greater than 83. The corresponding countries with females life expectancy greater than 83 would be Switzerland (S) and France (F). Set $A = \{S, F\}$. To determine set B, look at the

fourth column for numbers greater than 78. The corresponding countries with males life expectancy greater than 78 would be Switzerland (S) and Canada (C). Set $B = \{S, C\}$.

59. Answers may vary. Look at the third column for numbers less than 83. The corresponding countries with females life expectancy less than 83 would be USA (U) and Malta (M). Set $A = \{U, M\}$. To determine set B, look at the fourth column for numbers less than 78. The corresponding countries with males life expectancy less than 78 would be USA (U) and Luxenbourg (L). Set $B = \{U, L\}$.

61. An AB^+ person has all three antigens and thus may receive blood from any donor.

63. No, because the B^- person does not have the A antigen, which the AB^- does have.

65. No, because the O^- person does not have the Rh antigen, which the O^+ does have.

67. Add all the percents of people with the A antigen: $34 + 6 + 3 + 1 = 44\%$

69. Add all the percents of people without A or B antigens (that is, O type blood): $38 + 7 = 45\%$

71. Note the pattern that is developed in the text before the problem. We see that with one set, we have $2^1 = 2$ regions; with two sets, we have $2^2 = 4$ regions; and with three sets, $2^3 = 8$ regions. So, for n sets, we will have 2^n regions. Thus, four sets will have $2^4 = 16$ regions in the Venn diagram.

73. a. This requires everything common to A, B, and D, that is not in C. Thus, the answer is region 11.
    b. This requires everything that is outside of the union of A, B, and C. Thus, the answer is regions 8 and 16.

Section 2.4 The Number of Elements in a Set: A Problem-Solving Tool

Study Tips

Any natural number, $\{1, 2, 3, \ldots\}$, can be called a cardinal number when used for counting or showing how many. Thus, if you want to find the number of elements in set A, we are finding the *cardinal number of A*.

To find the cardinal number of A:
Think: How many elements are there in A? This can be done by simply counting the elements in the set.

Meaning of $n(A)$:
Think: Notation that "asks" you to determine the number of elements in the set, that is, the cardinal number.

To find the cardinal number of the union of A and B:
Think: Do the sets intersect?
If YES, you must be sure to not count an element twice. To find the number of elements in the union, you need the formula

$$n(A \cup B) = n(A) + n(B) - n(A \cap B).$$

If NO, the sets do not intersect, then the cardinal number is the sum of the elements in the two sets. You may use the formula

$$n(A \cup B) = n(A) + n(B)$$

Caution
When sets intersect, after adding all the elements in both sets, do not forget to subtract the number of elements in the intersection of the sets.

Application problems involving surveys:
Think: Venn diagram. Draw circles representing the sets, and be sure to label the region of the intersection first.

2.4 Exercises

1. Use Equation (1):
   $$n(A \cup B) = n(A) + n(B) - n(A \cap B).$$
   Then, since $n(A) = 15$, $n(B) = 20$, and $n(A \cap B) = 5$, it follows that
   $$n(A \cup B) = 15 + 20 - 5 = 30.$$

3. Use the same equation as in problem 1. Thus, $n(A) = 15$, $n(A \cap B) = 5$, and $n(A \cup B) = 30$, which we substitute into Equation (1) to get

$$30 = 15 + n(B) - 5$$
$$30 = 10 + n(B)$$
$$n(B) = 20$$

5. Let T be the set of families who subscribe to *Time* and N be the set who subscribe to *Newsweek*. Since 100 families were sur-veyed and 10 subscribe to neither maga-zine, 90 subscribe to one or both. Hence, we can use Equation (1) in the form:
$$n(T \cup N) = n(T) + n(N) - n(T \cap N)$$
With $n(T \cup N) = 90, n(T) = 75,$ and $n(N) = 55$, we get the equation
$$90 = 75 + 55 - n(T \cap N)$$
$$90 = 130 - n(T \cap N)$$
$$n(T \cap N) = 40$$
which says that 40 families subscribe to both.

7. To do this problem, first draw a Venn diagram. Let C be the set of students taking Chemistry, E be the set taking English, and M the set taking Mathematics. Draw the diagram with three circles as shown.

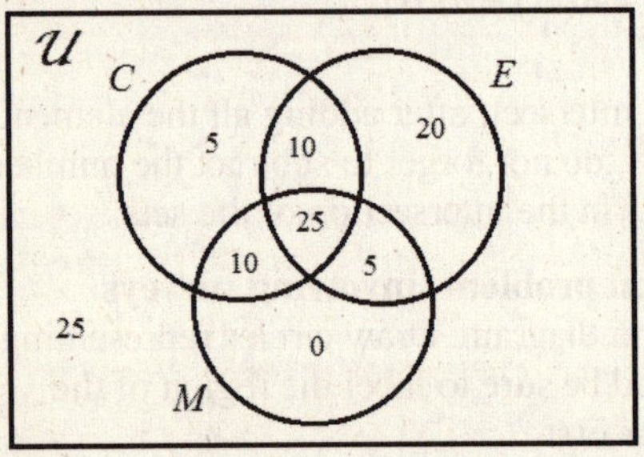

Start at the end of the given listing to fill in the proper numbers in the various regions. Because 25 are taking all three courses, write 25 in the region common to all three circles. Since 35 are taking Math and Chem, and we have accounted for 25 of these, write 10 in the remainder of the region common to M and C. Likewise, 35 are taking English and Chem, and we have accounted for 25 of these, write 10 in the remainder of the region common to E and C. Similarly, because 30 are taking English and Math, and we have accounted for 25 of these, write 5 in the remainder of the region common to E and M. Then, because 50 are taking Chem, and we have accounted for 10 + 25 + 10 = 45 of these, write 5 in the region of C that is outside of E and M. There are 40 in Math, and we have accounted for 5 + 10 + 25 = 40, so write 0 in the region of M that is outside of E and C. Since there are 60 taking English, and we have accounted for 5 + 10 + 25 = 40 of these, write 20 in the region of E that is outside of C and M. Up to this point, we have accounted for 20 + 5 + 5 + 10 + 10 + 25 = 75 students. Because 100 students were surveyed, write 25 in the region outside of the three circles. This completes the diagram and we can now read off the answers.

a. No students were taking Math and neither Chem nor English. (See the region of M that is outside of C and E.)
b. There were 10 students taking both Math and Chem, but not English. (See the region that is inside both M and C, but is outside of E.)
c. There were 10 students taking both English and Chem but not Math. (See the region inside both E and C, but outside of M.)

9. a. From the public 2-year college row, find the largest number, and then identify the corresponding column. The largest expense is Room and Board at \$7259. The total cost is the sum of all the expenses:
2713 + 7259 + 1135 + 1491 + 2041 = \$14,637.
b. From the public 4-year college (in-state) row, find the largest number, and then identify the corresponding column. The largest expense is Room and Board at \$8535. The total cost is the sum of all the expenses:
7605 + 8535 + 1137 + 1073 + 1959 = \$20,339.
c. From the public 4-year college (out-state) row, find the largest number, and then identify the corresponding column. The largest expense is Tuition and Fees at \$19,595. The total cost is the sum of all the expenses:
19,595 + 8535 + 1137 + 1073 + 1959 = \$32,329.

11. a. From the horizontal bar representing a public 2-year college, find the smallest number, and then identify the corresponding color with the legend. The least expense is Books and Supplies at \$1133.
b. From the horizontal bar representing a public 4-year college (in-state), find the

smallest number, and then identify the corresponding color with the legend. The least expense is Transportation at $1073.

c. From the horizontal bar representing a public 4-year college (out-state), find the smallest number, and then identify the corresponding color with the legend. The least expense is Transportation at $1073.

13. a. $C \cap M$, the set of cell phone (*C*) users who are Millennials (*M*), is the first item in the cell phone row: 95%

b. $D \cap X$, the set of desktop computer (*D*) users who are Gen X (*X*), is the second item in the desktop computer row: 69%

c. $L \cap Y$, the set of laptop computer (*L*) users who are Younger Baby Boomers (*Y*), is the third item in the laptop computer row: 49%

d. $X \cap I$, the set of Gen X (*X*) who are iPod/MP3 player (*I*) users, is the fourth item in the Gen X column: 56%

e. $Y \cap G$, the set of Younger Baby Boomers (*Y*) who are game console (*G*) users, is the fifth item in the Younger Baby Boomers column: 38%

15. Let *O* stand for onions, *M* for mustard, and *C* for catsup. Draw a Venn diagram as shown.

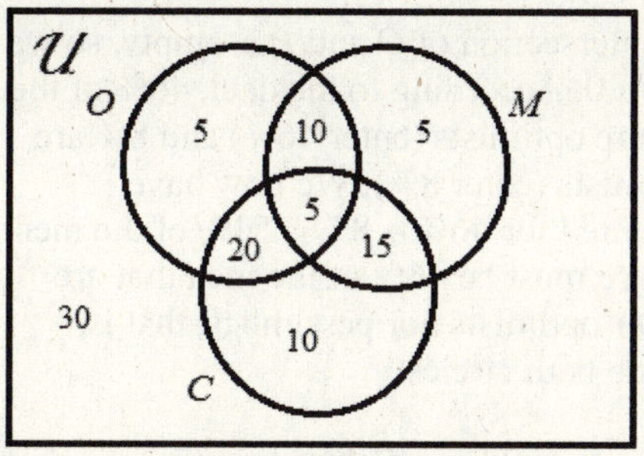

The numbers are obtained as follows: Start at the end of the list. Since 5 had all three condiments, write 5 in the region common to the three circles. Because 25 had onions and catsup, the region common to *O* and *C* must have a total of 25. We have already put 5 into this region, so 20 goes into the remainder of the region. Next, we see that 20 had mustard and catsup, so the region common to *M* and *C* must have a total of 20. Since 5 have already been put in this region, 15 must be put in the remainder of the region. Since 15 had onion and mustard, the region common to *O* and *M* must have a total of 15. Because 5 have already been put in this region, 10 must be in the remainder of the region. Now we see that 50 had catsup, so circle *C* must contain a total of 50. The diagram shows that we already have put 5 + 15 + 20 = 40 in *C*, so 10 must be put in the remainder of the region. Next, we see that 35 had mustard. Thus, circle *M* must have a total of 35. Since we have entered 5 + 15 + 10 = 30 in this circle, 5 must go into the remainder of this circle. Since 40 had onions, and the diagram shows that 5 + 10 + 20 = 35 have already been put in circle *O*, an additional 5 must go in this circle. Now, the diagram shows that 70 persons have been put into the three circles. Since 100 were surveyed, we must put 30 into the rectangle outside of the three circles. This completes the diagram and we can read off the answers to the questions.

a. 5 (See the region inside *O* and outside *M* and *C*.)

b. 30 (See the region in the rectangle outside of the three circles.)

c. 5 + 5 + 10 = 20 (See the non overlapping regions of the three circles.)

17. Let *G* be the set of people who liked ground coffee, and *I* be the set of people who liked instant coffee. Then we can use Equation (1) in the form

$n(G \cup I) = n(G) + n(I) - n(G \cap I)$ to find how many of those surveyed liked coffee (either ground or instant or both). We were given $n(G) = 200, n(I) = 270$, and

$n(G \cap I) = 70$, so that

$n(G \cup I) = 200 + 270 - 70 = 400$.

Since 50 people did not like coffee, 450 people were surveyed, and the company had to pay out $450.

19. The Venn diagram for the data reported in problem 18 is shown below.

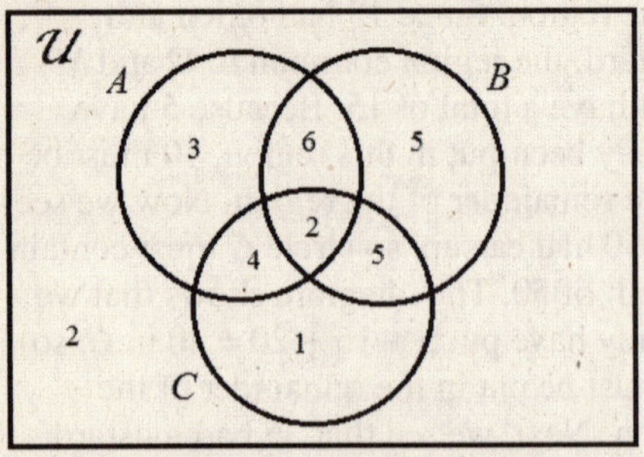

The sum of the numbers in the diagram is 28, which is the correct number of persons interviewed to give this data.

21. a. $n(A)$ is the sum of the numbers in A, $70 + 50 = 120$.
 b. $n(C)$ is the sum of the numbers in C, $30 + 50 = 80$.
 c. $n(A \cap C)$ is the number in the over-lapping portion of the two circles, 50.

23. a. $n(A')$ is the number of students in U and not in A, which is $200 - (70 + 50) = 80$.
 b. $n(C')$ is the number of students in U and not in C, which is $200 - (50 + 30) = 120$.
 c. $n(A' \cap C')$ is the number of students outside of both A and C, that is, $200 - (70 + 50 + 30) = 50$.

25. a. This is the sum of the numbers in E and T that are not in the region common to these two circles, $35 + 20 + 12 + 6 = 73$.
 b. This is the sum of the numbers in E and T that are not in M, $35 + 8 + 12 = 55$.
 c. This is the sum of all the numbers in the diagram except the number in the region common to all three circles. Thus, we get $35 + 8 + 12 + 20 + 6 + 10 = 91$.
 d. This is the sum of the numbers in all the regions common to at least two of the circles, $8 + 6 + 20 + 4 = 38$.
 e. Since the sum of all the numbers in the diagram is 95 and 100 persons were interviewed, 5 persons must have had none of these three types of investment.

27. Draw a diagram with two circles labeled C for cake and F for frosting.

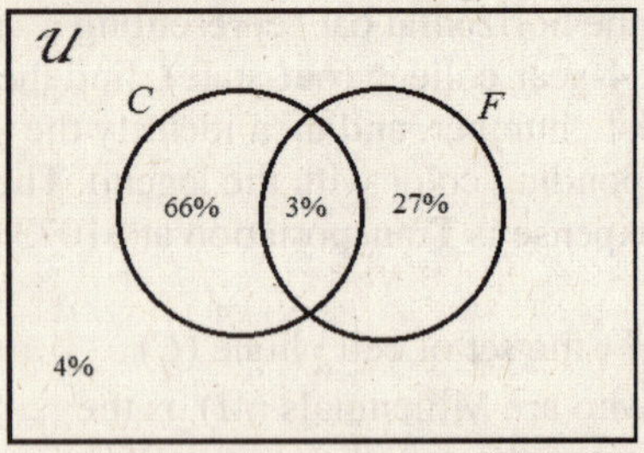

Since 3% eat them together, place 3% on the region common to C and F. The circle labeled C must contain 69% of the people, so enter 66% (69% – 3%) in the other region of C. The set F has 30% of the people, so we enter 27% (30% – 3%) on the other region of F. We now have accounted for 66% + 3% + 27% = 96% of the people, so 4% must be outside both circles and 27% + 4% = 31% are outside C. Thus, 31% do not eat cake.

29. Draw a diagram with two circles labeled O for optimist and P for pessimist.

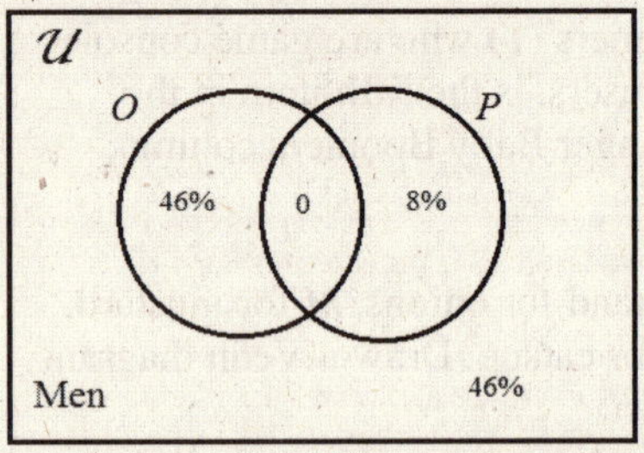

The intersection of O and P is empty, so we enter a 0. According to the data, 46% of the men are optimists (enter 46%) and 8% are pessimists (enter 8%). We now have accounted for 46% + 8% = 54% of the men, so there must be 46% of the men that are neither optimists nor pessimists, that is, outside both circles.

31. Answers may vary. False. A counterexample is $A = \{1, 2\}$ and $B = \{m, n\}$.

33. Answers may vary. False. A counterexample is $A = \{1, 2\}$ and $B = \{1, 2, 3\}$.

35. The Venn diagram shows that with the added information, the statistics in the cartoon are possible.

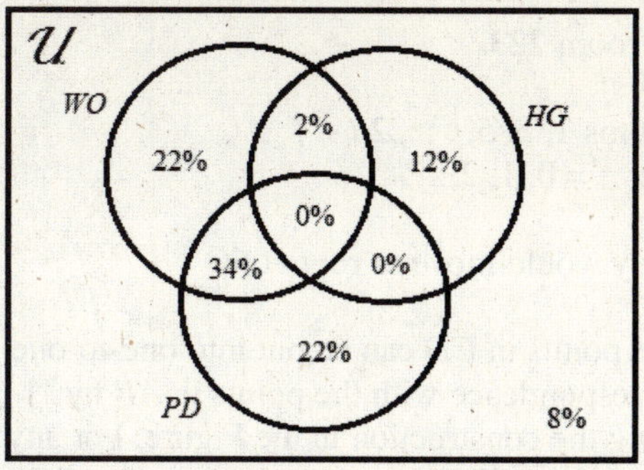

37. If there are 4 elements and each element can be either included or not included (two choices for each element), then there is a total of $2 \cdot 2 \cdot 2 \cdot 2 = 2^4 = 16$ different subsets.

Section 2.5 Infinite Sets

Study Tips

How do you know if a set is infinite? How can you determine if two finite sets are equivalent? If a set is infinite, does it have a cardinal number? To answer these questions, we will use cardinal numbers and one-to-one correspondences.

To place in one-to-one correspondence:
Think: Pair each element in one set with one element in another set.

To determine if finite sets are equivalent:
Think: Find the cardinal number of each, if they have the same cardinal number, they are equivalent.

To determine if infinite sets are equivalent:
Think: Can the sets be placed in a one-to-one correspondence with each other?

To determine if a set is infinite:
Think: Is it equivalent to one of its proper subset? That is, can the set be put into one-to-one correspondence with one of its proper subsets?

To determine the cardinal number of an infinite set A:
Think: $n(A) = \aleph_0$, the transfinite cardinal.

Mathematicians have even tried to do arithmetic with infinite numbers. To see how different this is, examine problems 24 – 27 and the *Using Your Knowledge*.

2.5 Exercises

1. $n(A) = 26$

3. $n(C) = 50$

5. The set E can be put into one-to-one correspondence with the set N of counting numbers: $n \leftrightarrow \frac{1}{n^2}$. So, $n(E) = \aleph_0$.

7. We can set up the correspondence

2	4	8	12
↕	↕	↕	↕
6	12	24	36

Thus, the sets P and Q are equivalent.

9. The correspondence

−1	−2	−3	. . .	−n	. . .
↕	↕	↕		↕	
1	2	3	. . .	n	. . .

shows that I^- and N are equivalent.

11. One such correspondence is

1	2	3	. . .	n	. . .
↕	↕	↕		↕	
1	3	5	. . .	2n − 1	. . .

Thus, the two sets N and O are equivalent.

13. One such correspondence is

2	4	6	. . .	2n	. . .
↕	↕	↕		↕	
102	104	106	. . .	100 + 2n	. . .

Thus, the two sets E and G are equivalent.

15. One such correspondence is

202	204	206	. . .	200 + 2n	. . .
↕	↕	↕		↕	
302	304	306	. . .	300 + 2n	. . .

Thus, the two sets G and T are equivalent.

17. The set {100, 200, 300, . . . } can be put Into one-to-one correspondence with a subset of itself, {200, 300, 400, . . .}.

$$\begin{array}{ccccc} 100 & 200 & 300 & \ldots & 100n \ \ldots \\ \updownarrow & \updownarrow & \updownarrow & & \updownarrow \\ 200 & 300 & 400 & \ldots & 100(n+1) \ \ldots \end{array}$$

This shows that the set is infinite.

19. The set $\left\{\frac{1}{3}, \frac{2}{3}, \frac{3}{3}, \ldots\right\}$ can be put into one-to-one correspondence with a subset of itself, $\left\{\frac{2}{3}, \frac{3}{3}, \frac{4}{3}, \ldots\right\}$.

$$\begin{array}{ccccc} \frac{1}{3} & \frac{2}{3} & \frac{3}{3} & \ldots & \frac{n}{3} \ \ldots \\ \updownarrow & \updownarrow & \updownarrow & & \updownarrow \\ \frac{2}{3} & \frac{3}{3} & \frac{4}{3} & \ldots & \frac{n+1}{3} \ \ldots \end{array}$$

Thus, the set is infinite.

21. Sets *B* and *D* are equal and equivalent.

23. Set *A* is neither equivalent nor equal to any of the other sets.

25. Answers may vary. Fill in the blank with $\aleph_0$. This is justified by considering the combination of two sets such as {1, 3, 5, . . .} and {2, 4, 6, . . .} into the set *N* = {1, 2, 3, 4, 5, 6, . . .}. Since all three of these sets have $\aleph_0$ as their cardinality, it follows that $\aleph_0 + \aleph_0 = \aleph_0$.

27. Answers may vary. Fill in the blank with $\aleph_0$. This is justified by considering the multiplication table for the set *N* of counting numbers. The set of products in this table can be put into one-to-one correspondence with the set *N* itself. This shows that $\aleph_0 \cdot \aleph_0 = \aleph_0$.

29. Answers may vary.
    a. The next two points are $\frac{7}{9}$ and $\frac{8}{9}$. (For the next cut, the points would be $\frac{1}{27}$ and $\frac{2}{27}$.)
    b. $\frac{1}{3} + \frac{2}{9} + \frac{4}{27} + \frac{8}{81} + \ldots$; the sum gets closer and closer to 1.

31. To room 223.

33. Rooms 1, 3, 5, . . ., 2*n* + 1, . . . (Let *n* = 0, 1, 2,...)

35. They would move to room 666.

37. The points in *CD* can be put into one-to-one correspondence with the points in *AB* by using the construction in the Figure. For any point *X* on *CD* draw a line through *O* and *X*. This line cuts *AB* at some point *Y*, so *X* and *Y* are corresponding points. Thus, there are as many points in *CD* as in *AB*.

Chapter 2 Practice Test

Study Tips

Being proactive about your test will help you to be successful. Take a look at this chapter and determine the concepts that will likely be on the exam. You might want to look at the ***Study Tips*** for a general guide as to what you would be expected to know. A helpful study technique is to make notecards of the concepts and procedures. Now use the practice test to determine what concepts need further work. Here are the solutions to the practice test.

1. b, c, and d define sets. Brilliant students (a) does not define a set.

2. The counting numbers between 2 and 10 are 3, 4, 5, 6, 7, 8, and 9, so the set is {3, 4, 5, 6, 7, 8, 9}.

3. a. This is the set of vowels in the English alphabet. In set-builder notation, it is {*x* | *x* is a vowel in the English alphabet}.
   b. This is the set of even counting numbers less than 10. In set builder notation, it is {*x* | *x* is an even counting number less than 10}.

4. The proper subsets are: ∅,{\$},{¢},{%},{\$,¢},{\$, %},{¢, %}

5. a. Both blanks take the symbol $\in$, because $A \cup B$ is the set of all elements that are in A or B.
   b. The first blank takes the symbol $\in$. The second blank takes the symbol $\notin$, because $A \cap B'$ is the set of elements that are in A and not in B.
   c. A' is the complement of A, that is, the set of elements in U, but not in A. Thus, the first blank takes the symbol $\in$, and the second blank takes the symbol $\notin$.
   d. Recall that $A - B$ is the set of elements in A with the elements that are also in B removed, that is, it is the set of elements that are in A and not in B. Thus, both blanks take the symbol $\in$.

6. a. A', the complement of A, = {King}.
   b. In this problem, $A \cup B = U$, so the complement of $A \cup B$ is the empty set, $\varnothing$.
   c. $A \cap B$ is the set of elements that are in both A and B, so the answer is {Queen}.
   d. The complement of $A \cap B$ is {Ace, King, Jack}. Taking this set away from U gives $U - (A \cap B)' = \{\text{Queen}\}$.

7. a. In problem 6c, we found that $A \cap B$ = {Queen}. The union of this set with the set C gives $(A \cap B) \cup C = \{\text{Ace, Queen, Jack}\}$.
   b. In problem 6a, we found that A' = {King}, so that $A' \cup C$ = {Ace, King, Jack}, which has only the element King in common with B. Thus, $(A' \cup C) \cap B = \{\text{King}\}$.

8. The two diagrams for this problem are shown:

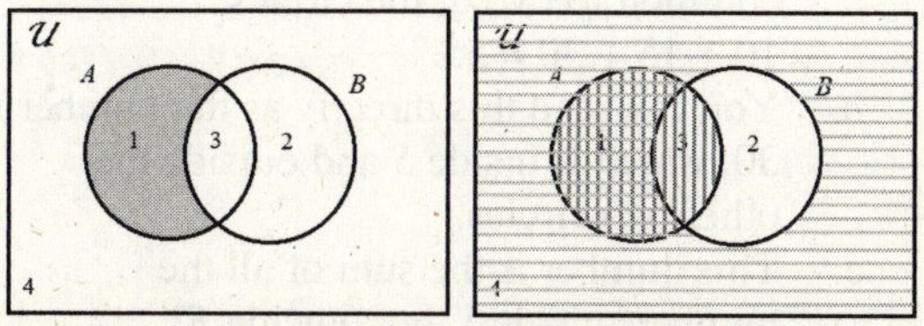

Refer to the diagram on the left. Since $A - B$ is the set A with the elements common to A and B removed, this corresponds to region 1, that is, circle A with region 3 removed. Shade region 1. Now, refer to the diagram on the right. $A \cap B'$ is the set of elements common to A and the complement of B. To show this, shade circle A one way and the region outside of B another way. The cross-hatched region corresponds to $A \cap B'$. The diagrams show that $A - B = A \cap B'$.

9. In the diagram below, first shade the region outside the circle C. This region corresponds to C'. Then shade darker that portion of the shaded region that lies inside both A and B. The darkly-shaded region corresponds to the set $A \cap B \cap C'$.

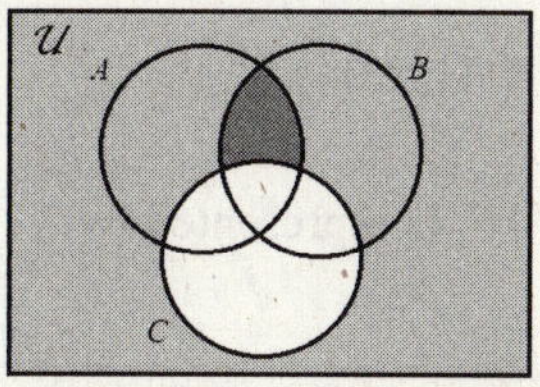

10. a. $A \cup B$ is the set of elements in A or in B, so the corresponding regions are 1, 2, 4, 5, 6, 7. C' is the set of elements in U but not in C, so the corresponding regions are 1, 2, 5, 8. The regions common to $A \cup B$ and C' are 1, 2, 5.
    b. A' is the complement of A, so the corresponding regions are those in U that are not in A; regions 2, 3, 6, 8. Similarly, the regions corresponding to B' are those in U that are not in B; regions 1, 3, 4, 8. The regions common to B' and C; regions 3, 4, correspond to $B' \cap C$. Thus, the regions corres-ponding to $A' \cup (B' \cap C)$ are those in A' or in $B' \cap C$; regions 2, 3, 4, 6, 8.

11. $A \cap B$ corresponds to the regions inside both circles A and B; regions 5, 7. $(A \cap B) \cup C$ corresponds to the regions in $A \cap B$ or in C; regions 3, 4, 5, 6, 7. Similarly, $A \cup C$ corresponds to the regions in A or C; regions 1, 3, 4, 5, 6, 7. $B \cup C$ corresponds to the regions in B or C; regions 2, 3, 4, 5, 6, 7. $(A \cup C) \cap (B \cup C)$ corresponds to the regions common to $A \cup C$ and $B \cup C$; regions 3, 4, 5, 6, 7.

This verifies the equation
$(A \cap B) \cup C = (A \cup C) \cap (B \cup C)$.

12. $(A \cap B)'$ is the complement of $A \cap B$, so the corresponding regions are those in U and not in $A \cap B$. This gives us regions 1, 2, 3, 4, 6, 8 (all except 5 and 7).
A' is the complement of A, so it corresponds to regions 2, 3, 6, 8; B' is the complement of B, so it corresponds to regions 1, 3, 4, 8. Since $A' \cup B'$ must correspond to the regions in A' or in B', we get regions 1, 2, 3, 4, 6, 8, as before. This verifies that $(A \cap B)' = A' \cup B'$.

13. Part (b) is correct: $(A \cup C) \cap B$

14. None of these. $B \cap C$ is represented by regions 6, 7.

15. a. $A \cup C = \{1,3,4,5,7,8\}$, so $n(A \cup C) = 6$.
    b. $B \cap C = \{4,8\}$, so $n(B \cap C) = 2 \neq 3$.

16. You can use the equation
$n(A \cup B) = n(A) + n(B) - n(A \cap B)$
for both parts of this problem.
    a. $n(A) = 25, n(B) = 35, n(A \cap B) = 0$
    Therefore,
    $n(A \cup B) = 25 + 35 - 0 = 60$
    b. $n(A) = 25, n(B) = 35$, and $n(A \cap B) = 5$. Therefore,
    $n(A \cup B) = 25 + 35 - 5 = 55$.

17. a. You can use the same equation as in problem 16, but with
    $n(A) = 15, n(B) = 25, n(A \cup B) = 35$.
    With these values, you get
    $35 = 15 + 25 - n(A \cap B)$, or
    $n(A \cap B) = 15 + 25 - 35 = 5$.
    b. $n(A' \cap B') = 8$ means there are 8 elements outside of both A and B.
    Since $n(A \cup B) = 35$,
    $n(U) = 35 + 8 = 43$.

18. The shaded region is the region inside of B and outside of A, so it may be described as $B - A$ or as $B \cap A'$. ($A' \cap B$ is also correct.)

19. First make a Venn diagram showing the various sets. Then start at the end of the list and work back through it.

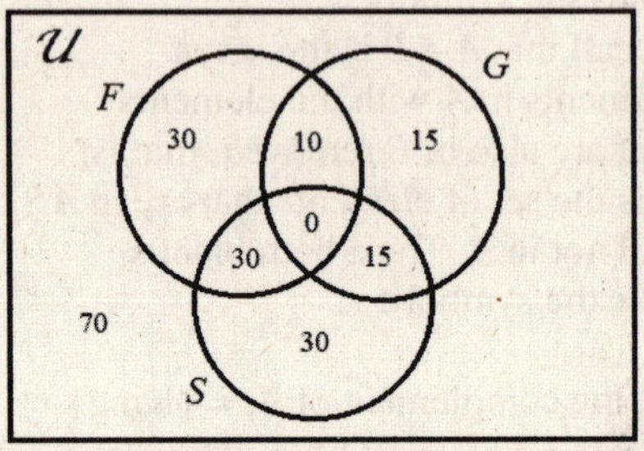

Since no students are taking all three of the courses, put a 0 in the region common to the circles F, G, and S. Because 70 are taking no language, write 70 in the rectangle outside of the three circles. As 15 are taking German and Spanish, write 15 in the region common to G and S, but outside of F. 30 are taking French and Spanish, so write 30 in the region common to F and S, but outside of G. 10 are taking French and German, so write 10 in the region common to F and G but outside of S. 75 are taking Spanish and of these, you have accounted for 0 + 30 + 15 = 45. (See the diagram.) Hence, you must write 30 in the region inside S but outside both F and G. 40 are taking German, and you have accounted for 0 + 10 + 15 = 25 of these, so write 15 in the region inside G, but outside F and S. 70 are taking French, and you have accounted for 0 + 10 + 30 = 40 of these, so write 30 in the region inside F, but outside G and S. This completes the diagram and you can get the answers from it.
    a. To get the number of students taking two languages, add the numbers common to two of the circles:
    10 + 15 + 30 = 55.
    b. You can read this directly as the number 30, which is inside S and outside the other two circles.
    c. This number is the sum of all the numbers inside S, but outside F:
    15 + 30 = 45.

20. a. This is the sum of all the numbers in the diagram:
$10+5+15+2+4+3+8=47$
b. This is the sum of the numbers that are in *A*, but not in *B*: $10+2=12$
c. This is the sum of the numbers that are in *B* or *C*, but not in *A*:
$15+3+8=26$.
d. This is the number that is in both *B* and *C*, but not in *A*, so the answer is 3.
e. This is the sum of all the numbers in the diagram that are not in either *B* or *C*, so the answer is 10.

21. The one-to-one correspondence

$$\begin{array}{cccccc} 1 & 3 & 5 & \ldots & 2n-1 & \ldots \\ \updownarrow & \updownarrow & \updownarrow & & \downarrow & \\ 2 & 4 & 6 & \ldots & 2n & \ldots \end{array}$$

shows that the two sets have the same cardinal number.

22. The one-to-one correspondence

$$\begin{array}{cccccc} 1 & 2 & 3 & \ldots & n & \ldots \\ \updownarrow & \updownarrow & \updownarrow & & \updownarrow & \\ 4 & 16 & 36 & \ldots & (2n)^2 & \ldots \end{array}$$

shows that the two sets are equivalent.

23. $n\{1,4,\ldots,n^2,\ldots,144\}=12$

24. Since you can set up a one-to-one correspondence between the given set and the set {1, 2, . . ., *n*, . . .}, the cardinality of the given set is the same as that of the set of counting numbers, $\aleph_0$.

25. The one-to-one correspondence

$$\begin{array}{cccccc} \frac{1}{2} & \frac{1}{3} & \frac{1}{4} & \ldots & \frac{1}{n} & \ldots \\ \updownarrow & \updownarrow & \updownarrow & & \updownarrow & \\ \frac{1}{2} & \frac{1}{4} & \frac{1}{6} & \ldots & \frac{1}{2n-1} & \ldots \end{array}$$

between the given set and a subset of itself shows that the given set is infinite.

Chapter 3 Logic

Section 3.1 Statements

Study Tips

In order to communicate, you can use statements, questions, commands, and exclamations. It is also hoped that you can reason logically, and determine if an argument is valid.

To determine if a sentence is a statement:
Think: Can the sentence be declared true or false, but not at the same time?
Note: Questions, commands, and exclamations are ***not*** statements.

To write a conjunction:
Think: simple statements connected by the word AND
Note: Symbolized by $p \wedge q$.

To write a disjunction:
Think: simple statements connected by the word OR
Note: Symbolized by $p \vee q$.

To write the negation of a simple statement:
Think: Opposite. If the given statement is true, then the negation will be false; if the given statement is false then the negation will be true.
Note: Symbolized by $\sim p$. Simple statements can be negated by inserting the word not or by writing "*It is not the case*" in front of the statement.

To negate conjunctions and disjunctions:
Think: De Morgan's Laws

1. $\sim(p \wedge q)$ is $\sim p \vee \sim q$.
   In words, not (*p* and *q*) is (not *p* **or** not *q*).
2. $\sim(p \vee q)$ is $\sim p \wedge \sim q$.
   In words, not (*p* or *q*) is (not *p* **and** not *q*).

The negation of…
All *p*'s are *q*'s *becomes* **Some** *p*'s **are not** *q*'s.
No *p*'s are *q*'s *becomes* **Some** *p*'s **are** *q*'s.

Word of Advice
Memorize De Morgan's Laws and the negations involving the words all, none, or some.

3.1 Exercises

1. This sentence cannot be classified as either true or false. It is not a statement.

3. This sentence is either true or false, so it is a statement. It is a compound statement with two components: *Jane is taking an English course. She has four themes to write.*

5. This sentence is a question and so is not a statement.

7. This sentence is either true or false, so it is a statement. It is a compound statement with the two components: *Students at Ohio State University are required to take a course in history. Students at Ohio State University are required to take a course in economics.*

9. This statement is a conjunction and is symbolized by $a \wedge f$.

11. This statement is a disjunction and is symbolized by $d \vee f$.

13. This statement is a conjunction and is symbolized by $b \wedge p$.

15. This statement is a disjunction and is symbolized by $a \vee m$.

17. The conjunction of *p* and *q*: $p \wedge q$

19. The negation of the disjunction of *p* and *q*: $\sim(p \vee q)$

21. Ricky loves Lucy, or Lucy does not love Ricky.

23. Ricky loves Lucy but Lucy does not love Ricky.

25. It is not the case that Ricky and Lucy love each other.

27. Bill's store is not making a good profit.

29. My dog is not a spaniel.

31. I like to work overtime.

33. These two are negations of each other because if either is true, the other is false.

35. This statement is negated by replacing the "All" by "Some… not": *Some men are not*

mortal.

37. This statement is negated by replacing the "Some … not" by "All": *All basketball players are 6 feet tall.*

39. Since this statement is a disjunction, you must negate both components: *It is not the case that he is bald or that he has a 10-in. forehead*, alternatively, *He is not bald, and he does not have a 10-inch forehead.*

41. This statement is negated by replacing the "Some" by "No": *No circles are round.*

43. This statement is negated by replacing the "Somebody" by "Nobody": *Nobody up there loves me.*

45. This statement is negated by replacing the "Everybody" by "Somebody does not": *Somebody does not like to go on a trip.*

47. This statement is negated by replacing the "All…" by "Some…. not": *Some persons occupying your covered auto are not insured.*

49. This statement is negated by replacing the "Some… not" by "All": *All expenses are subject to the 2% limit.*

51. Statement d. Conditions a, b, and c do not make the chairperson mistaken; only statement d does that.

53. $(d \wedge p) \vee r$ 55. $r \wedge (t \vee g)$

57. Answers may vary.
    a. The diagram is neither a square nor a rectangle. (The diagram is not a square and is not a rectangle.)
    b. The diagram is a square or not a rectangle.
    c. The diagram is a square and a rectangle.

59. Answers may vary. Sentence (1) is true, sentence (2) is false. If we assume that sentence (3) is true, there will be two sentences, sentence (1) and sentence (3), that are true, and thus, sentence (3) is false!
(Consider if sentence (3) is false as we have just determined, then there IS only one statement that is true (sentence (1)), thus making sentence (3) true. Paradox!)

61. False 63. False

65. True 67. False

69. a. (Hispanics and females and not high school graduates) or (white and females and not high school graduates): $16,500
    b. Males and Hispanics and not high school graduates: $22,200

Section 3.2 Truth Tables: A Problem-Solving Tool

Study Tips

Sometimes we do not know under what conditions compound statements are *true* or *false*.

To determine the truth value of a …

Conjunction $p \wedge q$:
Think: **true** only when both p and q are **true**.

Disjunction $p \vee q$:
Think: **false** only when both p and q are **false.**

Negation:
If a statement p is **true** its negation $\sim p$ is **false**.
If a statement p is **false** its negation $\sim p$ is **true.**

To make a truth table:
Think: Consistency. Columns 1 and 2 (for p and q) will always be

1	2	other columns follow
p	q	
T	T	
T	F	
F	T	
F	F	

To determine if two statements are equivalent:
Think: Use a truth table. The statements will have identical truth values.
Note: Symbolized by $p \Leftrightarrow q$.

3.2 Exercises

1. To write the disjunction, use the word "or": *Today is Friday or Monday.*

3. Here, you only have to insert the word "not":

Today is not Friday.

5. To write the disjunction, use the word "or": *He is a gentleman or a scholar.*

7. To write the conjunction, use the word "and": *He is a gentleman and a scholar.*

9. This statement is the conjunction of the given statements: $g \wedge s$

11. a. $p \wedge q$ b. $p \vee q$
    c. The statement in (a) is false, the one in (b) is true.

13. This is the conjunction of the negation of statement q and the statement p: $\sim q \wedge p$, or equivalently $\sim(q \vee \sim p)$

15. This is the disjunction of the two given statements: $p \vee q$

17. $p \vee q$. This is the disjunction of p and q. Since p is true, the disjunction is true.

19. $\sim p \wedge \sim q$. This is the conjunction of (not p) and (not q). Since (not p) is false, the conjunction is false.

21. $\sim q \wedge \sim p$. This is the conjunction of (not q) and (not p). Since (not p) is false, the conjunction is false.

23. $g \vee \sim j$. This is the disjunction of g and $\sim j$. Since g and $\sim j$ are both false, the disjunction is false.

25. $(g \vee j) \wedge \sim(g \wedge j)$. This is the conjunction of $g \vee j$ and $\sim(g \wedge j)$. Since j is true, the disjunction $g \vee j$ is true. Since g is false, the conjunction $g \wedge j$ is false, so that $\sim(g \wedge j)$ is true. Thus, the statement $(g \vee j) \wedge \sim(g \wedge j)$ is true.

27.

1	2	3	4
p	q	$\sim q$	$p \vee \sim q$
T	T	F	T
T	F	T	T
F	T	F	F
F	F	T	T

To construct the table, first fill in Columns 1 and 2: under p, 2 *T*s, 2 *F*s and under q, *T F T F*. This gives all the combinations of truth values for p and q. Next fill in Column 3, the negation of q: *F T F T*. Then fill in Column 4, the disjunction of p and $\sim q$. This combines Columns 1 and 3. Since the disjunction is false only when both components are false, write *F* in the third line and *T*s in the other lines. This completes the table and shows that *T T F T* are the truth values for $p \vee \sim q$.

29.

1	2	3	4
p	q	$\sim p$	$\sim p \wedge q$
T	T	F	F
T	F	F	F
F	T	T	T
F	F	T	F

First fill in Columns 1 and 2 just as in problem 27. Then fill in Column 3, the negation of p, *F F T T*. Then do Column 4, the conjunction of $\sim p$ and q. This combines Columns 3 and 2. Since the conjunction is true only when both components are true, write *T* in the third line and *F*s in the other lines. This completes the table and shows that *F F T F* are the truth values for $\sim p \wedge q$.

31.

1	2	3	4	5
p	q	$\sim q$	$p \vee \sim q$	$\sim(p \vee \sim q)$
T	T	F	T	F
T	F	T	T	F
F	T	F	F	T
F	F	T	T	F

First fill in Columns 1 and 2 just as in problem 27. Then do Column 3, the negation of *q*: *F T F T*. Next, do Column 4, the disjunction of *p* and ~*q* (Columns 1 and 3). Since the disjunction is false only when both components are false, write *F* in the third row, where *p* and *q* are both *F*s and write *T*s in the remaining rows. Column 5 is the negation of Column 4: *F F T F* are the truth values for $\sim(p\vee\sim q)$.

33.

1	2	3	4	5	6
p	q	$\sim p$	$\sim q$	$\sim p\wedge\sim q$	$\sim(\sim p\wedge\sim q)$
T	T	F	F	F	T
T	F	F	T	F	T
F	T	T	F	F	T
F	F	T	T	T	F

First fill in Columns 1 and 2 as in problem 27. Then do Column 3, the negation of *p*: *F F T T*; and Column 4, the negation of *q*: F T F T. Next, do Column 5, the conjunction of ~*p* and ~*q*. Since the conjunction is true only when both components are true, write *T* in the bottom row, where ~*p* and ~*q* are both *T*s, and write *F*s in the remaining rows. Column 6 calls for the negation of Column 5, so write *T* where Column 5 has *F*, and *F* where Column 5 has *T*. The final result gives *T T T F* for the truth values of $\sim(\sim p\wedge\sim q)$.

35.

1	2	3	4	5	6
p	q	$p\wedge q$	$\sim p$	$\sim p\wedge q$	$3\vee 5$
T	T	T	F	F	T
T	F	F	F	F	F
F	T	F	T	T	T
F	F	F	T	F	F

Fill in Columns 1 and 2 as in problem 27. Then do Column 3, the conjunction of *p* and *q*. The conjunction is true only when both components are true, so write *T* in the first row and *F* in the other rows. Next, do Column 4, which calls for the negation of *p*: *F F T T*; and then do Column 5, the conjunction of ~*p* and *q*. This takes a *T* in the third row, where both Columns 4 and 2 have *T*s, and takes *F*s in the other rows. Finally, Column 6, the disjunction of Columns 3 and 5, takes *T* in the rows where Column 3 or 5 is *T*, and *F* where both Columns 3 and 5 are *F*s. The required truth values are *T F T F*.

37.

1	2	3	4	5
p	q	r	$q\vee r$	$p\wedge(q\vee r)$
T	T	T	T	T
T	T	F	T	T
T	F	T	T	T
T	F	F	F	F
F	T	T	T	F
F	T	F	T	F
F	F	T	T	F
F	F	F	F	F

First, write four *T*s and then four *F*s in Column 1; then two *T*s, two *F*s, two *T*s, two *F*s in Column 2; in Column 3, write alternately *T* and *F*. This gives all the possible combinations of *T*s and *F*s for *p*, *q*, and *r*. Next, do Column 4, the disjunction of Columns 2 and 3. The disjunction is false only when both components are false, so write *F* in the 4th and 8th rows, where Columns 2 and 3 are both *F*s, and write *T* in the other rows. Column 5 calls for the conjunction of Columns 1 and 4, so write *T* in the first three rows, where Columns 1 and 4 are both *T*s, and write *F* in the remaining rows. This completes the table.

39.

1	2	3	4	5	6	7
p	q	r	$p\vee q$	$\sim q$	$r\wedge\sim q$	$4\vee 6$
T	T	T	T	F	F	T
T	T	F	T	F	F	T
T	F	T	T	T	T	T
T	F	F	T	T	F	T
F	T	T	T	F	F	T
F	T	F	T	F	F	T
F	F	T	F	T	T	T
F	F	F	F	T	F	F

First write four *T*s and then four *F*s in Column 1; then, two *T*s, two *F*s, two *T*s, two *F*s in Column 2; in Column 3, write *T* and *F* alternately. This gives all the possible combinations of *T*s and *F*s for *p*, *q*, and *r*. Next, do Column 4, by combining Columns 1 and 2 to form the disjunction of *p* and *q*. Thus, write *F* in the last two rows, where *p* and *q* are both *F*s, and write *T*s in the remaining rows.

Then do Column 5, the negation of q: *F F T T F F T T*. Column 6 is the conjunction of r and $\sim q$, Columns 3 and 5, so write *T* in the third and seventh rows, where r and $\sim q$ are both *T*, and write *F* in the other rows. Finally, Column 7 is the disjunction of Columns 4 and 6, so write *F* in the last row, where Columns 4 and 6 are both *F*, and write *T* in the remaining rows. This completes the table.

41. Let j be "I have a job." and let h be "I hate it."
  a. Since this is the conjunction of the two statements, it will be true when j and h are both true.
  b. This will be false if either j or h is false.
  c. This is the disjunction of the two statements, so it will be true if either j or h is true.
  d. This will be false only if both j and h are false.

43.

1	2	3	4	5	6	7	8
p	q	r	$q \wedge r$	$1 \vee 4$	$p \vee q$	$p \vee r$	$6 \wedge 7$
T	T	T	T	T	T	T	T
T	T	F	F	T	T	T	T
T	F	T	F	T	T	T	T
T	F	F	F	T	T	T	T
F	T	T	T	T	T	T	T
F	T	F	F	F	T	F	F
F	F	T	F	F	F	T	F
F	F	F	F	F	F	F	F

Columns 5 and 8 of the above truth table show that the two statements have the same truth values, so they are equivalent.

45.

1	2	3	4	5	6	7
p	q	$p \vee q$	$\sim(3)$	$\sim p$	$\sim q$	$5 \wedge 6$
T	T	T	F	F	F	F
T	F	T	F	F	T	F
F	T	T	F	T	F	F
F	F	F	T	T	T	T

Columns 4 and 7 of the table show that the two statements have the same truth values, so they are equivalent.

47.

1	2	3	5	6	7
p	q	$p \wedge q$	$\sim p$	$3 \vee 5$	$q \vee \sim p$
T	T	T	F	T	T
T	F	F	F	F	F
F	T	F	T	T	T
F	F	F	T	T	T

Columns 5 and 6 of the table show that the two statements have the same truth values, so they are equivalent.

49. a. $p \wedge q$ is true only when p and q are both true. Thus, the truth values are *T F F F* as in the table. $p \wedge \sim q$ is true only when p and $\sim q$ are both true, that is, when p is true and q is false. The truth values are *F T F F* as in the table. $\sim p \wedge q$ is true only when p is false and q is true, so that the truth values are *F F T F* as in the table. $\sim p \wedge \sim q$ is true only when p and q are both false, so the truth values are *F F F T* as in the table.
  b. The disjunction of two statements is true whenever at least one of the statements is true. Since $p \wedge q$ is true only in the first row and $\sim p \wedge \sim q$ is true only in the last row, the disjunction of these two has the truth values *T F F T* as stated.
  c. A statement with the truth values *F T T F* is $(p \wedge \sim q) \vee (\sim p \wedge q)$. Similarly, $(p \wedge \sim q) \vee (\sim p \wedge q) \vee (\sim p \wedge \sim q)$ has truth values *F T T T*. Since the given table has one F and three Ts, its negation will have one T and three Fs and will be a basic conjunction. Therefore a much simpler statement with truth values *F T T T* is $\sim(p \wedge q)$, the negation of $p \wedge q$, as can be seen from the table.

51. None of the applicants is eligible. Statement (*d*) is false for Joe, (*t*) is false for Mary and (*m*) is false for Ellen.

53. 7 is greater than or equal to 5

55. 0 is less than or equal to 3.

57. $\frac{1}{2}$ is greater than $\frac{1}{8}$.

59. I will not go fishing or the sun is not shining. This would be true if either or both of the statements: "I will not go fishing." and "The sun is not shining." is (are) true.

61. $\left[(e\wedge g)\vee a\right]\wedge h\wedge(c\vee n\vee \sim t)$

63. Mr. Baker is the carpenter.

Section 3.3 The Conditional and the Biconditional

Study Tips

How often have you said, to the effect, "If such and such, then blah and blah." We all have at one point or another. This is called a conditional, another important concept in logic.

To write a conditional:
Think: If p, then q. Statement p is the antecedent; statement q is the consequent.
Note: Symbolized by $p\rightarrow q$.

To write a biconditional:
Think: If p, then q AND if q, then p.
Note: Symbolized by $p\leftrightarrow q$, which is equivalent to $(p\rightarrow q)\wedge(q\rightarrow p)$.

To determine the truth value of a ...
Conditional $p\rightarrow q$:
Think: **false** only when p is **true** and q is **false**.
Note: $T\rightarrow F$ is <u>always</u> false!

Biconditional $p\leftrightarrow q$:
Think: **true** only when p and q have the **same** truth values.
Note: $T\leftrightarrow T$ or $F\leftrightarrow F$ is <u>always</u> true!

To write a conditional equivalence:
Think: negation and 'or'
Note: $(p\rightarrow q)\Leftrightarrow(\sim p\vee q)$

To write the negation of a conditional statement:
Think: negation of (negation and 'or')
Note: $\sim(p\rightarrow q)\Leftrightarrow\sim(\sim p\vee q)$

<u>Hint</u> Statements such as "All math students make A's" can be translated into the corresponding conditional "If you are a math student, then you make A's."

3.3 Exercises

1.

1	2	3	4	5	6
p	q	$\sim q$	$\sim p$	$\sim q\rightarrow\sim p$	$p\rightarrow q$
T	T	F	F	T	T
T	F	T	F	F	F
F	T	F	T	T	T
F	F	T	T	T	T

Columns 5 and 6 are identical, so $\sim q\rightarrow\sim p$ is equivalent to $p\rightarrow q$.

3.

1	2	3	4	5
p	q	$\sim p$	$\sim p\rightarrow q$	$p\vee q$
T	T	F	T	T
T	F	F	T	T
F	T	T	T	T
F	F	T	F	F

Since columns 4 and 5 are identical, $\sim p\rightarrow q$ and $p\vee q$ are equivalent.

5. This is a statement of the form $p\rightarrow q$, with p being "2 + 2 = 4," and q being "8 = 5." Since p is true and q is false, the conditional has the truth value *False*.

7. This is a statement of the form $p\rightarrow q$, with p being "2 + 2 = 22," and q being "4 = 26." Since p is false, the conditional has the truth value *True*.

9. Since the antecedent, $2+2=22$, is false, x may be any number.

11. Since the consequent, $2+2=32$, is false, x may be any number for which the antecedent is false, that is, any number except 4.

13. a. $\sim o$ is true when o is false.
    b. $\sim n\vee b$ is false when n is true and b is false.
    c. The statement $\sim o\rightarrow(\sim n\vee b)$ is false when o is false, n is true and b is false.

15.

1	2	3	4	5	6
p	q	r	$p \to q$	$p \vee r$	$4 \leftrightarrow 5$
T	T	T	T	T	T
T	T	F	T	T	T
T	F	T	F	T	F
T	F	F	F	T	F
F	T	T	T	T	T
F	T	F	T	F	F
F	F	T	T	T	T
F	F	F	T	F	F

The first three columns are filled out as usual to get all the possible combinations of *p*, *q* and *r*. The conditional, column 4, has *F*s in the rows where the antecedent *p* is true and the consequent *q* is false. The other rows are all *T*s. Column 5 has *T*s in the rows where either *p* or *r* is true and has *F*s in the remaining rows. Column 6 has *T*s in the rows where columns 4 and 5 both have *T*s and has *F*s in the remaining rows.

17.

1	2	3	4	5
p	q	r	$q \wedge r$	$p \to (q \wedge r)$
T	T	T	T	T
T	T	F	F	F
T	F	T	F	F
T	F	F	F	F
F	T	T	T	T
F	T	F	F	T
F	F	T	F	T
F	F	F	F	T

The first three columns are filled out as usual to get all the possible combinations of *p*, *q* and *r*. Column 4 is the conjunction of columns 2 and 3, so has *T*'s only in the rows where *q* and *r* are both *T*. Column 5, the conditional, has*F* in the rows where the antecedent *p* is true and the consequent $q \wedge r$ is false (second, third and fourth rows). The remaining entries are *T*s.

19. The final columns in the tables in problems 17 and 18 are identical, so the two statements are equivalent.

21. The antecedent is "I will adopt it," and the consequent is "It is a poodle." Thus, the symbolic form is $p \to q$.

23. The antecedent is "It is not a poodle," and the consequent is "I will not adopt it." Hence, the symbolic form is $\sim q \to \sim p$.

25. The antecedent is "It is a poodle," and the consequent is "I will not adopt it." Thus, the symbolic form is $q \to \sim p$.

27. The antecedent is "You are out of beer," and the consequent is "You are out of Schlitz." Thus, the symbolic form is $\sim b \to \sim s$

29. $\sim a \vee b$. The temperature is not above 80°, or I would go to the beach.

31. $\sim a \vee g$. Eva does not have a day off, or she would go to the beach.

33. Let *p* be "You got the time," and *q* be "We got the beer." Then $\sim p \vee q$ becomes "You do not have the time, or we've got the beer."

35. If it is a dog, then it is a mammal.

37. If it is a man, then it is created equal. (The equal refers to equal to each other, not to $10, not to a dog, etc.)

39. If it is a rectangle with perpendicular diagonals, then it is a square.

41.

1	2	3	4	5	6
p	q	$p \to q$	$\sim(3)$	$\sim q$	$p \wedge \sim q$
T	T	T	F	F	F
T	F	F	T	T	T
F	T	T	F	F	F
F	F	T	F	T	F

Since columns 4 and 6 are identical, $\sim(p \to q)$ and $p \wedge \sim q$ are equivalent.

43. Johnny does not play quarterback, and his team does not lose.

45. I kiss you once, but I do not kiss you again.

47. Evel Knievel is careless, but he will not lose his life.

49. If Johnny plays quarterback, then his team wins.

51. If Joe had not had an accident, then he would be able to get car insurance.

53. The antecedent "You eat the spinach and the liver" is false, so the conditional is true. She has not broken her original promise.

55. Let p be: "Mary is in Tampa," and q be: "Mary is in Florida." The given statement is $p \to q$, which is equivalent to $\sim p \vee q$. Statement (*d*) is symbolized by $\sim q \to \sim p$ which is equivalent to $\sim(\sim q) \vee \sim p$, that is, $q \vee \sim p$. Since $q \vee \sim p$ is equivalent to $\sim p \vee q$, statement (d) is logically equivalent to the given statement.

57. Let p be: "You studied hard," and q be: "You passed the course." The given statement is $p \to q$. The negation of $p \to q$ (See problem 17) is $p \wedge \sim q$, which is the symbolic form of statement (d).

59. Answers may vary. Let p be: "You stop here," and q be: "Your pain will stop." Then the given statement is $p \to q$, which is false only if p is true and q is false. Thus, if p is false, then q can be true or false. Hence, the statement: "If you did not stop, your pain will not stop either," does not follow logically from the given statement.

61. Answers may vary. The student has to take the placement examination only if the student is entering college for the first time even though this student may be admitted to sophomore standing due to satisfying freshman requirements (perhaps by advanced courses in high school, transfering from a different college, etc.).

63. Answers will vary.

65. Answers will vary.

67. $r \to a$

69. No. It only says that an adjustment will definitely be made if a report is made in 10 days but doesn't speak to the situation where a late report is made.

71. A must see at least one black hat, or she would know that her hat is black since they are not all white. B also must see at least one black hat, and further, that hat had to be on C, otherwise she would know that her hat was black (since she knows A saw at least one black hat). So C knows that her hat is black, without even seeing the others' hats.

73. The one who fell silent, presumably the quickest of the three, reasoned that his head must be painted also. The argument goes as follows. Let's call the quick one *Q*, and the other two *D* and *S*. Let's assume *Q*'s head is untouched. Then *D* is laughing because *S*'s head is painted, and vice versa. But eventually, *D* and *S* will realize that their head must be painted, because the other is laughing. So they will quit laughing as soon as they realize this. So, *Q* waits what he thinks is a reasonable amount of time for them to figure this out, and when they don't stop laughing, his worst fears are confirmed. He concludes that his assumption is invalid and he must be crowned in crimson too.

SKILL CHECKER

p	q	$\sim p$	$p \vee q$	$p \wedge q$
T	*T*	*F*	*T*	*T*
T	*F*	*F*	*T*	*F*
F	*T*	*T*	*T*	*F*
F	*F*	*T*	*F*	*F*

Section 3.4 Variations of the Conditional and Implications

Study Tips

Is there more than one way to say the same thing? Of course! As we continue our work with conditionals, $p \to q$, we will see that this is also true in logic.

To write the converse of a conditional:
Think: switch places; "if q, then p"
Note: Converse symbolized by $q \to p$

To write the inverse of a conditional:
Think: negate all; "if ~*p*, then ~*q*"
Note: Inverse symbolized by $\sim p \rightarrow \sim q$

To write the contrapositiveof a conditional:
Think: negate and switch; "if ~*q*, then ~*p*"
Note: Inverse symbolized by $\sim q \rightarrow \sim p$

To determine equalivalent to a conditional:
Think: contrapositive! (negate and switch)
Note: $p \rightarrow q$ is equivalent to $\sim q \rightarrow \sim p$

To write equivalent forms of a conditional:
Think: use sufficient or necessary
Note: Equivalent to $p \rightarrow q$ are the phrases…

p is sufficient for *q*
q is necessary for *p*
p only if *q*
q if *p*

To write equivalent forms of a biconditional:
Think: use sufficient and necessary
Note: Equivalent to $p \leftrightarrow q$ are the phrases…

p is necessary and sufficient for *q*
q is necessary and sufficient for *p*
p if and only if *q*

To determine tautology or contradiction:
Think: Use a truth table. All T's is a tautology; all F's is a contradiction.
Note: Statements are *equivalent* if the biconditional is a tautology.

To determine an implication, p implies q:
Think: Conditional $p \rightarrow q$ must be a tautology.
Note: Symbolized by $p \Rightarrow q$

Caution
Implication is a *relationship* between *p* and *q*, while a conditional is simply a *connective*.

3.4 Exercises

1. The *contrapositive* is: If *n* is divisible by 2, then *n* is an even number.

3. $q \rightarrow p$ 5. $q \rightarrow p$

7. If one is a mathematics major, then one takes calculus.

9. If the measure gets a two-thirds vote, then it carries.

11. If we have a stable economy, then we have low unemployment.

13. If birds are of a feather, then they flock together.

15.

p	q	$p \rightarrow q$	Converse $q \rightarrow p$	Inverse $\sim p \rightarrow \sim q$
T	T	T	T	T
T	F	F	T	T
F	T	T	F	F
F	F	T	T	T

The converse, $q \rightarrow p$, is true except when *q* is true and *p* is false (third row). The inverse, ~*p* is true and ~*q* is false (third row). Thus, the converse and the inverse have the same truth values, and hence are equivalent.

17. $u \rightarrow h$ 19. $u \rightarrow h$

21. $u \rightarrow h$ 23. $u \leftrightarrow h$

25. $u \rightarrow h$; $\sim h \rightarrow \sim u$; $\sim u \rightarrow \sim h$

27. $p \leftrightarrow s$

29. a. *Converse*: If you are not strong, then you do not eat your spinach.
Inverse: If you eat your spinach, then you are strong.
Contrapositive: If you are strong, then you eat your spinach.
b. *Converse*: If you are strong, then you eat your spinach.
Inverse: If you do not eat your spinach, then you are not strong.
Contrapositive: If you are not strong, then you do not eat your spinach.
c. *Converse*: If you eat your spinach, then you are strong.
Inverse: If you are not strong, then you do not eat your spinach.
Contrapositive: If you do not eat your spinach, then you are not strong.

31. If the square of an integer is divisible by 4, the integer is even. Always true.

33. If I am neat and well dressed, then I can get a date. Not always true.

35. If you pass this course, then you get passing grades on all the tests. Not always true.

37. If we cannot find a cure for cancer, then the research is inadequately funded.

39. If a person does not want to improve the world, then the person is not a radical.

41. Let p be: "n is even," and q be: "$3n$ is even." Then, we see that the equivalence given in (c) makes the desired transformation.

43. Let p be: "The day is cool," and q be: "I will go fishing." Then the given statement is $p \to q$. Statement (b) translates into $q \to p$, which is not logically equivalent to $p \to q$.

45.

1	2	3	4
p	q	$p \wedge q$	$(p \wedge q) \to p$
T	T	T	T
T	F	F	T
F	T	F	T
F	F	F	T

Column 3 is the conjunction of columns 1 and 2, so has T only in the first row, both p and q are T. Therefore, column 4 is all T's, which shows that $(p \wedge q) \to p$ is a tautology.

47.

1	2	3
p	$\sim p$	$p \leftrightarrow \sim p$
T	F	F
F	T	F

Since column 3 has all F's, the statement $p \leftrightarrow \sim p$ is a contradiction.

49.

1	2	3	4	5	6
p	q	$\sim p$	$\sim p \wedge q$	$p \to q$	$4 \to 5$
T	T	F	F	T	T
T	F	F	F	F	T
F	T	T	T	T	T
F	F	T	F	T	T

Since column 6 is all T's, the first statement, $\sim p \wedge q$, implies the second, $p \to q$. (Columns 4 and 5 are not identical, so the statements are not equivalent.)

51. If p is true, then the first statement is true, and if the first statement is true, then p is true. Hence, the two statements are equivalent as confirmed by the truth table.

1	2	3	4	5
p	q	$p \wedge q$	$p \vee 3$	$1 \leftrightarrow 4$
T	T	T	T	T
T	F	F	T	T
F	T	F	F	T
F	F	F	F	T

53. If $p \wedge \sim q$ is true, then p is true and $\sim q$ is true, so that $\sim p \vee \sim q$ is also true. Thus: $p \wedge \sim q$ implies $\sim p \vee \sim q$. However, $\sim p \vee \sim q$ is true if p is false and $\sim q$ is true, in which case $p \wedge \sim q$ is false. Hence: $\sim p \vee \sim q$ does not imply $p \wedge \sim q$. Hence, the two statements are not equivalent nor does one statement imply the other as confirmed by the truth table.

1	2	3	4	5	6	7	8
p	q	$\sim p$	$\sim q$	$3 \vee 4$	$p \wedge 4$	$5 \to 6$	$6 \to 5$
T	T	F	F	F	F	T	F
T	F	F	T	T	T	T	T
F	T	T	F	T	F	F	T
F	F	T	T	T	F	F	T

55. Answers will vary.

57. Answers may vary.
(1) For p to be true, it is necessary that q be true.
(2) A necessary condition for p to be true is that q be true.

59. Answers may vary. Should mention that the implication $\Rightarrow$ means that the conditional is a tautology.

61. Answers may vary. The conditional $p \to q$ is true whenever the antecedent p is false. Thus, if p is false, then $p \to q$ is a tautology and p implies q whether q is true or false.

63. a. The truth set of q is Q and the truth set of $\sim r$ is R'. Hence, the truth set of $q \wedge \sim r$ is $Q \cap R'$.

b. The truth set of $p \wedge q$ is $P \cap Q$ and the truth set of $\sim r$ is R'. Thus, the truth set of $(p \wedge q) \wedge \sim r$ is $(P \cap Q) \cap R'$.

65. The contrapositive of $\sim q \to \sim p$ is $\sim(\sim p) \to \sim(\sim q)$, that is $p \to q$.

67. The inverse of $p \to q$ is $\sim p \to \sim q$, and the contrapositive of $\sim p \to \sim q$ is $\sim(\sim p) \to \sim(\sim q)$, that is $q \to p$.

69. $(\sim r \wedge \sim s) \vee (p \vee q) \Leftrightarrow (r \vee s) \to (p \vee q)$
It is true because
$$(r \vee s) \to (p \vee q) \Leftrightarrow \sim(r \vee s) \vee (p \vee q)$$
$$\Leftrightarrow (\sim r \wedge \sim s) \vee (p \vee q)$$
and the converse of $(r \vee s) \to (p \vee q)$ is $(p \vee q) \to (r \vee s)$.

71. The direct statement

73. The contrapositive

Section 3.5 Euler Diagrams: A Problem-Solving Tool

Study Tips

Have you ever tried to reason with a defiant teenager? If so, you probably have been exposed to invalid arguments! However, it is not always easy to determine whether an argument is valid or invalid. This is where Euler diagrams are very helpful.

Definition of an Argument:
Think: set of statements; the premises and the conclusion
Note: The conclusion *follows* from the premises.

To determine validity with Euler diagram:
Think: Four basic types of statements

All P's are Q's

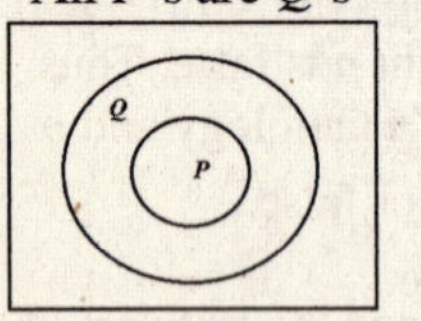

No P's are Q's

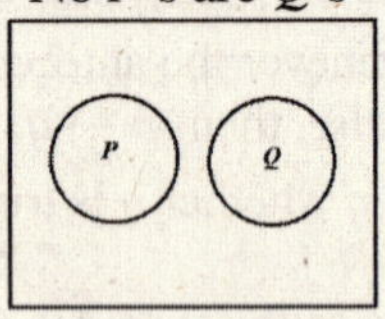

Some P's are Q's

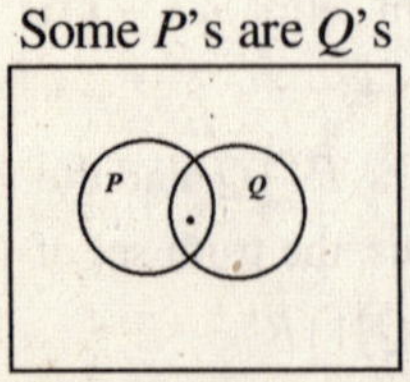

Some P's are not Q's

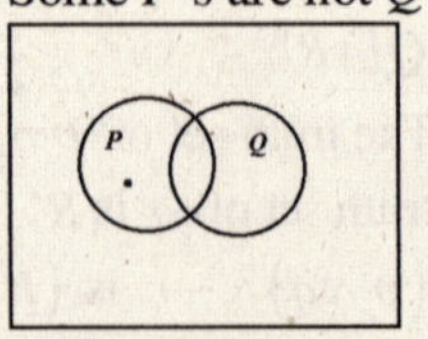

To determine validity of an argument:
Think: Euler diagram of the premises. In a valid argument, when all premises are true, the conclusion is true.
Note: The conclusion will be 'forced' by the premises when using a Euler diagram. The argument is *invalid* if you could show more than one conclusion.

Word of Advice
Memorize the diagrams of the *four basic types of statements*. If you do not know how to diagram these statements, you **will not** be able to do the problems.

3.5 Exercises

1. *Premises*: "No misers are generous," and "Some old persons are not generous."
Conclusion: "Some old persons are misers."

3. *Premises*: "All diligent students make A's," and "All lazy students are not successful."
Conclusion: "All diligent students are lazy."

5. *Premises*: "No kitten that loves fish is unteachable," and "No kitten without a tail will play with a gorilla."
Conclusion: "No unteachable kitten will play with a gorilla."

7. In the diagram, the circle P (professors) must be inside the circle W (wise), and the point B (Ms. Brown) must be inside the circle P, so the argument is valid.

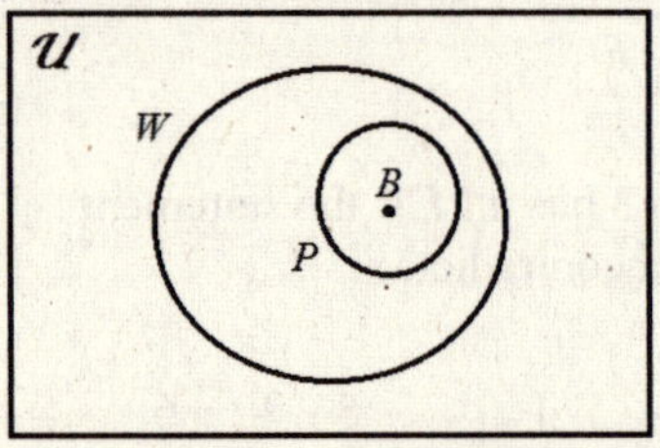

9. In the diagram, the circle D (drinkers) must be outside of both the circle H (healthy) and the circle J (joggers). The premises do not prevent the circles J and H from intersecting, so the argument is invalid.

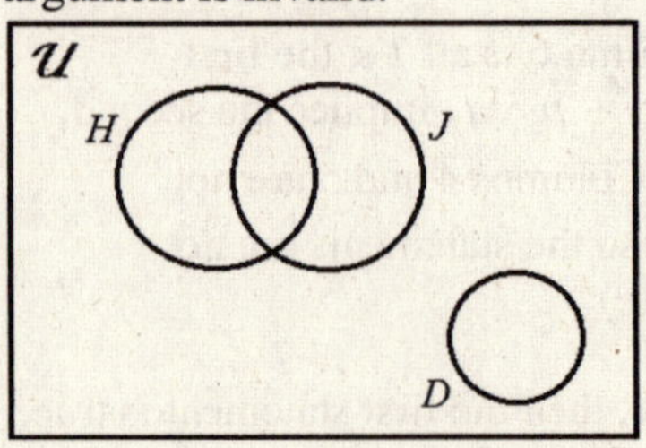

11. In the diagram, the circle *M* (men) must be inside the circle *F* (funny), and the dot *J* (Joey) must be inside the circle *M*, and so inside the circle *F*. Thus, the argument is valid.

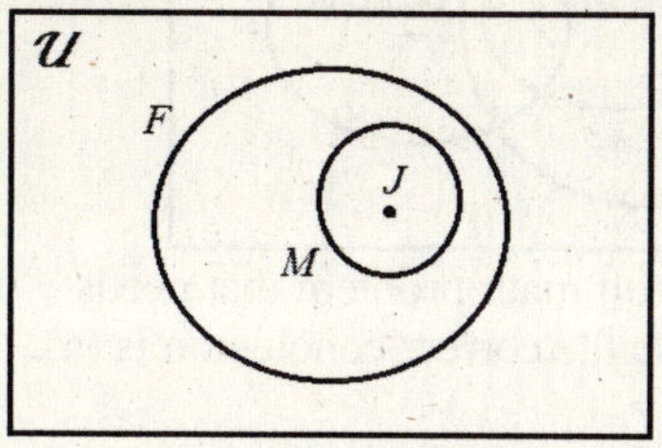

13. In the diagram, the circle *F* (felines) must be inside the circle *M* (mammals). The circle *D* (dogs) must not intersect the circle *F*, but the premises do not prevent circle *D* from intersecting circle *M*, so the argument is invalid.

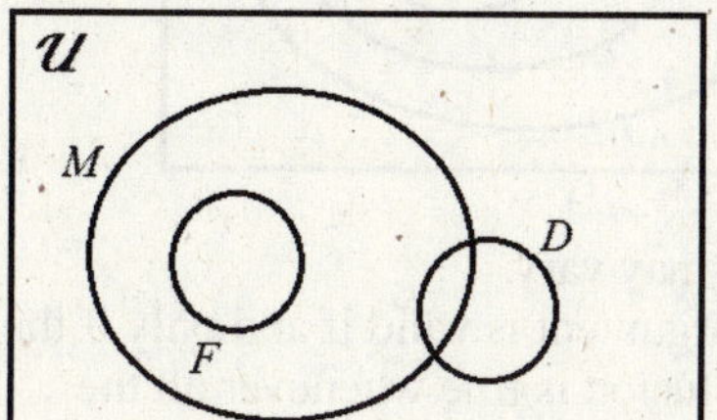

15. In the diagram, the circle *M* (Math teachers) must not intersect circle *W* (wealthy). The circle *P* (panthers) must not intersect the circle *M*, but may intersect the circle *W*, so the argument is invalid.

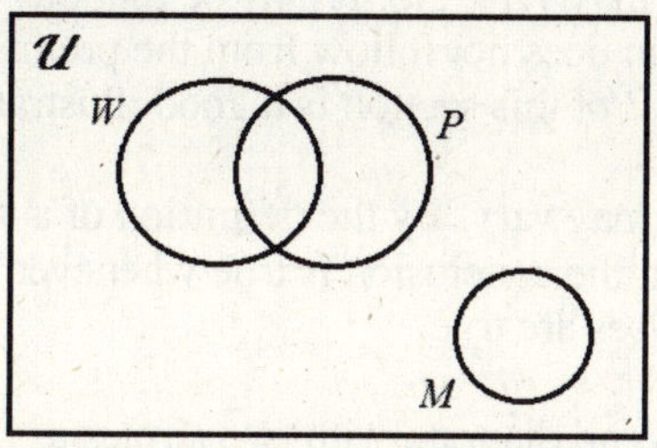

17. In the diagram, the circle *M* must be inside the circle *P* (publications). The *Ph* (Ph.D.) circle must intersect the circle *P*, but does not have to intersect the circle *M*, so the argument is invalid.

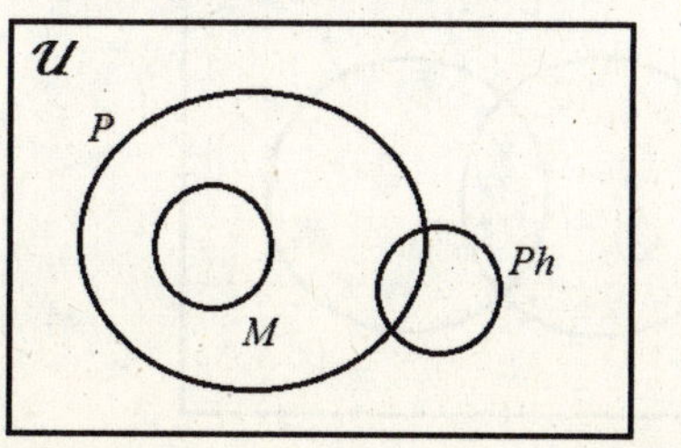

19. In the diagram, the circle *H* (heavy cars) must be inside the circle *C* (comfortable to ride in) and the circle *C* must be outside the circle *S* (shoddily built). Thus, circle *H* must lie outside of the circle *S*, so the argument is valid.

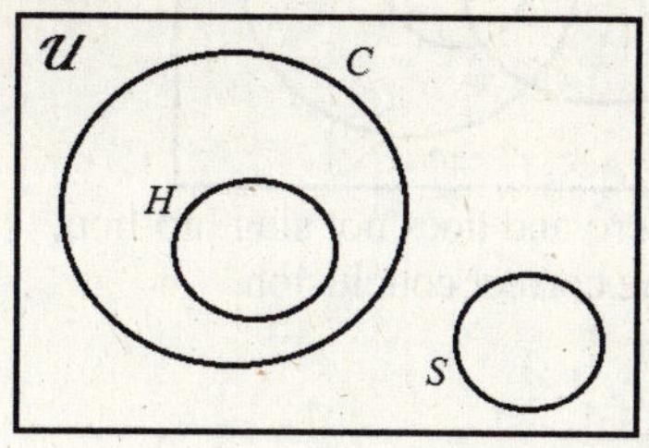

21. In the diagram, circle *B* (bulldogs) must be inside circle *U* (ugly). Point *D* (this dog) must be inside circle *U*, but not necessarily inside circle *B*, so the argument is invalid.

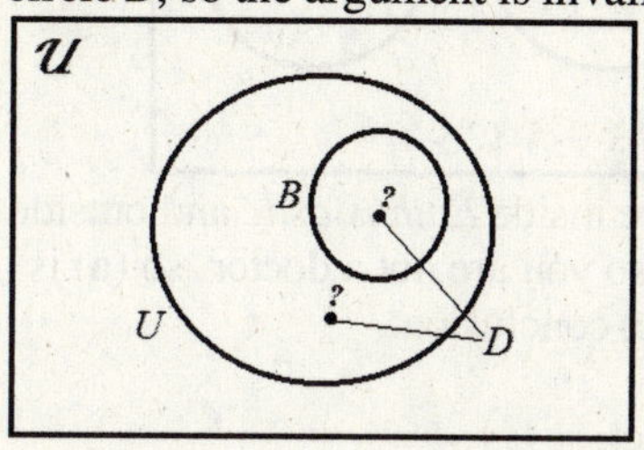

23. In the diagram, circle *A* (students who make A's) must be inside circle *S* (smart). The point *B* (student who made *B*) must be outside circle *A*, but can be inside or outside of circle *S*. Thus, the argument is invalid.

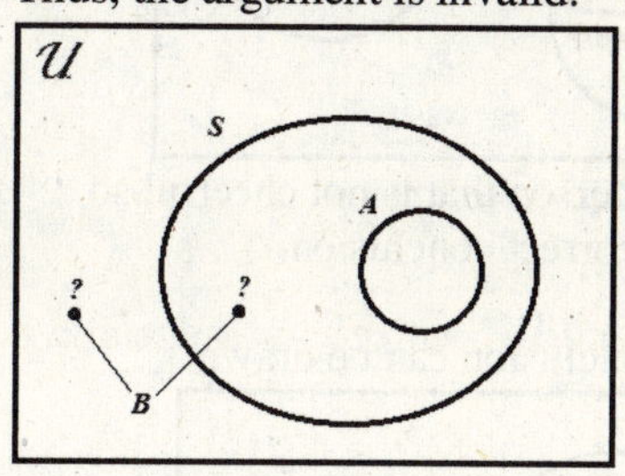

25. In the diagram, circle *F* (fishermen) must intersect circle *L* (lucky people) because some fishermen are lucky. Fred is unlucky, so he must be indicated outside of circle *L*, but the point Fred can be inside or outside the circle *F*. Thus, the argument is invalid.

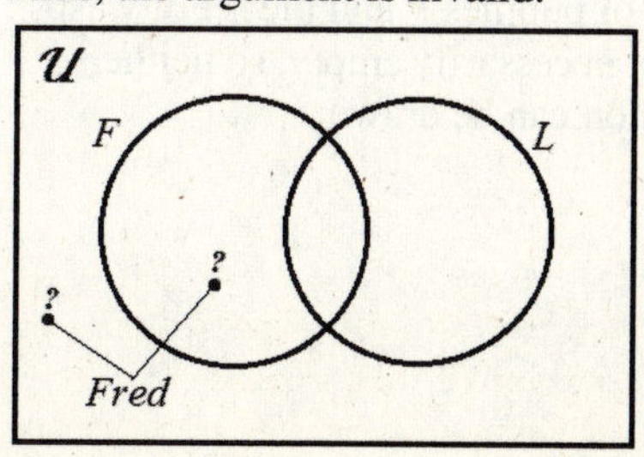

27. **b** is correct.

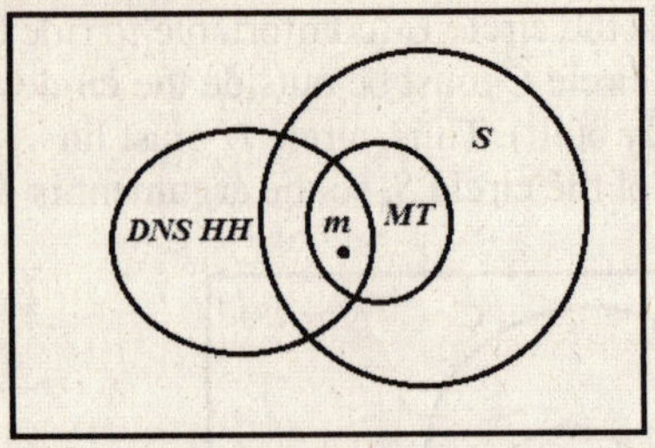

m is a salsero and does not sing hip-hop, so **(b)** is the correct conclusion.

29. **a** is correct.

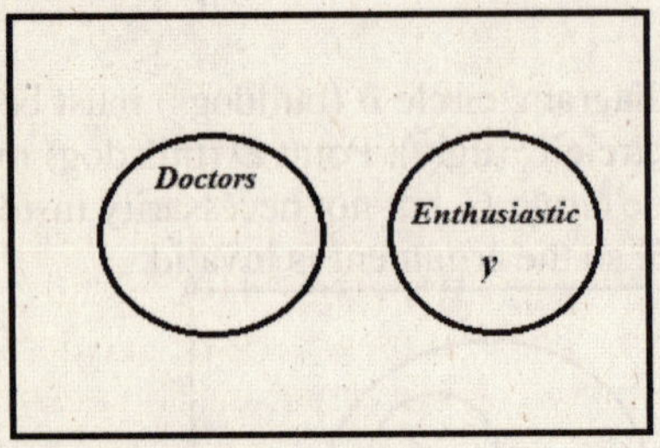

y (you) are inside *Enthusiastic* and outside *Doctors*, so you are not a doctor, so **(a)** is the correct conclusion.

31. **b** is correct.

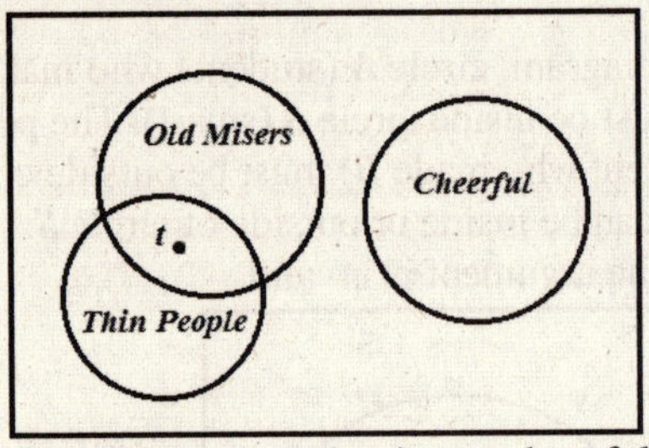

t is a thin person that is not cheerful so **(b)** is the correct conclusion.

33. Neither conclusion can be drawn.

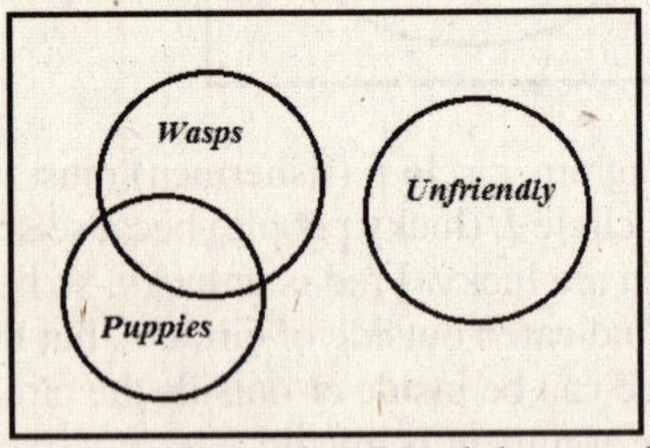

The set of puppies ***P*** and the set of wasps ***W*** is not necessarily empty, so neither conclusion can be drawn.

35. **a** is correct.

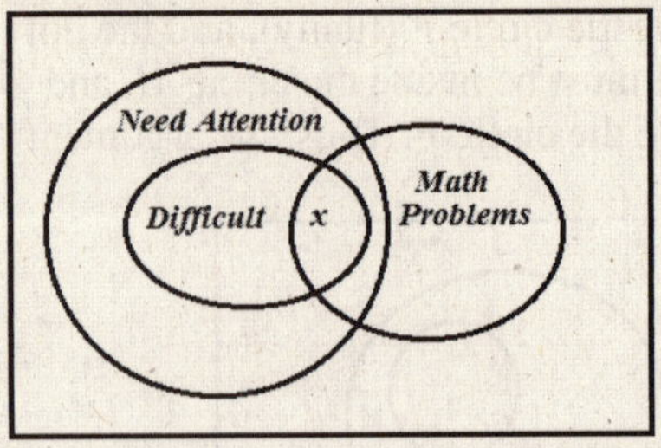

x is a difficult math problem that needs attention, so the correct conclusion is **(a).**

37. Statement **(b)** can be logically deduced from the diagram.

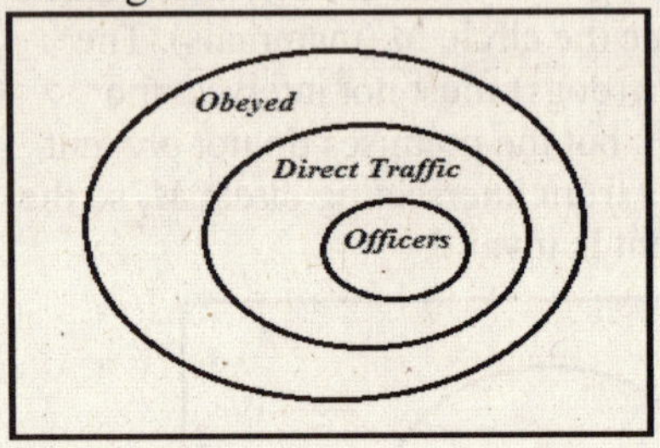

39. Answers may vary.
   a. An argument is valid if and only if the conclusion is true whenever all the premises are true. Thus, a valid argument reaches a true conclusion from true premises.
   b. No. By the definition stated in (a), if the premises are all true and the conclusion is false, then the argument is invalid.

41. Answers may vary. No. It may be that the conclusion does not follow from the premises. Example 2 of this section is a good illustration.

43. Answers may vary. By the definition of a valid argument, the conclusion is true whenever all the premises are true.

45. Draw a Venn diagram with two circles as shown: *M* for men and *F* for funny. The statement. "All men are funny," means that region 1 is empty. Thus, the point *J* (for Joey) must be in region 3. This shows that the argument is valid.

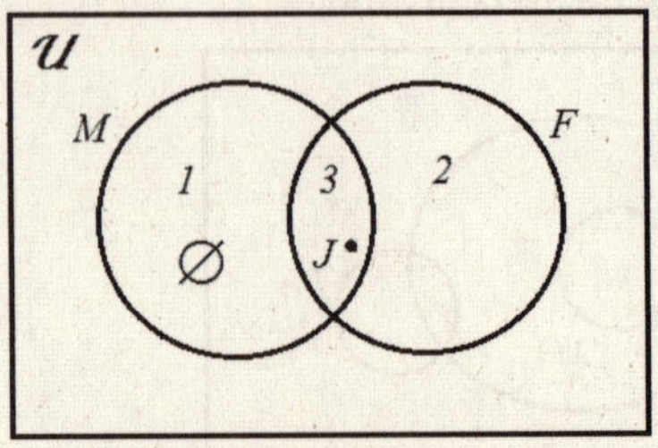

47. Draw a Venn diagram with three circles as shown: F for feline, M for mammals, and D for dogs. Because "All felines are mammals," regions 1 and 6 are empty. "No dog is a feline," makes region 7 empty. "This dog" can be in region 3 or 5, so the argument is invalid.

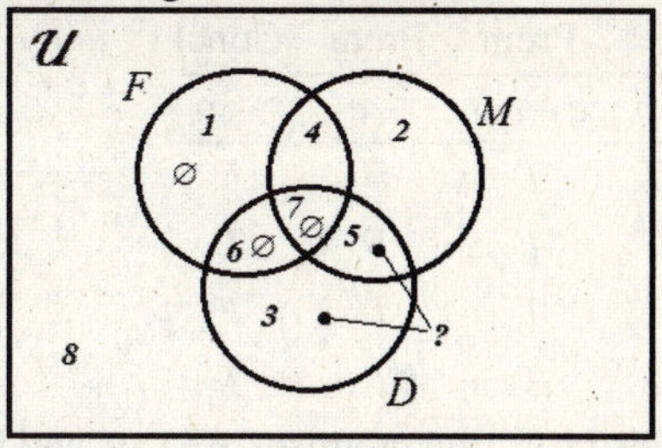

49. Draw a Venn diagram with three circles as shown: M for Math teachers, W for wealthy, and P for panthers. The first premise makes regions 4 and 7 empty. The second premise makes region 6 also empty. Since a given panther could be in region 3 or 5, the argument is invalid.

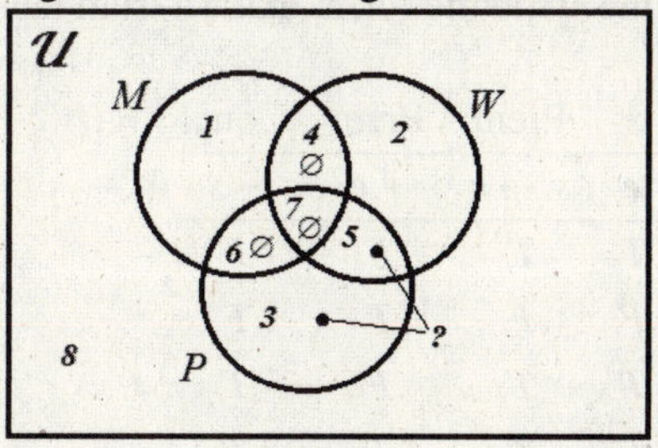

51. Draw a Venn diagram with three circles as shown: M for Math teachers, P for publications, and Ph for Ph.D's. The first premise makes regions 1 and 5 empty. The second premise means regions 6 and 7 can't both be empty. But region 7 might be empty, so the argument is invalid.

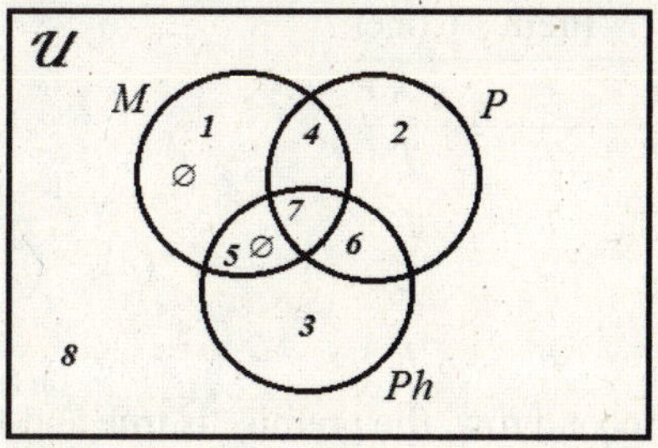

53. Draw a Venn diagram with three circles as shown: H for heavy cars, C for comfortable cars, and S for poorly-built cars. The first premise makes regions 1 and 5 empty, and the second premise makes regions 6 and 7 empty. Since the regions common to H and S (regions 6 and 7) are empty, the argument is valid.

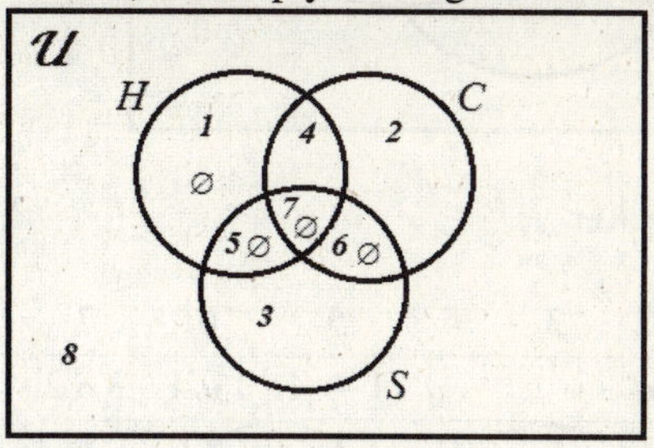

55. Draw a Venn diagram with three circles as shown: W (won't study), A (get A's), and S (students). The first premise makes regions 4 and 7 empty. The second means that region 5 is not empty, so the first conclusion is valid. Nothing in the premises makes region 6 or region 3 empty, so conclusions (b) and (c) are both invalid. Only conclusion **a** is valid.

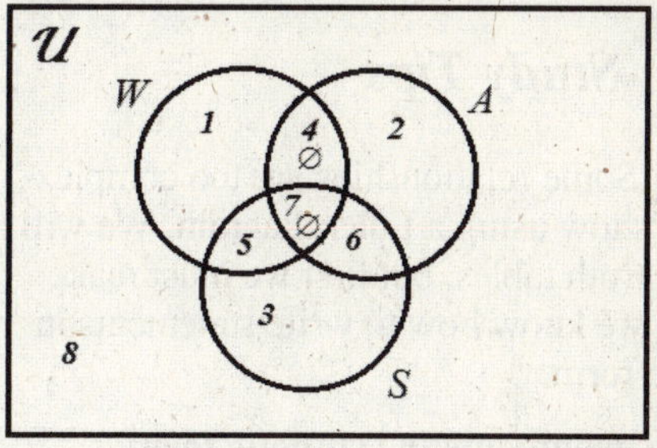

57. Yes, see the diagram. (If x is the empty set, then y and z could be disjoint.)

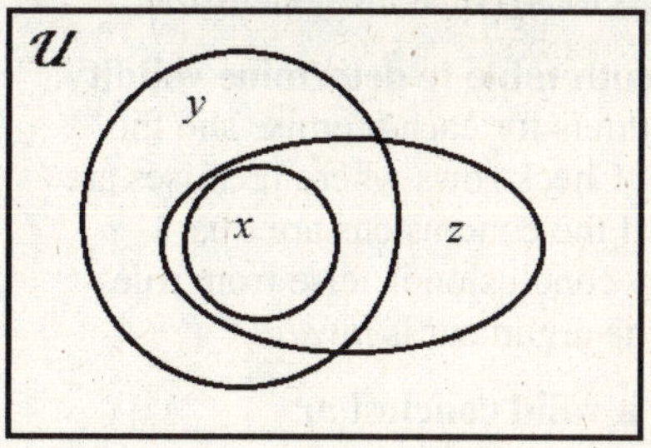

59. Some z's are y's. See the diagram.

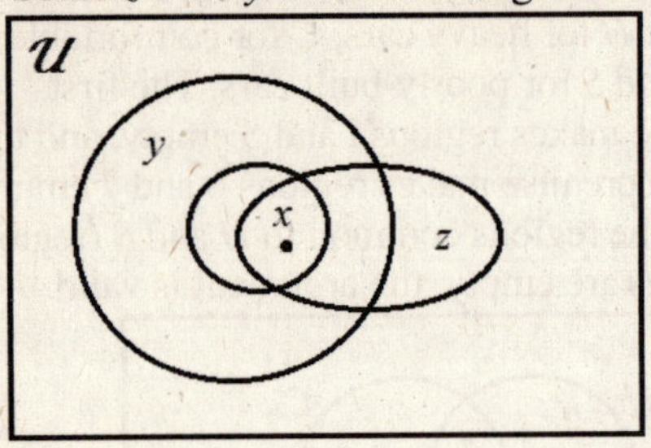

Skill Checker

1.

1	2	3	4	5	6	7
p	q	$\sim p$	$\sim q$	$1 \rightarrow 4$	$1 \vee 4$	$3 \wedge 2$
T	T	F	F	F	T	F
T	F	F	T	T	T	F
F	T	T	F	T	F	T
F	F	T	T	T	T	F

Section 3.6 Truth Tables and Validity of Arguments

Study Tips

Some relationships are too complex to easily show using a Euler diagram. We will return to truth tables, but first we must make sure that we know how to write statements in symbolic form.

To write arguments in symbolic form:
Think: $p, q, r, s, \wedge, \vee, \sim, \rightarrow, \leftrightarrow, \therefore$
Note: Identify the statements; express relationships to arrive at the conclusion.

To use a truth table to determine validity:
Think: Columns for each premise and the conclusion. Check rows where premises are true to see if the conclusions are true.
Note: If any conclusion is false from true premises, the argument is *invalid.*

To supply a valid conclusion:
Think: Rearrange premises.
Note: A premise may need to be rewritten as its contrapositive.

3.6 Exercises

1. $e \rightarrow p$

$\sim e$

$\therefore \sim p$

Using the symbolic form, make a truth table:

		Prem	Prem	Concl
e	p	$e \rightarrow p$	$\sim e$	$\sim p$
T	T	T	F	F
T	F	F	F	T
F	T	T	T	F
F	F	T	T	T

In the third row, the premises are true and the conclusion is false. Thus, the argument is invalid.

3. $s \rightarrow e$

$\sim e$

$\therefore \sim s$

Using the symbolic form, make a truth table:

		Prem	Prem	Concl
s	e	$s \rightarrow e$	$\sim e$	$\sim s$
T	T	T	F	F
T	F	F	T	F
F	T	T	F	T
F	F	T	T	T

In the fourth row, the premises are true and the conclusion is true. Thus, the argument is valid.

5. g

$\therefore g \wedge r$

Using the symbolic form, make a truth table:

		Prem	Concl
g	r	g	$g \wedge r$
T	T	T	T
T	F	T	F
F	T	F	F
F	F	F	F

In the second row, the premise is true and the conclusion is false, so the argument is invalid.

7. $w \rightarrow m$

$\sim w \rightarrow g$

$\therefore m \vee g$

Using the symbolic form, make a truth table:

			Prem	Prem	Concl
w	m	g	$w \to m$	$\sim w \to g$	$m \vee g$
T	T	T	T	T	T
T	T	F	T	T	T
T	F	T	F	T	T
T	F	F	F	T	F
F	T	T	T	T	T
F	T	F	T	F	T
F	F	T	T	T	T
F	F	F	T	F	F

Rows 1, 2, 5, and 7 are the only rows where both premises are true. In these rows the conclusion is also true, so the argument is valid.

9. $t \to b$

$\underline{t}$

$\therefore b$

Using the symbolic form, make a truth table:

		Prem	Prem	Concl
t	b	$t \to b$	t	b
T	T	T	T	T
T	F	F	T	F
F	T	T	F	T
F	F	T	F	F

The first row is the only one in which both premises are true. In that row, the conclusion is also true, so the argument is valid.

11. $s \to f$

$\underline{s}$

$\therefore f$

Using the symbolic form, make a truth table:

		Prem	Prem	Concl
s	f	$s \to f$	s	f
T	T	T	T	T
T	F	F	T	F
F	T	T	F	T
F	F	T	F	F

The first row is the only one where both premises are true. In that row, the conclusion is also true, so the argument is valid.

13. $m \to e$

$\underline{\sim m}$

$\therefore \sim e$

Using the symbolic form, make a truth table:

		Prem	Prem	Concl
m	e	$m \to e$	$\sim m$	$\sim e$
T	T	T	F	F
T	F	F	F	T
F	T	T	T	F
F	F	T	T	T

In the third row, both premises are true and the conclusion is false, so the argument is invalid.

15. $f \to s$

$\underline{\sim f}$

$\therefore \sim s$

Using the symbolic form, make a truth table:

		Prem	Prem	Concl
f	s	$f \to s$	$\sim f$	$\sim s$
T	T	T	F	F
T	F	F	F	T
F	T	T	T	F
F	F	T	T	T

In the third row, both premises are true and the conclusion is false, so the argument is invalid.

17. Using the symbolic form, make a truth table:

		Prem	Prem	Concl
p	q	$p \vee q$	$\sim p$	q
T	T	T	F	T
T	F	T	F	F
F	T	T	T	T
F	F	F	T	F

In the third row, both premises are true and the conclusion is true, so the argument is valid.

19. Using the symbolic form, make a truth table:

		Prem	Prem	Concl
p	q	$p \to q$	$\sim p$	q
T	T	T	F	T
T	F	F	F	F
F	T	T	T	T
F	F	T	T	F

In rows 3 and 4, both premises are true but the conclusion in row 4 is false, so the argument is invalid.

21. Using the symbolic form, make a truth table

					Prem	Prem	Concl
p	q	r	$\sim p$	$\sim r$	$p \to q$	$q \to r$	$\sim r \to \sim p$
T	T	T	F	F	T	T	T
T	T	F	F	T	T	F	F
T	F	T	F	F	F	T	T
T	F	F	F	T	F	T	F
F	T	T	T	F	T	T	T
F	T	F	T	T	T	F	T
F	F	T	T	F	T	T	T
F	F	F	T	T	T	T	T

Rows 1, 5, 7, and 8 are the only rows where both premises are true. In these rows the conclusion is also true, so the argument is valid.

23. Use the hint and replace $\sim q \vee r$ by $q \to r$. The premises are then $p \to q$ and $q \to r$, so that a valid conclusion is $p \to r$.

25. Replace the premise $\sim s \to \sim r$ by its contrapositive, $r \to s$. The premises can then be written p, $p \to q$, $q \to r$, and $r \to s$. Thus, a valid conclusion using all the premises is s.

27. Replace the disjunction $\sim p \vee r$ by its equivalent, $\sim r \to \sim p$. The premises can then be written ~r, $\sim r \to \sim p$, and $\sim p \to q$. Thus, a valid conclusion using all the premises is q.

29. $p \to ro \to \sim r$ (p: politicians; ro: run for office; ~r: not reliable)
Conclusion: If you are a politician you are not reliable or equivalently, No politicians are reliable.

31. Since there is at least one student that is intelligent, that student has to be a snob (because all intelligent people are snobs). Thus, we can conclude: Some students are snobs.

33. Let n be "I do not recommend them," h be "The books are healthy in tone," b be "The books are bound," w be "The books are well written," and r be "The books are romances." Then the symbolic forms of the given premises are: $n \to \sim h$, $b \to w$, $r \to h$, and $\sim b \to n$. Replace the first and last of these premises by their contrapositives, $h \to \sim n$, and $\sim n \to b$. Then arrange the premises in the order: $r \to h$, $h \to \sim n$, $\sim n \to b$, $b \to w$.
You can now read a conclusion that uses all the premises: $r \to w$, or in words, "All romances are well written."

35. Let p be "You are patriotic," v be "You vote," a be "You are an aardvark," and e be "You have emotions." Then, the symbolic forms of the given premises are: $\sim p \to \sim v$, $a \to \sim e$, and $\sim e \to \sim p$. Now arrange these in the order: $a \to \sim e$, $\sim e \to \sim p$, $\sim p \to \sim v$. A conclusion that uses all the premises is: $a \to \sim v$, or in words, "Aardvarks do not vote."

37. There is at least a college student that does not go to class but that student will not pass the course (because no college student who doesn't go to class will pass the course). Thus, we can conclude that: Some college students will not pass the course, that is conclusion (c).

39. Let t: you are a teacher, tc: you teach classes and d: you are dumb. The symbolic form of the premises is: $tc \to \sim d$, $t \to tc$.
We can conclude that $t \to \sim d$. This means that if you are a teacher you are not dumb or equivalently "No teacher is dumb," that is, conclusion (b).

41. Conclusion (c) makes the argument valid. This can be seen by letting p be "I drive to work," q be "I will not be late," and n be "I do not lose any pay." Then the premises can be written as $p \to q$ and $q \to n$, from which it follows that $p \to n$, which is conclusion (c).

43. Let p be "All persons pay their bills on time," and q be "Some collection agencies are needed." Then the first premise is $p \to \sim q$ and the second premise is q. The contrapositive of the first premise is $q \to \sim p$. Thus, we have q and $q \to \sim p$, so a valid conclusion is ~p. This is conclusion (b).

45. Let e be "Bill studies economics," g be "Bill makes good money," b be "Bill studies business procedures." The premises are

$e \to g, b \to g$, and $e \wedge \sim b$. If $e \wedge \sim b$ is true, then e is true and b is false. If b is false, then the second premise $b \to g$ with b false, allows g to be either true or false. However, if g is false $e \to g$ is also false, so g must be true. Thus, a logical conclusion is stated in (b). Note that statement (c) cannot be deduced from the given premises as a logical conclusion.

47. You read X magazine.

49. Answers may vary. "Affirming the consequent" means that the "then" statement is affirmed and the "if" statement is taken as a valid conclusion. This is a fallacy because the statement $p \to q$ is true if p is false and q is either true or false.

51. Let f be "A kitten loves fish," u be "A kitten is unteachable," t be "A kitten has a tail," p be "A kitten will play with a gorilla," w be "A kitten has whiskers," and g be "A kitten has green eyes." Then, the symbolic forms of the given statements are: (1) $f \to \sim u$, (2) $\sim t \to \sim p$, (3) $w \to f$, (4) $\sim u \to \sim g$, (5) $\sim w \to \sim t$. Now replace statements (2) and (5) by their respective contra-positives: (2') $p \to t$ (5') $t \to w$. Then arrange the premises in the order (2'), (5'), (3), (1), (4): $p \to t$. $t \to w$, $w \to f$, $f \to \sim u$, $\sim u \to \sim g$. A conclusion using all these premises is: $p \to \sim g$ or, in words, "Kittens that will play with a gorilla do not have green eyes." (An equivalent conclusion is, "No kitten with green eyes will play with a gorilla.")

Chapter 3 Practice Test

Study Tips

In preparing for a test, not only is it important to be mentally prepared but also physically prepared.

1. Be sure to get plenty of sleep the night before the test. That means do not try to cram for your test til dawn!

2. Eat something before testing, but not heavy foods. Your body will function best when properly nurished.

3. Wear comfortable clothes. Dress appropriately so you will not be too hot or too cold. You do not want your focus to be diverted from the test .

Here are the solutions to the practice test.

1. (b), (c), (d), and (e) are statements. Each is either true or false. (a) is not a statement because people do not agree on what is good. (f) is a question, so is neither true nor false.

2. a. d: The number of years is divisible by 4. p: The year is a presidential election year. The logical connective is *if ... then ...*, $d \to p$.
   b. b: I love Bill. $\sim m$: Bill does not love me. The logical connective is *and*, $b \wedge \sim m$.
   c. e: A candidate is elected president of the United States. m: He receives a majority of the electoral college votes. The logical connective is *if and only if*, $e \leftrightarrow m$.
   d. s: Janet can make sense out of symbolic logic. f: She fails this course. The logical connective is *or*, $s \vee f$.
   e. s: Janet can make sense out of symbolic logic. The logical modifier is *not*, $\sim s$.

3. a. It is not the case that he is a gentleman and a scholar.
   b. He is not a gentleman, but he is a scholar.

4. a. I will go neither to the beach nor to the movies.
   b. I will either not stay in my room or not do my homework.
   c. Pluto is a planet.

5. a. Some cats are not felines.
   b. No dog is well trained.
   c. Some dogs are afraid of a mouse.

6. a. Joey does not study, and he will not fail this course.
   b. Sally studies hard, but she does not get an A in this course.

7. a. $p \leftrightarrow q$ b. $p \wedge q$ c. $\sim p$
   d. $p \to q$ e. $p \vee q$

8.

1	2	3	4	5	6	7
p	q	$\sim p$	$\sim q$	$p \vee q$	$3 \vee 4$	$5 \wedge 6$
T	T	F	F	T	F	F
T	F	F	T	T	T	T
F	T	T	F	T	T	T
F	F	T	T	F	T	F

9.

1	2	3	4	5
p	q	$\sim p$	$p \vee q$	$(p \vee q) \to \sim p$
T	T	F	T	F
T	F	F	T	F
F	T	T	T	T
F	F	T	F	T

10. Make a truth table for the given statement and statements a and b:

1	2	3	4	5	6	7	8
p	q	$\sim p$	$3 \vee 2$	$\sim q$	$3 \wedge 5$	$1 \wedge 5$	~ 7
T	T	F	T	F	F	F	T
T	F	F	F	T	F	T	F
F	T	T	T	F	F	F	T
F	F	T	T	T	T	F	T

Since columns 4 and 8 show the same truth values, *T F T T*, statement (b) is equivalent to the given statement.

11. When at least one of the statements "Sally is naturally beautiful," and "Sally knows how to use makeup," is true.

12. The premise, "2 + 2 = 5," is false, so the statement (a conditional) is true.

13. In making the truth table, keep in mind that a biconditional is true when both components have the same truth value, and is false in the other cases.

1	2	3	4	5	6
p	q	$p \to q$	$\sim p$	$2 \vee 4$	$3 \leftrightarrow 5$
T	T	T	F	T	T
T	F	F	F	F	T
F	T	T	T	T	T
F	F	T	T	T	T

14. You should keep in mind that, starting with $p \to q$, the converse is $q \to p$, the inverse is $\sim p \to \sim q$, and the contrapositive is $\sim q \to \sim p$.

a. If you make a golf score of 62 again, then you made it once.

b. If you do not make a golf score of 62 once, then you will not make it again.

c. If you do not make a golf score of 62 again, then you did not make it once.

15. a. $m \to p$ b. $p \to m$ c. $p \leftrightarrow m$

16. a. $b \to c$ b. $c \to b$ c. $b \leftrightarrow c$

17. *b* implies *a*; *b* implies *c*; *c* implies *a*.

18. Statement (b), $p \vee \sim p$, is always true, so it is a tautology.
Statement (a), $p \wedge \sim p$, is always false, so it is not a tautology, but is known as a contradiction.
Statement (c), $(p \to q) \leftrightarrow (\sim q \vee p)$, is not always true. For example, if *p* is false and *q* is true, then $p \to q$ is true, but $\sim q \vee p$ is false, so the biconditional is false. Thus, it is not a tautology.

19. Nothing in the premises tells whether the *J* (for John) is inside or outside of the circle *H*. Thus, the argument is invalid.

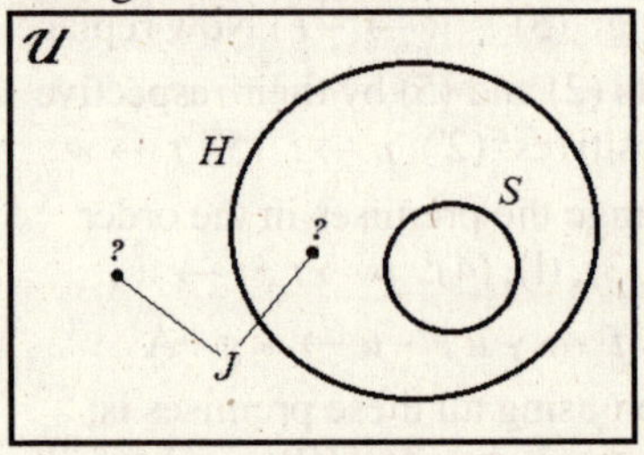

20. Nothing in the premises tells whether *S* (for Sally) goes inside the circle *L* or not. Thus, the argument is invalid.

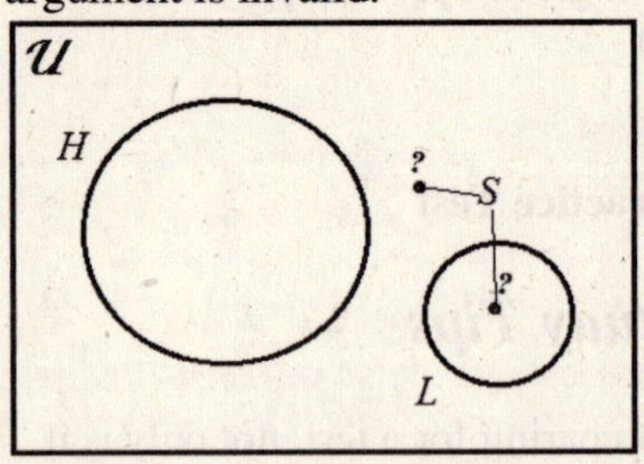

21. With s for "He is a student," and h for "He studies hard," the argument can be symbolized and a truth table constructed.

$$\begin{array}{l} s \to h \\ \underline{\sim s} \\ \therefore \sim h \end{array}$$

		Prem	Prem	Concl
s	h	$s \to h$	$\sim s$	$\sim h$
T	T	T	F	F
T	F	F	F	T
F	T	T	T	F
F	F	T	T	T

In the third row, both premises are true and the conclusion is false, so the argument is invalid.

22. With f for "Sally is a loafer," and h for "Sally works hard," the argument can be symbolized and a truth table constructed.

$$\begin{array}{l} f \to \sim h \\ \underline{\sim h} \\ \therefore f \end{array}$$

		Prem	Prem	Concl
f	h	$f \to \sim h$	$\sim h$	f
T	T	F	F	T
T	F	T	T	T
F	T	T	F	F
F	F	T	T	F

In the fourth row of the truth table, the premises are both true and the conclusion is false, so the argument is invalid.

23. With w for "You win the race," and r for "You are a good runner", the argument can be symbolized and a truth table constructed.

$$\begin{array}{l} w \to r \\ \underline{w} \\ \therefore r \end{array}$$

		Prem	Prem	Concl
r	w	$w \to r$	w	r
T	T	T	T	T
T	F	T	F	T
F	T	F	T	F
F	F	T	F	F

The first row of the table is the only row where the premises are both true. In this row the conclusion is also true, so the argument is valid.

24. Write the contrapositive of 2 which is $r \to p$, 1 which is $p \to q$ and the contrapositive of 3 which is $q \to \sim s$. We then have: $r \to p$, $p \to q$, $q \to \sim s$.

A valid conclusion is $r \to \sim s$, or its contrapositive $s \to \sim r$.

Chapter 4 Numeration Systems

Section 4.1 Egyptian, Babylonian, and Roman Numeration Systems

Study Tips

This chapter will give you an insight into the development of our number system by studying the Egyptian, Babylonian, and Roman systems. What are the differences and similarities of these systems?

The Egyptian System:
Think: base of 10, additive system
Note: Symbols and value

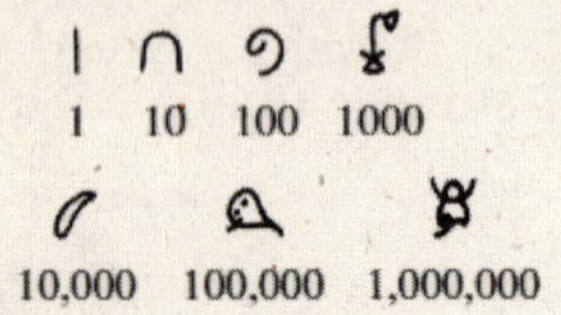

The Babylonian System:
Think: base of 60, positional, additive system
Note: Symbols and value

▼	◀	▼
1	10	60

The Roman System:
Think: positional; addition, subtraction, and multiplication principles
Note: Symbols and value

I	V	X	L	C	D	M
1	5	10	50	100	500	1000

FACT
The numeral system we use is the Hindu-Arabic system, which has base of 10 and is positional.

4.1 Exercises

1. ∩∩|||| 3. 9∩∩∩∩||

5. 9 999∩∩∩|||
   9999 ||

7. 113   9. 322

11. 11,232

13. ∩∩∩||||
    +∩∩ |||
    ∩∩∩∩∩|||||||

15. Step 1. Subtract one coil, two heel bones, and two strokes:

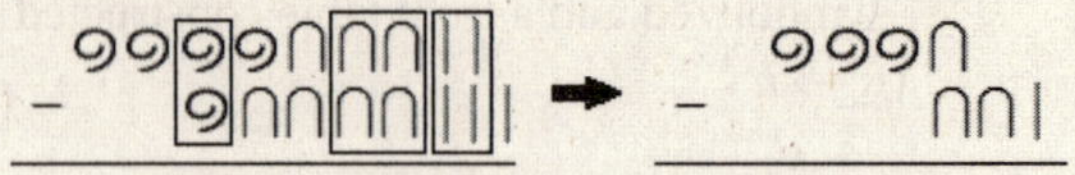

Step 2. In the top line, replace one coil by ten heel bones and one heel bone by ten strokes. This lets you do the remainder of the subtraction. Subtract two heel bones and one stroke. The answer is given as two coils, eight heel bones, and nine strokes.

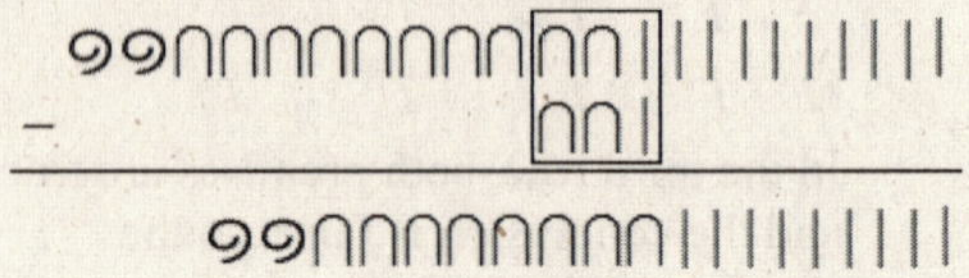

17.

\1	40
\2	80
\4	160
\8	320
15	600

19.

1	51
\2	102
\4	204
8	408
\16	816
22	1122

21.

18	32
9	[64]
4	128
2	256
1	[512]
	576

23.

12	51
6	102
3	[204]
1	[408]
	612

25. ▼▼▼▼▼▼

27. $32 = 3\times10+2 =$ ◀◀◀▼▼

29. $123 = 2\times60+3 =$ ▼▼ ▼▼▼

31. $258 = 4\times60+1\times10+8 =$
▼▼▼▼ ◀▼▼▼▼▼▼▼▼

33. $3733 = 1\times3600+2\times60+1\times10+3 =$
▼ ▼▼ ◀▼▼▼

35. One 60 plus three 10's plus two 1's = 92

37. Three 60's plus one 10 plus two 1's = 192

39. One 3600 plus twelve 60's plus two 1's = 4322

41. First add, then replace six 10's by one 60.

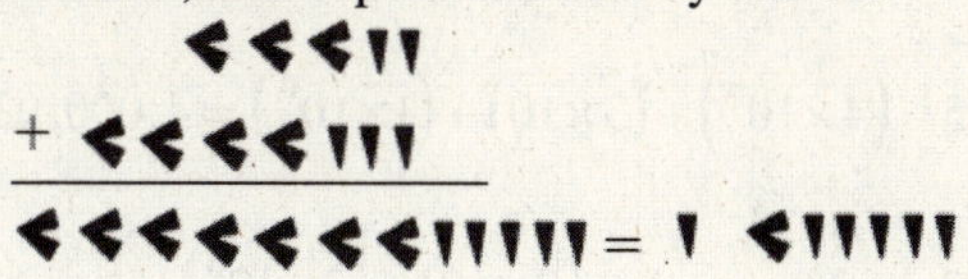

43. First add, then replace ten 1's by one 10.

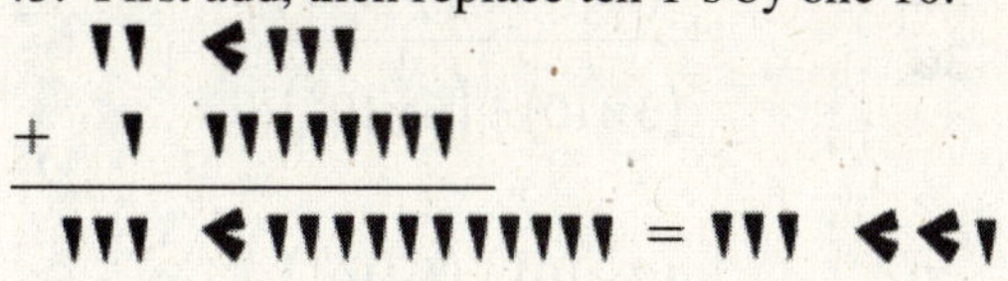

45. $\text{CXXVI} = 100 + 2\times10 + 5 + 1 = 126$

47. $\overline{\text{XLII}} = [(50-10)+2]\times1000 = 42{,}000$

49. $\overline{\text{XC}}\text{CDV} = \overline{\text{XC}}(\text{CD})\text{V}$
$= (100-10)\times1000 + (500-100) + 5$
$= 90{,}405$

51. $72 = 50+10+10+2 = \text{LXXII}$

53. $145 = 100 + (50-10) + 5 = \text{CXLV}$

55. $32{,}503 = (3\times10+2)\times1000 + 500 + 3$
$= \overline{\text{XXXII}}\text{DIII}$

57. J o h n
$1+6+8+5 = 20 \to 2+0 = 2$
F i t z g e r a l d
$6+9+2+8+7+5+9+1+3+4 = 54$
$\to 5+4 = 9$
K e n n e d y
$2+5+5+5+5+4+7 = 33 \to 3+3 = 6$
$2+9+6 = 17 \to 1+7 = 8$
$\to\to$ Lonely, misunderstood

59. R i n g o
$9+9+5+7+6 = 36 \to 3+6 = 9$
S t a r r
$1+2+1+9+9 = 22 \to 2+2 = 4$
$4+9 = 13 \to 1+3 = 4$
$\to\to$ Rebels, unconventional

61. Answers may vary. No. C is more than two steps larger than I, so this subtraction is not allowed. I may be subtracted from V or X only.

63. Answers may vary. The Babylonian system is a base 60 system and our decimal system is a base 10 system. Another important difference is the lack of a symbol for zero in the Babylonian system. The Babylonian system was not a good place system, it depended on spacing. The symbol for 1 was the same as that for 60, and only the spacing could show which was intended.

65. Answers may vary. The Egyptian system was based on 10 and the Babylonian on 60. The Egyptian system was not a positional system; it depended essentially on the addition of the symbol values. The Babylonian system used spacing to change symbol values.

67. Assume the answer is 6. $6+\frac{1}{6}\cdot6 = 7$, and $21\div7 = 3$. Hence, the correct answer is $3\cdot6 = 18$.

69. Assume the answer is 3. $3+\frac{2}{3}\cdot3 = 5$, and $5-\frac{1}{3}\cdot5 = \frac{10}{3}$ and $10\div\frac{10}{3} = 3$. Hence, the correct answer is $3\times3 = 9$.
(To avoid fractions, assume the answer is a multiple of 3, so start with 9. Then $9+\frac{2}{3}\cdot9 = 15$, and $15-\frac{1}{3}\cdot15 = 10$ and $10\div10 = 1$. Hence, the correct answer is $9\times1 = 9$.)

71. If you calculated $n^3 + n^2$ for $n = 8$, you found the result to be 5121 + 64 = 576. Thus, the answer is $n = 8$.

Section 4.2 The Hindu-Arabic (Decimal) System

Study Tips

How familiar are you really with the Hindu-Arabic system? To understand how to read, write and operate with numbers you need to know how to write numbers in expanded form using exponents.

To write in expanded form:
Think: place value names, either in word or value, and how many of each.
Note: An example,
312 = 3 hundreds + 1 ten + 2 ones
$= (3\cdot 100) + (1\cdot 10) + (2\cdot 1)$
The number 312 is in *decimal form*.

The meaning of exponents:
Think: shorthand for addition, a^n is a power with a the base and n the exponent.
Note: n tells you how many times to multiply a to itself. For example, $3^4 = 3\cdot 3\cdot 3\cdot 3$.

To multiply powers $a^m \times a^n$ (Product Law):
Think: bases the same, add the exponents
Note: Never multiply the bases together, only add exponents: a^{m+n}

To divide powers $a^m \div a^n$ (Quotient Law):
Think: bases the same, subtract the exponents
Note: Never divide the bases, only subtract exponents: a^{m-n}

To raise a power to a power $\left(a^m\right)^n$ (Power Law):
Think: shorthand for multiplication, multiply the exponents
Note: $\left(a^m\right)^n$ means $a^m \cdot a^m \cdot \ldots$ n times, so we multiply the exponents: $a^{m\times n}$

Exercise 4.2

1. $432 = \left(4\times 10^2\right) + \left(3\times 10^1\right) + \left(2\times 10^0\right)$

3. $2307 = \left(2\times 10^3\right) + \left(3\times 10^2\right) + \left(0\times 10^1\right) + \left(7\times 10^0\right)$

5. $12{,}349 = \left(1\times 10^4\right) + \left(2\times 10^3\right) + \left(3\times 10^2\right) + \left(4\times 10^1\right) + \left(9\times 10^0\right)$

7. $5^0 = 1$

9. $(4\times 10) + \left(5\times 10^0\right) = 45$

11. $\left(9\times 10^3\right) + (7\times 10) + \left(1\times 10^0\right) = 9071$

13. $\left(7\times 10^5\right) + \left(4\times 10^4\right) + \left(8\times 10^3\right) + \left(3\times 10^2\right) + \left(8\times 10^0\right) = 748{,}308$

15. $\left(4\times 10^6\right) + (3\times 10) + \left(1\times 10^0\right) = 4{,}000{,}031$

17.
$$\begin{array}{r} 23 \\ \underline{+13} \\ 36 \end{array} \qquad \begin{array}{r} (2\times 10) + \left(3\times 10^0\right) \\ \underline{+(1\times 10) + \left(3\times 10^0\right)} \\ (3\times 10) + \left(6\times 10^0\right) \end{array}$$

19.
$$\begin{array}{r} 71 \\ \underline{+23} \\ 94 \end{array} \qquad \begin{array}{r} (7\times 10) + \left(1\times 10^0\right) \\ \underline{+(2\times 10) + \left(3\times 10^0\right)} \\ (9\times 10) + \left(4\times 10^0\right) \end{array}$$

21.
$$\begin{array}{r} 76 \\ \underline{-54} \\ 22 \end{array} \qquad \begin{array}{r} (7\times 10) + \left(6\times 10^0\right) \\ \underline{-\left[(5\times 10) + \left(4\times 10^0\right)\right]} \\ (2\times 10) + \left(2\times 10^0\right) \end{array}$$

23.
$$\begin{array}{r} 84 \\ \underline{-31} \\ 53 \end{array} \qquad \begin{array}{r} (8\times 10) + \left(4\times 10^0\right) \\ \underline{-\left[(3\times 10) + \left(1\times 10^0\right)\right]} \\ (5\times 10) + \left(3\times 10^0\right) \end{array}$$

25. $7^8 \times 7^3 = 7^{8+3} = 7^{11}$

27. $6^{19} \times 6^{21} = 6^{19+21} = 6^{40}$

29. $6^{10} \div 6^3 = 6^{10-3} = 6^7$

31. $6^{12} \div 6^0 = 6^{12-0} = 6^{12}$

33. $\left(5^4\right)^3 = 5^{4\times 3} = 5^{12}$

35. $\left(10^3\right)^{10} = 10^{3\times 10} = 10^{30}$

37. $\begin{array}{r} 25 \\ \underline{\times 51} \\ 25 \\ \underline{125} \\ 1275 \end{array}$

$$\begin{array}{r} (2\times10)+(5\times10^0) \\ \underline{\times(5\times10)+(1\times10^0)} \\ (2\times10)+(5\times10^0) \\ \underline{(10\times10^2)+(25\times10)} \\ 10^3+(27\times10)+(5\times10^0) \end{array}$$

$=1\times10^3+2\times10^2+7\times10+5\times10^0=1275$

39. $\begin{array}{r} 62 \\ \underline{\times 25} \\ 310 \\ \underline{124} \\ 1550 \end{array}$

$$\begin{array}{r} (6\times10)+(2\times10^0) \\ \underline{\times(2\times10)+(5\times10^0)} \\ (30\times10)+(10\times10^0) \\ \underline{(12\times10^2)+(4\times10)} \\ 12\times10^2+(34\times10)+(10\times10^0) \end{array}$$

$=1\times10^3+5\times10^2+5\times10+0\times10^0=1550$

41. $\begin{array}{r} 8 \\ 8\overline{)64} \\ \underline{64} \\ 0 \end{array}$

$$\begin{array}{r} 8\times10^0 \\ 8\times10^0\overline{)(6\times10)+(4\times10^0)} \\ \underline{(6\times10)+(4\times10^0)} \\ 0 \end{array}$$

43. $\begin{array}{r} 12 \\ 6\overline{)72} \\ \underline{6} \\ 12 \\ \underline{12} \\ 0 \end{array}$

$$\begin{array}{r} (1\times10)+(2\times10^0) \\ 6\times10^0\overline{)(7\times10)+(2\times10^0)} \\ \underline{(6\times10)} \\ (1\times10)+(2\times10^0) \\ \underline{(1\times10)+(2\times10^0)} \\ 0 \end{array}$$

45. $4\times75{,}000=300{,}000=3\times10^5$ trees saved

47. $1.7\times(45.5\times10^6)=77.35\times10^6$

$=7.735\times10^7$ dogs

49. $\dfrac{4.8\times10^7}{30}=0.16\times10^7=1.6\times10^6$ searches/day

51. $200\times(3.20\times10^8)=640\times10^6=6.4\times10^8$ lb/yr

53. $58.8\times(3.20\times10^8)=188.16\times10^8$

$=1.8816\times10^{10}$ lb/yr

55. $24\times(3.20\times10^8)=76.8\times10^8$

$=7.68\times10^9$ gal/yr

57. $\dfrac{(2.7\times10^8)\times(2\times10^3)}{(3\times10^8)\times(3.6\times10^2)}=\dfrac{5.4\times10^{11}}{10.8\times10^{10}}$

$=0.5\times10=5$ lb

59. $(1.8\times10^3)\times(4.2\times10^6)=7.56\times10^9$

$=7{,}560{,}000{,}000$ lb

61. Answers may vary. You must add the exponents to obtain a^{m+n}.

63. Answers may vary. You must multiply the exponent m by n to obtain a^{mn}.

65. 7 women, $7^2=49$ mules. $7^3=343$ sacks $7^4=2401$ loaves. $7^5=16{,}807$ knives. $7^6=117{,}649$ sheaths. Thus there were: $7+49+343+2401+16{,}807+117{,}649$ $=137{,}256$ in all on the road to Rome.

Section 4.3 Number Systems with Bases Other Than 10

Study Tips

The *base* of a system tells you how the numbers are grouped in that system, so in binary base we group by 2's but in base 5 we group by 5's. Numbers in bases other than **10** are used in science and computers. These ideas are used in the *bar codes* used to price and identify merchandise in stores and the *Postnet code* used by the U.S. Postal Service in delivering your mail.

To change to base 10:
Think: expanded notation, use the base given
Note: $101_{two} = 1 \cdot 2^2 + 0 \cdot 2^1 + 1 \cdot 2^0$ is equal to 5 in base 10.

Caution
When writing expanded notation, the exponent on the base furthest to the right is always zero (b^0), not one (b^1).

To change from base 10:
Think: divide by target base, read remainders upward
Note: The remainder will make up your answer.

Word of Advice
Just remember that to change a number from base 10 to any base *b*, you must divide the base 10 number by *b*.

4.3 Exercises

1. 22_{three}

3. 31_{four}

5. [********] ******* $= 17_{eight}$

7. [*******] [*******] * $= 21_{seven}$

9. $42_{five} = 4 \times 5 + 2 = 22$

11. $213_{eight} = 2 \times 8^2 + 1 \times 8 + 3$
$= 128 + 8 + 3 = 139$

13. $11011_{two} = 2^4 + 2^3 + 2 + 1$
$= 16 + 8 + 2 + 1 = 27$

15. $123_{sixteen} = 1 \times 16^2 + 2 \times 16 + 3$
$= 256 + 32 + 3 = 291$

17. 5)15
3|0 $15 = 30_{five}$

19. 2)28
2)14|0
2) 7|0
2) 3|1
1|1 $28 = 11100_{two}$

21. 16)25
1|9 $25 = 19_{sixteen}$

23. 6)25
4|1 $25 = 41_{six}$

25. 7)64
7) 9|1
1|2 $64 = 121_{seven}$

27. 8)38
4|6 $38 = 46_{eight}$

29. 16)1467
16) 91|11
5|11 $1467 = 5BB_{sixteen}$

31. 2)73
2)36|1
2)18|0
2) 9|0
2) 4|1
2) 2|0
1|0 $73 = 1001001_{two}$

8)73
8) 9|1
1|1 $73 = 111_{eight}$

33. 00110 01010 00101 00011 01100
3 5 2 1 6

35. 01001 00101 00011 00110 01010
    4 2 1 3 5

37.

39.

41. Answers may vary. Yes. The Zip Code + 4 consists of 9 numbers and each number is represented by 5 bars, so there should only be 45 bars and the two long bars at the beginning and the end of the Zip Code + 4 numbers for a total of 47 bars. As it turns out, the other 5 bars are the *checking* digit.

43. Answers may vary. One of the meanings of binary is "based on two." The prefix bi means "two."

45. Answers may vary. Hexadecimal means "based on 16." The prefix hexa means "six" and the "deci" means ten to make sixteen.

47. The trick works because the columns correspond to the binary digits in the number. For instance, $6 = 110_{two}$ and this corresponds to the number 6 = 2 + 4, the numbers that head columns *B* and *C*. Note that 6 occurs in columns *B* and *C*, but not in *A*.

49. Use the same procedure as for the numbers from l to 7, but with five columns instead of three. If you label the columns *A, B, C, D, E*, then the numbers l, 2, 4, 8, and 16 would head these columns. The number 13, for example, would be put into columns *A, C*, and *D* because $13 = 01101_{two}$ which means that $13 = 1 + 4 + 8$, the numbers that head these three columns.

Calculator Corner

1. Key in the following and the calculator will show the answer 13:
   [1] [×] [2] [+] [1] [=] [×] [2] [+] [0] [=] [×] [2] [+] [1] [=]

3. Key in the following and the calculator will show the answer 113:
   [4] [×] [5] [+] [2] [=] [×] [5] [+] [3] [=]

5. Key in the following and the calculator will show the answer 1914:
   [3] [×] [8] [+] [5] [=] [×] [8] [+] [7] [=] [×] [8] [+] [2] [=]

Section 4.4 Binary, Octal, and Hexadecimal Arithmetic

Study Tips

If you are familiar with computers, you may know that computer characters (like the letter A) are written in ASCII code using binary values. For example, the letter A has the value 01000001_2. In this section we study *binary*, *octal*, and *hexadecimal* arithmetic.

To add, multiply, divide in base 2, 8, and 16:
Think: similar to base 10, refer to tables
Note: You are also to subtract in base 2 and 8.

Word of Advice
Go back in time and recall how you learned the addition tables for base 10 so you could do arithmetic. It's the same idea here. You must know the addition and multiplication facts (**Tables 4.6, 4.7, 4.8, 4.9, 4.10** and **4.11**) but this time, instead of doing base 10, we will be doing binary arithmetic (base 2), octal arithmetic (base 8), or hexadecimal arithmetic (base 16). Start this section by studying these tables.

Caution
Be aware that in base 16 the symbols A, B, C, D, E and F are used for the decimal numbers 10, 11, 12, 13, 14, and 15. Just as in the decimal system the number 12 is 10 + 2, the hexadecimal number $F2 = 15 \times 16 + 2 = 242$ in the decimal system.

4.4 Exercises

1. $^{1}111_2$
$+10_2$
1001_2

3. $^{1}1101_2$
$+110_2$
10011_2

5. $^{1}1^{1}10_2$
101_2
$+111_2$
10010_2

7. $^{1}5^{1}31_8$
$+47_8$
600_8

9. $^{1}7^{1}2^{1}56_8$
$+634_8$
10112_8

11. $^{1}3^{1}CB_{16}$
$+4C_{16}$
417_{16}

13. $9^{1}8D_{16}$
$+2B_{16}$
$9B8_{16}$

15. 111_2
-10_2
101_2

17. $\not{1}^{1}0^{1}0^{1}0_2$
-111_2
1_2

19. 1111_2
-101_2
1010_2

21. $5\,^{2}\not{3}\,^{1}4_8$
-25_8
507_8

23. $^{2}\not{3}\,^{1}2\,^{5}\not{6}\,^{1}4_8$
-756_8
2306_8

25. 110_2
$\times 11_2$
110_2
1100_2
10010_2

27. 1111_2
$\times 11_2$
1111_2
11110_2
101101_2

29. 1011_2
$\times 101_2$
1011_2
101100_2
110111

31. 57_8
$\times 6_8$
432_8

33. 216_8
$\times 32_8$
434
652
7154_8

35. $2C5_{16}$
$\times 3B_{16}$
37
84
16
F
24
6
$A367_{16}$

37. $6F3_{16}$
$\times AB_{16}$
21
A5
42
1E
96
3C
$4A451_{16}$

39. $10_2 \overline{)1101_2}$, quotient $110_2\,R1_2$
10
10
10
01

41. $11_2 \overline{)1110_2}$, quotient $100_2\,R10_2$
11
010

43. $101_2 \overline{)111011_2}$, quotient $1011_2\,R100_2$
101
1001
101
1001
101
100

45. $7_8 \overline{)317_8}$, quotient $35_8\,R4_8$
25
47
43
4

47. $15_8 \overline{)4215_8}$, quotient $250_8\,R5_8$
32
101
101
05

49. Answers will vary. The main reason is that there are only two digits in the binary system contrasted with 16 digits in the hexadecimal system. Compare the addition and multiplication tables for the two systems.

51. Answers will vary depending on the numbers chosen.

53. $110111_2 = 2^5 + 2^4 + 2^2 + 2 + 1$

$= 32 + 16 + 4 + 2 + 1 = 55$

55. Since white is represented by 000000_2, the next smallest number, 000001_2, would represent the lightest shade of gray that is not white.

Chapter 4 Practice Test

Study Tips

For your convenience, a list of most of the symbols used in the Babylonian, Egyptian and Roman numeration systems appears in the *Summary*. Consult your instructor to see if you have to memorize all these symbols or if they will be available to you when you are taking the test. Make sure you are familiar with exponents and their uses, since they are needed in many of the problems in the practice test. Here are the solutions to the practice test.

1. a. 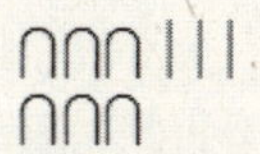

   b. 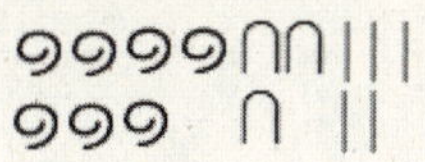

2. a. $10 + 10 + 1 + 1 + 1 = 23$
   b. $100 + 10 + 10 + 1 = 121$

3. a. 

   b. 

4. a. $60 + 10 + 10 + 1 + 1 = 82$
   b. $60 + 60 + 10 + 1 = 131$

5. a.

\1	21
\2	42
\4	84
8	168
\16	336
23	483

b.

23	[21]
11	[42]
5	[84]
2	168
1	[336]
	483

6. a. 67 b. 48,000

7. a. LIII b. XLII c. $\overline{\text{XXII}}$

8. a. $2\times10^3 + 5\times10^2 + 0\times10 + 7\times10^0$
   b. $1\times10^2 + 8\times10 + 9\times10^0$

9. a. 3702 b. 59,040

10. a. $75 + 32 = 107$

$$\begin{array}{r} (7\times10)+(5\times10^0) \\ +(3\times10)+(2\times10^0) \\ \hline (10\times10)+(7\times10^0) \end{array}$$

$= (1\times10^2) + (7\times10^0) = 107$

b. $56 - 24 = 32$

$$\begin{array}{r} (5\times10)+(6\times10^0) \\ -(2\times10)+(4\times10^0) \\ \hline (3\times10)+(2\times10^0) = 32 \end{array}$$

11. a. $3^4 \times 3^8 = 3^{4+8} = 3^{12}$
    b. $2^9 \div 2^3 = 2^{9-3} = 2^6$

12. a.

$$\begin{array}{r} 83 \\ \times21 \\ \hline 83 \\ 166 \\ \hline 1743 \end{array}$$

$$\begin{array}{r} (8\times10)+(3\times10^0) \\ \times(2\times10)+(1\times10^0) \\ \hline (8\times10)+(3\times10^0) \\ (16\times10^2)+(6\times10) \\ \hline 16\times10^2+(14\times10)+(3\times10^0) \end{array}$$

$= 1\times10^3 + 7\times10^2 + 4\times10 + 3\times10^0 = 1743$

b.

$$\begin{array}{r} 7\text{R}5 \\ 7\overline{)54} \\ 49 \\ \hline 5 \end{array}$$

$$\begin{array}{r} (7\times10^0)\text{R}(5\times10^0) \\ 7\times10^0\overline{)(5\times10)+(4\times10^0)} \\ (4\times10)+(9\times10^0) \\ \hline 5\times10^0 \end{array}$$

13. a. $203_4 = 2\times4^2 + 0\times4 + 3 = 32 + 3 = 35$
    b. $143_5 = 1\times5^2 + 4\times5 + 3$
    $= 25 + 20 + 3 = 48$
    c. $1101_2 = 1\times2^3 + 1\times2^2 + 0\times2 + 1$
    $= 8 + 4 + 1 = 13$

14. a. $152_8 = 1\times8^2 + 5\times8 + 2$
    $= 64 + 40 + 2 = 106$

b. $A2C_{16} = 10 \times 16^2 + 2 \times 16 + 12$
$= 2560 + 32 + 12 = 2604$

15. a.
$$\begin{array}{l} 5)\underline{33} \\ 5)\ \underline{6|3} \\ \quad 1|1 \qquad 33 = 113_{\text{five}} \end{array}$$

b.
$$\begin{array}{l} 6)\underline{33} \\ \quad 5|3 \qquad 33 = 53_{\text{six}} \end{array}$$

16. a.
$$\begin{array}{l} 2)\underline{39} \\ 2)\underline{19|1} \\ 2)\ \underline{9|1} \\ 2)\ \underline{4|1} \\ 2)\ \underline{2|0} \\ \quad 1|0 \qquad 39 = 100111_{\text{two}} \end{array}$$

b.
$$\begin{array}{l} 2)\underline{527} \\ 2)\underline{263|1} \\ 2)\underline{131|1} \\ 2)\ \underline{65|1} \\ 2)\ \underline{32|1} \\ 2)\ \underline{16|0} \\ 2)\ \underline{8|0} \\ 2)\ \underline{4|0} \\ 2)\ \underline{2|0} \\ \quad 1|0 \qquad 527 = 1000001111_{\text{two}} \end{array}$$

17. a.
$$\begin{array}{l} 8)\underline{47} \\ \quad 5|7 \qquad 47 = 57_{\text{eight}} \end{array}$$

b.
$$\begin{array}{l} 16)\underline{47} \\ \quad 2|15 \qquad 47 = 2F_{\text{sixteen}} \end{array}$$

18. a.
$$\begin{array}{r} {}^{1}1{}^{1}1{}^{1}01_2 \\ +\ 101_2 \\ \hline 10010_2 \end{array}$$

b.
$$\begin{array}{r} {}^{1}1\,{}^{0}\not{1}\,{}^{1}01_2 \\ -\ 111_2 \\ \hline 110_2 \end{array}$$

19. a.
$$\begin{array}{r} 1101_2 \\ \times 11_2 \\ \hline 1101_2 \\ 1101_2 \\ \hline 11010_2 \\ \hline 100111_2 \end{array}$$

b.
$$\begin{array}{r} 10110_2 \\ \times 101_2 \\ \hline 10110_2 \\ \hline 1011000_2 \\ \hline 1101110_2 \end{array}$$

20. a.
$$\begin{array}{r} 111_2\ R1_2 \\ 11_2\overline{)10110_2} \\ \underline{11} \\ 101 \\ \underline{11} \\ 100 \\ \underline{11} \\ 1 \end{array}$$

b.
$$\begin{array}{r} 1001_2\ R1_2 \\ 110_2\overline{)110111_2} \\ \underline{110} \\ 0111 \\ \underline{110} \\ 1 \end{array}$$

21.
$$\begin{array}{r} {}^{1}6\,{}^{1}32_8 \\ +\ 46_8 \\ \hline 700_8 \end{array}$$

22.
$$\begin{array}{r} 37_8 \\ \times 5_8 \\ \hline 43 \\ \underline{17} \\ 233_8 \end{array}$$

23.
$$\begin{array}{r} {}^{5}\not{6}\,{}^{12}\not{3}\,{}^{1}2_8 \\ -\ 4\ 6_8 \\ \hline 564_8 \end{array}$$

24.

$$\begin{array}{r} 77_8 \\ 6_8 \overline{)572_8} \\ \underline{52} \\ 52 \\ \underline{52} \\ 0 \end{array}$$

25. a.

$$\begin{array}{r} {}^{1}2\,{}^{1}BC_{16} \\ \underline{+\ 5D_{16}} \\ 319_{16} \end{array}$$

b.

$$\begin{array}{r} 3C4_{16} \\ \underline{\times\ 2B_{16}} \\ 2C \\ 84 \\ 21 \\ 8 \\ 18 \\ \underline{6} \\ A1EC_{16} \end{array}$$

Chapter 5 Number Theory and the Real Numbers

Section 5.1 Number Theory: Primes and Composites

Study Tips

Remember counting blocks as a young child? These first numbers that you used naturally are called the natural numbers, or counting numbers.

To classify natural numbers:
Think: 3 dogs (how many?: *cardinal*), 1st place (assign order: *ordinal*), (Social Security number: *identification*)

To determine a prime natural number:
Think: two divisors only; itself and 1
Note: A *composite* number has more divisors. The *number 1* is neither prime nor composite.

Hint
To find the prime numbers, you can use the *Sieve of Eratosthenes* given in the text.

To find the prime factorization of a number:
Think: factor tree, only prime number factors
Note: Try dividing the given number by successive primes (2, 3, 5, 7, and so on).

Fact
The *Fundamental Theorem of Arithmetic* tells us that any composite number can be written (factored) as a product of primes.

To determine divisibility by 2, 3, or 5:
Think: divisibility rules, Table 5.1
Note: A number is divisible …
by 2 if number ends in 0, 2, 4, 6, or 8
by 3 if sum of digits is divisible by 3
by 5 if number ends in 0 or 5

Word of Advice
Know these divisibility rules to help you find the prime factorization of a number, which you'll need when you find the LCM or GCF.

To find the GCF of two numbers:
Think: largest factor that *divides into* both numbers (greatest common factor)
Note: After prime factorization of numbers, take the *smallest* exponent of only common *factors*.

Application
You will use the GCF when reducing fractions.

To find the LCM of two numbers:
Think: smallest number that is a *multiple* of both numbers (least common multiple)
Note: After prime factorization of numbers, take every prime factor to its highest power.

Application
When working with fractions, this is referred to as LCD (least common denominator).You will use the LCD when adding fractions.

5.1 Exercises

1. For identification only

3. A cardinal number (used for counting)

5. The "First" is for identification; the "one" is an ordinal number.

7. The sieve is constructed in the same way as it was for the numbers from l to 50. We cross out all the multiples of 2, of 3, of 5, and of 7. The remaining numbers are all primes. We don't have to go beyond 7, because the next prime is 11 and its square is greater than 100. The primes in the table are boxed.

~~51~~	~~52~~	53	~~54~~	~~55~~	~~56~~	~~57~~	~~58~~	59	~~60~~
61	~~62~~	~~63~~	~~64~~	~~65~~	~~66~~	67	~~68~~	~~69~~	~~70~~
71	~~72~~	73	~~74~~	~~75~~	~~76~~	~~77~~	~~78~~	79	~~80~~
~~81~~	~~82~~	83	~~84~~	~~85~~	~~86~~	~~87~~	~~88~~	89	~~90~~
~~91~~	~~92~~	~~93~~	~~94~~	~~95~~	~~96~~	97	~~98~~	~~99~~	~~100~~

9. There are 6 primes between 25 and 50: 29, 31, 37, 41, 43, 47.

11. There are 4 primes between 75 and 100: 79, 83, 89, 97.

13. a. 2 and 3 are both primes and are consecutive counting numbers.
    b. No. If any pair of consecutive counting numbers greater than 2 is selected, one of the pair must be an even number (divisible by 2) and hence, not a prime. Thus, there cannot be a second pair of primes that are consecutive counting numbers .

15. a. The product part of m is exactly divisible by 2, so that m divided by 2 would have a remainder of 1.
    b. The product part of m is exactly divisible by 3, so that m divided by 3 would have a remainder of 1.

c. and d. Exactly the same reasoning as in parts a and b applies here. If m is divided by any prime from 2 to P, there is remainder of 1.

e. Because P was assumed to be the largest prime.

f. Because m is not divisible by any of the primes from 2 to P.

17. $50 = 2 \cdot 5 \cdot 5$. Thus, the divisors of 50 are 1, 2, 5, or 10 (or $2 \cdot 5$), 25 (or $5 \cdot 5$), and 50 (or $2 \cdot 5 \cdot 5$).

19. $128 = 2 \cdot 2 \cdot 2 \cdot 2 \cdot 2 \cdot 2 \cdot 2 = 2^7$. Thus, the divisors of 128 are 1, 2, 4 (or $2 \cdot 2$), 8 (or $2 \cdot 4$), 16 (or $2 \cdot 8$), 32 (or $2 \cdot 16$), 64 (or $2 \cdot 32$), and 128 (or $2 \cdot 64$).

21. $1001 = 7 \cdot 11 \cdot 13$. Thus, the divisors of 1001 are 1, 7, 11, 13, 77 (or $7 \cdot 11$), 91 (or $7 \cdot 13$), 143 (or $11 \cdot 13$), and 1001 (or $7 \cdot 11 \cdot 13$).

23. 41 is a prime.

25. $91 = 7 \cdot 13$

27. $148 = 2^2 \cdot 37$

29. $2 \cdot 5 \cdot 7^2 = 490$

31. $2^4 \cdot 3 \cdot 5^2 = 1200$

33. a. 6345 is not an even number, so is not divisible by 2. The sum of the digits is 18, so the number is divisible by **3**. The number ends in 5, so is divisible by **5**.

b. 8280 ends in 0, so is divisible by **2** and by **5**. The sum of the digits is 18, so the number is divisible by **3**.

c. 11,469,390 ends in 0, so is divisible by **2** and by **5**. The sum of the first three digits is 6, which is divisible by 3, and the other digits are all multiples of 3; thus, the number is divisible by **3**.

35. In order to leave a remainder of 1 when divided into 23, the divisor must be an exact divisor of 22; 2, 11 or 22. All three of these leave a remainder of 1 when divided into 45. Thus, there are three whole numbers, 2, 11 and 22, that satisfy the given condition.

37. $135 = 3^3 \cdot 5;\ 351 = 3^3 \cdot 13$.
GCF (135, 351) $= 3^3 = 27$

39. $147 = 3 \cdot 7^2;\ 260 = 2^2 \cdot 5 \cdot 13$.
GCF (147, 260) = 1, so 147 and 260 are relatively prime.

41. $282 = 2 \cdot 3 \cdot 47;\ 329 = 7 \cdot 47$.
GCF (282, 329) = 47

43. $12 = 2^2 \cdot 3;\ 15 = 3 \cdot 5;\ 20 = 2^2 \cdot 5$.
GCF (12, 15, 20) = 1, so 12, 15 and 20 are relatively prime.

45. $100 = 2^2 \cdot 5^2;\ 200 = 2^3 \cdot 5^2;\ 320 = 2^6 \cdot 5$.
GCF (100, 200, 320) $= 2^2 \cdot 5 = 20$.

47. $\dfrac{62}{88} = \dfrac{2 \cdot 31}{2 \cdot 44} = \dfrac{31}{44}$

49. $156 = 2^2 \cdot 3 \cdot 13;\ 728 = 2^3 \cdot 7 \cdot 13$.
GCF (156, 728) $= 2^2 \cdot 13 = 52$. So,
$\dfrac{156}{728} = \dfrac{52 \cdot 3}{52 \cdot 14} = \dfrac{3}{14}$

51. $96 = 2^5 \cdot 3;\ 384 = 2^7 \cdot 3$.
GCF (96, 384) $= 2^5 \cdot 3 = 96$. Thus,
$\dfrac{96}{384} = \dfrac{96 \cdot 1}{96 \cdot 4} = \dfrac{1}{4}$

53. $15 = 3 \cdot 5;\ 55 = 5 \cdot 11$.
LCM (15, 165) $= 3 \cdot 5 \cdot 11 = 165$.
$\dfrac{1}{15} + \dfrac{1}{55} = \dfrac{11}{165} + \dfrac{3}{165} = \dfrac{14}{165}$

55. $32 = 2^5;\ 124 = 2^2 \cdot 31$.
LCM $= 2^5 \cdot 31 = 992$.
$\dfrac{3}{32} + \dfrac{1}{124} = \dfrac{3 \cdot 31}{992} + \dfrac{1 \cdot 8}{992} = \dfrac{93}{992} + \dfrac{8}{992}$
$= \dfrac{101}{992}$

57. $180 = 2^2 \cdot 3^2 \cdot 5;\ 240 = 2^4 \cdot 3 \cdot 5.$

LCM $(180, 240) = 2^4 \cdot 3^2 \cdot 5 = 720.$

$$\frac{1}{180} + \frac{1}{240} = \frac{4}{720} + \frac{3}{720} = \frac{7}{720}$$

59. $12 = 2^2 \cdot 3;\ 18 = 2 \cdot 3^2;\ 30 = 2 \cdot 3 \cdot 5$

LCM $(12, 18, 30) = 2^2 \cdot 3^2 \cdot 5 = 180$

$$\frac{1}{12} + \frac{1}{18} + \frac{1}{30} = \frac{15}{180} + \frac{10}{180} + \frac{6}{180} = \frac{31}{180}$$

61. $285 = 3 \cdot 5 \cdot 19;\ 315 = 3^2 \cdot 5 \cdot 7;$

$588 = 2^2 \cdot 3 \cdot 7^2$

LCM $(285, 315, 588) = 2^2 \cdot 3^2 \cdot 5 \cdot 7^2 \cdot 19$

$= 167{,}580$

$$\frac{1}{285} + \frac{1}{315} + \frac{1}{588}$$
$$= \frac{2^2 \cdot 3 \cdot 7^2}{167{,}580} + \frac{2^2 \cdot 7 \cdot 19}{167{,}580} + \frac{3 \cdot 5 \cdot 19}{167{,}580}$$
$$= \frac{588 + 532 + 285}{167{,}580} = \frac{1405}{167{,}580} = \frac{281}{33{,}516}$$

63. $200 = 2^3 \cdot 5^2;\ 300 = 2^2 \cdot 3 \cdot 5^2;$

$420 = 2^2 \cdot 3 \cdot 5 \cdot 7$

LCM $(200, 300, 420) = 2^3 \cdot 3 \cdot 5^2 \cdot 7$

$= 4200$

$$\frac{1}{200} + \frac{1}{300} + \frac{1}{420} = \frac{21}{4200} + \frac{14}{4200} + \frac{10}{4200}$$
$$= \frac{45}{4200} = \frac{3}{280}$$

65. To determine in how many hours you will need to take all four medications again, find the LCM of 6, 2, 4, and 12.

$6 = 2 \cdot 3;\ 2 = 2;\ 4 = 2^2; 12 = 2^2 \cdot 3$

LCM $(6, 2, 4, 12) = 2^2 \cdot 3 = 12$

You will take all our medications again in 12 hours.

67. To determine in how many hours there will be fresh tamales and fresh pastries, find the LCM of 5 and 4.

$5 = 5;\ 4 = 2^2$

LCM $(5, 4) = 2^2 \cdot 5 = 20$

You will fresh items again in 20 days.

69. $1 - \frac{3}{10} - \frac{1}{5} - \frac{1}{20} - \frac{3}{20}$

$$= \frac{20}{20} - \frac{6}{20} - \frac{4}{20} - \frac{1}{20} - \frac{3}{20} = \frac{6}{20} = \frac{3}{10}$$

$\frac{3}{10}$ of the water is used for toilet flushing.

71. a. This can be done in several ways:
$100 = 3 + 97 = 11 + 89 = 17 + 83$
$= 29 + 71 = 41 + 59 = 47 + 53.$

b. This can also be done in several ways:
$200 = 3 + 197 = 7 + 193 = 19 + 181$
$= 37 + 163 = 43 + 157 = 61 + 139$
$= 73 + 127 = 97 + 103$

73. Answers may vary. The number 1 has only one divisor, itself. It is not a prime because a prime must have exactly *two distinct divisors*, 1 and itself. It is not a composite number because it has only one divisor.

75. Answers may vary. The largest prime that you need to try is 13, because the next prime is 17 and $17^2 = 289$ which is greater than 211.

77. All the other digits are multiples of 3, so their sum is divisible by 3. Hence, only the sum of 2 and 7 needs to be checked.

79. Since 999 and 99 and 9 are all divisible by 9, only the sum $2 \cdot 1 + 8 \cdot 1 + 5 \cdot 1 + 3$, which is exactly the sum of the digits, needs to be checked. If this sum is divisible by 9, the original number is divisible by 9 and not otherwise.

81. a. 1436 is divisible by 4 because 36 is divisible by 4. 1436 is not divisible by 8 because 436 is not divisible by 8.

b. 21,408 is divisible by 8 because 408 is divisible by 8. Being divisible by 8, the number is also divisible by 4.

c. 347,712 is divisible by 8 because 712 is divisible by 8. Since the number is divisible by 8, it is also divisible by 4.

d. 40,924 is divisible by 4 because 24 is divisible by 4. The number is not divisible by 8 because 924 is not divisible by 8.

83. None of the numbers 1, 2, 3, 4, 5 is the sum of its proper divisors. Therefore, 6 is the smallest perfect number.

85. $496 = 2^4 \cdot 31$ Thus, the proper divisors of 496 are 1, 2, 4, 8, 16, 31, 62, 124, and 248. The sum of these divisors is 496, so 496 is a perfect number.

87. The only proper divisor of a prime is the number 1. Thus, a prime is always greater than the sum of its proper divisors. Consequently, all primes are deficient.

89. They end in 6 or 28. Also, they are sums of powers of 2. $6 = 2 + 2^2$, $28 = 2^4 + 2^3 + 2^2$, and so on.

Section 5.2 Whole Numbers, Integers, and Order of Operations

Study Tips

You are introduced to two new sets of numbers: the set of *whole numbers* {0, 1, 2, 3...} (almost the same as the natural numbers, only **0** has been added), and the set of *integers* {...–2, –1, 0, 1, 2... }.

Important Properties to Remember...

Identity Property for Multiplication:
Think: the number <u>1</u> can multiply a number and not change the number
Note: $a \cdot 1 = 1 \cdot a = a$

Identity Property for Addition:
Think: the number <u>0</u> can be added to any number and not change the number
Note: $a + 0 = 0 + a = a$

Properties of 0:
Think: 0 can be subtracted, multiplied, and divided by another number
Note: Never divide by zero!

Additive Inverse Property:
Think: opposites, their sum is zero
Note: $n + (-n) = (-n) + n = 0$. Opposites are also called *additive inverses*.

Distributive Property:
Think: multiply each term in the parentheses
Note: The factor will be distributed to each term in the parentheses: $a(b + c) = ab + ac$.

<u>Fact</u>
Integers can be *positive* (1, 2, 3 and so on), *negative* (–1, –2, –3 and so on) or 0, which is neither positive, nor negative.

<u>Caution</u>
When we say *non-negative* it means 0, 1, 2, 3 and so on, since these numbers are *not* negative.

To find the absolute value of a number:
Think: distance from zero, always positive
Note: Notation is $|a|$. Again, absolute value will never be a negative number.

To add integers:
Think: same sign, add, keep sign
different sign, subtract absolute values, sign of number furthest from zero (greatest absolute value)
Note: Rewrite *subtraction* as the addition of the additive inverse, then use the addition rules for adding: $a - b = a + (-b)$

To multiply integers:
Think: same sign, answer always positive; different sign, answer always negative
Note: For *division* of integers, the same rules apply.

To apply the order of operations:
Think: order in which addition, subtraction, multiplication, exponents, and division are done
Note: **PEMDAS** is a good way to remember the order:

1. remove **P**arentheses
2. do **E**xponents
3. do **M**ultiplications and **D**ivisions as they occur from left to right
4. do **A**ddition and **S**ubtraction also as they occur from left to right.

<u>Hint</u>
Can also use this phrase to remember:
Please Excuse My Dear Aunt Sally

5.2 Exercises

1. Because 9 + (–9) = 0, the additive inverse of 9 is –9.

3. Because –10 + 10 = 0, the additive inverse of –10 is 10.

5. Because 0 + 0 = 0, the additive inverse of 0 is 0.

7. $|27| = 27$ because 27 is 27 units from zero.

9. $|-16| = 16$ because –16 is 16 units from zero.

11. –3 + 5 = 5 – 3 = 2

13. 8 + (–1) = 8 – 1 = 7

15. $-8 + 13 = 13 - 8 = 5$

17. $-17 + 4 = -(17 - 4) = -13$

19. Add the positives 8 and 6 and the negatives (–4) and (–2); then add the results.
$$-4 + 8 + 6 + (-2) = 8 + 6 + (-4) + (-2) = 14 + (-6) = 14 - 6 = 8$$

21. $3 - 8 = 3 + (-8) = -5$

23. $3 - 4 = 3 + (-4) = -1$

25. $-5 - 2 = -5 + (-2) = -7$

27. $5 - (-6) = 5 + (+6) = 11$

29. $-3 - (-4) = -3 + (+4) = 1$

31. The product of like-signed numbers is a positive number: $16 \cdot 2 = 32$

33. The product of unlike-signed numbers is a negative number: $-7 \cdot 8 = -(7 \cdot 8) = -56$

35. The product of unlike-signed numbers is a negative number:
$2 \cdot (-5) = -(2 \cdot 5) = -10$

37. The product of like-signed numbers is a positive number: $-4 \cdot (-5) = 4 \cdot 5 = 20$

39. The product of like-signed numbers is a positive number: $-7 \cdot (-10) = 7 \cdot 10 = 70$

41. The quotient of like-signed numbers is a positive number: $10 \div 2 = 5$

43. The quotient of unlike-signed numbers is a negative number: $\dfrac{-20}{5} = -\left(\dfrac{20}{5}\right) = -4$

45. The quotient of unlike-signed numbers is a negative number:
$-40 \div 8 = -(40 \div 8) = -5$

47. The quotient of unlike-signed numbers is a negative number:
$150 \div (-15) = -(150 \div 15) = -10$

49. The quotient of unlike-signed numbers is a negative number: $\dfrac{140}{-7} = -\left(\dfrac{140}{7}\right) = -20$

51. The quotient of like-signed numbers is a positive number: $-98 \div (-14) = 98 \div 14 = 7$

53. The quotient of like-signed numbers is a positive number: $\dfrac{-98}{-7} = \dfrac{98}{7} = 14$

55. Division by zero is not allowed.
$\dfrac{3}{0}$ is not defined.

57. Zero divided by any number is 0.
$\dfrac{0}{-33} = 0$

59. Division by zero is not allowed.
$-19 \div 0$ is not defined.

61. $(-8)^2 = (-8) \cdot (-8) = 64$

63. $-6^2 = -(6 \cdot 6) = -36$

65. $-6^3 = -(6 \cdot 6 \cdot 6) = -216$

67. $(-3)^4 = (-3) \cdot (-3) \cdot (-3) \cdot (-3)$
$= 9 \cdot (-3) \cdot (-3) = -27 \cdot (-3) = 81$

69. $(-4)^3 = (-4) \cdot (-4) \cdot (-4) = 16 \cdot (-4) = -64$

71. a. $-3(4+5) = -3(9) = -27$
b. $-4(4-5) = -4(-1) = 4$

73. a. $-5 + (-5+1) = -5 + (-4) = -9$
b. $-8 + (-2+5) = -8 + (3) = -5$

75. $(-2-4)(-3) - 8(5-4) = -6(-3) - 8(1)$
$= 18 - 8 = 10$

77. $6 \times 2 \div 3 + 6 \div 2 \times (-3) = 12 \div 3 + 3 \times (-3)$
$= 4 + (-9) = -5$

79. $4\times9\div3\times10^3-2\times10^2$

$$=4\times9\div3\times1000-2\times100$$
$$=36\div3\times1000-200$$
$$=12\times1000-200$$
$$=12{,}000-200$$
$$=11{,}800$$

81. $20\div5+\{3\cdot4-[4+(5-3)]\}$

$$=20\div5+\{3\cdot4-[4+(2)]\}$$
$$=20\div5+\{3\cdot4-6\}$$
$$=20\div5+\{12-6\}$$
$$=20\div5+6$$
$$=4+6$$
$$=10$$

83. $(20-15)\cdot[20\div2-(2\cdot2+2)]$

$$=(20-15)\cdot[20\div2-(4+2)]$$
$$=5\cdot[20\div2-6]$$
$$=5\cdot[10-6]$$
$$=5\cdot4=20$$
$$=20$$

85. -14; $506-14=492$

87. $+19$; $506+19=525$

89. $+53$; $506+53=559$

91.

Altitude in (1000 ft)	Altitude × Rate of Change	Temperature Change
5	5(–4)	–20

93.

Altitude in (1000 ft)	Altitude × Rate of Change	Temperature Change
15	15(–4)	–60°F

95.

Altitude in Kilometers	Altitude × Rate of Change	Temperature Change
5	5(–7)	–35°C

97. 14,000 ft is 14 thousands and the rate of change is $-4°F$ for each 1000 ft of altitude. The temperature change is $70+14(-4)=70-56=14°F$.

99. 6000 m is 6 kilometers and the rate of change is $-7°C$ for each kilometer of altitude. The temperature at the top is $20+6(-7)=20+(-42)=-22°C$.

101. Answers may vary. The product of two positive numbers is a positive number.

103. Answers may vary. The product of two negative numbers is a positive number.

105. $2\cdot1+2\cdot6+7\cdot(-2)=2+12-14=0$

Section 5.3 Operations with Rational Numbers, Expanded and Scientific Notation

Study Tips

Rational numbers include integers, fractions, decimals and percents. However, would that be enough to define a rational number?

To define a rational number:
Think: fraction, integers only
Note: The key here is that a rational number *can be* written in the form $\frac{a}{b}$, where *a* and *b* are integers.

Fact
The *Fundamental Property of Rational Numbers*, $\frac{a}{b}=\frac{ak}{bk}$, enables us to reduce fractions!

Operations with rational numbers in fraction form....

To multiply:
Think: multiply numerators, multiply denominators
Note: Reduce, if possible, before you multiply.

To add or subtract:
Think: LCD, write equivalent fractions, add or subtract numerators

Caution
Only fractions with *like* denominators can be added or subtracted.

To divide:
Think: rewrite as multiplication by its reciprocal

Note: To find the reciprocal of a number, invert the number, keep the sign. Reciprocal of $\frac{a}{b}$ is $\frac{b}{a}$.

Word of Advice
Be consistent with answers. If the problem has proper fractions, give the answer as a proper fraction when possible. If it has mixed numbers, use mixed numbers for the answer.

To write decimals in expanded form:
Think: how many of each place value, add
Note: All terminating decimals are rational numbers. Writing decimals in expanded form requires *negative* exponents. Just remember that $a^{-n} = \frac{1}{a^n}$, the exponent n remains the same!

To write numbers in scientific notation:
Think: special form of $m \times 10^n$, $m \geq 1$ and $m < 10$, n shows direction and number of places to move
Note: Example, $345.85 = 3.4585 \times 10^2$. Negative exponents are also used with scientific notation: $0.00537 = 5.37 \times 10^{-3}$.

Operations with decimals….

To add or subtract:
Think: align decimal points, add or subtract as normal
Note: You may need to *round* your answer to a certain number of decimal places, follow instructions in the problem.

To multiply:
Think: multiply first, count total places, move decimal point to the left
Note: Multiply as if they were whole numbers, then move the decimal point, starting at the right, left the appropriate number of places. Round to the correct number of decimal places.

To divide:
Think: divisor must be made into a whole number
Note: Be sure to multiply both the divisor and dividend by the same power of 10.

5.3 Exercises

1. The numerator is 3; the denominator is 4.

3. The numerator is 3; the denominator is –5.

5. $\frac{17}{41} \stackrel{?}{=} \frac{289}{697}$; $17 \cdot 697 = 11{,}849 = 41 \cdot 289$ and
$\frac{17}{41} \stackrel{?}{=} \frac{714}{1682}$; $17 \cdot 1682 = 28{,}594$
$\neq 29{,}274 = 41 \cdot 714$
Thus, $\frac{17}{41} = \frac{289}{697}$.

7. $\frac{11}{91} \stackrel{?}{=} \frac{253}{2093}$; $11 \cdot 2093 = 23{,}023 = 91 \cdot 253$ and
$\frac{11}{91} \stackrel{?}{=} \frac{111}{911}$; $11 \cdot 911 = 10{,}021 \neq 10{,}101 = 91 \cdot 111$
Thus, $\frac{11}{91} = \frac{253}{2093}$.

9. $\frac{95}{38} = \frac{5 \cdot 19}{2 \cdot 19} = \frac{5}{2}$

11. $\frac{21}{48} = \frac{3 \cdot 7}{3 \cdot 16} = \frac{7}{16}$

13. $\frac{30}{28} = \frac{2 \cdot 15}{2 \cdot 14} = \frac{15}{14}$

15. $\frac{52}{78} = \frac{2 \cdot 26}{3 \cdot 26} = \frac{2}{3}$

17. $\frac{2}{9} + \frac{1}{6} + \frac{7}{18} = \frac{4}{18} + \frac{3}{18} + \frac{7}{18} = \frac{14}{18}$

19. $\frac{1}{3} + \frac{1}{6} + \frac{1}{9} = \frac{6}{18} + \frac{3}{18} + \frac{2}{18} = \frac{11}{18}$

21. $\frac{1}{7} + \frac{1}{9} = \frac{9}{63} + \frac{7}{63} = \frac{16}{63}$

23. $\frac{3}{4} + \frac{5}{6} = \frac{9}{12} + \frac{10}{12} = \frac{19}{12}$

25. $\frac{3}{17} + \frac{7}{19} = \frac{57}{323} + \frac{119}{323} = \frac{176}{323}$

27. $\frac{1}{7} - \frac{1}{9} = \frac{9}{63} - \frac{7}{63} = \frac{2}{63}$

29. $\frac{3}{4} - \frac{5}{6} = \frac{9}{12} - \frac{10}{12} = -\frac{1}{12}$

31. $\frac{7}{19} - \frac{3}{17} = \frac{119}{323} - \frac{57}{323} = \frac{62}{323}$

33. $\frac{2}{5}\times\frac{5}{3}=\frac{2}{\cancel{5}_{1}}\times\frac{\cancel{5}^{1}}{3}=\frac{2}{3}$

35. $\frac{3}{4}\div\frac{2}{7}=\frac{3}{4}\times\frac{7}{2}=\frac{21}{8}$

37. $\frac{7}{9}\div\frac{3}{8}=\frac{7}{9}\times\frac{8}{3}=\frac{56}{27}$

39. $\left(-\frac{6}{7}\right)\times\left(-\frac{3}{11}\right)=\frac{18}{77}$

41. $\left(-\frac{3}{4}\right)\div\left(-\frac{7}{6}\right)=\left(-\frac{3}{\cancel{4}_{2}}\right)\cdot\left(-\frac{\cancel{6}^{3}}{7}\right)=\frac{9}{14}$

43. $\frac{1}{8}\div\left(-\frac{3}{4}\right)=\frac{1}{\cancel{8}_{2}}\cdot\left(-\frac{\cancel{4}^{1}}{3}\right)=-\frac{1}{6}$

45. $\left(-\frac{1}{8}\right)+\frac{1}{4}=-\frac{1}{8}+\frac{2}{8}=\frac{1}{8}$

47. $\frac{3}{8}-\left(\frac{1}{4}-\frac{1}{8}\right)=\frac{3}{8}-\left(\frac{2}{8}-\frac{1}{8}\right)=\frac{3}{8}-\frac{1}{8}$

$=\frac{2}{8}=\frac{1}{4}$

49. $\frac{1}{2}\times\left(\frac{7}{8}\times\frac{7}{5}\right)=\frac{1}{2}\times\frac{49}{40}=\frac{49}{80}$

51. $\left(\frac{1}{2}\div\frac{1}{8}\right)\div\frac{1}{4}=\left(\frac{1}{2}\cdot\frac{8}{1}\right)\cdot\frac{4}{1}=4\cdot4=16$

53. $\frac{2}{3}\left(\frac{1}{2}+\frac{3}{4}\right)+\frac{2}{3}=\frac{2}{3}\left(\frac{2}{4}+\frac{3}{4}\right)+\frac{2}{3}$

$=\frac{2}{3}\left(\frac{5}{4}\right)+\frac{2}{3}=\frac{5}{6}+\frac{4}{6}=\frac{9}{6}=\frac{3}{2}$

55. $\frac{1}{3}\left(\frac{3}{2}-\frac{1}{5}\right)-\frac{1}{30}=\frac{1}{3}\left(\frac{15}{10}-\frac{2}{10}\right)-\frac{1}{30}$

$=\frac{1}{3}\left(\frac{13}{10}\right)-\frac{1}{30}=\frac{13}{30}-\frac{1}{30}=\frac{12}{30}=\frac{2}{5}$

57. $1\frac{1}{2}+\frac{1}{7}=\frac{3}{2}+\frac{1}{7}=\frac{21}{14}+\frac{2}{14}=\frac{23}{14}=1\frac{9}{14}$

59. $\frac{1}{4}\times1\frac{1}{7}=\frac{1}{4}\cdot\frac{8}{7}=\frac{2}{7}$

61. $3\frac{1}{4}+\frac{1}{6}=3+\frac{1}{4}+\frac{1}{6}=3+\frac{3}{12}+\frac{2}{12}=3\frac{5}{12}$

63. $\frac{1}{5}\cdot2\frac{1}{7}=\frac{1}{5}\cdot\frac{15}{7}=\frac{3}{7}$

65. $-3+2\frac{1}{4}=-3+2+\frac{1}{4}=-1+\frac{1}{4}=-\frac{3}{4}$

67. $(-8)\cdot2\frac{1}{4}=(-8)\cdot\frac{9}{4}=(-2)\cdot9=-18$

69. $-2+1\frac{1}{5}=-2+1+\frac{1}{5}=-1+\frac{1}{5}=-\frac{4}{5}$

71. $(-9)\times3\frac{1}{3}=(-9)\cdot\frac{10}{3}=(-3)\times10=-30$

73. $7\frac{1}{4}+\left(-\frac{1}{8}\right)=7+\frac{1}{4}-\frac{1}{8}=7+\frac{2}{8}-\frac{1}{8}$

$=7+\frac{1}{8}=7\frac{1}{8}$

75. $\left(-1\frac{1}{4}\right)\times\left(-2\frac{1}{10}\right)=\left(-\frac{\cancel{5}^{1}}{4}\right)\times\left(-\frac{21}{\cancel{10}_{2}}\right)$

$=\left(-\frac{1}{4}\right)\times\left(-\frac{21}{2}\right)=\frac{21}{8}=2\frac{5}{8}$

77. $\frac{1}{2}\times\frac{1}{6}-\frac{1}{3}+\frac{1}{4}=\frac{1}{12}-\frac{4}{12}+\frac{3}{12}=\frac{0}{12}=0$

79. $\frac{1}{3}-\frac{1}{3}\times\frac{2}{3}\div\frac{2}{5}=\frac{1}{3}-\frac{2}{9}\div\frac{2}{5}=\frac{1}{3}-\frac{2}{9}\times\frac{5}{2}$

$=\frac{3}{9}-\frac{5}{9}=-\frac{2}{9}$

81. $\left(2\frac{1}{2}\right)\times\left(-3\frac{1}{4}\right)-\left(-7\frac{1}{8}\right)\div 3$
$=\frac{5}{2}\cdot\left(-\frac{13}{4}\right)-\left(-\frac{57}{8}\right)\times\frac{1}{3}$
$=-\frac{65}{8}-\left(-\frac{19}{8}\right)=-\frac{65}{8}+\frac{19}{8}$
$=-\frac{46}{8}=-\frac{23}{4}=-5\frac{3}{4}$

83. $12\div 6-\left(\frac{1}{3}+\frac{1}{2}\right)$
$=12\div 6-\left(\frac{2}{6}+\frac{3}{6}\right)$
$=12\div 6-\left(\frac{5}{6}\right)=\frac{12}{6}-\frac{5}{6}=\frac{7}{6}=1\frac{1}{6}$

85. $\frac{1}{3}\cdot\frac{1}{4}\div\frac{1}{2}+\left(\frac{5}{6}-\frac{1}{2}\right)=\frac{1}{3}\cdot\frac{1}{4}\div\frac{1}{2}+\left(\frac{5}{6}-\frac{3}{6}\right)$
$=\frac{1}{3}\cdot\frac{1}{4}\div\frac{1}{2}+\frac{2}{6}=\frac{1}{12}\cdot\frac{2}{1}+\frac{2}{6}$
$=\frac{1}{6}+\frac{2}{6}=\frac{3}{6}=\frac{1}{2}$

87. $\frac{1}{6}\div\frac{1}{3}\cdot\frac{1}{3}\cdot\frac{1}{3}+\left(\frac{1}{4}-\frac{1}{9}\right)$
$=\frac{1}{6}\div\frac{1}{3}\cdot\frac{1}{3}\cdot\frac{1}{3}+\left(\frac{9}{36}-\frac{4}{36}\right)$
$=\frac{1}{6}\div\frac{1}{3}\cdot\frac{1}{3}\cdot\frac{1}{3}+\frac{5}{36}$
$=\frac{1}{6}\cdot\frac{3}{1}\cdot\frac{1}{3}\cdot\frac{1}{3}+\frac{5}{36}$
$=\frac{1}{18}+\frac{5}{36}=\frac{2}{36}+\frac{5}{36}=\frac{7}{36}$

89. $8\div\frac{1}{2}\cdot\frac{1}{2}\cdot\frac{1}{2}-\left(\frac{1}{3}+\frac{1}{5}\right)$
$=8\div\frac{1}{2}\cdot\frac{1}{2}\cdot\frac{1}{2}-\left(\frac{5}{15}+\frac{3}{15}\right)$
$=8\div\frac{1}{2}\cdot\frac{1}{2}\cdot\frac{1}{2}-\frac{8}{15}$
$=8\cdot\frac{2}{1}\cdot\frac{1}{2}\cdot\frac{1}{2}-\frac{8}{15}=4-\frac{8}{15}=3\frac{7}{15}$

91. $692.087=6\times10^2+9\times10+2\times10^0$
$+0\times10^{-1}+8\times10^{-2}+7\times10^{-3}$

93. $0.00107=1\times10^{-3}+7\times10^{-5}$

95. 5020.39

97. 0.004702

99. $935=9.35\times10^2$

101. $0.0012=1.2\times10^{-3}$

103. $8.64\times10^4=86,400$

105. $6.71\times10^{-3}=0.00671$

107. $0.0346\div1,730,000$
$=\left(3.46\times10^{-2}\right)\div\left(1.73\times10^6\right)$
$=\left(3.46\div1.73\right)\times\left(10^{-2}\div10^6\right)$
$=2\times10^{-8}$

109. $\left(3.1\times10^5\right)\times\left(2.2\times10^{-6}\right)$
$=\left(3.1\times2.2\right)\times\left(10^5\times10^{-6}\right)$
$=6.82\times10^{-1}$

111. $\frac{\left(2\times10^6\right)\left(6\times10^{-5}\right)}{4\times10^3}=\frac{(2\cdot6)\cdot\left(10^6\times10^{-5}\right)}{4\times10^3}$
$=\frac{12\times10^1}{4\times10^3}=3\times10^{-2}$

113. a. $3.81+0.93=4.74$
b. $-3.81+(-0.93)=-4.74$

115. a. $2.08-6.238=-4.158$
b. $3.07-8.934=-5.864$

117. a. $(-0.03)\times(-1.5)=0.045$

b. $(-3.2)\times(-0.04)=0.128$

119. a. $(-0.07)\div 1.4=-0.05$

b. $(-0.09)\div(-4.5)=0.02$

121. $\frac{1}{12}$ of the time was spent eating.

123. $\frac{1}{12}$ of the time was spent doing homework.

125. a. $4.8\times\left(5.6\times10^{6}\right)\times\left(3.65\times10^{2}\right)$

$=\left(4.8\times5.6\times3.65\right)\times\left(10^{6}\times10^{2}\right)$

$=98.112\times10^{8}=9.8112\times10^{1}\times10^{8}$

$=9.8112\times10^{9}$

b. From part a,

$9.8112\times10^{9}=9,811,200,000$

127. a. To determine minimum pulse rate for a 20-year old, use $0.72(220-A)$, $A = 20$.
$0.72(220 - 20) = 0.72(200) = 144$
The minimum pulse rate is 144.

b. To determine minimum pulse rate for a 45-year old, use $0.72(220-A)$, $A = 20$.
$0.72(220 - 45) = 0.72(175) = 126$
The minimum pulse rate is 126.

129. 2007 CO_2 concentration is 384.42 ppm and 1967 CO_2 concentration is 322.16 ppm. The difference is 384.42 – 322.16 = 62.26 ppm.

131. 2009 CO_2 concentration is 388.79 ppm and 1969 CO_2 concentration is 324.62 ppm. The difference is 388.79 – 324.62 = 64.17 ppm.

133. Answers may vary.

135. Special K provides $\frac{4}{70}$ of the Recommended daily protein needs.

Rounded to the hundredths, $\frac{4}{70}\approx 0.06$.

137. One egg provides $\frac{7}{70}=\frac{1}{10}$ of the recommended daily protein needs. Expressed as a decimal, $\frac{1}{10}=0.10$.

Section 5.4 Rationals and Irrationals as Decimals: Percents

Study Tips

Some numbers, like $\sqrt{2}$ and π are not rational. Such numbers are called *irrational* numbers. In particular, the irrational number π is used in the formula for the circumference of a circle.

To write a fraction as a decimal:
Think: divide the numerator (top) by the denominator (bottom)
Note: If the decimal terminates or repeats, you have a *rational* number. If the decimal is nonterminating and nonrepeating, it is an *irrational* number.

To write a repeating decimal as a fraction:
Think: write equation ($x =$), multiply by power of 10, subtract original equation, solve for x
Note: Entire procedure is given in your text.

Shortcut
To write $0.\overline{37}$ as a fraction: $0.\overline{37}$ has two digits, 3 and 7, under the bar, so write $0.\overline{37}=\frac{37}{99}$. If the number were $0.\overline{371}$, which has *three* digits under the bar, write $0.\overline{371}=\frac{371}{999}$, with *three nines* in the denominator this time.

To write a percent as a decimal:
Think: move decimal point two places to left, drop the % sign
Note: To change a decimal to a percent reverse the procedure, that is, move decimal point two places to the right, write the % sign.

Fact
The % sign means '$\div 100$,' so
$37\% = 37\div 100 = 0.37$.

To write a fraction as a percent:
Think: change to decimal, then to percent

To classify numbers:
Think: rational: repeating, terminating decimals, square roots of perfect squares $\left(\sqrt{121}=11\right)$, fractions with integers in both numerator and denominator
irrational: nonrepeating, nonterminating decimals, square roots of nonperfect squares $\left(\sqrt{24}\right)$, π

To find numbers between decimals and fractions:
Think: write as fractions with a common denominator, choose a number for the numerator between the numbers in the other numerators

Fact
The set of rational numbers is dense.

5.4 Exercises

1. $\frac{9}{10}=0.9$

3. $\frac{11}{10}=1.1$

5. $\frac{17}{100}=0.17$

7. $\frac{121}{100}=1.21$

9. $\frac{3}{1000}=0.003$

11. $\frac{1243}{1000}=1.243$

13. $\frac{3}{5}=0.6$

15. $\frac{9}{16}=0.5625$

17. $\frac{5}{8}=0.625$

19. $\frac{5}{7}=0.714285714285...=0.\overline{714285}$

21. $\frac{4}{15}=0.2666...=0.2\overline{6}$

23. $7\frac{1}{7}=7.142857142857...=7.\overline{142857}$

25. $\frac{3}{16}=0.1875$

27. $\frac{1}{64}=0.015625$

29. $\frac{31}{3125}=0.00992$

31. $0.555555\cdots=0.\overline{5}$

33. $0.646464\cdots=0.\overline{64}$

35. $0.235235\cdots=0.\overline{235}$

37. $0.215555\cdots=0.21\overline{5}$

39. $0.079353535\cdots=0.079\overline{35}$

41. $5.070707\cdots=5.\overline{07}$

43. To change $0.\overline{8}$ to a fraction:
$10x=8.88888\cdots$
$-\ x=.88888\cdots$
$9x=8$
$x=\frac{8}{9}$

45. To change $0.\overline{31}$ to a fraction:
$100x=31.313131\cdots$
$-\ x=\ .313131\cdots$
$99x=31$
$x=\frac{31}{99}$

47. To change $0.\overline{114}$ to a fraction:
$1000x=114.114114\cdots$
$-\ x=\ .114114\cdots$
$999x=114$
$x=\frac{114}{999}=\frac{38}{333}$

49. To change $2.\overline{31}$ to a fraction, rewrite as $2+0.\overline{31}$ and then
$100x=31.313131\cdots$
$-\ x=\ .313131\cdots$
$99x=31$
$x=\frac{31}{99}$
$2.\overline{31}=2+\frac{31}{99}=\frac{229}{99}$

51. To change $1.\overline{234}$ to a fraction, rewrite as $1+0.\overline{234}$ and then

$$\begin{array}{rl} 1000x = & 234.234234\cdots \\ - \quad x = & .234234\cdots \\ \hline 999x = & 234 \end{array}$$

$$x = \frac{234}{999}$$

$$1.\overline{234} = 1 + \frac{234}{999} = \frac{1233}{999} = \frac{137}{111}$$

53. To change $1.\overline{27}$ to a fraction, rewrite as $1+0.\overline{27}$ and then

$$\begin{array}{rl} 100x = & 27.272727\cdots \\ - \quad x = & .272727\cdots \\ \hline 99x = & 27 \end{array}$$

$$x = \frac{27}{99} = \frac{3}{11}$$

$$1.\overline{27} = 1 + \frac{3}{11} = \frac{14}{11}$$

55. To change $0.45\overline{75}$ to a fraction:

$$\begin{array}{rl} 10,000x = & 4575.757575\cdots \\ - \quad 100x = & 45.757575\cdots \\ \hline 9900x = & 4530 \end{array}$$

$$x = \frac{4530}{9900} = \frac{151}{330}$$

57. To change $0.\overline{2016}$ to a fraction:

$$\begin{array}{rl} 10,000x = & 2016.20162016\cdots \\ - \quad x = & 0.20162016\cdots \\ \hline 9999x = & 2016 \end{array}$$

$$x = \frac{2016}{9999} = \frac{224}{1111}$$

For problems 59 – 67, to change a percent to a decimal, move the decimal point two places to the left and omit the % symbol.

59. $29\% = 0.29$

61. $0.9\% = 0.009$

63. $45.69\% = 0.4569$

65. $34.15\% = 0.3415$

67. $0.0234\% = 0.000234$

For problems 69 – 75, to change a decimal to a percent, move the decimal point two places to the right and attach the % symbol.

69. $3.45 = 345\%$

71. $0.567 = 56.7\%$

73. $9.003 = 900.3\%$

75. $0.0045 = 0.45\%$

77. $\frac{3}{5} = 0.6 = 60\%$

79. $\frac{5}{6} \approx 0.8333 \approx 83.3\%$

81. $\sqrt{120}$ is an irrational number because 120 is not a perfect square.

83. $\sqrt{125}$ is an irrational number because 125 is not a perfect square.

85. $\sqrt{\frac{9}{16}} = \frac{3}{4}$ is a rational number because 9 and 16 are perfect squares.

87. $\frac{3}{5}$ is a rational number.

89. $-\frac{5}{3}$ is a rational number.

91. $0.2323\cdots$ is a repeating decimal, so is a rational number.

93. 0.121231234 . . . is a nonterminating, nonrepeating decimal, so is an irrational number.

95. $6\frac{1}{4} = \frac{25}{4} = 6.25$ is a terminating decimal, so is a rational number.

97. $\sqrt{3\frac{1}{4}} = 1.802995638\cdots$ is a nonrepeating, nonterminating decimal, so is an irrational number.

99. 0.1122334455 is a terminating decimal, so is a rational number.

101. $\pi = 3.141592654\cdots$ is a nonrepeating, nonterminating decimal, so is an irrational

number.

103. $17 > 11$ 105. $\frac{12}{19} < \frac{11}{17}$

107. $1\frac{2}{3} > \frac{8}{6}$

109. $3.777\cdots < \sqrt{15}$

111. $0.101001000 < 0.1101001000\cdots$

113. $0.333\cdots + 0.666\cdots = 1$

115. $0.112233 > 0.111222333\cdots$

For problems 117 – 129, other answers are possible.

117. 0.2825 119. $0.28101001000\cdots$

121. 0.3031 123. $0.303103003000\cdots$

125. $\frac{79}{99}$; This is the average (one-half the sum) of the given numbers.

127. $\frac{2}{11} = 0.1818\cdots$ and $\frac{3}{11} = 0.2727\cdots$, so $0.2273101001000\cdots$

129. $0.1 = \frac{1}{10}$ and $0.\bar{1} = 0.111\cdots = \frac{1}{9}$, so $\frac{19}{180}$. This is the average (one-half the sum) of the two fractions.

131. Slightly wicked: $(0.54)(-2.5) = -1.35$

133. Rather lovable: $(0.84)(2.4) = 2.016$

135. $0.\bar{4} = 0.444\cdots = \frac{4}{9}$

Section 5.5 Radicals and Real Numbers

Study Tips

If you rounded an irrational number, that is, a nonrepeating, nonterminating decimal, it would be an appoximation. So, how would you write it exactly? Write it as a radical! That is, in the form $\sqrt[n]{m}$, a radical expression with index n and radicand m. Square roots have the index 2, so we usually write them as $\sqrt{m}$.

To simplify a radical:
Think: no perfect squares under the radical sign
Note: Use this fact: $\sqrt{a \cdot b} = \sqrt{a} \cdot \sqrt{b}$

To rationalize the denominator:
Think: multiply both numerator and denominator by a radical that makes a perfect square in the radical in the denominator, then simplify
Note: The reason for making the denominator rational is so that we are not dividing by an irrational number.

To multiply radicals:
Think: rewrite as the product of radicals, then simplify
Note: You will use this fact: $\sqrt{a \cdot b} = \sqrt{a} \cdot \sqrt{b}$

To divide radicals:
Think: rewrite as the quotient of radicals, then simplify
Note: You will use this fact: $\sqrt{\frac{a}{b}} = \frac{\sqrt{a}}{\sqrt{b}}$

To add or subtract radicals:
Think: simplify each radical, add/subtract the coeffiecients of like radicals
Note: After simplification, it is as easy as adding apples to apples. For example, 3 apples + 4 apples = 7 apples. Similarly, $3\sqrt{10} + 4\sqrt{10} = 7\sqrt{10}$.

5.5 Exercises

1. $\sqrt{90} = \sqrt{9 \cdot 10} = 3\sqrt{10}$

3. $\sqrt{122}$ is in simplest form.

5. $\sqrt{180} = \sqrt{36 \cdot 5} = 6\sqrt{5}$

7. $\sqrt{200} = \sqrt{100 \cdot 2} = 10\sqrt{2}$

9. $\sqrt{384} = \sqrt{64 \cdot 6} = 8\sqrt{6}$

11. $\sqrt{588} = \sqrt{196 \cdot 3} = 14\sqrt{3}$

13. $\frac{3}{\sqrt{7}} = \frac{3}{\sqrt{7}} \cdot \frac{\sqrt{7}}{\sqrt{7}} = \frac{3\sqrt{7}}{7}$

15. $-\frac{\sqrt{2}}{\sqrt{5}}=-\frac{\sqrt{2}}{\sqrt{5}}\cdot\frac{\sqrt{5}}{\sqrt{5}}=-\frac{\sqrt{10}}{5}$

17. $\frac{4}{\sqrt{8}}=\frac{4}{\sqrt{8}}\cdot\frac{\sqrt{2}}{\sqrt{2}}=\frac{4\sqrt{2}}{\sqrt{16}}=\frac{4\sqrt{2}}{4}=\sqrt{2}$

19. $\sqrt{\frac{3}{49}}=\frac{\sqrt{3}}{\sqrt{49}}=\frac{\sqrt{3}}{7}$

21. $\sqrt{\frac{4}{3}}=\frac{\sqrt{4}}{\sqrt{3}}=\frac{2}{\sqrt{3}}\cdot\frac{\sqrt{3}}{\sqrt{3}}=\frac{2\sqrt{3}}{3}$

23. $\sqrt{\frac{8}{49}}=\frac{\sqrt{8}}{\sqrt{49}}=\frac{\sqrt{4\cdot 2}}{\sqrt{49}}=\frac{2\sqrt{2}}{7}$

25. $\sqrt{\frac{18}{50}}=\sqrt{\frac{9}{25}}=\frac{3}{5}$

27. $\sqrt{\frac{32}{125}}=\frac{\sqrt{16\cdot 2}}{\sqrt{25\cdot 5}}=\frac{4\sqrt{2}}{5\sqrt{5}}=\frac{4\sqrt{2}}{5\sqrt{5}}\cdot\frac{\sqrt{5}}{\sqrt{5}}$

$=\frac{4\sqrt{10}}{5\cdot 5}=\frac{4\sqrt{10}}{25}$

29. $\sqrt{5}\cdot\sqrt{50}=\sqrt{250}=\sqrt{25\cdot 10}=5\sqrt{10}$

31. $\frac{\sqrt{28}}{\sqrt{2}}=\sqrt{\frac{28}{2}}=\sqrt{14}$

33. $\frac{\sqrt{10}}{\sqrt{250}}=\sqrt{\frac{10}{250}}=\sqrt{\frac{1}{25}}=\frac{1}{5}$

35. $\frac{\sqrt{33}}{\sqrt{22}}=\sqrt{\frac{33}{22}}=\sqrt{\frac{3}{2}}=\frac{\sqrt{3}}{\sqrt{2}}\cdot\frac{\sqrt{2}}{\sqrt{2}}=\frac{\sqrt{6}}{2}$

37. $\sqrt{3}+\sqrt{12}=\sqrt{3}+\sqrt{4\cdot 3}=\sqrt{3}+2\sqrt{3}$

$=3\sqrt{3}$

39. $\sqrt{125}+\sqrt{80}=\sqrt{25\cdot 5}+\sqrt{16\cdot 5}$

$=5\sqrt{5}+4\sqrt{5}=9\sqrt{5}$

41. $\sqrt{3^2+4^2}=\sqrt{9+16}=\sqrt{25}=5$

43. $\sqrt{13^2-12^2}=\sqrt{169-144}=\sqrt{25}=5$

45. $6\sqrt{7}+\sqrt{7}-2\sqrt{7}=(6+1-2)\sqrt{7}=5\sqrt{7}$

47. $5\sqrt{7}-3\sqrt{28}-2\sqrt{63}$

$=5\sqrt{7}-3\sqrt{4\cdot 7}-2\sqrt{9\cdot 7}$

$=5\sqrt{7}-3\cdot 2\sqrt{7}-2\cdot 3\sqrt{7}$

$=5\sqrt{7}-6\sqrt{7}-6\sqrt{7}=-7\sqrt{7}$

49. $-3\sqrt{45}+\sqrt{20}-\sqrt{5}$

$=-3\sqrt{9\cdot 5}+\sqrt{4\cdot 5}-\sqrt{5}$

$=-3\cdot 3\sqrt{5}+2\sqrt{5}-\sqrt{5}$

$=-9\sqrt{5}+2\sqrt{5}-\sqrt{5}=-8\sqrt{5}$

51. a. To determine how long (t) a storm 6 mi in diameter will last, let $d = 6$ in the formula $t=\sqrt{\left(\frac{d}{6}\right)^3}$:

$t=\sqrt{\left(\frac{d}{6}\right)^3}=\sqrt{\left(\frac{6}{6}\right)^3}=\sqrt{(1)^3}=1$

The storm will last 1 hour.

b. To determine how long (t) a storm 10 mi in diameter will last, let $d = 10$ in the formula $t=\sqrt{\left(\frac{d}{6}\right)^3}$:

$t=\sqrt{\left(\frac{d}{6}\right)^3}=\sqrt{\left(\frac{10}{6}\right)^3}=\sqrt{\left(\frac{5}{3}\right)^3}=\sqrt{\frac{5^3}{3^3}}$

$=\frac{\sqrt{5^3}}{\sqrt{3^3}}=\frac{5\sqrt{5}}{3\sqrt{3}}$

$=\frac{5\sqrt{5}}{3\sqrt{3}}\cdot\frac{\sqrt{3}}{\sqrt{3}}=\frac{5\sqrt{15}}{9}\approx 2.15$

The storm will last about 2.15 hours.

53. a. To determine the average speed S of a tsunami when the average depth d of the water is 40 m, let $d = 40$ in the formula $S=\sqrt{10}\cdot\sqrt{d}$:

$S=\sqrt{10}\cdot\sqrt{d}=\sqrt{10}\cdot\sqrt{40}=\sqrt{400}=20$

The average speed of the tsunami is 20 m/sec.

b. To determine the average speed S of this tsunami when the average depth d of the water is 4267 m, let $d = 4267$ in the formula $S=\sqrt{10}\cdot\sqrt{d}$, round to the nearest meter:

$S = \sqrt{10} \cdot \sqrt{d} = \sqrt{10} \cdot \sqrt{4267}$

$= \sqrt{42670} \approx 207$

The average speed of this tsunami is 207 m/sec.

55. $r = \sqrt{\frac{144}{100}} - 1 = \frac{12}{10} - 1 = 1.2 - 1 = 0.20$

$= 20\%$

57. $v = \sqrt{20 \cdot 10 + 25} = \sqrt{225} = 15$ m/sec

59. $v = \sqrt{64 \cdot 12 + 16} = \sqrt{784} = 28$ ft/sec

61. $\sqrt{5^2 - 4 \cdot 1 \cdot 4} = \sqrt{25 - 16} = \sqrt{9} = 3$

63. $\sqrt{(-3)^2 - 4 \cdot 2 \cdot (-20)} = \sqrt{9 + 160}$

$= \sqrt{169} = 13$

65. $\sqrt{\left(\frac{1}{3}\right)^2 - 4 \cdot \frac{1}{12} \cdot (-1)} = \sqrt{\frac{1}{9} + \frac{1}{3}}$

$= \sqrt{\frac{4}{9}} = \frac{2}{3}$

67. 4.2: check Rational numbers, Real numbers

69. 0: check Whole numbers, Integers, Rational numbers, Real numbers

71. $\sqrt{9}$: check Natural numbers, Whole numbers, Integers, Rational numbers, Real numbers

73. $0.\overline{66}$: check Rational numbers, Real numbers

75. $\sqrt{20}$: check Irrational numbers, Real numbers

77. Answers may vary. Try an example. Suppose $a = -2$ and $b = -3$. Then, by definition, $\sqrt{-2}$ and $\sqrt{-3}$ are not real numbers but $\sqrt{-2}\sqrt{-3} = \sqrt{(-2)(-3)} = \sqrt{6}$, which is a real number. How can the product of two undefined numbers yield a real number?

79. $40 - 36 = 4$

$\sqrt{36} = 6$

$\sqrt{40} = \boxed{6\frac{4}{13} \approx 6.31}$

$\sqrt{49} = 7$

$49 - 36 = 13$

(Calculator gives 6.32.)

81. $85 - 81 = 4$

$\sqrt{81} = 9$

$\sqrt{85} = \boxed{9\frac{4}{19} \approx 9.21}$

$\sqrt{100} = 10$

$100 - 81 = 19$

(Calculator gives 9.22.)

Section 5.6 Number Sequences

Study Tips

We study two types of sequences: *arithmetic* and *geometric*. How do you tell them apart? Look at successive terms.

To determine an arithmetic sequence:
Think: subtract successive terms, if difference always the same, then it is arithmetic
Note: The general term a_n of an arithmetic sequence is $a_n = a_1 + (n-1) \cdot d$.

To find the sum of an arithmetic sequence:
Think: note the number *n* of terms to add, then identify the first term a_1, the last term a_n , use formula below
Note: The sum of a finite arithmetic sequence with *n* terms is $S_n = \dfrac{n(a_1 + a_n)}{2}$.

To determine an geometric sequence:
Think: divide successive terms, if quotient always the same, then it is geometric
Note: The general term a_n of a geometric sequence is $a_n = a_1 r^{n-1}$.

To find the sum of an geometric sequence:
Think: note the number *n* of terms to add, then identify the common ratio *r*, and the first term a_1,

use formula below
Note: The sum of a finite arithmetic sequence with n terms is $S_n = \dfrac{a_1(1-r^n)}{1-r}$.

To find the sum of an inifinite geometric sequence:
Think: identify the common ratio r and the first term a_1, use formula below
Note: The sum of an infinite geometric sequence is $S_n = \dfrac{a_1}{1-r}$.

Fact
An important application of the sum of an infinite geometric sequence is to write a given repeating decimal as a fraction.

To write a repeating decimal as a fraction:
Think: write as a geometric sequence, identify the common ratio r and the first term a_1, use formula $S_n = \dfrac{a_1}{1-r}$

Word of Advice
Memorize these formulas. Notecards are a helpful tool.

5.6 Exercises

1. a. $a_1 = 7$ b. $d = 6$
   c. $a_{10} = a_1 + 9d = 7 + 9 \cdot 6 = 7 + 54 = 61$
   d. $a_n = a_1 + (n-1)d = 7 + (n-1)6$
   $= 7 + 6n - 6 = 6n + 1$

3. a. $a_1 = 43$ b. $d = -9$
   c. $a_{10} = a_1 + 9d = 43 + 9 \cdot (-9)$
   $= 43 - 81 = -38$
   d. $a_n = a_1 + (n-1)d = 43 + (n-1)(-9)$
   $= 43 - 9n + 9 = 52 - 9n$

5. a. $a_1 = 2$ b. $d = -5$
   c. $a_{10} = a_1 + 9d = 2 + 9 \cdot (-5)$
   $= 2 - 45 = -43$
   d. $a_n = a_1 + (n-1)d = 2 + (n-1)(-5)$
   $= 2 - 5n + 5 = 7 - 5n$

7. a. $a_1 = -\dfrac{5}{6}$ b. $d = \dfrac{1}{2}$
   c. $a_{10} = a_1 + 9d = -\dfrac{5}{6} + 9 \cdot \dfrac{1}{2}$
   $= -\dfrac{5}{6} + \dfrac{9}{2} = \dfrac{22}{6} = \dfrac{11}{3}$
   d. $a_n = a_1 + (n-1)d = -\dfrac{5}{6} + (n-1)\left(\dfrac{1}{2}\right)$
   $= -\dfrac{5}{6} + \dfrac{1}{2}n - \dfrac{1}{2} = \dfrac{1}{2}n - \dfrac{4}{3} = \dfrac{3n-8}{6}$

9. a. $a_1 = 0.6$ b. $d = -0.4$
   c. $a_{10} = a_1 + 9d = 0.6 + 9 \cdot (-0.4)$
   $= 0.6 - 3.6 = -3$
   d. $a_n = a_1 + (n-1)d = 0.6 + (n-1)(-0.4)$
   $= 0.6 - 0.4n + 0.4 = 1 - 0.4n$

11. $S_{10} = \dfrac{10(7+61)}{2} = 5(68) = 340$
    $S_n = \dfrac{n(7+6n+1)}{2} = \dfrac{n(6n+8)}{2}$
    $= n(3n+4) = 3n^2 + 4n$

13. $S_{10} = \dfrac{10(43+(-38))}{2} = 5(5) = 25$
    $S_n = \dfrac{n(43+52-9n)}{2} = \dfrac{n(95-9n)}{2}$
    $= \dfrac{95n - 9n^2}{2}$

15. $S_{10} = \dfrac{10(2+(-43))}{2} = 5(-41) = -205$
    $S_n = \dfrac{n(2+7-5n)}{2} = \dfrac{n(9-5n)}{2}$
    $= \dfrac{9n - 5n^2}{2}$

17. $S_{10} = \dfrac{10\left(-\frac{5}{6} + \frac{11}{3}\right)}{2} = 5\left(\dfrac{17}{6}\right) = \dfrac{85}{6} = 14\dfrac{1}{6}$
    $S_n = \dfrac{n\left(-\frac{5}{6} + \frac{3n-8}{6}\right)}{2} = \dfrac{n\left(\frac{3n-13}{6}\right)}{2}$
    $= \dfrac{n(3n-13)}{12} = \dfrac{3n^2 - 13n}{12}$

19. $S_{10} = \frac{10(0.6+(-3))}{2} = 5(-2.4) = -12$

$S_n = \frac{n(0.6+1-0.4n)}{2} = \frac{n(1.6-0.4n)}{2}$

$= n(0.8-0.2n)$

$= 0.8n - 0.2n^2$, or $\frac{n}{5}(4-n)$

21. a. $a_1 = 3$ b. $r = 2$

c. $a_{10} = a_1 \cdot r^{n-1} = 3 \cdot 2^9 = 3 \cdot 512 = 1536$

d. $a_n = 3 \cdot 2^{n-1}$

23. a. $a_1 = \frac{1}{3}$ b. $r = 3$

c. $a_{10} = a_1 \cdot r^{n-1} = \frac{1}{3} \cdot 3^9 = 3^8 = 6561$

d. $a_n = \frac{1}{3} \cdot 3^{n-1} = 3^{n-2}$

25. a. $a_1 = 16$ b. $r = -\frac{1}{4}$

c. $a_{10} = a_1 \cdot r^{n-1} = 16 \cdot \left(-\frac{1}{4}\right)^9$

$= -4^2 \cdot 4^{-9} = -4^{-7} = -\frac{1}{16{,}384}$

d. $a_n = 16 \cdot \left(-\frac{1}{4}\right)^{n-1}$

$= 4^2 (-1)^{n-1} (4^{1-n})$

$= (-1)^{n-1} (4^{3-n}) = \frac{(-1)^{n-1}}{4^{n-3}}$

27. $S_{10} = \frac{3(1-2^{10})}{1-2} = \frac{3(-1023)}{-1} = 3069$

$S_n = \frac{3(1-2^n)}{1-2} = \frac{3(1-2^n)}{-1} = 3(2^n - 1)$

29. $S_{10} = \frac{\frac{1}{3}(1-3^{10})}{1-3} = \frac{\frac{1}{3}(1-3^{10})}{-2}$

$= \frac{1}{6}(3^{10}-1) = \frac{1}{6}(59{,}048) = 9841\frac{1}{3}$

$S_n = \frac{\frac{1}{3}(1-3^n)}{1-3} = \frac{\frac{1}{3}(1-3^n)}{-2} = \frac{1}{6}(3^n - 1)$

31. $S_{10} = \frac{16\left(1-\left(-\frac{1}{4}\right)^{10}\right)}{1-\left(-\frac{1}{4}\right)} = \frac{16\left(1-\frac{1}{4^{10}}\right)}{\frac{5}{4}}$

$= \frac{64}{5}\left(\frac{4^{10}-1}{4^{10}}\right) = \frac{4^3}{5}\left(\frac{4^{10}-1}{4^{10}}\right) = \frac{4^{10}-1}{5 \cdot 4^7}$

$= \frac{1{,}048{,}575}{81{,}920} = \frac{209{,}715}{16{,}384}$

$S_n = \frac{16\left(1-\left(-\frac{1}{4}\right)^n\right)}{1-\left(-\frac{1}{4}\right)} = \frac{16\left(1-\frac{(-1)^n}{4^n}\right)}{\frac{5}{4}}$

$= \frac{64}{5}\left(\frac{4^n-(-1)^n}{4^n}\right) = \frac{4^3}{5}\left(\frac{4^n-(-1)^n}{4^n}\right)$

$= \frac{4^n-(-1)^n}{5 \cdot 4^{n-3}}$

33. $a_1 = 6,\ r = \frac{1}{2}$

$S = \frac{a_1}{1-r} = \frac{6}{1-\frac{1}{2}} = \frac{6}{\frac{1}{2}} = 12$

35. $a_1 = -8,\ r = \frac{1}{2}$

$S = \frac{a_1}{1-r} = \frac{-8}{1-\frac{1}{2}} = \frac{-8}{\frac{1}{2}} = -16$

37. $a_1 = 0.7,\ r = 0.1$

$S = \frac{a_1}{1-r} = \frac{0.7}{1-0.1} = \frac{0.7}{0.9} = \frac{7}{9}$

39. First write this as $2+0.101010\cdots$. Then for the decimal part:
$a_1 = 0.10,\ r = 0.01$

$$S = \frac{a_1}{1-r} = \frac{0.1}{1-0.01} = \frac{0.1}{0.99} = \frac{10}{99}$$

The number is: $2+\frac{10}{99} = \frac{208}{99}$

41. This results in an arithmetic sequence with $a_1 = 1380$ and $d = -40$.

a. $a_{10} = a_1 + (n-1)d = 1380 + 9(-40)$
$= 1380 - 360 = 1020$
Thus, the depreciation the tenth year will be $1020.

b. The total depreciation for the ten years is obtained as the sum of the first ten terms of the sequence:

$$S_{10} = \frac{n}{2}(a_1 + a_{10}) = \frac{10}{2}(1380+1020)$$
$$= 5(2400) = 12{,}000$$

Thus, the value of the property at the end of ten years will be 30,000 – 12,000 = $18,000.

43. The costs per foot form an arithmetic progression with $a_1 = 50$ and $d = 5$.

a. The cost of the tenth foot is
$a_{10} = a_1 + (n-1)d = 50 + 9(5)$
$= 50 + 45 = 95$
So the cost is $95.

b. $a_{50} = a_1 + (n-1)d = 50 + 49(5)$
$= 50 + 245 = 295$
The total cost for the 50 ft is obtained from the sum

$$S_{50} = \frac{n}{2}(a_1 + a_{10}) = \frac{50}{2}(50+295)$$
$$= 25(345) = 8625$$

Thus, the total cost for a 50 foot well is $8625.

45. Here, the sequence is geometric with $a_1 = 100$ and $r = 1.10$. The sum of the first five terms is

$$S_5 = 100\left(\frac{1-1.10^5}{1-1.10}\right) = 100\left(\frac{1-1.10^5}{-0.10}\right)$$
$$= 100\left(\frac{1.10^5-1}{0.10}\right) = 1000\left(1.10^5-1\right)$$
$$= 1000(1.61051-1) = 1000(0.61051)$$
$$= 610.51$$

Thus, the final amount is $610.51.

47. Assume that n is an even number. Then there are $\frac{n}{2}$ pairs and the sum of each pair is $n + 1$.

Thus, the sum of the sequence is $\frac{n}{2}(n+1)$. If n is an odd number, first consider the sequence $1+2+3+\cdots+(n-3)+(n-2)+(n-1)$. Here there are $\frac{n-1}{2}$ pairs with n as the sum of each pair. So the sum is $\frac{n-1}{2}(n) = \frac{n(n-1)}{2}$.
Adding the term n that was omitted, the final sum is

$$\frac{n(n-1)}{2} + n = \frac{n(n-1)+2n}{2} = \frac{n(n-1+2)}{2}$$
$$= \frac{n}{2}(n+1) \text{ as before.}$$

49. Answers may vary. In an arithmetic sequence, each term after the first is obtained by adding the constant difference d to the preceding term. In a geometric sequence, each term after the first is obtained by multiplying the preceding term by the constant ratio r.

51. $a_n = 0 + (n-1)4.5 = 4.5n - 4.5$

53. $a_n = 45 + (n-1)0.5 = -0.5n + 45.5$

Chapter 5 Practice Test

Study Tips

If you have memorized the formulas and tips we have given you, write them on your test paper as soon as you receive it and before you forget them.
Here are the solutions to the practice test.

1. a. The <u>third</u> is an ordinal number.
   b. The <u>two</u> is a cardinal number.

c. The 270–891 is used for identification only.

2. a. A prime number is a natural number with exactly two distinct divisors: itself and 1.
   b. A composite number is a natural number with more than two distinct divisors.

3. $2\,\overline{)1220}$
   $2\,\overline{)\ 610}$
   $5\,\overline{)\ 305}$
   61
   Thus, $1220 = 2^2 \cdot 5 \cdot 61$

4. 53, 59, 61 and 67

5. $143 = 11 \cdot 13$, so 143 is composite.

6. a. 436 and 1530 are divisible by **2**.
   b. 3 + 8 + 7 = 18, which is divisible by 3, so 387 is divisible by **3**.
   1 + 5 + 3 = 9, which is divisible by 3, so 1530 is divisible by **3**.
   c. 2345 ends in a 5 and 1530 ends in a 0, so these are both divisible by **5**.

7. $216 = 2 \cdot 2 \cdot 2 \cdot 3 \cdot 3 \cdot 3 = 2^3 \cdot 3^3$
   $254 = 2 \cdot 127$
   The GCF is 2.
   $\frac{216}{254} = \frac{2 \cdot 108}{2 \cdot 127} = \frac{108}{127}$

8. $18 = 2 \cdot 3^2$
   $54 = 2 \cdot 3^3$
   $60 = 2^2 \cdot 3 \cdot 5$
   $\text{LCM} = 2^2 \cdot 3^3 \cdot 5 = 4 \cdot 27 \cdot 5 = 540$
   $\frac{1}{18} + \frac{1}{54} - \frac{1}{60} = \frac{30}{540} + \frac{10}{540} - \frac{9}{540} = \frac{31}{540}$

9. $1 - \frac{1}{4} - \frac{1}{2} - \frac{1}{8} = \frac{8-2-4-1}{8} = \frac{1}{8}$

10. a. –27 b. 41

11. a. $8 - 19 = 8 + (-19) = -11$
    b. $8 - (-19) = 8 + (+19) = 27$
    c. $-8 - 19 = -8 + (-19) = -27$
    d. $-8 - (-19) = -8 + (+19) = 11$

12. a. $6 \cdot 7 = 42$ b. $-8 \cdot 9 = -72$
    c. $-3 \cdot (-7) = 21$ d. $5 \cdot (-8) = -40$

13. a. $(-7)^2 = 49$ b. $-7^2 = -49$
    c. $(-4)^3 = -64$ d. $-4^3 = -64$

14. a. $81 \div 9 = 9$ b. $\frac{-15}{5} = -3$
    c. $12 \div (-2) = -6$ d. $\frac{-20}{-10} = 2$
    e. $\frac{0}{15} = 0$ f. $\frac{15}{0}$ is not defined.

15. a. $4 \times 12 \div 3 \times 10^3 - 2(-6+4) \times 10^4$
    $= 4 \times 12 \div 3 \times 10^3 - 2(-2) \times 10^4$
    $= 4 \times 12 \div 3 \times 1000 - 2(-2) \times 10{,}000$
    $= 48 \div 3 \times 1000 + 4 \times 10{,}000$
    $= 16 \times 1000 + 4 \times 10{,}000$
    $= 16{,}000 + 40{,}000 = 56{,}000$
    b. $\left(\frac{1}{2}\right)^3 \div \frac{1}{4} \cdot \frac{1}{2} + \frac{1}{3}\left(\frac{7}{2} - \frac{3}{2}\right) - \frac{1}{2} \cdot \frac{1}{3}$
    $= \left(\frac{1}{2}\right)^3 \div \frac{1}{4} \cdot \frac{1}{2} + \frac{1}{3}\left(\frac{4}{2}\right) - \frac{1}{2} \cdot \frac{1}{3}$
    $= \frac{1}{8} \div \frac{1}{4} \cdot \frac{1}{2} + \frac{1}{3}(2) - \frac{1}{2} \cdot \frac{1}{3}$
    $= \frac{1}{8} \cdot \frac{4}{1} \cdot \frac{1}{2} + \frac{1}{3}(2) - \frac{1}{2} \cdot \frac{1}{3}$
    $= \frac{1}{4} + \frac{2}{3} - \frac{1}{6} = \frac{3}{12} + \frac{8}{12} - \frac{2}{12}$
    $= \frac{9}{12} = \frac{3}{4}$

16. $\frac{3}{4} = \frac{3 \cdot 4}{4 \cdot 4} = \frac{12}{16}$

17. a. $\frac{3}{2}$ b. $-\frac{7}{4}$ c. $\frac{8}{21}$ d. $-\frac{1}{8}$

18. a. $\frac{7}{8}\times\left(-\frac{5}{16}\right)=-\frac{35}{128}$

b. $-\frac{7}{8}\div\left(-\frac{5}{16}\right)=-\frac{7}{8}\cdot\left(-\frac{16}{5}\right)=\frac{14}{5}$

19. a. 23.508

$=2\times10+3\times10^0+5\times10^{-1}+8\times10^{-3}$

b. 803.04

20. $\left(6\times10^4\right)\times\left(8\times10^{-6}\right)$

$=6\times8\times10^{4-6}=48\times10^{-2}=4.8\times10^{-1}$

21. a. $6.73+2.8=9.53$
b. $9.34-4.71=4.63$
c. $0.29\times6.7=1.943$
d. $17.36\div3.1=5.6$

22. a. $\frac{3}{4}=0.75$ b. $\frac{1}{15}=0.0\overline{6}$

23. a. $0.\overline{12}=\frac{12}{99}=\frac{4}{33}$

b. $2.6555\ldots=\frac{26}{10}+\frac{5}{90}=\frac{234+5}{90}=\frac{239}{90}$

24. a. $21\%=0.21$
b. $9.35\%=0.0935$
c. $0.26\%=0.0026$

25. a. $0.52=52\%$
b. $2.765=276.5\%$
c. $\frac{3}{5}=0.60=60\%$
d. $\frac{2}{11}=0.1818\ldots\approx18.2\%$

26. a. $\sqrt{49}=7$, a rational number.
b. $\sqrt{45}$ is an irrational number because 45 is not a perfect square.
c. $\sqrt{121}=11$, a rational number.
d. 0.41252525 . . . is a nonterminating, repeating decimal, so is a rational number.
e. 0.212112111 . . . is a nonterminating, nonrepeating decimal, so is an irrational number.
f. 0.246810 . . . is a nonterminating, nonrepeating decimal, so is an irrational number.

27. a. $\sqrt{59}<7.7$ b. $\sqrt{31}>5.5$

28. $0.\overline{2}=0.222\ldots$ so 0.24 is a rational number between $0.\overline{2}$ and 0.25. (Other answers are possible.)

29. 0.23456 . . . is a nonterminating, nonrepeating decimal, so is an irrational number. Also, it is between 0.222 . . . and 0.25. (Other answers are possible.)

30. a. $\sqrt{96}=\sqrt{16\cdot6}=4\sqrt{6}$

b. $\sqrt{58}$ is in simplest form.

31. a. $\frac{4}{\sqrt{20}}=\frac{4}{\sqrt{4\cdot5}}=\frac{4}{2\sqrt{5}}\cdot\frac{\sqrt{5}}{\sqrt{5}}=\frac{2\sqrt{5}}{5}$

b. $\sqrt{\frac{48}{49}}=\frac{\sqrt{16\cdot3}}{\sqrt{49}}=\frac{4\sqrt{3}}{7}$

32. a. $\sqrt{8}\cdot\sqrt{6}=\sqrt{48}=\sqrt{16\cdot3}=4\sqrt{3}$

b. $\frac{\sqrt{56}}{\sqrt{7}}=\sqrt{\frac{56}{7}}=\sqrt{8}=\sqrt{4\cdot2}=2\sqrt{2}$

33. a. $\sqrt{90}-\sqrt{40}=\sqrt{9\cdot10}-\sqrt{4\cdot10}$

$=3\sqrt{10}-2\sqrt{10}=\sqrt{10}$

b. $\sqrt{32}+\sqrt{18}-\sqrt{50}$

$=\sqrt{16\cdot2}+\sqrt{9\cdot2}-\sqrt{25\cdot2}$

$=4\sqrt{2}+3\sqrt{2}-5\sqrt{2}=2\sqrt{2}$

34. $c=\sqrt{4^2+6^2}=\sqrt{16+36}=\sqrt{52}$

$=\sqrt{4\cdot13}=2\sqrt{13}$

35. a. Check Natural numbers, Whole numbers, Integers, Rational numbers and Real Numbers.
b. Check Rational numbers and Real numbers.
c. Check Irrational numbers and Real numbers.
d. Check Rational numbers and Real numbers.
e. Check Rational numbers and Real numbers.

36. a. Each term is 2 times the preceding term, so this is a geometric sequence.
b. Each term is 3 plus the preceding term so this is an arithmetic sequence.

37. This is an arithmetic sequence with $a_1 = 9$ and $d = 4$. So,

$$a_{10} = 9 + 9(4) = 9 + 36 = 45$$

$$S_{10} = \frac{10}{2}(9 + 45) = 5(54) = 270$$

38. This is a geometric sequence with $a_1 = 1$ and $r = \frac{1}{2}$. So,

$$S_5 = \frac{1\left(1 - \left(\frac{1}{2}\right)^5\right)}{1 - \frac{1}{2}} = \frac{1 - \frac{1}{32}}{\frac{1}{2}} = 2\left(1 - \frac{1}{32}\right)$$

$$= 2 - \frac{1}{16} = \frac{31}{16}$$

39. a. $a_5 = \frac{1}{3}\left(\frac{1}{2}\right)^4 = \frac{1}{3} \cdot \frac{1}{16} = \frac{1}{48}$

b. $S_5 = \frac{\frac{1}{3}\left(1 - \left(\frac{1}{2}\right)^5\right)}{1 - \frac{1}{2}} = \frac{\frac{1}{3}\left(1 - \frac{1}{32}\right)}{\frac{1}{2}}$

$$= \frac{2}{3}\left(1 - \frac{1}{32}\right) = \frac{2}{3} \cdot \frac{31}{32} = \frac{31}{48}$$

40. a. $S = \frac{0.4}{1 - 0.1} = \frac{0.4}{0.9} = \frac{4}{9}$

b. $S = \frac{0.21}{1 - 0.01} = \frac{0.21}{0.99} = \frac{21}{99} = \frac{7}{33}$

c. $S = 2 + \frac{0.5}{1 - 0.1} = 2 + \frac{0.5}{0.9} = 2 + \frac{5}{9} = \frac{23}{9}$

Chapter 6 Equations, Inequalities, and Problem Solving

Section 6.1 Solutions of First-Degree Sentences

Study Tips

In application problems, a question needs to be answered. We use equations and inequalities to answer questions.

To find the solution set of an equation or inequality given the replacement set:
Think: substitute each number from the replacement set into the equation or inequality, if it produces a true statement, then that number is part of the solution set
Note: The *replacement set is* given and you *replace* the variable in the equation by the elements of this set until a true statement results. (The equation is *satisfied.*) When this happens, the element becomes part of the *solution set.*

To solve an equation:
Think: get variable by itself on one side of the equation
Note: The elementary operations you can perform on an equation and the procedure to actually *solve* an equation are given in part B in this section. *Study and use this procedure in every problem.*

Fact
Linear equations can have *one* solution (the solution set has one element), *infinitely many* solutions (the solution set may be the set of real numbers or the set of integers), or *no* solution (the solution set is the empty set.)

Hint
When you have fractions in the equation, find the LCD and multiply both sides of the equation. This will *eliminate* all fractions.

To solve a formula for a variable:
Think: goal is to get the specified variable by itself on one side of the equation
Note: The procedures will be similar to solving an equation, see part C in this section. You might have to use the distributive property if your specified variable shows up more than once.

Hint
To distinguish the specified variable, underline or highlight it so that you do not confuse it with the other variables.

To solve an inequality:
Think: get variable by itself on one side of the inequality
Note: The procedure for solving *inequalities* is the same as for equations with *one* exception: *If you multiply or divide an inequality by a negative number, you must reverse the sense of the inequality.*

Fact
An inequality of the first degree in x is of the form $ax + b < 0$, or $ax + b > 0$. The solution set can be expressed in the form of an inequality.

6.1 Exercises

1. $2 + x \le 2 - x$
   a. Substitute 2 for x: $2 + 2 = 4$; $2 - 2 = 0$ and $4 > 0$. So 2 is not a solution.
   b. Substitute -2 for x: $2 + (-2) = 0$, $2 - (-2) = 4$ and $0 < 4$. So -2 is a solution.
   c. Substitute 0 for x: $2 + 0 = 2$; $2 - 0 = 2$ and $2 = 2$. So 0 is a solution.
   d. Substitute 5 for x: $2 + 5 = 7$; $2 - 5 = -3$ and $7 > -3$. So 5 is not a solution.

   Thus, the only solutions are -2 and 0.

3. $3x - 2 \ge 2x - 1$
   a. Substitute 0 for x: $3 \cdot 0 - 2 = -2$; $2 \cdot 0 - 1 = -1$ and $-2 < -1$. So 0 is not a solution.
   b. Substitute 3 for x: $3 \cdot 3 - 2 = 7$; $2 \cdot 3 - 1 = 5$ and $7 > 5$. So 3 is a solution.
   c. Substitute -2 for x: $3(-2) - 2 = -8$; $2(-2) - 1 = -5$ and $-8 < -5$. So -2 is not a solution.
   d. Substitute 1 for x: $3 \cdot 1 - 2 = 1$; $2 \cdot 1 - 1 = 1$ and $1 = 1$. So 1 is a solution.

   Thus, the only solutions are 3 and 1.

5. a. Substitute -1 for x: $2(-1) - 1 = 3$; $-2 - 1 = 3$; and $-3 = 3$. So -1 is not a solution.
   b. Substitute 1 for x: $2(1) - 1 = 3$; $2 - 1 = 3$; and $1 = 3$. So 1 is not a solution.
   c. Substitute 2 for x: $2(2) - 1 = 3$; $4 - 1 = 3$; and $3 = 3$. So 2 is a solution.
   d. Substitute 0 for x: $2(0) - 1 = 3$; $0 - 1 = 3$; and $-1 = 3$. So 0 is not a solution.

   2 is the only solution of $2x - 1 = 3$.

For problems 7 – 31, checking the answer (step 7 of the procedure to solve linear equations) is left to the student.

7. The given equation is: $x+10=15$
Subtract 10 from both sides:
$x+10-10=15-10$
Simplify: $x=5$

9. The given equation is: $2x-1=5$
Add 1 to both sides: $2x-1+1=5+1$
Simplify: $2x=6$
Divide both sides by 2: $x=3$

11. The given equation is: $2x+2=x+4$
Subtract x from both sides:
$2x+2-x=x+4-x$
Simplify: $x+2=4$
Subtract 2 from both sides: $x=2$

13. The given equation is: $3x+1=4x-8$
Subtract $3x$ from both sides:
$3x+1-3x=4x-8-3x$
Simplify: $1=x-8$
Add 8 to both sides: $9=x$

15. The given equation is: $7=3x+4$
Subtract 4 from both sides:
$7-4=3x+4-4$
Simplify: $3=3x$
Divide both sides by 3: $1=x$

17. The given equation is: $4=3x-2$
Add 2 to both sides: $4+2=3x-2+2$
Simplify: $6=3x$
Divide both sides by 3: $2=x$

19. The given equation is:
$7n+10-2n=4n-2+3n$
Simplify: $5n+10=7n-2$
Subtract $5n$ from both sides:
$5n+10-5n=7n-2-5n$
Simplify: $10=2n-2$
Add 2 to both sides: $10+2=2n-2+2$
Simplify: $12=2n$
Divide both sides by 2: $6=n$

21. The given equation is: $2(x+5)=13$
Simplify: $2x+10=13$
Subtract 10 from both sides:
$2x+10-10=13-10$
Simplify: $2x=3$

Divide both sides by 3: $x=\frac{3}{2}=1\frac{1}{2}$

23. The given equation is: $\frac{1}{2}(x-2)=5$
Multiply both sides by 2: $x-2=10$
Add 2 to both sides: $x=12$

25. The given equation is: $3(x+1)-x=2(9-x)$
Multiply: $3x+3-x=18-2x$
Simplify: $2x+3=18-2x$
Add $2x$ to both sides: $4x+3=18$
Subtract 3 from both sides: $4x=15$
Divide both sides by 4: $x=\frac{15}{4}=3\frac{3}{4}$

27. The given equation is: $8(x-1)=x+2$
Multiply: $8x-8=x+2$
Subtract x from both sides: $7x-8=2$
Add 8 to both sides: $7x=10$
Divide both sides by 7: $x=\frac{10}{7}=1\frac{3}{7}$

29. The given equation is: $\frac{1}{4}(x-2)=\frac{1}{3}(x-4)$
Multiply both sides by 12: $3(x-2)=4(x-4)$
Simplify: $3x-6=4x-16$
Subract $3x$ from both sides: $-6=x-16$
Add 16 to both sides: $10=x$

31. The given equation is: $3\times 2p+5=37-\frac{4p^2}{2p}$
Multiply and divide: $6p+5=37-2p$
Add $2p$ to both sides: $8p+5=37$
Subtract 5 from both sides: $8p=32$
Divide both sides by 8: $p=4$

33. Given: $V=\pi r^2 h$; solve for h:
Divide both sides by πr^2: $h=\frac{V}{\pi r^2}$

35. Given: $V=LWH$; solve for W:
Divide both sides by LH : $W=\frac{V}{LH}$

37. Given: $P=s_1+s_2+b$; solve for b:

Subtract $s_1 + s_2$ from both sides:
$b = P - s_1 - s_2$

39. Given: $D = RT$
   a. Solve for T:
   Divide both sides by R: $T = \frac{D}{R}$
   b. $D = 220$ mi, $R = 55$ mph
   $T = \frac{220}{55} = 4$ hr

41. Given: $H = 17 - \frac{A}{2}$
   a. Solve for A:
   Multiply each term by 2:
   $2H = 34 - A$
   Add A and subtract $2H$: $A = 34 - 2H$
   b. $H = 8$ hr
   $A = 34 - 2 \cdot 8 = 34 - 16 = 18$ years old

43. The given inequality is: $x - 3 < 1$
   Add 3 to both sides: $x < 4$
   The solution is $\{x | x < 4\}$.

45. The given inequality is: $x - 4 > -1$
   Add 4 to both sides: $x > 3$
   The solution is $\{x | x > 3\}$.

47. The given inequality is: $2x - 1 > x + 2$
   Subtract x from both sides: $x - 1 > 2$
   Add 1 to both sides: $x > 3$
   The solution is $\{x | x > 3\}$.

49. The given inequality is: $2x + 3 \le 9 + 5x$
   Subtract $2x$ from both sides: $3 \le 9 + 3x$
   Subtract 9 from both sides: $-6 \le 3x$
   Divide both sides by 3: $-2 \le x$
   The solution is $\{x | x \ge -2\}$.

51. The given inequality is: $x + 1 > \frac{1}{2}x - 1$
   Multiply both sides by 2: $2x + 2 > x - 2$
   Subtract x from both sides: $x + 2 > -2$
   Subtract 2 from both sides: $x > -4$
   The solution is $\{x | x > -4\}$.

53. The given inequality is: $x \ge 4 + 3x$
   Subtract x from both sides: $0 \ge 4 + 2x$
   Subtract 4 from both sides: $-4 \ge 2x$
   Divide both sides by 2: $-2 \ge x$
   The solution is $\{x | x \le -2\}$.

55. The given inequality is: $\frac{1}{3}x - 2 \ge \frac{2}{3}x + 1$
   Multiply both sides by 3: $x - 6 \ge 2x + 3$
   Subtract x from both sides: $-6 \ge x + 3$
   Subtract 3 from both sides: $-9 \ge x$
   The solution is $\{x | x \le -9\}$.

57. The given inequality is: $2x - 2 > x + 1$
   Subtract x from both sides: $x - 2 > 1$
   Add 2 to both sides: $x > 3$
   The solution is $\{x | x > 3\}$.

59. The given inequality is: $x + 3 > \frac{1}{2}x + 1$
   Multiply both sides by 2: $2x + 6 > x + 2$
   Subtract x from both sides: $x + 6 > 2$
   Subtract 6 from both sides: $x > -4$
   The solution is $\{x | x > -4\}$.

61. The given inequality is: $x \ge 2 + 4x$
   Subtract x from both sides: $0 \ge 2 + 3x$
   Subtract 2 from both sides: $-2 \ge 3x$
   Divide both sides by 3: $-\frac{2}{3} \ge x$
   The solution is $\left\{x \middle| x \le -\frac{2}{3}\right\}$.

63. The given inequality is: $2x + 1 < 2x$
   Subtract $2x$ from both sides: $1 < 0$
   which is false for any value of x. There is no value of x for which the given inequality is true. The solution is $\varnothing$.

65. The given inequality is: $8x + 2 \le 3(x + 4)$
   Multiply: $8x + 2 \le 3x + 12$
   Subtract $3x$ from both sides: $5x + 2 \le 12$
   Subtract 2 from both sides: $5x \le 10$
   Divide both sides by 5: $x \le 2$
   The solution is $\{x | x \le 2\}$.

67. The given inequality is: $3(x + 4) > -5x - 4$
   Multiply: $3x + 12 > -5x - 4$
   Add $5x$ to both sides: $8x + 12 > -4$

Subtract 12 from both sides: $8x > -16$
Divide both sides by 8: $x > -2$
The solution is $\{x|x > -2\}$.

69. The given inequality is:
$$-2(x+1) \geq 3x - 4$$
Multiply: $-2x - 2 \geq 3x - 4$
Add 2x to both sides: $-2 \geq 5x - 4$
Add 4 to both sides: $2 \geq 5x$
Divide both sides by 5: $\frac{2}{5} \geq x$
The solution is $\left\{x \middle| x \leq \frac{2}{5}\right\}$.

71. The given inequality is:
$$a(x-1) \leq a(2x+3)$$
Divide by a: $x - 1 \geq 2x + 3$
(reverse inequality since $a < 0$)
Subtract x from both sides: $-1 \geq x + 3$
Subtract 3 from both sides: $-4 \geq x$
The solution is $\{x|x \leq -4\}$.

73. Use the equation $0.01 \cdot r \cdot n = p$. In this problem, $r = 40, n = 80$, and p is unknown. Thus, the equation becomes
$0.01 \cdot 40 \cdot 80 = p$
$32 = p$
The number is 32.

75. Use the equation $0.01 \cdot r \cdot n = p$. In this problem, $n = 3150, p = 315$, and r is unknown. Thus, the equation becomes
$0.01 \cdot r \cdot 3150 = 315$
$31.50r = 315$ Divide by 31.50.
$r = 10$
The percent is 10%.

77. Use the equation $0.01 \cdot r \cdot n = p$. In this problem, $n = 40, p = 5$, and r is unknown. Thus, the equation becomes
$0.01 \cdot r \cdot 40 = 5$
$0.4r = 5$ Divide by 0.4.
$r = 12.5$
The percent is 12.5%.

79. Use the equation $0.01 \cdot r \cdot n = p$. In this problem, $r = 30, p = 60$, and n is unknown. Thus, the equation becomes
$0.01 \cdot 30 \cdot n = 60$
$0.30n = 60$ Divide by 0.30.
$n = 200$
The number is 200.

81. a. Determine 10% of 20 mpg: $0.1 \cdot 20 = 2$
Since this is an increase, add to 20 mpg:
$2 + 20 = 22$ mpg
b. Currently, you drive $20 \cdot 600 = 12,000$ miles per year. Divide by the new mpg:
$12,000 \div 22 \approx 545$ gal/yr

83. Use the equation $0.01 \cdot r \cdot n = p$. In this problem, $n = 60, p = 40$, and r is unknown. Thus, the equation becomes
$0.01 \cdot r \cdot 60 = 40$
$0.6r = 40$ Divide by 0.6.
$r \approx 67$
The students score was 67%.

85. Use the equation $0.01 \cdot r \cdot n = p$. In this problem, $r = 25, n = 140$, and p is unknown. Thus, the equation becomes
$0.01 \cdot 25 \cdot 140 = p$
$35 = p$
The discount is $35, so the sale price at store A is $105. Store B has the lower price.

87. Find when $B = M$. So solve the equation:
$1.73x + 30 = 0.53x + 33$
$1.20x + 30 = 33$
$1.20x = 3$
$x = \frac{3}{1.20} = 2.5$
The amounts are the same in 2.5 years.

89. a. Evaluate the equation when $N = 13$.
$E = 0.15(13) + 7.77$
$= 1.95 + 7.77 = 9.72$ million
This value is within 0.18 million, or 180,000, of the value of the graph.
b. Evaluate the equation when $N = 25$.
$E = 0.15(25) + 7.77$
$= 3.75 + 7.77 = 11.52$ million
This value is within 0.02 million,

or 20,000, of the value of the graph.

c. Evaluate the equation when $N = 21$.

$$E = 0.15(21) + 7.77$$
$$= 3.15 + 7.77 = 10.92 \text{ million}$$

91. Solve the men's equation for $S = 11$.

$$11 = 3L - 22$$
$$33 = 3L$$
$$11 = L$$

Tyrone's foot is 11 in. long.

93. Solve the men's equation for $S = 7$.

$$7 = 3L - 22$$
$$29 = 3L$$
$$L = \frac{29}{3} = 9\frac{2}{3}$$

Sam's foot is $9\frac{2}{3}$ in. long. So Sue's foot is also $9\frac{2}{3}$ in. long. Solve the women's equation for S:

$$S = 3\left(\frac{29}{3}\right) - 21 = 29 - 21 = 8$$

Sue's shoe size is 8.

95. Solve the men's equation for $S = 14$.

$$14 = 3L - 22$$
$$36 = 3L$$
$$12 = L$$

The foot is 12 in. long.

97. a. For a snowman, solve the men's equation for $L = 23$:

$$S = 3 \cdot 23 - 22 = 69 - 22 = 47$$

The Abominable Snowman needs size 47 shoes.

b. For a snowwoman, solve the women's equation for $L = 23$:

$$S = 3 \cdot 23 - 21 = 69 - 21 = 48$$

The Abominable Snowwoman needs size 48 shoes.

99. The U.S. Post Office price is cheaper when $0.44 + 0.20(x - 1) < 5.00$

$$0.44 + 0.20x - 0.20 < 5.00$$
$$0.20x + 0.24 < 5.00$$
$$0.20x < 4.76$$
$$x < 23.8$$

The Post Office is cheaper when the package weighs less than 23.8 oz, or 1 lb 7.8 oz.

101. $N > A$ when

$$0.165x + 4.68 > -0.185x + 5.38$$
$$0.35x + 4.68 > 5.38$$
$$0.35x > 0.70$$
$$x > 2$$

This will occur afer 1996 + 2 or 1998.

103. Answers will vary. The solution set for an equation is the set of numbers taken from the replacement set that will make an open sentence a true statement when the number is substituted for the variable in the open sentence.

105. Answers will vary. Yes. In order to be in the solution set a number must be part of the replacement set.

107. Since 50% of the total calories c is 125 fat calories, we have:

$$0.50c = 125$$
$$c = 250 \quad \text{Divide by 0.50.}$$

Thus, there are 250 calories in the apple pie.

109. Since 50% of the calories is 355 of the total calories c, we have:

$$0.50c = 355$$
$$c = 710 \quad \text{Divide by 0.50.}$$

Thus, there are 710 calories in a Whopper.

111. $N = 4(77.5 - 40) = 4(37.5) = 150$ chirps per minute.

113. $S = \frac{1}{6}(10 - 4) = \frac{1}{6}(6) = 1$ cm/sec

Section 6.2 Graphs of Algebraic Sentences

Study Tips

Sometimes the elements of the solution set of an equation are not easy to write, but a picture of the solution set, its *graph*, is easily obtainable.

Comparing graph and solution set:
The *graph of the solution set* is similar to the solution set itself.

1. If the solution set is a point, the graph is a point.
2. If the solution set consists of all real numbers such that $x > 7$, the graph is

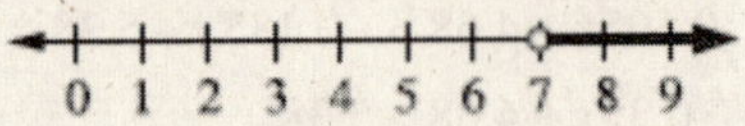

with the arrow pointing right just like in the > sign. The 7 is marked by an open circle and is not included.

3. For $x \le -1$, the graph is

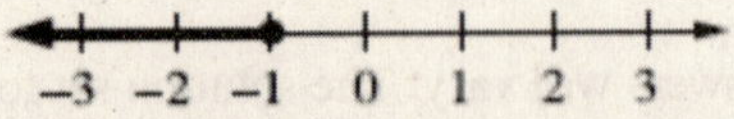

pointing left, like the < and with −1 included as indicated by the shaded circle.

Word of Advice
Remember to look at your graph and see if it is similar to the solution itself.

To graph inequalities with *and*:
Think: intersection of the graphs
Note: When the replacement set is the set of real numbers and the compound sentence has an *and*, the answer is either the *empty set* or a *single line segment*.

To graph inequalities with *or*:
Think: union of graphs
Note: If the compound sentence contains an *or* the answer consists of *two pieces* called *rays*.

Application Problems
Sometimes the replacement set is *not* the set of real numbers, then the answer is the *intersection* of the two sets when ***and*** is used or the *union* of the two sets when ***or*** is used.

6.2 Exercises

1. Subtract 2 from both sides to find the solution $x = 2$. The solution set is thus $\{2\}$, which is shown on the graph.

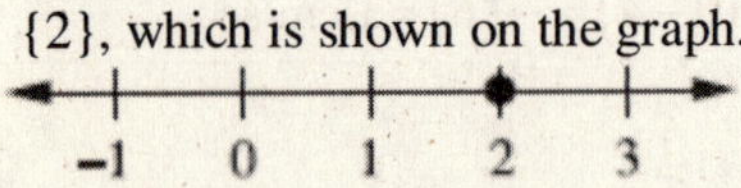

3. Subtract 1 from both sides to get $x \ge 1$. Thus, the solution set consists of all the positive integers as indicated on the graph.

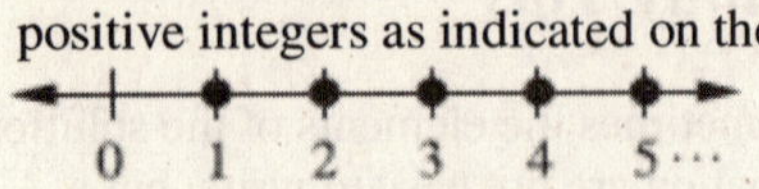

5. Add 3 to both sides to get $x \ne 4$. This means that the solution consists of all the integers except 4, as indicated on the graph.

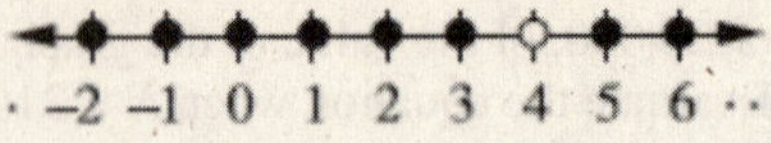

7. Since x is to be an integer, the inequality $-2 \le x \le 4$ has the solution set $\{-2, -1, 0, 1, 2, 3, 4\}$, as shown on the graph.

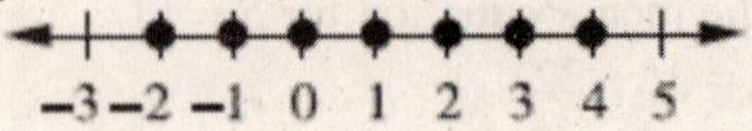

9. The solution set of $x < 4$ is the set of all real numbers less than 4, as shown on the graph.

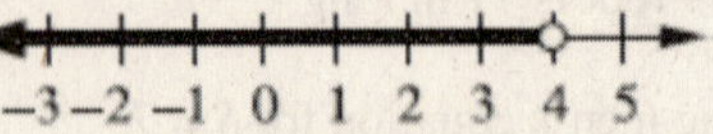

11. Adding 2 to both sides gives the inequality $x \le 2$ so the solution set consists of all the real numbers less than or equal to 2.

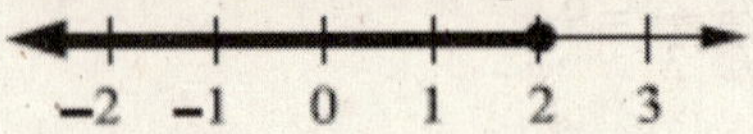

13. The solution set of $-2 \le x \le 4$ consists of all the real numbers between −2 and +4, inclusive.

15. Subtract 2 from both sides to get $x > 3$. The solution set is the set of all real numbers greater than 3.

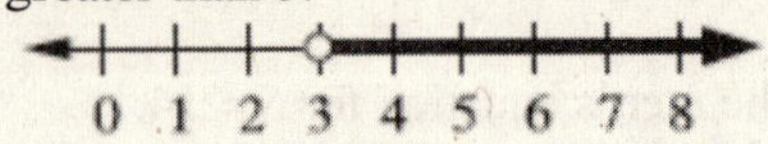

17. The solution set of $-1 < x < 2$ consists of all the real numbers between −1 and +2 with neither endpoint included.

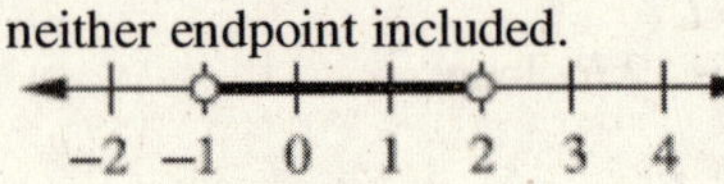

19. Subtracting 4 from both sides gives $x < 1$, so that the solution set consists of all the real numbers less than 1.

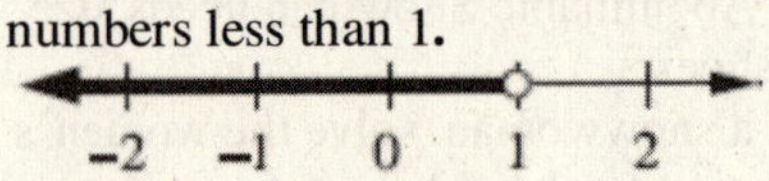

21. $x + 1 < x$ is not true for any real number. Thus, the solution set is $\varnothing$.

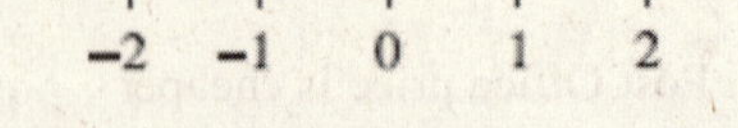

23. Subtract x from both sides to get $x + 3 < 1$. Then subtract 3 from both sides to get $x < -2$. Thus, the solution set consists of all the real numbers less than −2.

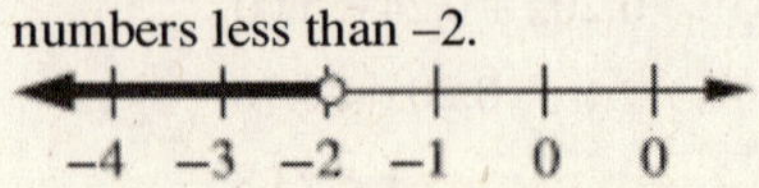

25. Add 7 to both sides to get $3x \geq 0$. Then divide both sides by 3 to get $x \geq 0$. The solution set consists of all the real numbers that are greater than or equal to 0.

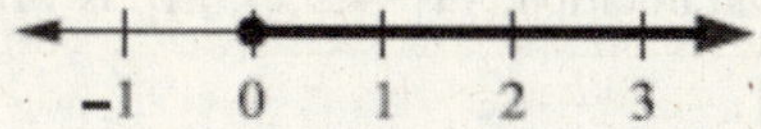

27. Add 1 to both sides of the second inequality to get $x \geq -1$. The solution set of this inequality is the set of integers greater than or equal to –1. The solution set of $x \leq 4$ is the set of integers less than or equal to 4. So the solution set of $x \leq 4$ **and** $x \geq -1$ is the set of integers greater than or equal to –1 and less than or equal to 4, that is, {–1, 0, 1, 2, 3, 4}.

29. Subtract 1 from both sides of the first inequality to get $x \leq 6$. The solution set of this inequality is the set of integers less than or equal to 6. The solution set of $x > 2$ is the set of integers greater than 2. Thus, the solution set of $x > 2$ **and** $x \leq 6$ is the set of integers greater than 2 but less than or equal to 6, that is, {3, 4, 5, 6}.

31. Add 1 to both sides of the first inequality to get $2x > 2$. Then divide both sides by 2 to obtain $x > 1$. The solution set of this inequality is the set of integers greater than 1. Now, subtract 1 from both sides of the second inequality to get $x < 3$. The solution set of this inequality is the set of integers less than 3. Thus, the solution set of $x > 1$ **and** $x < 3$ consists of the integer 2, that is, {2}.

33. The solution set of $x < -5$ **or** $x > 5$ consists of all the integers except those between –5 and +5, inclusive, that is the set {. . ., –8, –7, –6, 6, 7, 8, . . .}.

35. Subtract 1 from both sides of the first inequality to get $x \geq 1$. The solution set of $x \geq 1$ **and** $x \leq 4$ is the set of all real numbers between 1 and 4, inclusive. This is shown on the graph.

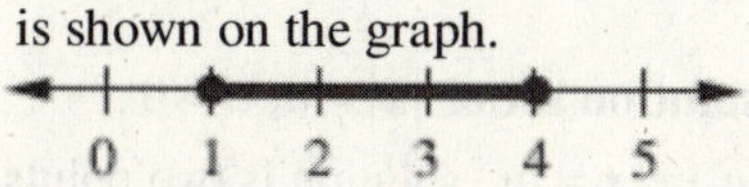

37. There are no numbers that are both greater than 2 and less than –2, so the solution set is the empty set, $\varnothing$.

39. Add 1 to both sides of the first inequality to get $x > 1$. Then, subtract 1 from both sides of the second inequality to get $x < 4$. The solution set of $x > 1$ **and** $x < 4$ is the set of all real numbers between 1 and 4, not including the 1 and the 4. This is shown on the graph.

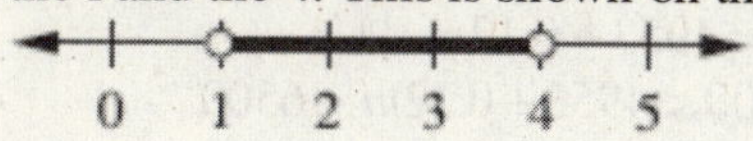

41. The first inequality, $x \leq x+1$, is satisfied by all real numbers. Therefore, only the inequality $x \geq 2$ must be satisfied, and the solution set of this inequality is the set of all real numbers greater than or equal to 2. The graph shows this set.

43. Add 2 to both sides of the first inequality to get $x \geq 4$. Since there is no real number that is both greater than or equal to 4 and less than 0, the solution set is the empty set, $\varnothing$.

45. Add 1 to both sides of the second inequality to get $x \geq 3$. Then, the solution set of $x \geq 0$ **and** $x \geq 3$ is the set of all real numbers that are greater than or equal to 3. This set is shown on the graph.

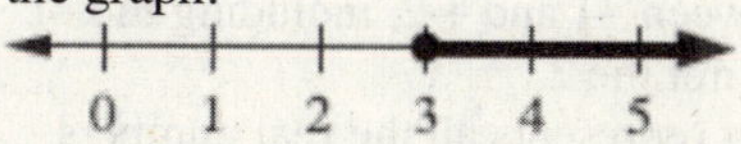

47. Add 1 to both sides of the second inequality to get $x < 3$. Then, the solution set of $x < 0$ **or** $x < 3$ is the set of all real numbers that are less than 3. (Note that each number of this set satisfies at least one of the two inequalities.) The graph is shown below.

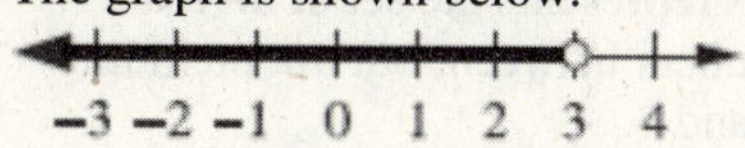

49. Subtract 1 from both sides of the first inequality to get $x > 1$. Then, add 2 to both sides of the second inequality to get $x < 5$. The solution set of $x > 1$ **and** $x < 5$ is the set of all real numbers between 1 and 5, not including the 1 or the 5. The graph shows this set.

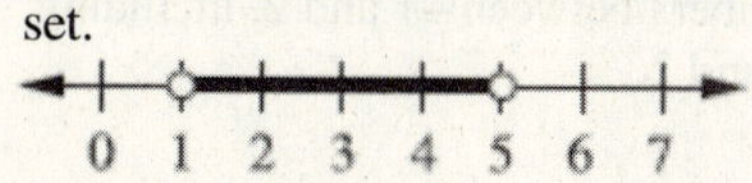

51. Add 1 to both sides of the first inequality and subtract 2 from both sides of the second inequality to get the system $x > 1$ **or** $x < 2$. Every real number is greater than 1 or less

than 2, so the solution set consists of all the real numbers, as indicated by the graph.

53. a. $C = 4953 + 0.12m$
    b. $6000 < 4953 + 0.12m < 6500$
    c. $6000 < 4953 + 0.12m < 6500$
    $1047 < 0.12m < 1547$
    $8725 < m < 12{,}892$
    To stay within the budget the car can be driven between 8725 mi and 12,892 mi annually.

55. Solve for N when $E > 36$ and $E = 1.5N + 27$.
    $1.5N + 27 > 36$
    $1.5N > 9$
    $N > 6$
    $N > 6$ years after 2010, that is, 2016.

57. a. This represents all the real numbers between −1 and +2, including the 2 but not the −1.
    b. This represents all the real numbers between −1 and + 2, including the −1 but not the 2.
    c. This represents all the real numbers between −1 and + 2, not including the endpoints.
    d. This represents all the real numbers between −1 and + 2, including the endpoints.

59. a. This represents the set of all real numbers between, but not including −1 and 2.
    b. This represents the set of all real numbers between −1 and 2, including 2 but not −1.
    c. This represents the set of all real numbers between −1 and 2, including −1 but not 2.
    d. This represents the set of all real numbers between −1 and 2, including −1 and 2.

61. In interval notation, $\{x|x<5\}$ is written $(-\infty,5)$.

63. In interval notation, $\{x|x>9\}$ is written $(9,\infty)$.

65. In interval notation, $\{x|-4 \le x < -1\}$ is written $[-4,-1)$.

67. In interval notation, $\{x|-1 \le x \le 10\}$ is written $[-1,10]$.

69. Let J in. be Joe's height. Then $J = 60$.

71. Let F in. be Frank's height and S in. be Sam's height. Then, $F = S - 3$.

73. Let S in. be Sam's height. Then $S = 77$.

75. Let B be Bill's height. Then $B > 74$ in. (6 ft 2 in.)

Section 6.3 Sentences Involving Absolute Values

Study Tips

The best way to understand this section is with examples you can construct yourself. First, graph $|x| = 1$. The graph consists of two points one unit away from 0.

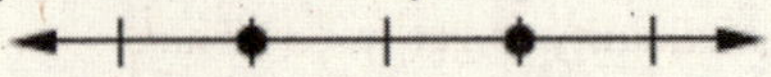

Then graph $|x| < 1$. The graph consists of all the points less than one unit away from 0, that is, all real numbers x such that $-1 < x < 1$.

Finally, $|x| > 1$ has to consist of all elements more than one unit from 0, that is, $|x| > 1$ consists of all real numbers x such that $x > 1$ or $x < -1$.

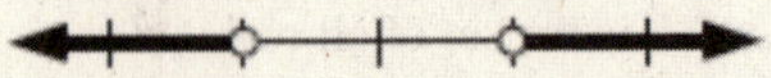

To find the solution set of $|x| = a,\ a > 0$:
Think: $x = -a$ or $x = a$, solution is two points
Note: Treat a problem of the form $|x + b| = a$ in the same way: $x + b = -a$ or $x + b = a$. You will need to get x by itself, then you will have the solution.

To find the solution set of $|x| < a$:
Think: $-a < x < a$, solution is line segment between two points
Note: Treat a problem of the form $|x+b| < a$ in the same way: $-a < x+b < a$. You will need to get x by itself, then you will have the solution.

To find the solution set of $|x| > a$:
Think: $x < -a$ or $x > a$, solution is two rays going in opposite directions from the points
Note: Treat a problem of the form $|x+b| > a$ in the same way: $x+b < -a$ or $x+b > a$. You will need to get x by itself, then you will have the solution.

Remember

1. graph of $|x| = a$: **two points**
2. the graph of $|x| \le a$: **line segment**
3. graph of $|x| \ge a$: **two pieces** called rays

6.3 Exercises

1. $|-10| = 10$ 3. $\left|-\frac{1}{8}\right| = \frac{1}{8}$

5. $|5-8| = |-3| = 3$ 7. $|0| + |-2| = 0 + 2 = 2$

9. $-|8| = -8$

11. a. For $x = 2$,
$|1-3\cdot 2| = |1-6| = |-5| = 5 > 3$,
a solution.

b. For $x = -\frac{1}{2}$,
$\left|1-3\left(-\frac{1}{2}\right)\right| = \left|1+\frac{3}{2}\right| = \left|\frac{5}{2}\right| = \frac{5}{2} \not> 3$,
not a solution.

c. For $x = \frac{5}{3}$,
$\left|1-3\cdot\frac{5}{3}\right| = |1-5| = |-4| = 4 > 3$,
a solution.

d. For $x = 0$, $|1-3\cdot 0| = |1-0| = |1| = 1 \not> 3$,
not a solution.

13. $\{0\}$ 15. $\{-5, 5\}$

17. $\{\ldots, -3, -2, -1, 1, 2, 3, \ldots\}$

19. $|x| = 1$ is true for $x = -1$ and $x = 1$. These points are shown on the graph. There is no interval that satisfies this equation.

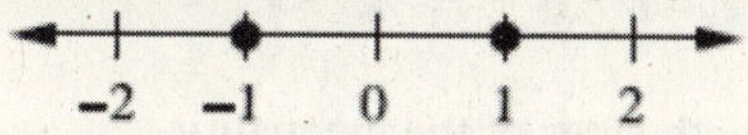

21. $|x| \le 4$ is true for all the real numbers between –4 and 4, inclusive. These numbers are shown on the graph. In interval notation, the solution is [–4, 4].

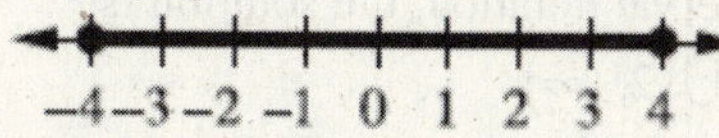

23. $|x+1| < 3$ is equivalent to $-3 < x+1 < 3$. Thus, by subtracting 1 from each side of this inequality, you get $-4 < x < 2$. The last inequality is true for all real numbers between –4 and 2, not including the –4 or the 2. This gives the required solution set, which is shown on the graph. In interval notation, the solution is (–4, 2).

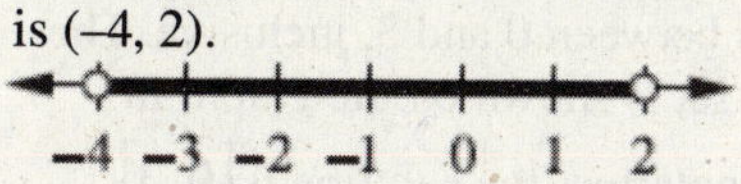

25. $|x| \ge 1$ is true if $x \ge 1$ or $x \le -1$. Thus, the solution set consists of all real numbers that are less than or equal to –1 or that are greater than or equal to 1. This set is shown on the graph. In interval notation, the solution is $(-\infty, -1] \cup [1, +\infty)$.

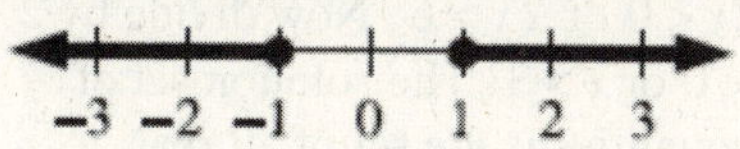

27. $|x-1| > 2$ is true if $x-1 > 2$ or $x-1 < -2$. By adding 1 to both sides of the last two inequalities, you get $x > 3$ or $x < -1$. Thus, the solution set of the given inequality consists of all the real numbers that are less than –1 or are greater than 3. This set is shown on the graph. In interval notation, the solution is $(-\infty, -1) \cup (3, +\infty)$.

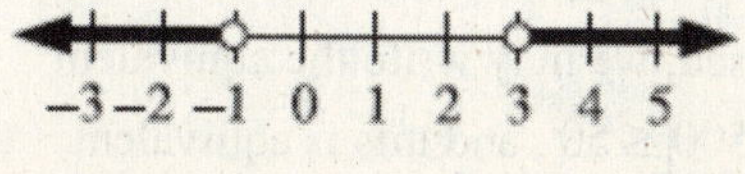

29. Divide both sides of the inequality $|2x| < 4$ by 2 to get $|x| < 2$. This inequality is satisfied by all the real numbers between –2 and 2, not including the –2 or the 2. This set is shown on the graph. In interval notation, the solution is (–2, 2).

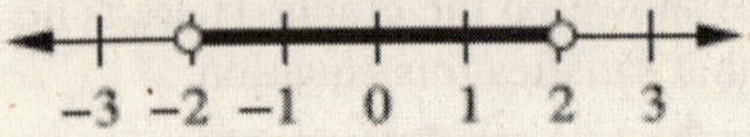

31. Divide both sides of the inequality $|3x| \geq 6$ by 3 to get $|x| \geq 2$. This inequality is satisfied by all the real numbers that are less than or equal to –2 or that are greater than or equal to 2. The graph shows this set. In interval notation, the solution is $(-\infty,-2] \cup [2,+\infty)$.

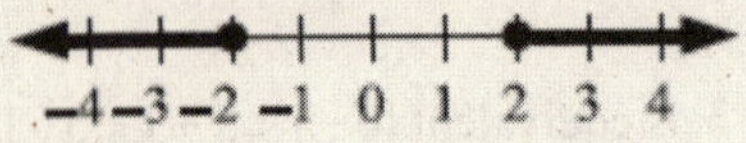

33. $|2x-3| \leq 3$ is equivalent to $-3 \leq 2x-3 \leq 3$. Add 3 to each member of this inequality to get $0 \leq 2x \leq 6$. Now divide by 2 to get the inequality $0 \leq x \leq 3$. This inequality is satisfied by all the real numbers between 0 and 3, inclusive. The solution set is shown on the graph. In interval notation, the solution is $[0, 3]$.

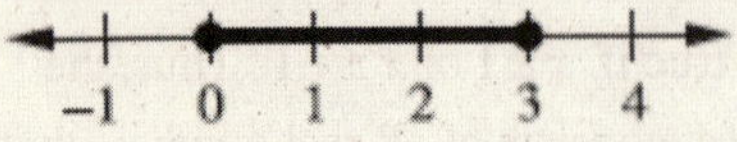

35. $|2x-3| > 3$ is equivalent to the system $2x-3 < -3$ or $2x-3 > 3$. Add 3 to both sides of these inequalities to get the system $2x < 0$ or $2x > 6$. Now divide by 2 to get $x < 0$ or $x > 3$. The solution set of these inequalities is the set of all real numbers that are less than 0 or that are greater than 3. The graph shows this set. In interval notation, the solution is $(-\infty,0) \cup (3,+\infty)$.

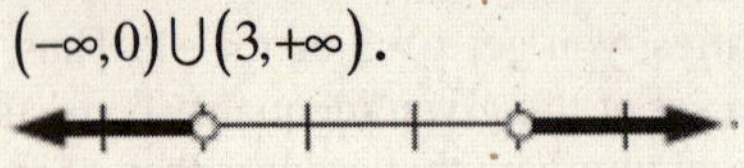

37. Substitute $b = 500$ and $c = 50$ to get $|500-a| \leq 50$. Since the absolute value is being used, we may write the equivalent as $|a-500| \leq 50$, and this is equivalent to $-50 \leq a-500 \leq 50$. Add 500 to each member to get $450 \leq a \leq 550$. Thus, the company can spend any amount between \$450 and \$550, inclusive.

39. Yes. Substitute $b = 300$ and $a = 290$ to get $|\$300-\$290| = \$10$. Since 5% of the budgeted $\$300 = 0.05 \cdot \$300 = \$15$, George would have to spend within \$15 of his budget to be within the 5% variance, and George only spent \$10.

41. a. $|w-137| \leq 7$ b. $-7 \leq w-137 \leq 7$

$130 \leq w \leq 144$

43. a. $|L-12| \leq 0.02 \cdot 12 = 0.24$

b. $-0.24 \leq L-12 \leq 0.24$

$11.76 \leq L \leq 12.24$

45. a. $|s-505| \leq 4$ b. $-4 \leq s-505 \leq 4$

$501 \leq s \leq 509$

47. Substitute $b = 500$ and $c = 50$ to get $|500-a| \leq 50$. This is equivalent to $-50 \leq 500-a \leq 50$. Subtract 500 from each member to get $-550 \leq -a \leq -450$. Multiply each member by –1, and reverse the inequality, to get $\$450 \geq a \geq \$550$.

49. A 5% variance of \$300 is $\$300 = 0.05 \cdot 300 = \$15$. Substitute $a = 290$, $b = 300$ and $c = 15$ into the inequality $|300-a| \leq 15$.

$$|300-290| \overset{?}{\leq} 15$$

$$|10| \overset{?}{\leq} 15$$

$$10 \leq 15$$

Yes, the actual expenses are within a 5% budget variance.

51. Answers may vary. "The absolute value of x is less than a." is equivalent to "x is between $-a$ and a."

53. Answers may vary. The solution is the set of all real numbers, since the absolute value of an expression is always non-negative, the expression $|x-2|$ is always non-negative and has to be greater than –5.

Skill Checker

1. $\sqrt{(-2)^2-4(1)(-1)}=\sqrt{4+4}$
$=\sqrt{8}=2\sqrt{2}$

3. $\sqrt{(-2)^2-4(2)(-1)}=\sqrt{4+8}$
$=\sqrt{12}=2\sqrt{3}$

5. $\sqrt{3^2-4(2)(-5)}=\sqrt{9+40}=\sqrt{49}=7$

7. $\sqrt{(-6)^2-4(9)(-2)}=\sqrt{36+72}=\sqrt{108}$
$=\sqrt{36\cdot 3}=6\sqrt{3}$

9. $\sqrt{8^2-4(4)(-5)}=\sqrt{64+80}=\sqrt{144}=12$

Section 6.4 Quadratic Equations

Study Tips

Frequently used in application problems is the *quadratic equation*. Here we learn some ways to solve quadratic equations.

To write a quadratic equation in standard form:
Think: equation must equal zero
Note: To solve a quadratic, it is helpful to be in the standard form $ax^2+bx+c=0$.

To solve $ax^2+bx+c=0$ when $a = 1$:
Think: factor, set each factor equal to zero, solve the two equations
Note: If unable to factor, remember the quadratic formula.

To solve quadratics with no x term:
Think: get x^2 by itself, take the square root of both sides, answer will be + and –
Note: Equation might look like this $3x^2+2=50$. This is called solving quadratics by root extraction.

To solve $ax^2+bx+c=0$ by quadratic formula:
Think: identify a, b, and c; substitute into formula
Note: All quadratic equations can be solved using the quadratic formula:

$$x=\frac{-b\pm\sqrt{b^2-4ac}}{2a}$$

Caution
Be sure to extend the division line under all of $-b\pm\sqrt{b^2-4ac}$ as all of this expression is divided by $2a$.

6.4 Exercises

1. $x^2+6x+8=(x+2)(x+4)$

3. $x^2-x-12=(x-4)(x+3)$

5. $x^2+7x-18=(x+9)(x-2)$

7. $x^2-10x+25=(x-5)(x-5)=(x-5)^2$

9. $x^2+10x+25=(x+5)(x+5)=(x+5)^2$

11. $2x^2+x-3=(2x+3)(x-1)$

13. $6x^2-5x+1=(3x-1)(2x-1)$

15. This equation is satisfied if and only if $x-2=0$ or $x-4=0$, that is, if and only if $x=2$ or $x=4$. The solution set is {2, 4}.

17. This equation is satisfied if and only if $x+2=0$ or $x-3=0$, that is, if and only if $x=-2$ or $x=3$. The solution set is {-2, 3}.

19. This equation is satisfied if and only if $x=0$ or $x-1=0$ or $x+1=0$, that is, if and only if $x=0$ or $x=1$ or $x=-1$. The solution set is {-1, 0, 1}.

21. This equation is satisfied if and only if $2x-1=0$ or $x+2=0$. Add 1 to both sides of the first equation to get $2x=1$. Then, divide both sides by 2 to get $x=\frac{1}{2}$. The second equation is satisfied by $x=-2$. The solution set of the given equation is $\left\{-2,\frac{1}{2}\right\}$.

23. Add 16 to both sides of the given equation to get $x^2=16$. Take the square root of both sides.

The solution set of this equation is {–4, 4}, because $(-4)^2 = 16$ and $4^2 = 16$.

25. Divide both sides of the given equation by 5 to get $x^2 = 25$. Take the square root of both sides. The solution set of this equation is {–5, 5}, because $(-5)^2 = 25$ and $5^2 = 25$.

27. The equation $(3x-6)(2x+3)(5x-8) = 0$ is satisfied if and only if $3x-6=0$ or $2x+3=0$ or $5x-8=0$. The equation $3x-6=0$ can be solved by adding 6 to both sides and then dividing both sides by 3 to get $x=2$. Similarly, the second equation can be solved by subtracting 3 from both sides and then dividing by 2 to get $x=-\frac{3}{2}$. The third equation is solved by adding 8 to both sides and then dividing by 5 to get $x=\frac{8}{5}$. The solution set of the given equation is $\left\{-\frac{3}{2}, \frac{8}{5}, 2\right\}$.

29. To solve the equation $6x^2 - 1 = 215$, add 1 to both sides to get $6x^2 = 216$, and then divide both sides by 6 to get $x^2 = 36$. Take the square root of both sides. This equation is satisfied by –6 and by 6, because $(-6)^2 = 36$ and $6^2 = 36$. The solution set is {–6, 6}.

31. Since $x^2 - 12x + 27 = (x-3)(x-9)$, the given equation can be rewritten as $(x-3)(x-9) = 0$, which is satisfied if and only if $x = 3$ or $x = 9$. The solution set of the given equation is {3, 9}.

33. Since $x^2 - 8x - 20 = (x+2)(x-10)$, the given equation can be rewritten as $(x+2)(x-10) = 0$, which is satisfied if and only if $x = -2$ or $x = 10$. The solution set is {–2, 10}.

35. Since $10x^2 + 7x + 1 = (2x+1)(5x+1)$, the given equation can be rewritten as $(2x+1)(5x+1) = 0$, which is satisfied if and only if $x = -\frac{1}{2}$ or $x = -\frac{1}{5}$. The solution set is $\left\{-\frac{1}{2}, -\frac{1}{5}\right\}$.

37. First subtract 5 from both sides to get $3x^2 + 2x - 5 = 0$. Since $3x^2 + 2x - 5 = (3x+5)(x-1)$ the given equation can be rewritten as $(3x+5)(x-1) = 0$, which is satisfied if and only if $x = -\frac{5}{3}$ or $x = 1$. The solution set is $\left\{-\frac{5}{3}, 1\right\}$.

39. For the equation $2x^2 + 3x - 5 = 0$, $a = 2$, $b = 3$, and $c = -5$. By substituting these numbers in the quadratic formula $x = \frac{-b \pm \sqrt{b^2 - 4ac}}{2a}$ we get

$$x = \frac{-3 \pm \sqrt{3^2 - 4\cdot 2(-5)}}{2\cdot 2} = \frac{-3 \pm \sqrt{9+40}}{4}$$

$$= \frac{-3 \pm \sqrt{49}}{4} = \frac{-3 \pm 7}{4} = -\frac{10}{4} \text{ or } \frac{4}{4}$$

that is, $-\frac{5}{2}$ or 1. The solution set is $\left\{-\frac{5}{2}, 1\right\}$.

41. For the equation $2x^2 + 5x - 7 = 0$, $a = 2$, $b = 5$, and $c = -7$. By substituting these numbers in the quadratic formula (see problem 39), we get

$$x = \frac{-5 \pm \sqrt{5^2 - 4\cdot 2(-7)}}{2\cdot 2} = \frac{-5 \pm \sqrt{25+56}}{4}$$

$$= \frac{-5 \pm \sqrt{81}}{4} = \frac{-5 \pm 9}{4} = -\frac{14}{4} \text{ or } \frac{4}{4}$$

that is, $-\frac{7}{2}$ or 1. The solution set is $\left\{-\frac{7}{2}, 1\right\}$.

43. For the equation $x^2 + 5x + 3 = 0$, $a = 1$,

$b = 5$, and $c = 3$. By substituting these numbers in the quadratic formula (see problem 39), we get

$$x = \frac{-5 \pm \sqrt{5^2 - 4 \cdot 1(3)}}{2 \cdot 1} = \frac{-5 \pm \sqrt{25-12}}{2}$$
$$= \frac{-5 \pm \sqrt{13}}{2}$$

The solution set is $\left\{\frac{-5-\sqrt{13}}{2}, \frac{-5+\sqrt{13}}{2}\right\}$.

45. For the equation $5x^2 - 8x + 2 = 0$, $a = 5$, $b = -8$, and $c = 2$. By substituting these numbers in the quadratic formula (see problem 39), we get

$$x = \frac{-(-8) \pm \sqrt{(-8)^2 - 4 \cdot 5 \cdot 2}}{2 \cdot 5}$$
$$= \frac{8 \pm \sqrt{64-40}}{10} = \frac{8 \pm \sqrt{24}}{10}$$
$$= \frac{8 \pm 2\sqrt{6}}{10} = \frac{4 \pm \sqrt{6}}{5}$$

The solution set is $\left\{\frac{4-\sqrt{6}}{5}, \frac{4+\sqrt{6}}{5}\right\}$.

47. By adding 1 to both sides, we can write the given equation in the standard form: $7x^2 - 6x + 1 = 0$, and $a = 7$, $b = -6$, and $c = 1$. By substituting these numbers in the quadratic formula (see problem 39), we get

$$x = \frac{-(-6) \pm \sqrt{(-6)^2 - 4 \cdot 7 \cdot 1}}{2 \cdot 7}$$
$$= \frac{6 \pm \sqrt{36-28}}{14} = \frac{6 \pm \sqrt{8}}{14}$$
$$= \frac{6 \pm 2\sqrt{2}}{14} = \frac{3 \pm \sqrt{2}}{7}$$

The solution set is $\left\{\frac{3-\sqrt{2}}{7}, \frac{3+\sqrt{2}}{7}\right\}$.

49. For the equation $9x^2 - 6x - 2 = 0$, $a = 9$, $b = -6$, and $c = -2$. By substituting these numbers in the quadratic formula (see problem 39), we get

$$x = \frac{-(-6) \pm \sqrt{(-6)^2 - 4 \cdot 9(-2)}}{2 \cdot 9}$$
$$= \frac{6 \pm \sqrt{36+72}}{18} = \frac{6 \pm \sqrt{108}}{18}$$
$$= \frac{6 \pm 6\sqrt{3}}{18} = \frac{1 \pm \sqrt{3}}{3}$$

The solution set is $\left\{\frac{1-\sqrt{3}}{3}, \frac{1+\sqrt{3}}{3}\right\}$.

51. First subtract 1 from both sides to get $2x^2 + 2x - 1 = 0$, for which $a = 2$, $b = 2$, and $c = -1$. By substituting these numbers in the quadratic formula (see problem 39), we get

$$x = \frac{-2 \pm \sqrt{2^2 - 4 \cdot 2(-1)}}{2 \cdot 2} = \frac{-2 \pm \sqrt{4+8}}{4}$$
$$= \frac{-2 \pm \sqrt{12}}{4} = \frac{-2 \pm 2\sqrt{3}}{4} = \frac{-1 \pm \sqrt{3}}{2}$$

The solution set is $\left\{\frac{-1-\sqrt{3}}{2}, \frac{-1+\sqrt{3}}{2}\right\}$.

53. First, add $8x - 5$ to both sides to get $4x^2 + 8x - 5 = 0$, $a = 4$, $b = 8$, and $c = -5$. By substituting these numbers in the quadratic formula (see problem 39), we get

$$x = \frac{-8 \pm \sqrt{8^2 - 4 \cdot 4(-5)}}{2 \cdot 4} = \frac{-8 \pm \sqrt{64+80}}{8}$$
$$= \frac{-8 \pm \sqrt{144}}{8} = \frac{-8 \pm 12}{8} = -\frac{20}{8} \text{ or } \frac{4}{8}$$

that is, $-\frac{5}{2}$ or $\frac{1}{2}$. The solution set is $\left\{-\frac{5}{2}, \frac{1}{2}\right\}$.

55. $$(2ax+b)^2 = (2ax+b)(2ax+b)$$
$$= 2ax(2ax+b) + b(2ax+b)$$
$$= (4a^2x^2 + 2axb) + (2abx + b^2)$$
$$= 4a^2x^2 + 4abx + b^2$$

Which is the left side of the equation, as stated. If we subtract b from both sides of

$$2ax + b = \pm\sqrt{b^2 - 4ac}$$

we get

$$2ax = -b \pm \sqrt{b^2 - 4ac}$$

If we divide both sides of this last equation by $2a$, we get the quadratic formula as given.

57. Let the shortest side be x cm long. Then the hypotenuse is $(x + 4)$ cm long and the other leg is $(x + 2)$ cm long. By the Pythagorean theorem,

$$(x+4)^2 = x^2 + (x+2)^2$$

$$x^2 + 8x + 16 = x^2 + x^2 + 4x + 4 \quad \text{Multiply.}$$

$$0 = x^2 - 4x - 12 \quad \text{Subtract } x^2 + 8x + 16.$$

$$0 = (x-6)(x+2) \quad \text{Factor.}$$

$$x = 6 \text{ or } x = -2 \quad \text{Solve.}$$

Since this is to be the side of a triangle, we discard the negative solution. The sides of the triangle are, respectively, 6 cm, 6 + 2 = 8 cm, and 6 + 4 = 10 cm long.

59. Let the shortest side be x in. long. Then the hypotenuse is $(x + 8)$ in. long and the other leg is $(x + 7)$ in. long. By the Pythagorean theorem

$$(x+8)^2 = x^2 + (x+7)^2$$

$$x^2 + 16x + 64 = x^2 + x^2 + 14x + 49 \quad \text{Multiply.}$$

$$0 = x^2 - 2x - 15 \quad \text{Subtract } x^2 + 16x + 64.$$

$$0 = (x-5)(x+3) \quad \text{Factor.}$$

$$x = 5 \text{ or } x = -3 \quad \text{Solve.}$$

Since this is to be a side of a triangle, we discard the negative solution. The sides of the triangle are, respectively, 5 in., 5 + 7 = 12 in. and 5 + 8 = 13 in. long.

61. Here $v_0 = 5, h = 10$ so we have:

$$10 = 5t^2 + 5t$$

$$2 = t^2 + t \quad \text{Divide by 5.}$$

$$0 = t^2 + t - 2 \quad \text{Subtract 2.}$$

$$0 = (t+2)(t-1) \quad \text{Factor.}$$

$$t = -2 \text{ or } t = 1 \quad \text{Solve.}$$

Since time cannot be negative, we discard $t = -2$ and use $t = 1$ second.

63. Here $v_0 = 10, h = 15$ so we have:

$$15 = 5t^2 + 10t$$

$$3 = t^2 + 2t \quad \text{Divide by 5.}$$

$$0 = t^2 + 2t - 3 \quad \text{Subtract 3.}$$

$$0 = (t+3)(t-1) \quad \text{Factor.}$$

$$t = -3 \text{ or } t = 1 \quad \text{Solve.}$$

Since time cannot be negative, we discard $t = -3$ and use $t = 1$ second.

65. a. $t^2 + 26t + 169$

$$= (t+13)(t+13) = (t+13)^2$$

b. In 2020, the value of $t = 2020 - 1980 = 40$.

$$E(40) = (40+13)^2 = 53^2 = 2809$$

In 2020, there will be 2809 endangered species.

67. a. $P - 2Pr + Pr^2 = P(1 - 2r + r^2)$

$$= P(1-r)(1-r) = P(1-r)^2$$

b. Let P = 25,000 and $r = 0.15$.

$$P(1-r)^2 = 25{,}000(1-0.15)^2$$

$$= 25{,}000(0.85)^2$$

$$= 25{,}000(0.7225)$$

$$= 18{,}062.5$$

In two years, the Prius will be worth $18,062.50.

69. a. The $9,000 is lower.

b. Let $t = 15$ in the equation

$$H(t) = -3t^2 + 300t + 6300$$

$$H(15) = -3(15)^2 + 300(15) + 6300$$

$$= \$10{,}125$$

This is lower than the $13,100 approximation in the graph.

71. Answers will vary. 73. Answers will vary.

75. If $b^2 - 4ac = 0$, the two answers coincide. The only solution is $-\dfrac{b}{2a}$.

77. If $b^2 - 4ac < 0$, there are no real number solutions.

79. From problem 78, $t = 3.25$ so

$$D(3.25) = -5(3.25)^2 + 115(3.25) - 110$$

$$= 210.9375 \approx 211 \text{ ft}$$

81. In the equation $x^2 = y^2 + 13^2$, we want y as large as possible relative to x, so we try putting $y = x - 1$. The equation becomes

$x^2 = (x-1)^2 + 13^2$

$x^2 = x^2 - 2x + 1 + 169$ Multiply.

$0 = -2x + 170$ Subtract x^2.

$2x = 170$ Add $2x$.

$x = 85$ Divide by 2.

We can check this by noting that $85^2 = 7225$ and $84^2 = 7056$. Thus, $85^2 - 84^2 = 7225 - 7056 = 169 = 13^2$.

So the maximum number of letters that Charlie could have received is $85^2 = 7225$.

Section 6.5 Modeling and Problem Solving

Study Tips

Before you start this section it is a good idea to review the **RSTUV** procedure covered in Chapter 1.

Know the problem-solving procedure:
Think: **RSTUV**
Note: In short, **RSTUV** stands for
1. *Read the problem...*
2. *Select the unkown...*
3. *Think of a plan...*
4. *Use the techniques you are studying...*
5. *Verify your answer.*

Word of Advice
You may find it helpful to write these steps on a notecard for easy reference. Try covering the solution in the right hand column of the *Problem Solving* section with a 3 by 5 card and see if you can figure out how to proceed.

Know key words and translation:
Think: study Table 6.1 in text
Note: Expressions for integers include:
Two consecutive integers: *n*, *n* + 1
Two consecutive even integers: *n*, *n* + 2
Two consecutive odd integers: *n*, *n* + 2

6.5 Exercises

1. $4m = m + 18$

3. If n is the first consecutive even integer, then $n + 2$ would be the next consecutive integer. The equation is $n(n+2) = 10(n+2) - 20$.

5. If x is the number, then the required equation is: $4x + 5 = 29$

$4x = 24$ Subtract 5.

$x = 6$ Divide by 4.

Thus, the number is 6.

7. If x is the number, then the required equation is: $3x + 8 = 29$

$3x = 21$ Subtract 8.

$x = 7$ Divide by 3.

Thus, the number is 7.

9. If x is the number, then the required equation is: $3x - 2 = 16$

$3x = 18$ Add 2.

$x = 6$ Divide by 3.

Thus, the number is 6.

11. If x is the number, then the required equation is: $2x^2 = 2x + 12$

$2x^2 - 2x - 12 = 0$ Subtract $2x + 12$.

$x^2 - x - 6 = 0$ Divide by 2.

$(x-3)(x+2) = 0$ Factor.

$x = 3$ or $x = -2$ Solve.

Thus, the number is –2 or 3.

13. If x is the number, then the required equation is: $\frac{1}{3}x^2 - 2 = 10$

$\frac{1}{3}x^2 = 12$ Add 2.

$x^2 = 36$ Multiply by 3.

$x = \pm 6$ Take square roots.

Thus, the number is –6 or 6.

15. Let x million lb be the weight of the orbiter and fuel. Then the weight of the tank and the boosters is $(x - 1.26)$ million lb. The total weight of these components is 4.16 million lb, so the required equation is:

$x + (x - 1.26) = 4.16$

$2x - 1.26 = 4.16$ Simplify.

$2x = 5.42$ Add 1.26.

$x = 2.71$ Divide by 2.

Check:
$2.71 - 1.26 = 1.45$, and
$2.71 + 1.45 = 4.16$.
Thus, the weight of the orbiter and fuel is 2.71 million lb.

17. Let x be Maria's age. Then Pedro is $2x$ years old. The sum of their ages is 30 less than 5 times Maria's, so the required equation is: $x + 2x = 5x - 30$

$3x = 5x - 30$ Simplify.

$-2x = -30$ Subtract $5x$.

$x = 15$ Divide by -2.

Check: $15 + 2(15) \overset{?}{=} 5(15) - 30$

$15 + 30 \overset{?}{=} 75 - 30$

$45 = 45$

Thus, Pedro is 30 years old and Maria is 15 years old.

19. Let m be the number of miles that Margie traveled. Then, $0.20m$ is the number of dollars that her travel cost. Since the total cost was \$44, the required equation is:
$0.20m + 18 = 44$

$0.20m = 26$ Subtract 18.

$m = 130$ Divide by 0.20.

Check: $0.20(130) = 26$ and $26 + 18 = 44$.
Thus, Margie traveled 130 mi.

21. Let r be the annual interest rate in $I = Prt$. Then, the interest on \$10,000 for 3 months (one-fourth of a year) is $\dfrac{10000r}{4} = 2500r$.
Since the interest was \$50, the required equation is: $2500r = 50$

$r = 0.02$ Divide by 2500.

Check: $(10{,}000)(0.02) = 200$
$200 \div 4 = 50$
Thus, the interest rate was 2% per year.

23. a. If the total cost is \$71, then the required equation is:
$0.25m + 20 = 71$

$0.25m = 51$ Subtract 20.

$m = 204$ Divide by 0.25.

Check: $(0.25)(204) = 51$, and
$51 + 20 = 71$.
Thus, the answer is 204 miles.

b. Using the mileage rate, the cost would be (0.25)(60) + 20, which comes to 35 dollars. The flat rate would be \$40, so use the mileage rate.

25. Substitute $P = 1000$ and $A = 1040.40$ into the equation $A = P(1+r)^2$ to get the required equation:

$1040.40 = 1000(1+r)^2$

$1.0404 = (1+r)^2$ Divide by 1000.

$1.02 = 1 + r$ Take square root.

$0.02 = r$ Subtract 1.

Thus, the rate is 0.02 or 2%.
You can check this answer as follows: The interest for 1 yr would be 2% of \$1000, which is \$20. So the total amount at the end of the first year would be \$1000 + \$20 = \$1020. The interest for the second year would be 2% of \$1020, that is, \$20.40. Adding this interest to the \$1020 gives \$1040.40, which shows that the answer is correct.

27. Substitute $b = 54$ into the given formula to get the required equation:

$54 = 0.06v^2$

$900 = v^2$ Divide by 0.06.

$30 = v$ Take positive square root.

Thus, the answer is 30 mph.

29. Substitute $v = 30$ and $t = 0.5$ into the given formula to get:

$d = 1.5(0.5)(30) + 0.06(30)^2$

$= 22.5 + 54 = 76.5$

Thus, the stopping distance is 76.5 ft.

31. Substitute $v = 20$ and $d = 42$ into the given formula to get:

$42 = 1.5(20)t + 0.06(20)^2$

$42 = 30t + 24$ Simplify.

$18 = 30t$ Subtract 24.

$0.6 = t$ Divide by 30.

Thus, the driver's reaction time is 0.6 sec.

33. Substitute $t = \dfrac{2}{3}$ and $d = 44$ into the given formula to get the equation:

$$44 = 1.5\left(\frac{2}{3}\right)v + 0.06v^2$$

$44 = v + 0.06v^2$ Simplify.

$4400 = 100v + 6v^2$ Multiply by 100.

$2200 = 50v + 3v^2$ Divide by 2.

$0 = 3v^2 + 50v - 2200$ Standard form

In the last equation, you can read off: $a = 3$, $b = 50$, and $c = -2200$. Substitute these numbers into the quadratic formula to get:

$$v = \frac{-50 \pm \sqrt{50^2 - 4 \cdot 3(-2200)}}{2 \cdot 3}$$

$$= \frac{-50 \pm \sqrt{2500 + 26400}}{6}$$

$$= \frac{-50 \pm \sqrt{28900}}{6} = \frac{-50 \pm 170}{6}$$

Since a positive number is needed for the answer, use the plus sign to find $v = \frac{-50 + 170}{6} = \frac{120}{6} = 20$. Thus, Loren was going 20 mph at the instant of the stop signal.

35. Let the two consecutive integers be x and $x + 1$. Then the given condition says that $6x < 5(x+1)$. Simplify this inequality to get $6x < 5x + 5$. Then subtract $5x$ from both sides to get $x < 5$. Since 4 is the largest integer less than 5, take $x = 4$, so that $x + 1 = 5$. Thus, 4 and 5 are the largest integers for which 6 times the smaller is less than 5 times the larger.

37. Let x billion cans be the number that are recycled. Then $\frac{2}{3}x$ billion cans are thrown away. Since the total number of billion cans is 100, we have the equation

$$x + \frac{2}{3}x = 100$$

$$\frac{5}{3}x = 100$$

$$x = \frac{3}{5} \cdot 100 = 60$$

So 60 billion cans are recycled and 40 billion cans are not recycled.

39. a. Tuition $= T$

State Contribution $= = \frac{3N}{40}(2800)$
$= 210N$

Revenue to Institution $= T + 210N$

b. Costs to Institution:

(s) (40)($100)	=	$4,000
(U) (40)($150)	=	$6,000
(G) (40)($1000)	=	$40,000
(S) (3)($1000)	=	$3,000
Total Cost	=	$53,000

c. $T + 210N = 53{,}000$

d. $N = \frac{53000 - T}{210}$

e. If N is the number of students, then the total tuition revenue is $T = 40 \cdot 3 \cdot N = 120N$.
To break even, $T + 210N = 53{,}000$, that is:

$$120N + 210N = 53000$$

$$330N = 53000$$

$$N = \frac{53000}{330} = 160.6060\ldots$$

Thus, to insure no loss, the least number of students required is 161.

41. a. Identical Row Alignment

	Unit A	Unit B	Unit C
Cost per month	$25	$90	$128
Number per row	5	10	10
Rows	5	5	10
Layers	3	5	5
Total	75	250	500
Units Needed	5	2	1
Cost for 2 months	$250	$360	$256

Staggered Row Alignment

	Unit A	Unit B	Unit C
Cost per month	$25	$90	$128
Number per row	2-4s & 3-5s	3-10s & 2-9s	6-10s & 5-9s
Rows	5	5	11
Layers	3	5	5
Total	69	240	525
Units Needed	6	2	1
Cost for 2 months	$300	$360	$256

b. Rent 5 units A for a cost of $250.

c. Rent one unit C for a cost of $256.

d. Rent one unit A and one unit B for a cost of $230.

43. HOURS SPENT AND GRADE EARNED

	Math	Science	English	GPA
I	9(A)	2(C)	1(C)	$\frac{30}{11} = 2.73$
II	6(B)	4(B)	2(B)	$\frac{33}{11} = 3.00$
III	3(C)	6(A)	3(A)	$\frac{36}{11} = 3.23$

Computations:

I: $\frac{16+8+6}{11} = \frac{30}{11} = 2.73$

II: $\frac{12+12+9}{11} = \frac{33}{11} = 3.00$

III: $\frac{8+16+12}{11} = \frac{36}{11} = 3.23$

The table shows that you should spend 3 hr on Math, 6 hr on Science and 3 hours on English since the last row gives the best GPA.

45. a. **Blockbusters:**

Returned after

	2 days	3 days	4 days
New	$3+2=\$5$	$3+4=\$7$	$3+6=\$9$
Old	$3+0=\$3$	$3+2=\$5$	$3+4=\$7$

Red Rabbit:

Returned after

	2 days	3 days	4 days
New	$3+3=\$6$	$3+6=\$9$	$3+9=$ \$12
Old	$1.6+0$ $=\$1.60$	$1.6+1.5$ $=\$3.10$	$1.6+3$ $=\$4.60$

b. Returned after n days:

Blockbusters:

New $3+2(n-1)$

Old $3+2(n-2)$

New + Old $6+2n-2+2n-4=\$4n$

Red Rabbit:

New $3+3(n-1)$

Old $1.60+1.50(n-2)$

New + Old $4.60+3n-3+1.5n-3$

$=4.5n-1.40$

c. Charges when a new movie and an old movie is returned after 5 days at Blockbuster: $\$4(5)=\$20$

Charges when a new movie and an old movie is returned after 5 days at Red Rabbit:

$\$4.5(5)-\$1.40=\$21.10$

47. a. Average stopping distance in feet:

$d=1.5(0.6)v+0.06v^2$

$=0.9v+0.06v^2$

b. $d=0.9(45)+0.06(45)^2$

$=40.5+121.5=162$ ft

c. 162 ft + 80 ft = 242 ft

d. $D=RT$

$$242=\frac{45(5280)T}{3600}$$

$$T=\frac{242(3600)}{45(5280)}\approx 3.67 \text{ sec}$$

49. Answers may vary. You should try to determine the unknown, that is, what is wanted.

51. $$y=\frac{32(12)(28{,}000)}{(32-12)(30{,}000)(4.00)}$$

$$=\frac{32(12)(28)}{20(30)(4)}=\frac{10{,}752}{2400}=4.48 \text{ yr}$$

53. If M is greater than G, then y would be negative, which is unrealistic. If $M=G$, then there is a zero denominator, which could be interpreted to mean that you should keep the old car and not buy the new one.

Section 6.6 Ratio, Proportion, and Variation

Study Tips

This section covers ratio, proportion and variation. A *ratio* is an indicated quotient and can be written in three ways: a to b, a:b, or the actual quotient $\frac{a}{b}$. A *proportion* is an equality between ratios.

To solve a proportion:

Think: cross multiply, then solve for variable

Note: Proportions are usually given in words, so be prepared to use the **RSTUV** procedure when solving proportions.

To solve a direct variation problem:
Think: $y = kx$, need to find k, put value of k into $y = kx$, then solve for the requested variable

Note: If you think of y as $\frac{y}{1}$, and y varies *directly* as x, ($y = kx$) x and y are both in the numerator.

To solve a inverse variation problem:
Think: $y=\frac{k}{x}$, need to find k, put value of k into $y=\frac{k}{x}$, then solve for the requested variable

Note: If y varies **inversely** as x, $\left(y=\frac{k}{x}\right)$ y is in the numerator and x is in the denominator.

6.6 Exercises

1. 7000 to 2000; 7000 : 2000; $\frac{7000}{2000}$; or 7 to 2; 7:2; $\frac{7}{2}$

3. 70 to 4260; 70 : 4260; $\frac{70}{4260}$ or 7 to 426; 7:426; $\frac{7}{426}$

5. $\frac{2000}{600}=\frac{10}{3}$

7. $\frac{12{,}000}{700}=\frac{120}{7}\approx 17$ mpg

9. a. $\frac{1.31}{22}=0.0595\approx \$0.06=6$ cents

   b. $\frac{1.75}{32}=0.0547\approx \$0.05=5$ cents

   c. Based on price per oz alone, White Magic is the better buy.

11. From the proportion $\frac{x}{9}=\frac{4}{3}$, you get $3x = 36$, by using "cross-products." Divide both sides by 3 to find $x = 12$.

13. Use "cross-products" for the proportion $\frac{8}{x}=\frac{4}{3}$ to get $4x = 24$. Then, divide both sides by 4 to find $x = 6$.

15. Use "cross-products" for the proportion $\frac{3}{8}=\frac{9}{x}$ to get $3x = 72$. Then, divide both sides by 3 to find $x = 24$.

17. $\frac{n}{40}=\frac{9}{2}$

19. $\frac{\frac{1}{2}\text{ hr}}{1\text{ tank}}=\frac{3\text{ hr}}{x\text{ tanks}}$

    Cross multiply: $\frac{x}{2}=3$

    Solve for x: $x = 6$

    Thus, you need 6 tanks for a 3 hr flight.

21. Let x inches be the length. Then, you must have the proportion: $\frac{35}{x}=\frac{10}{19}$.
    Now, use "cross-products" to get $10x = 35\cdot 19 = 665$. Then divide both sides by 10 to find $x = 66.5$. Thus, the flag should be 66.5 in. long.

23. Let t be the number of trees. Then, you must have the proportion: $\frac{t\text{ trees}}{7600\text{ lb}}=\frac{1\text{ tree}}{48\text{ lb}}$.
    Now, use "cross-products" to get $48t = 1\cdot 7600 = 7600$. Then divide both sides by 48 to find $t = 158.\overline{3}$. Thus, 159 trees are needed to absorb 7600 lb of CO_2.

25. Let x be the number of fish in the lake. Assuming that the ratio of the total number of fish to the total number tagged is the same as the corresponding ratio for the sample of 53 fish, we have the proportion $\frac{x}{250}=\frac{53}{5}$. Now, multiply both sides by 250 to get
    $x=\frac{250\cdot 53}{5}=50\cdot 53=2650.$
    Thus, there are about 2650 fish in the lake.

27. a. $S = kn$
    b. Substitute $S = 307{,}200$ and $n = 2$ into the equation in part a to get $307{,}200 = k \cdot 2$. Then divide both sides by 2 to find $k = 153{,}600$ bytes.
    c. Use $k = 153{,}600$ and $n = 20$ to get the equation $S = 153{,}600 \cdot 20$. Then simplify to find $t = 3{,}072{,}000$ bytes.

29. a. $T = kh^3$
    b. Substitute $T = 196$ and $h = 70$ into the equation in part a to get $196 = k \cdot 70^3$. Then divide both sides by 70^3 to find k: $k = \frac{196}{70^3} = \frac{196}{343000} \approx 0.0005714$.
    c. $T = \frac{75^3}{1750} = \frac{421875}{1750} \approx 241.07$
    Thus, the threshold weight is about 241 lb.

31. a. $f = \frac{k}{d}$
    b. Substitute $f = 8$ and $k = \frac{1}{2}$ into the equation in part a to get $8 = \frac{k}{\frac{1}{2}}$. Then multiply both sides by $\frac{1}{2}$ to find $k = 4$.
    c. Put $d = \frac{1}{4}$ and $k = 4$ into the equation in part a. This gives $f = \frac{4}{\frac{1}{4}} = 16$. So the f number is 16.

33. Since the pressure P varies inversely as the volume V, we write $P = \frac{k}{V}$ or $PV = k$.
    To find k, substitute $P = 24$ and $V = 18$ to find $k = 18 \cdot 24 = 432$. This gives the equation $PV = 432$. For $P = 40$, we get $40V = 432$. Divide by 40 to find $V = \frac{432}{40} = 10.8 \text{ in}^3$.

35. a. $w = \frac{k}{s}$
    b. Substitute $w = 600$ and $s = 12$ into the equation in part a to get $600 = \frac{k}{12}$. Then multiply both sides by 12 to find k: $k = 7200$.
    c. $w = \frac{7200}{10} = 720$
    Thus, 720 words can be typed on a page if a 10-point font were used.

37. a. $b = \frac{k}{a}$
    b. Substitute $b = 110$ and $a = 27$ into the equation in part a to get $110 = \frac{k}{27}$. Then multiply both sides by 27 to find k: $k = 2970$.
    c. $b = \frac{2970}{33} = 90$; you would expect 90 births (per 1000 women) for 33-year old women.

39. a. $d = ks$
    b. Substitute $b = 500$ and $a = 28.41$ into the equation in part a to get $500 = k \cdot 28.41$. Then divide both sides by 28.41 to find k: $501 = k \cdot 28.41$ or $k \approx 17.63$.
    c. Since d is the distance, s the average speed and $k \approx 17.63$, $d = 17.63s$ and k is the number of hours needed to travel the distance d at the speed s.

41. a. $\text{BAC} = k(N-1)$
    b. According to the problem, the BAC is 0.052 when $N = 3$. Thus, $0.052 = k(3-1)$, that is, $0.052 = 2k$. Dividing by 2, $k = \frac{0.052}{2} = 0.026$.
    c. When $k = 0.026, \text{BAC} = k(N-1)$ becomes $\text{BAC} = 0.026(N-1)$. To find the BAC for $N = 5$ beers, $\text{BAC} = 0.026(5-1) = 0.104$.
    d. 4 beers: when $N = 4$, $\text{BAC} = 0.026(4-1) = 0.078$. Note that for $N = 5$, BAC = 0.104, which is over the limit!

43. a. $C = k(F - 37)$ According to the problem, when $C = 80$, $F = 57$ so we have $80 = k(57 - 37) = 20k$. Dividing by 20, $4 = k$. Thus, $C = 4(F - 37)$.

b. Substituting $F = 90$, $C = 4(90 - 37) = 4 \cdot 53 = 212$.

45. a. $T = \frac{k}{n}$

b. Substitute $T = 35$ and $n = 2$ into the equation in part a to get $35 = \frac{k}{2}$. Then multiply both sides by 2 to find k: $k = 70$.

c. $T = \frac{70}{48} = 1.458333...$ Thus, it would take approximately 1.4583 minutes to completely read a 48*X* CD.

47. $C = k\frac{w}{L}$. For the Rhodesian man, $w = 15$ and $L = 21$ and $C = 98$, so $98 = k\frac{15}{21}$. Multiplying by 21 and dividing by 15 gives $k = 137.2$ so we rewrite the equation as $C = 137.2\frac{w}{L}$. For Cro-Magnon, $L = 20$, $w = 15$, $C = 137.2 \cdot \frac{15}{20} = 102.9$.

49. a. $\text{BAC} = \frac{k}{W}$. When $W = 130$, and BAC $= 0.066$, $0.066 = \frac{k}{130}$. This means that $k = 130 \cdot 0.066 = 8.58$ and now $\text{BAC} = \frac{8.58}{W}$.

b. $\text{BAC} = \frac{8.58}{W}$ where $W = 260$. Thus, $\text{BAC} = \frac{8.58}{260} \approx 0.033$.

c. According to the problem, $\text{BAC} = 0.08 = \frac{8.58}{W}$. Multiplying both sides by W and then dividing by 0.08 we get $W = \frac{8.58}{0.08} = 107.25$ lb.

d. If you weigh more than 107.25 lb, BAC would be less than 0.08.

51. Answers may vary. In direct variation, as one quantity increases, a related quantity increases. In inverse variation as one quantity increases, a related quantity decreases.

53. Answers may vary. The proportion $a/b = c/d$ can be solved by cross-multiplication means that $a/b = c/d$ can be rewritten as $ad = cb$.

55. Answers may vary. As the number of miles that you drive increases, the number of gallons of gas that the car uses will increase.

57. a. $T = 15S$

b. 1 hour is 60 minutes, so substitute $T = 60$. $60 = 15S$, then divide both sides by 15 to get $S = 4$.

Chapter 6 Practice Test

Study Tips

You won't have "mental blocks" on a test if you are thoroughly prepared and confident you will get an A. Mental blocks usually occur when you are not prepared! To be really prepared:

1. Begin studying at least a week in advance of your test. Determine your *study plan* during for the days prior to the test. It is helpful to make a specific plan of *when* you will study and *what* topics or sections you will study.
2. Ask questions about material you do not understand.

Here are the solutions to the practice test.

1. a. $x + 7 = 2$; subtract 7 from both sides to get $x = -5$.

b. $x - 4 = 9$; add 4 to both sides to get $x = 13$.

2. a. $x + 5 > 4$; subtract 5 from both sides to get $x > -1$. The set of integers satisfying this inequality is $\{0, 1, 2, 3, \ldots\}$.

b. $2 + x \geq -x - 1$

$2 + 2x \geq -1$ Add x to both sides.

$2x \geq -3$ Subtract 2 from both sides.

$x \geq -\dfrac{3}{2}$ Divide both sides by 2.

Thus, the solution set is $\{-1, 0, 1, 2, 3, \ldots\}$.

3. a. $2x + 2 = 3x - 2$

$2 = x - 2$ Subtract $2x$ from both sides.

$4 = x$ Add 2 to both sides.

b. $\dfrac{x+3}{5} = \dfrac{x-1}{3} + \dfrac{6}{5}$

$15 \cdot \dfrac{x+3}{5} = 15 \cdot \dfrac{x-1}{3} + 15 \cdot \dfrac{6}{5}$

$3(x+3) = 5(x-1) + 18$ Multiply by 15.

$3x + 9 = 5x - 5 + 18$ Simplify.

$3x + 9 = 5x + 13$

$9 = 2x + 13$

$-4 = 2x$

$x = -2$

4. a. Given: $S = 3L - 21$

$S + 21 = 3L$ Add 21.

$\dfrac{S+21}{3} = L$ Divide by 3.

b. Here $S = 6$, $L = \dfrac{6+21}{3} = \dfrac{27}{3} = 9$ in.

5. $\dfrac{x}{2} + 2 \geq \dfrac{-(x+1)}{4}$ Multiply by 4.

$2x + 8 \geq -x - 1$ Add x.

$3x + 8 \geq -1$ Subtract 8.

$3x \geq -9$ Divide by 3.

$x \geq -3$

The solution set consists of all real numbers greater than or equal to –3.

6. a. $x - 3 \leq 0$ Add 3 to both sides to get $x \leq 3$. This inequality is satisfied by all real numbers less than or equal to 3. The solution set is shown on the graph.

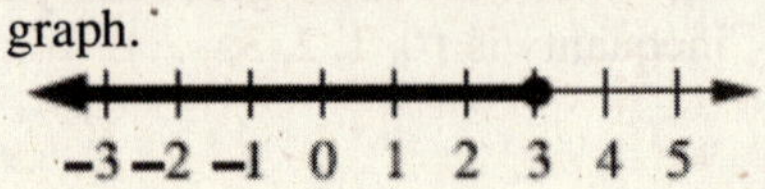

b. $-2x + 4 > x + 1$

$4 > 3x + 1$ Add $2x$.

$3 > 3x$ Subtract 1.

$1 > x$ Divide by 3.

The last inequality is satisfied by all real numbers that are less than 1. This solution set is shown on the graph.

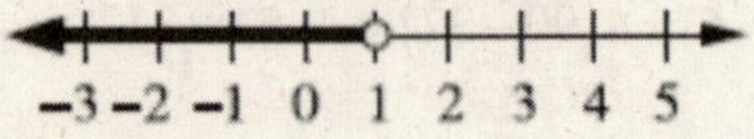

7. a. $x + 2 \geq 3$ and $x \leq 4$

$x \geq 1$ and $x \leq 4$ Subtract 2.

The solution set of this system of inequalities is the set of all real numbers between 1 and 4, inclusive. This set is shown on the graph.

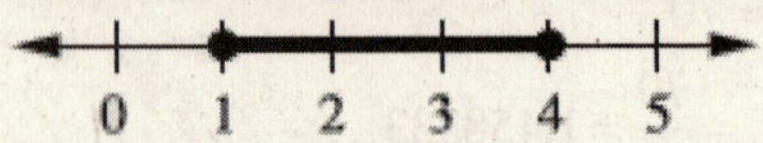

b. $x - 3 \geq 1$ and $x \leq 0$

$x \geq 4$ and $x \leq 0$ Add 3.

There are no numbers that are greater than or equal to 4 and that are less than or equal to zero, so the solution set is empty.

8. a. $x < 0$ or $x - 2 < 1$

$x < 0$ or $x < 3$ Add 2.

The solution set of this system of inequalities is the set of all real numbers that are less than 3. This set is shown on the graph.

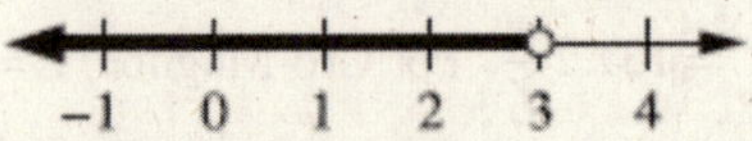

b. $x + 2 < 3$ or $x - 1 > 2$

$x < 1$ or $x > 3$

(Subtract 2 from both sides of the first inequality and add 1 to both sides of the second inequality.) The solution set of this system of inequalities is the set of all real numbers that are less than 1 or that are greater than 3. This set is shown on the graph.

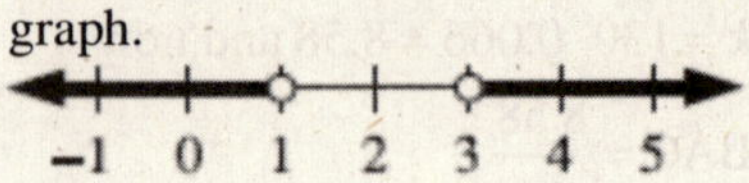

9. Since $|3| = 3$ and $|-3| = 3$, the solutions of $|x| = 3$ are $x = \pm 3$.

10. $|x| < 2$ means that $-2 < x < 2$, so the solution set consists of all the real numbers between –2 and 2, not including the –2 or the 2. The graph shows this set.

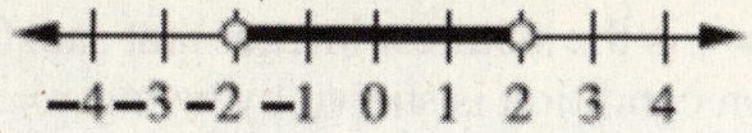

11. $|x-2|<2$ means that $-2<x-2<2$. By adding 2 to each side, we get $0<x<4$. The solution set of this system is the set of all real numbers that are between 0 and 4. The endpoints are not included.

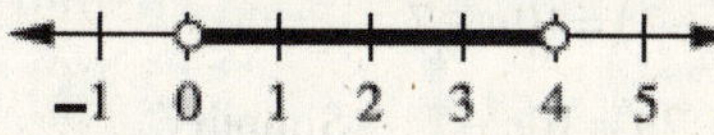

12. $|x|>2$ is equivalent to the system $x<-2$ or $x>2$. The solution set of this system is the set of all real numbers that are less than –2 or that are greater than +2. The graph shows this set.

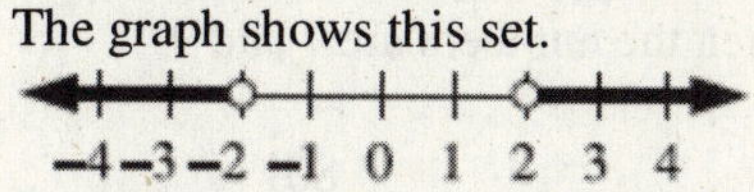

13. $|x-2|>3$ is equivalent to the system $x-2<-3$ or $x-2>3$. By adding 2 to both sides of these inequalities, we get the system $x<-1$ or $x>5$. The solution set of this system is the set of all real numbers that are either less than –1 or that are greater than 5. The graph shows this set.

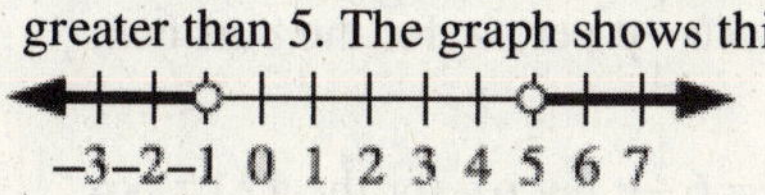

14. a. Since $1\cdot 2=2$ and $1+2=3$, $x^2+3x+2=(x+1)(x+2)$.

    b. Since $1\cdot(-4)=-4$ and $1+(-4)=-3$, $x^2-3x-4=(x+1)(x-4)$.

15. a. $(x-1)(x+2)=0$ is satisfied if and only if $x-1=0$ or $x+2=0$, that is, if and only if $x=1$ or $x=-2$.

    b. $x(x-1)=0$ is satisfied if and only if $x=0$ or $x-1=0$, that is, if and only if $x=0$ or $x=1$.

16. Since $x^2+7x+10=(x+5)(x+2)$, The given equation can be written as $(x+5)(x+2)=0$ which is satisfied if and only if $x+5=0$ or $x+2=0$, that is, if and only if $x=-5$ or $x=-2$.

17. Since $x^2-3x-10=(x-5)(x+2)$, the given equation can be rewritten as $(x-5)(x+2)=0$, which is satisfied if and only if $x-5=0$ or $x+2=0$, that is, if and only if $x=5$ or $x=-2$.

18. a. $9x^2-16=0$

$9x^2=16$ Add 16.

$x^2=\dfrac{16}{9}$ Divide by 9.

$x=\pm\dfrac{4}{3}$ Take square roots.

b. $25x^2-4=0$

$25x^2=4$ Add 4.

$x^2=\dfrac{4}{25}$ Divide by 25.

$x=\pm\dfrac{2}{5}$ Take square roots.

19. In the equation $2x^2+3x-5=0$, $a=2$, $b=3$, and $c=-5$. Substitute these values into the quadratic formula

$$x=\frac{-b\pm\sqrt{b^2-4ac}}{2a}$$ to get

$$x=\frac{-3\pm\sqrt{3^2-4\cdot 2(-5)}}{2\cdot 2}=\frac{-3\pm\sqrt{9+40}}{4}$$

$$=\frac{-3\pm\sqrt{49}}{4}=\frac{-3\pm 7}{4}=-\frac{10}{4}\text{ or }\frac{4}{4}$$

that is, $-\frac{5}{2}$ or 1. Thus, the solution set is $\left\{-\frac{5}{2},1\right\}$.

20. Write the equation $x^2-2x=2$, in the standard form $x^2-2x-2=0$ with $a=1$, $b=-2$, and $c=-2$. Substitute these values in the quadratic formula (see problem 18) to get

$$x=\frac{-(-2)\pm\sqrt{(-2)^2-4\cdot 1(-2)}}{2\cdot 1}$$

$$=\frac{2\pm\sqrt{4+8}}{2}=\frac{2\pm\sqrt{12}}{2}=\frac{2\pm 2\sqrt{3}}{2}$$

$$=1\pm\sqrt{3}$$

Thus, the solution set is $\{1-\sqrt{3}, 1+\sqrt{3}\}$.

21. Let x cm be the length of the shorter leg. Then the hypotenuse is of length $(x + 9)$ cm and the other leg is of length $(x + 7)$ cm. Thus,

$$(x+9)^2 = x^2 + (x+7)^2$$
$$x^2 + 18x + 81 = x^2 + x^2 + 14x + 49$$
$$0 = x^2 - 4x - 32$$
$$0 = (x-8)(x+4)$$
$$x = 8 \text{ or } x = -4$$

Since x must be positive, we discard the –4. If the shorter leg is 8 cm long, then the hypotenuse is 8 + 9 = 17 cm long and the other leg is 8 + 7 = 15 cm long. You can check that $17^2 = 8^2 + 15^2$.

22. Let m be the number of miles you could drive for a charge of \$63. Then the required equation is

$21 + 0.21m = 63$

$0.21m = 42$ Subtract 21.

$21m = 4200$ Multiply by 100.

$m = 200$ Divide by 21.

200 miles could be driven for \$63.

23. Let x and $x + 1$ be the integers. Then the required equation is

$3(x + x + 1) = 45$

$3(2x+1) = 45$ Simplify.

$2x + 1 = 15$ Divide by 3.

$2x = 14$ Subtract 1.

$x = 7$ Divide by 2.

Thus, the two integers are 7 and 8.
Check: 7 + 8 = 15 and $3 \cdot 15 = 45$.

24. Let x and $x + 1$ be the integers.

$12x > 9(x+1)$

$12x > 9x + 9$ Simplify.

$3x > 9$ Subtract $9x$.

$x > 3$ Divide by 3.

Thus, the smaller integer must be greater than 3. Since the smallest such integer is 4, the two integers must be 4 and 5.
Check: $12 \cdot 4 = 48, 9 \cdot 5 = 45$ and 48 > 45.

That $x = 4$ is the smallest integer that satisfies the given condition is shown by trying $x = 3$. This gives $12 \cdot 3 = 36,\ 9 \cdot 4 = 36$ so that $12 \cdot 3$ equals, but is not greater than, $9 \cdot 4$.

25. Let n be an odd number, then the next consecutive odd number is $n + 2$. Solve the equation $n(n+2) = 10n - 7$.

$n(n+2) = 10n - 7$

$n^2 + 2n = 10n - 7$ Simplify.

$n^2 - 8n + 7 = 0$ Subtract $10n$ and add 7.

$(n-1)(n-7) = 0$ Factor.

$n = 1$ or $n = 7$ Solve.

If $n = 1$, then the numbers are 1 and 3 or if $n = 7$, then the numbers are 7 and 9.

26. a. 801 to 688; 801:688; $\frac{801}{688}$

b. 688 + 801 + 501 = 1990 = total number of stocks. So the ratio of losers to total number is $\frac{801}{1990}$.

27. a. $\frac{50}{8} = 6\frac{1}{4}$¢ per oz for the 8 oz box,

$\frac{76}{12} = 6\frac{1}{3}$¢ per oz for the 12 oz box

b. The 8 oz box is the better buy.

28. a. $\frac{x}{8} = \frac{9}{5}$

b. Multiply both sides by 8 to get $x = \frac{72}{5} = 14.4$. So the second side is 14.4 ft long.

29. a. $C = kx^2$

b. Put $C = 100$ and $x = 150$ in the equation of part a to obtain $100 = k(150)^2 = 22{,}500k$. Divide both sides by 22,500 to get $k = \frac{1}{225}$.

c. Put $k = \frac{1}{225}$ and $x = 180$ in the equation of part a to get

$C = \frac{180^2}{225} = \frac{32400}{225} = 144$, so the cost is \$144 per hr.

30. a. $t = \frac{k}{I}$

b. Put $t = \frac{1}{30}$ and $I = 300$ in the equation of part a. This gives $\frac{1}{30} = \frac{k}{300}$. Now multiply both sides by 300 to find $k = 10$.

c. Put $k = 10$ and $I = 600$ in the equation of part a to get $t = \frac{10}{600} = \frac{1}{60}$. Thus, the correct exposure time is $\frac{1}{60}$ sec.

Chapter 7 Functions and Graphs

Section 7.1 Graphing Relations and Functions

Study Tips

The best way of learning about relations and functions is to study some examples that you understand. Say you are making $5 per hour. The amount A of money you earn is a *function* of the number of hours H that you work. We write $A = f(H) = 5H$. If you work 1, 2 or 3 hours, you make 5, 10 or 15 dollars, respectively. The set of ordered pairs $\{(1, 5), (2, 10), (3, 15)\}$ gives the relationship between the number of hours worked and your earnings. The set $\{1, 2, 3\}$ is called the *domain* and the set $\{5, 10, 15\}$ is the *range*.

To find the domain of a relation:
Think: what are the x values of the relation?

To find the range of a relation:
Think: what are the y values of the relation?
Note: As a matter of fact, any set of ordered pairs is a relation. In general, if $y = f(x)$ the x's are in the domain, the y's are in the range.

To determine a function:
Think: all the x values are different
Note: A relation is a function if for every x in the domain there is only one y.
Example
If you look at the ordered pairs, there are never two distinct pairs with the same first coordinate, such as (3, 4) and (3, 5).

To determine if a relation is a function from its graph:
Think: vertical line test
Note: Geometrically, a function will never have two points on the same vertical line.

Fact
Since both functions and relations are sets of ordered pairs, graphs are points of ordered pairs satisfying the function or relation.

To graph a function:
Think: make a table of values, plot points on Cartesian coordinate plane, if domain is the real numbers, then connect the points
Note: The ordered pairs needed to find the graph are found by giving different values to x and finding the corresponding y values.

Caution
Keep in mind that $f(x)$ and y are always the same. If $f(x) = 2x + 4$, do not panic because there is no y, this means that $y = f(x) = 2x + 4$.

Fact
If the function is of the form $f(x) = ax + b$, the result is a line. If $f(x) = ax^2 + bx + c$, it is a parabola opening up (like a soup bowl) when a is positive, and opening down (invert the soup bowl) when a is negative.

7.1 Exercises

1. The domain is the set of first numbers in the pairs of the relation: $\{1, 2, 3\}$. The range is the set of second numbers in these pairs: $\{2, 3, 4\}$.

3. The domain is the set of first numbers in the pairs of the relation: $\{1, 2, 3\}$. The range is the set of second numbers in these pairs: $\{1, 2, 3\}$.

5. The domain is the set of x values: $\{x | x \text{ is a real number}\}$. The range is the set of y values: $\{y | y \text{ is a real number}\}$.

7. The domain is the set of x values: $\{x | x \text{ is a real number}\}$. The range is the set of y values: $\{y | y \text{ is a real number}\}$.

9. The domain is the set of x values: $\{x | x \text{ is a real number}\}$. The range is the set of y values: $\{y | y \geq 0\}$.

11. The domain is the set of x values: $\{x | x \geq 0\}$. The range is the set of y values: $\{y | y \text{ is a real number}\}$.

13. The domain is the set of x values: $\{x | x \neq 0\}$. The range is the set of y values: $\{y | y \neq 0\}$.

15. This is a function because one real value of y corresponds to each real value of x.

17. This is not a function, because two values of y correspond to each positive value of x. For instance, (4, 2) and (4, –2) both belong to this relation.

19. This is a function. The domain is $\{x | x \geq 0\}$, and to each value of x in this domain there corresponds one y value, the nonnegative square root of the x-value.

21. This is a function. The domain consists of the set of real numbers, and to each number in this domain there corresponds one y value, the cube root of x.

23. a. $f(0)=3\cdot 0+1=0+1=1$
    b. $f(2)=3\cdot 2+1=6+1=7$
    c. $f(-2)=3(-2)+1=-6+1=-5$

25. a. $F(1)=\sqrt{1-1}=\sqrt{0}=0$
    b. $F(5)=\sqrt{5-1}=\sqrt{4}=2$
    c. $F(26)=\sqrt{26-1}=\sqrt{25}=5$

27. a. $f(x+h)=3(x+h)+1=3x+3h+1$
    b. $f(x+h)-f(x)$
    $=3(x+h)+1-(3x+1)$
    $=3x+3h+1-3x-1=3h$
    c. $\dfrac{f(x+h)-f(x)}{h}=\dfrac{3h}{h}=3$ for $h\neq 0$

29. $g(x)=x^2$. The missing numbers are $\frac{1}{16}$, 4.41, and ± 8.

31. a. $g(0)=2\cdot 0^3+0^2-3\cdot 0+1=1$
    b. $g(-2)=2(-2)^3+(-2)^2-3(-2)+1$
    $=-16+4+6+1=-5$
    c. $g(2)=2(2)^3+(2)^2-3(2)+1$
    $=16+4-6+1=15$

33. a. Put $a=50$ in the function $U(a)=-a+190$. This gives $U(50)=-50+190=140$. Thus, the upper limit is 140 beats per min.
    b. Put $a=60$ in the function $U(a)=-a+190$. This gives $U(60)=-60+190=130$. Thus, the upper limit is 130 beats per min.

35. a. Put $h=70$ in the function $w(h)=5h-190$. This gives $w(70)=5\cdot 70-190=350-190=160$, so his weight should be 160 lb.
    b. Put $w(h)=200$ to get the equation $200=5h-190$. Then add 190 to both sides to get $390=5h$. Now divide both sides by 5 to find $h=78$. Thus, his height should be 78 in.

37. a. Put $d=10$ in the function $P(d)=63.9d$ to get $P(10)=63.9\cdot 10=639$. The pressure is 639 lb per sq ft.
    b. Put $d=100$ in the function $P(d)=63.9d$ to get $P(100)=63.9\cdot 100=6390$. The pressure is 6390 lb per sq ft.

39. $S(t)=\frac{1}{2}gt^2=\frac{1}{2}\cdot 32t^2=16t^2$
    a. $S(3)=16\cdot 3^2=16\cdot 9=144$ ft
    b. $S(5)=16\cdot 5^2=16\cdot 25=400$ ft

41. The table of values and the graph for the relation $\{(x,y)\,|\,y=x, x$ an integer between -1 and 4, inclusive$\}$.

x	y
−1	−1
0	0
1	1
2	2
3	3
4	4

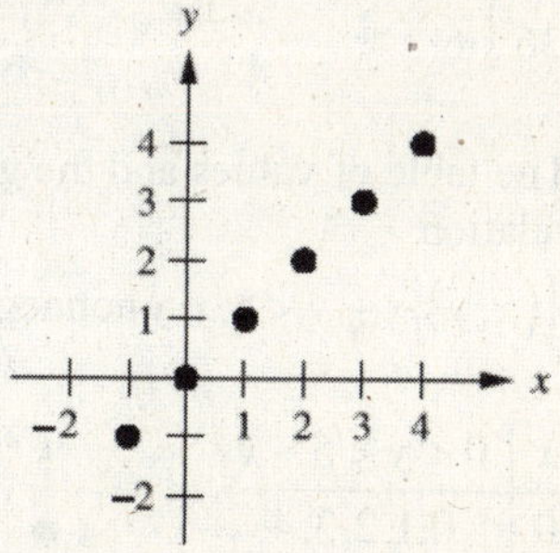

43. The table of values and the graph for the relation $\{(x,y)\,|\,y=2x+1, x$ an integer between 0 and 5, inclusive$\}$.

x	y
0	1
1	3
2	5
3	7
4	9
5	11

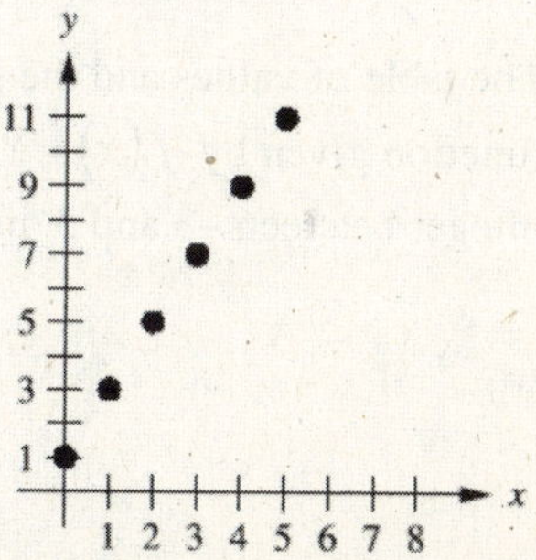

45. The table of values and the graph for the relation $\{(x,y)|2x-y=4, x \text{ an integer between } -2 \text{ and } 2, \text{ inclusive}\}$.

x	y
−2	−8
−1	−6
0	−4
1	−2
2	0

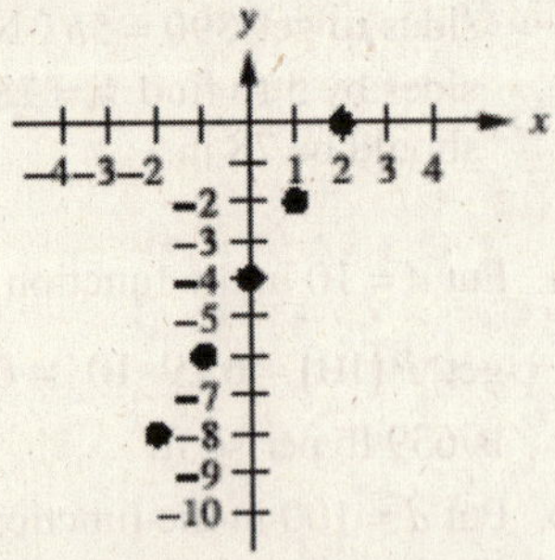

47. The table of values and the graph for the relation

$\{(x,y)|y=\sqrt{x}, x=0,1,4,9,16,25,36\}$.

x	y
0	0
1	1
4	2
9	3
16	4
25	5
36	6

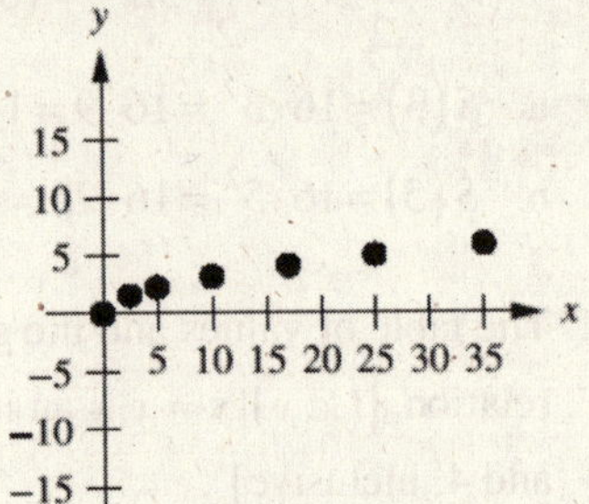

49. The table of values and the graph for the relation

$\{(x,y)|x+y<5, x, y \text{ nonnegative integers}\}$.

x	$0<y<5-x$
0	0,1,2,3,4
1	0,1,2,3
2	0,1,2
3	0,1
4	0

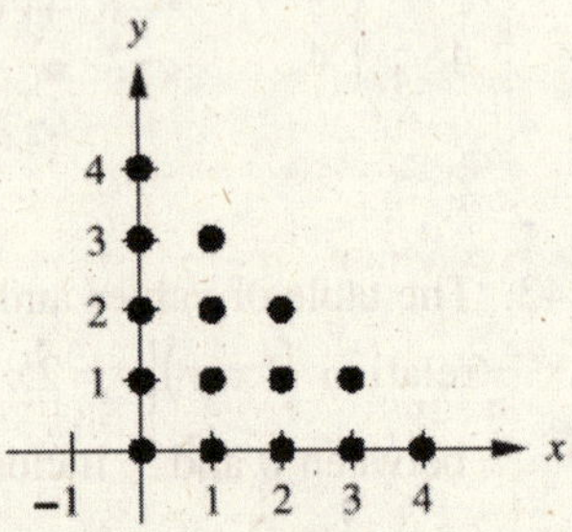

51. The table of values and the graph for the function given by $f(x)=x+1$, x an integer between −3 and 3, inclusive.

x	$f(x)=x+1$
−3	−2
−2	−1
−1	0
0	1
1	2
2	3
3	4

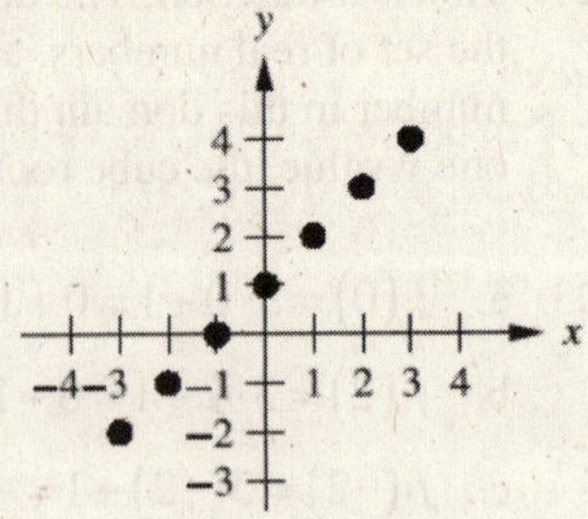

53. The table of values and the graph for the function given by $g(x)=x^2+1$, x an integer between −3 and 3, inclusive.

x	$g(x)=x^2+1$
−3	10
−2	5
−1	2
0	1
1	2
2	5
3	10

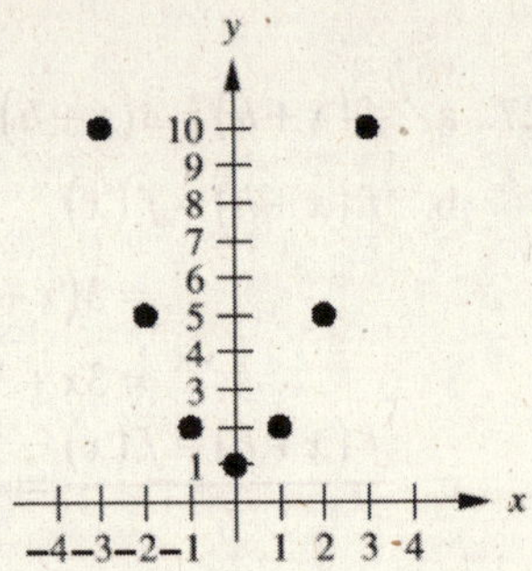

55. a. $h(x)=2.89x+70.64$

b. $h(34)=2.89(34)+70.64=168.9$, so the height was about 169 cm.

57. a. Let m be monthly data allowance and f be monthly access fee. The table of values and the graph for the data plan is

m	f
75-0.1	10
2	30
5	50
10	80

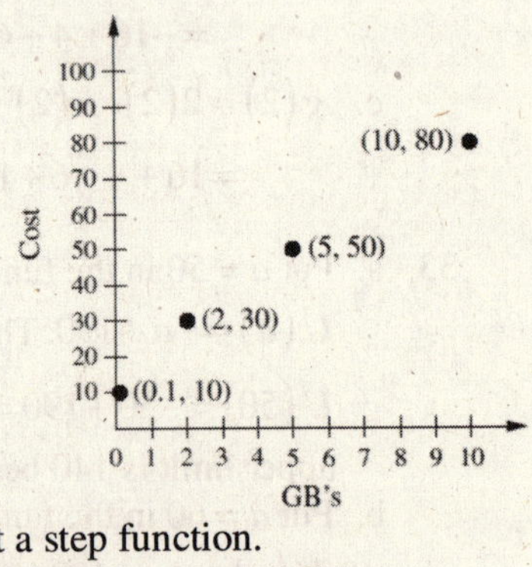

The graph is not a step function.

b. The access fee is $50 if you use 5 GB.

c. No, you cannot determine the fee for 8 GB.

d. The monthly allowance of 5 GB has a monthly access fee of $50. The access fee for 8 GB is determined by adding $10 for each GB over 5 GB, which is 8 – 5 = 3 GB, to the monthly access fee of $50: 50 + 3(10) = $80. For 8.6 GB, we have 8.6 – 5 = 3.6 GB; since even a fraction will incur an additional $10, the monthly access fee is

50 + 4(10) = \$90.

59. a. Put $n = 100$ in the function $b(n) = 2.520n$ to get $b(100) = 2.520(100) = 252$. Thus, there were 252 million gal spilled in 100 days.
b. For a leak of 4.2 million gal per day, the function is $e(n) = 4.2n$ million gal per day.
c. Put $n = 60$ in the function $e(n) = 4.2n$ to get $e(60) = 4.2(60) = 252$. Thus, there were 252 million gal spilled in 60 days.

61. a. Write the function $b(n) = 2.520n$ million gal per day as $b(n) = 2{,}520{,}000n$ gal. The actual amount of oil that spilled daily is $a(n) = 2{,}520{,}000n - 441{,}000n$ $= 2{,}079{,}000n$ gal per day.
b. Put $n = 60$ in the function $a(n) = 2{,}079{,}000n$ to get $a(n) = 2{,}079{,}000(60) = 124{,}740{,}000$. Thus, there were 124,740,000 gal spilled in 60 days.

63. Answers may vary. $g(x) = \sqrt{x-1}$ is real if and only if $x - 1 \geq 0$, that is, if $x \geq 1$. Thus, we would exclude all values of x less than 1 if g is to have real values.

65. Answers may vary. The graph of a function $f(x)$ is a picture of the set of points $\{(x, y) \mid y = f(x), x$ an element of the domain of $f\}$.

67. Answers may vary. Yes. A function is a relation in which there is exactly one value of y for each value of x in the domain.

69. Notice that if the units digit in the temperature reading is multiplied by 4, the result is the number of chirps per minute. Thus, the desired function is $c = f(x) = 4(x - 40)$, where x is the temperature in degrees Fahrenheit.

71. The table shows that the distance in feet is given by multiplying the square of the time (in sec) by 16. Thus, the desired function is $f(t) = 16t^2$.

73. For each table entry, the number of seconds is the square root of the number of units of length, so the suggested rule is $f(x) = \sqrt{x}$.

75. If $f(x) = \sqrt{x} = 100$, then, squaring both sides of $\sqrt{x} = 100$, we obtain $x = 100^2 = 10{,}000$. Thus, the length would be 10,000 units, or since 1 unit is about 25 cm, the length is about 250,000 cm.

Section 7.2 Linear Functions, Relations, and Applications

Study Tips

A graph of a linear function is a straight line, and is written as $f(x) = ax + b$. Two special lines are the horizontal line and the vertical line.

To graph a horizontal line:
Think: the line will cross the y-axis, points will all have the same y value
Note: The equation of the horizontal line is $y = f(x) = c$. No matter what value you give x, y is always the constant c.

To graph a vertical line:
Think: the line will cross the x-axis, points will all have the same x value
Note: The equation of the vertical line is $x = c$, where no matter what value you give y, x is always the constant c.

Hint
You should recognize both of these as special lines because they involve one variable only.

To graph an equation using intercepts:
Think: find the points that a) cross the x-axis and b) cross the y-axis, plot, then draw a line through the points
Note: To find the x-intercept, let $y = 0$ in the equation and solve for x; to find the y-intercept, let $x = 0$ in the equation and solve for y.
Hint
Fill in this table to find the x- and y-intercepts.

	x	y	
	0		**y-intercept**
x-intercept		0	

To find the distance between two points:
Think: (x_1, y_1), (x_2, y_2), and distance formula
Note: The distance formula between point A (x_1, y_1) and $B(x_2, y_2)$ is

$$|AB| = \sqrt{(x_2 - x_1)^2 + (y_2 - y_1)^2}$$

Hint
It will not matter which point you label (x_1, y_1) or (x_2, y_2).

7.2 Exercises

1. According to the table, when $x = 3$, $y = 2\frac{1}{4} = 2.25$ and $y = ax$. Thus, $2.25 = 3a$ and $a = 0.75 = \frac{3}{4}$. Now, $y = ax = \frac{3}{4}x$. If $y = 3$, then $3 = \frac{3}{4}x$, so that $x = 4$.

3. Since $f(x) = 3x + 6, f(0) = 6$ and $f(-2) = 0$. Thus, (0, 6) and (–2, 0) are two points on the line. Graph these points and draw the line through them, as shown on the graph:

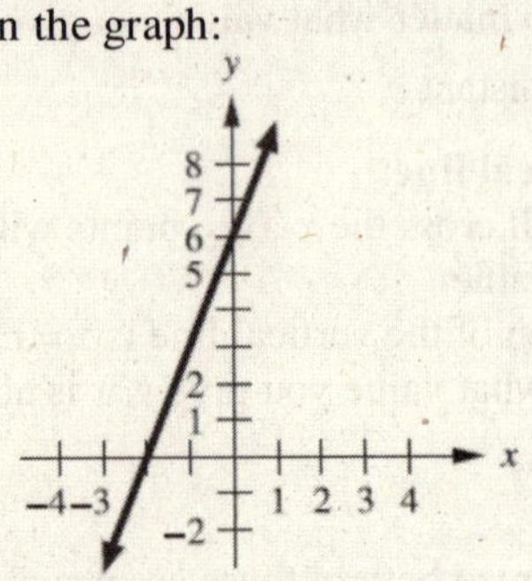

5. Since $f(x) = 3$ for all values of x, the graph is a straight line parallel to the x-axis and 3 units above this axis as shown in the figure:

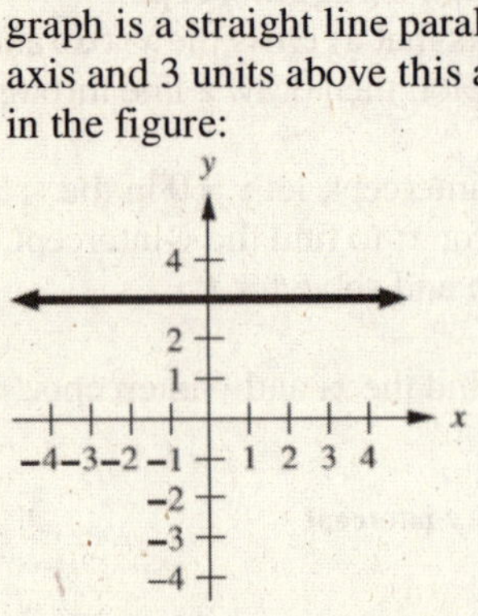

7. Here we are given $x = -1$ for all values of y. Thus, the graph is a straight line parallel to the y-axis and 1 unit to the left of this axis as shown in the figure:

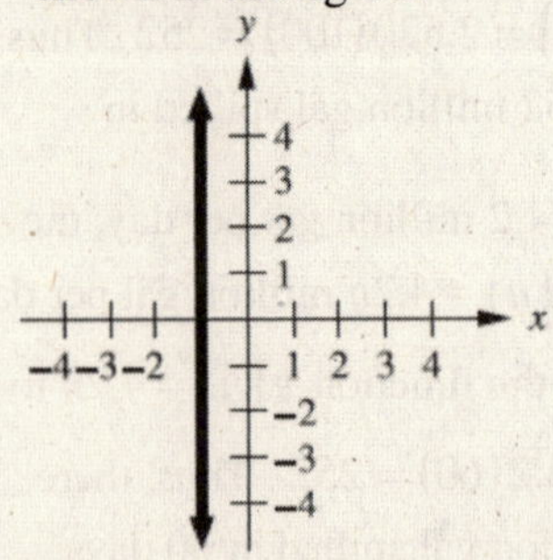

9. Since $f(x) = -x + 2$, then $f(0) = 2$ and $f(2) = 0$. Thus, (0, 2) and (2, 0) are two points on the line. Graph these points and draw the straight line through them as in the figure:

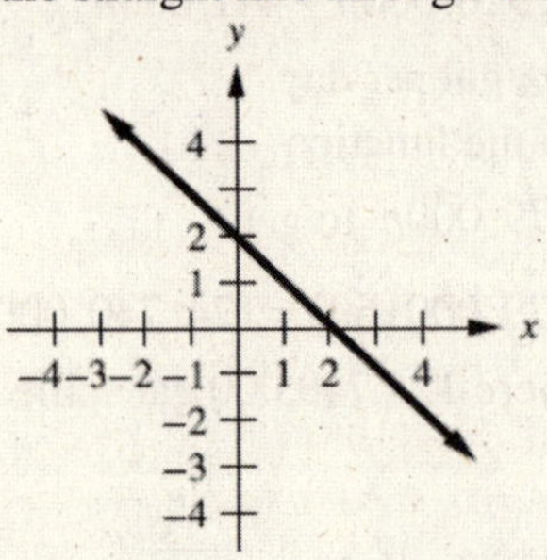

11. With $g(x) = -3x - 6$, we get $g(0) = -6$ and $g(-2) = 0$. So (0, –6) and (–2, 0) are two points on the line. Graph these two points and draw a straight line through them as shown in the figure:

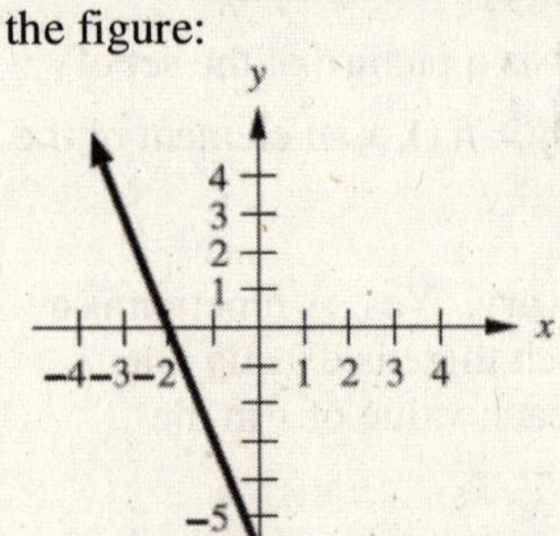

13. The equation $3x + 2y = 6$ is satisfied by the pairs (0, 3) and (2, 0), so these are two points on the line. Graph these points and draw the straight line through them as in the figure:

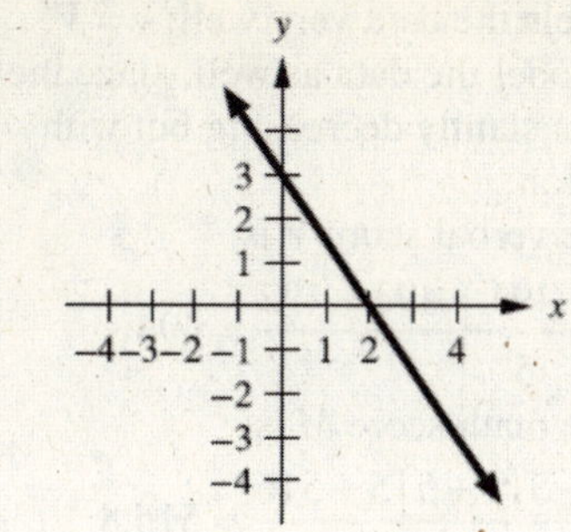

15. The equation $-2x + 3y = 6$ is satisfied by the pairs (0, 2) and (–3, 0), so these are two points on the line. Graph these points and draw the straight line through them as in the figure:

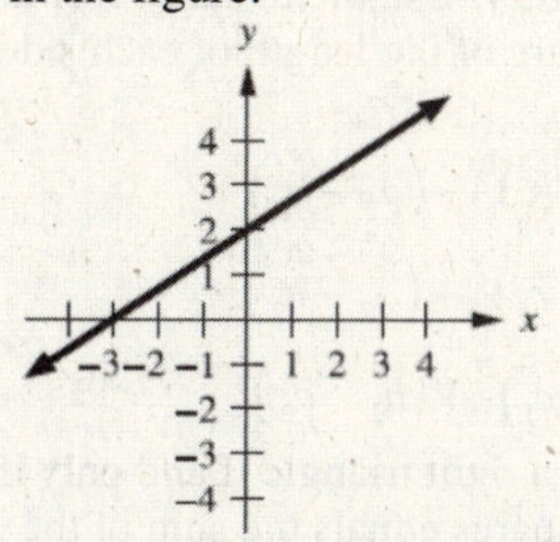

17. Two of the solutions of the equation $4x - 3y = 12$ are (3, 0) and (0, –4), and these are two points on the graph. Graph these two points and draw the straight line through them as shown in the figure:

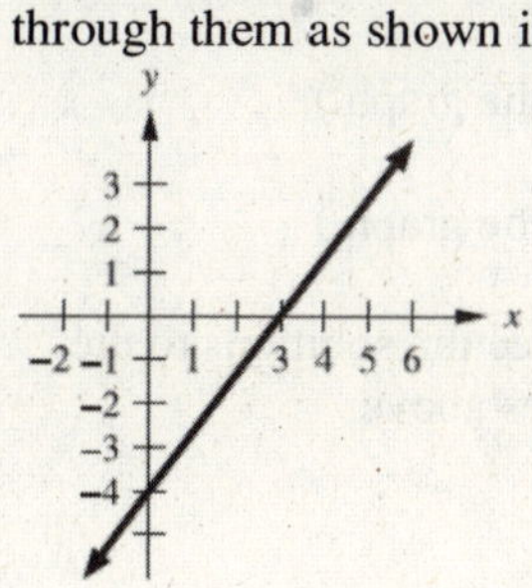

19. $d = \sqrt{(-1-2)^2 + (0-4)^2}$

$= \sqrt{(-3)^2 + (-4)^2} = \sqrt{9+16} = \sqrt{25}$

$= 5$ units

21. $d = \sqrt{(-1-(-4))^2 + (3-(-5))^2}$

$= \sqrt{(3)^2 + (8)^2} = \sqrt{9+64} = \sqrt{73}$

≈ 8.54 units

23. $d = \sqrt{(1-4)^2 + (-1-(-8))^2}$

$= \sqrt{(-3)^2 + 7^2} = \sqrt{9+49} = \sqrt{58}$

≈ 7.62 units

25. These points are on a line parallel to the y-axis, so the distance is
$d = |y_2 - y_1| = |-2-0| = 2$ units

27. These points are on a line parallel to the y-axis, so the distance is
$d = |y_2 - y_1| = |7-3| = 4$ units

29. a. $E(x) = 500 + 25x$; $S(x) = 1000 + 20x$

b.

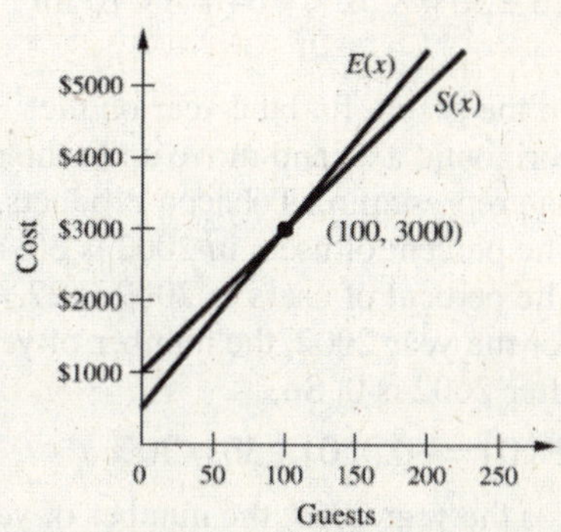

c. The two lines in the graph intersect at (100, 3000). This point checks in both equations. Thus, the cost is the same for 100 persons.

31. a. Refer to the table:
$m(2) = \$30$; $m(5) = \$50$.

b. The monthly access fee would be the base $m(2) = \$30$ and $d = 3$, 1 unit past the allowance for 2 GB, so fee of \$10. Thus for $2 < d \le 3$, $m(d) = \$30 + \$10 = \$40$.

c. The monthly access fee would be the base $m(2) = \$30$ plus d = 4, 2 units past the allowance for 2 GB. For $3 < d \le 4$, $m(d) = \$30 + 2(\$10) = \$50$.

d. The monthly access fee would be the base $m(2) = \$30$ plus d = 5, 3 units past the allowance for 2 GB. For $4 < d < 5$, $m(d) = \$30 + 3(\$10) = \$60$. (The economical choice for this customer would have been to choose the 5GB plan.)

e.

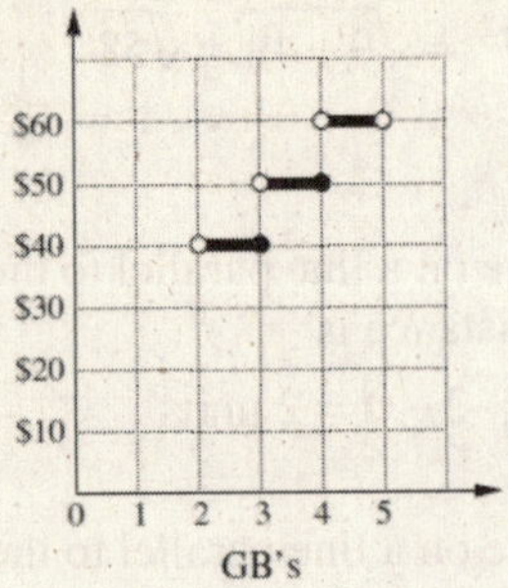

33. Yes. The distance d from (2, –3) to (–2, 3) is:

$$\sqrt{(-2-2)^2+(3-(-3))^2}=\sqrt{(-4)^2+6^2}$$

$$=\sqrt{16+36}=\sqrt{52}<\sqrt{64}=8<10 \text{ mi}$$

35. a. On the graph, find the year on the horizontal axis and move up to the red line representing Tobacco Products. The percent of users in 2002 is 30.4%. The percent of users in 2009 is 27.7%.
b. For the year 2002, the number of years after 2002 is 0. So,
$P(0)=-0.3(0)+30=30\%$.
For the year 2009, the number of years after 2002 is 7. So,
$P(7)=-0.3(7)+30=27.9\%$.
c. For the year 2012, the number of years after 2002 is 10. So,
$P(10)=-0.3(10)+30=27\%$.
d. Put $P(t)=24$ in the equation
$P(t)=-0.3t+30$ to get
$24=-0.3t+30$. Subtract 30, then divide by –0.3 to find $t=20$. Thus, 20 years after 2002, the year 2022, is when the percent will reach 24%.

37. a., c.The graph for the given data and for $y = V$ and $y = M$ are shown:

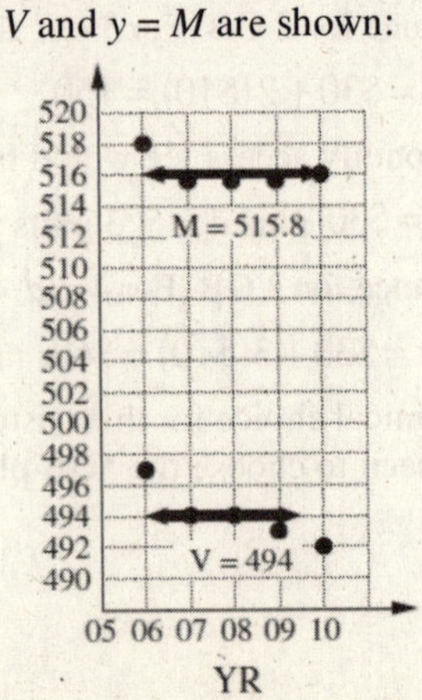

$y = M$ models the data very well; $y = V$ does not model the data as well, since the verbal is constantly decreasing but with one exception.
b. The average verbal score V is:
$$\frac{497+494+494+493+492}{5}=494$$
The average math score M is:
$$\frac{518+515+515+515+516}{5}=515.8$$
d. Yes, they are functions. You can tell by using the vertical line test (any vertical line touches the graph only once).

39. Answers may vary. Use the distance formula to find the square of the length of each side of the triangle:
$$(AB)^2=(a_1-b_1)^2+(a_2-b_2)^2$$
$$(AC)^2=(a_1-c_1)^2+(a_2-c_2)^2$$
$$(BC)^2=(b_1-c_1)^2+(b_2-c_2)^2$$
The triangle is a right triangle if and only if one of these squares equals the sum of the other two squares. So we can check this.

41. Answers may vary. See the Problem Solving procedure following Example 1 and follow it step by step for $x = c$.

43. About $940 (See the graph.)

45. About $800 (See the graph.)

For problems 47–57, see the solutions to the corresponding exercises above.

Calculator Corner

1. Steps may vary due to type of calculator used to solve the problem.
Problem 19:
[√] [(] [–1] [–] [2] [)] [x^2] [+] [(] [0] [–] [4] [)] [x^2] [)]
[ENTER] Answer: 5
Problem 21:
[√] [(] [–1] [–] [(–)] [4] [)] [x^2] [+] [(] [3] [–] [(–)] [5]
[x^2] [)] [ENTER] Answer: 8.54
Problem 23:
[√] [(] [1] [–] [4] [)] [x^2] [+] [(] [–1] [–] [(–)] [8] [)] [x^2]

ENTER Answer: 7.62

Problem 25:

√ (3 − 3) x^2 + (−2 − 0) x^2)

ENTER Answer: 2

Problem 27:

√ (−2 − (−) 2) x^2 + (7 − 3) x^2)

ENTER Answer: 4

Section 7.3 Slopes and Equations of a Line

Study Tips

We now shift emphasis from graphs to equations. We want to explore the different ways in which the equation of a line can be written, as well as the information necessary to write these equations.

To find the slope of a line:

Think: identify two points: (x_1, y_1), (x_2, y_2) then use slope formula

Note: The inclination of a line is called the slope. The formula for the slope of a line is

$$m = \frac{y_2 - y_1}{x_2 - x_1}.$$

Caution

While it does not matter which point you label (x_1, y_1) or (x_2, y_2), it is critical that you place the x and y values in the correct place. After you have put the values into the formula, a quick check is that you should be able to see that the points could be read vertically.

To find the equation of the line given the slope and the y-intercept:

Think: what form shows only the slope and y-intercept? The slope-intercept form:

$y = mx + b$

Note: Substitute the slope m and the y-intercept b into the formula. You now have the equation of the line!

To find the equation of the line given the slope and a point (x_1, y_1):

Think: what form shows only the slope and a point? The point-slope form:

$y - y_1 = m(x - x_1)$

Note: Substitute the slope m and the point (x_1, y_1) into the formula. You now have the equation of the line!

Fact

Once you have made the substitutions into the form, you *have* the equation of the line; it just is not in a 'nice' form. You will often be asked to put the equation of the line into the slope-intercept form $y = mx + b$ or standard form $Ax + By = C$.

Note

If you are given just *two points*, find the slope between the points. Use that slope and either one of the points in the point-slope formula. This is a summary of what you do when using the two-point form.

Word of Advice

Do not worry about which formula you have to use to find the equation of a line because the name of the formula will tell you when to apply it. For your convenience, all this information is in Table 7.8. Familiarize yourself with it.

Relationships that you must know:

1. If two lines are parallel, they have the same slope.

2. If two lines are perpendicular, the slopes are negative reciprocals. If a line has slope 4, a line perpendicular to it must have slope $-\frac{1}{4}$, the negative reciprocal of 4.

7.3 Exercises

1. $m = \frac{y_2 - y_1}{x_2 - x_1} = \frac{4-2}{3-1} = \frac{2}{2} = 1$

3. $m = \frac{y_2 - y_1}{x_2 - x_1} = \frac{0-5}{5-0} = \frac{-5}{5} = -1$

5. $m = \frac{y_2 - y_1}{x_2 - x_1} = \frac{-4-(-3)}{7-(-1)} = \frac{-1}{8} = -\frac{1}{8}$

7. $m = \frac{y_2 - y_1}{x_2 - x_1} = \frac{3-0}{12-0} = \frac{3}{12} = \frac{1}{4}$

9. $m = \frac{y_2 - y_1}{x_2 - x_1} = \frac{5-5}{-2-3} = \frac{0}{-5} = 0$

11. Use the point-slope form $y - y_1 = m(x - x_1)$ to get:

$$y-2=\frac{1}{2}(x-1)$$
$$y-2=\frac{1}{2}x-\frac{1}{2}$$
$$y=\frac{1}{2}x+\frac{3}{2}$$

13. Use the point-slope form $y - y_1 = m(x - x_1)$ to get:

$$y-4=-1(x-2)$$
$$y-4=-1x+2$$
$$y=-x+6$$

15. Use the point-slope form $y - y_1 = m(x - x_1)$ to get:

$$y-5=0(x-4)$$
$$y-5=0$$
$$y=5$$

17. By comparing $y = x + 2$ with the equation $y = mx + b$, where m is the slope and b the y-intercept we find that:
a. $m = 1$ b. $b = 2$

19. First solve $3y = 4x$ for y to get $y=\frac{4}{3}x$ then compare with $y = mx + b$, where m is the slope and b the y-intercept to find:

a. $m=\frac{4}{3}$ b. $b = 0$

21. First solve $x + y = 14$ for y to get $y = -x + 14$ then compare with $y = mx + b$ to find:
a. $m = -1$ b. $b = 14$

23. Compare the equation $y = 6$ with $y = mx + b$ to find:
a. $m = 0$ b. $b = 6$

25. The line whose equation is $x = 3$ is parallel to the y-axis. Thus,
a. The slope is not defined.
b. The line does not intersect the y-axis.

27. For the line through the points (1, –1) and (2, 2), find the slope:

$$m=\frac{2-(-1)}{2-1}=\frac{3}{1}=3$$

Then use the point-slope form to get:

$$y-(-1)=3(x-1)$$
$$y+1=3x-3$$
$$3x-y=4$$

29. For the line through the points (3, 2) and (2, 3), find the slope:

$$m=\frac{3-2}{2-3}=\frac{1}{-1}=-1$$

Then use the point-slope form to get:

$$y-2=-1(x-3)$$
$$y-2=-x+3$$
$$x+y=5$$

31. For the line through the points (0, 0) and (1, 10), find the slope:

$$m=\frac{10-0}{1-0}=\frac{10}{1}=10$$

Then use the point-slope form to get:

$$y-0=10(x-0)$$
$$y=10x$$
$$10x-y=0$$

33. Use the first two entries of the table: (62, 134) and (63,139) to find the slope m, then with h in place of x and w in place of y, substitute in the point-slope form for the line and solve for w:

$$m=\frac{139-134}{63-62}=\frac{5}{1}=5$$
$$w-134=5(h-62)$$
$$w-134=5h-310$$
$$w=5h-176$$

35. Use the first two entries of the table: (62, 123) and (63,128) to find the slope m, then with h in place of x and w in place of y, substitute in the point-slope form for the line and solve for w: $m=\frac{128-123}{63-62}=\frac{5}{1}=5$

$$w-123=5(h-62)$$
$$w-123=5h-310$$
$$w=5h-187$$

37. The cost C is 2 dollars for the first mile and \$1.70 for each additional mile. That is: $C = 2 + 1.70(m-1)$

$= 2 + 1.70m - 1.70$

$= 1.70m + 0.30$

For $m = 30$,

$C = 1.70(30) + 0.30 = \$51.30.$

39. From problem 37: $C = 1.70m + 0.30$
From problem 38: $C = 1.50m + 1.50$
If the cost C is the same:

$1.70m + 0.30 = 1.50m + 1.50$

$0.20m + 0.30 = 1.50$

$0.20m = 1.20$

$m = 6$ miles

41. a. The cost of the first mile is \$2.
b. After the first mile, you travel $m - 1$ miles.
c. The cost per mile after the first mile is \$1.70.
d. The cost of the miles beyond the first is $C = 1.70(m-1)$.
e. $C = 1.70(m-1) + 2$

$= 1.70m - 1.70 + 2 = 1.70m + 0.30$

Yes, this is the same as the answer to problem 37.

43. Since the cost C is equal to the rental fee R plus \$2 for each minute m, we get: $C = R + 2m$.
Knowing that cost was $C = \$175$ and $60 = m$ minutes were used, we get:

$175 = R + 2 \cdot 60$

$175 = R + 120$

$55 = R$

For $R = 55$, $C = R + 2m$ becomes $C = 55 + 2m$ or $C = 2m + 55$.

45. To find the slopes, compare with the slope-intercept form $y = mx + b$:
For $y = 2x + 5, m = 2$.. Solve $4x - 2y = 7$ for y to get $y = 2x - \frac{7}{2}$, which shows that $m = 2$ for the second line. Since the slopes are equal and the y-intercepts are different, the lines are parallel.

47. To find the slopes, compare with the slope-intercept form $y = mx + b$. Solve the first equation, $2x + 5y = 8$, for y to get $y = -\frac{2}{5}x + \frac{8}{5}$. This result shows that $m_1 = -\frac{2}{5}$. Solve the second equation, $5x - 2y = -9$ for y to get $y = \frac{5}{2}x + \frac{9}{2}$, which shows that $m_2 = \frac{5}{2}$ for the second line. Since the slopes are not equal, the lines are not parallel.

49. To find the slopes, compare with the slope-intercept form $y = mx + b$. Solve the first equation, $x + 7y = 7$, for y to get $y = -\frac{1}{7}x + 1$. This result shows that $m_1 = -\frac{1}{7}$. Solve the second equation, $2x + 14y = 21$ for y to get $y = -\frac{1}{7}x + \frac{3}{2}$, which shows that $m_2 = -\frac{1}{7}$ for the second line. Since $m_1 = m_2$, and the y-intercepts are different, the lines are parallel.

51. Since the required line is to be parallel to the line $4x - y = 7$ or, equivalently, $y = 4x - 7$, the new line must have slope $m = 4$. Since the line passes through the point (1, –2), use the point-slope formula $y - y_1 = m(x - x_1)$ with $m = 4$ and $(x_1, y_1) = (1, -2)$ to obtain $y - (-2) = 4(x-1)$. Solve for y to obtain $y = 4x - 6$, which is an equation with slope 4, so is parallel to the given line and passes through (1, –2). In standard form, we have $4x - y = 6$.

53. a. In order to be perpendicular to the line $2x + 5y = 7$, the desired line must have an equation of the form $5x - 2y = c$, with c chosen so that the line goes through (2, 0). By substituting (2, 0) into the equation, we find that $5 \cdot 2 - 2 \cdot 0 = C$, so $C = 10$. Thus, the equation of the desired line is $5x - 2y = 10$.
b. The equation $y = 2x - 3$ can be written in the form $2x - y = 3$. In order to be

perpendicular to the line $2x - y = 3$, the desired line must have an equation of the form $x + 2y = c$, with c chosen so that the line goes through (1, 1). By substituting (1, 1) into the equation, we find that $1 + 2 \cdot 1 = c$, so $c = 3$. Thus, an equation of the desired line is $x + 2y = 3$.

c. In order to be perpendicular to the line $x - 2y = 3$, the desired line must have an equation of the form $2x + y = c$, with c chosen so that the line goes through (2, –2). By substituting (2, –2) into the equation, we find that $2 \cdot 2 + (-2) = c$, so $c = 2$. Thus, an equation of the desired line is $2x + y = 2$.

d. In order to be perpendicular to the line $4x + 5y = 9$, the desired line must have an equation of the form $5x - 4y = c$, with c chosen so that the line goes through (1, 1). By substituting (1, 1) into the equation, we find that $5 \cdot 1 - 4 \cdot 1 = c$, so $c = 1$. Thus, an equation of the desired line is $5x - 4y = 1$.

55. For the first line, $m_1 = \frac{200 - 0}{100 - 0} = \frac{200}{100} = 2$, and for the second line, $m_2 = \frac{-405 - 10}{790 - 0} = \frac{-415}{790} \neq -\frac{1}{2}$. Since $m_2 \neq -\frac{1}{m_1}$, the two lines are not perpendicular.

57. a. For (*f*, *F*), we have the points (0, 0) and (1000, 1120) to find the slope:
$$m = \frac{1120 - 0}{1000 - 0} = \frac{1120}{1000} = \frac{28}{25} = 1.12$$
The slope is the number of friends *F* for every follower *f*.

b. The point (0, 0) is on the line and so the *y*-intercept is 0.

c. $F(f) = 1.12f$

d. Put f = 12,000,000 in the equation $F(f) = 1.12f$.
$$F(12{,}000{,}000) = 1.12(12{,}000{,}000)$$
$$= 13{,}440{,}000$$
Lady Gaga has 13.44 million friends.

59. a. Compare $T = 3.87 - 0.13t$ with $y = mx + b$, where m is the slope and b the y-intercept to find that the slope is –0.13. Since the slope is negative, the consumption of tuna is decreasing.

b. Compare $F = -0.29t + 15.46$ with $y = mx + b$, where m is the slope and b the y-intercept to find that the slope is –0.29. Since the slope is negative, the consumption of fish and shellfish is decreasing.

c. The steepest slope is –0.29, which means that fish and shellfish consumption is decreasing faster.

61. a. Compare $y = 0.19t + 74$ with $y = mx + b$, where m is the slope and b the y-intercept to find that the slope is 0.19.

b. Since the slope is positive, the life span of American men is increasing.

c. The slope represents the annual increase in life span for American men, which is 0.19 yr per year.

63. a. Compare $f = 165 + 0.4t$ with $y = mx + b$, where m is the slope and b the y-intercept to find that the slope is 0.4.

b. Since the slope is positive, the consumption of fat is increasing .

c. The slope represents the annual increase in fat consumption by Americans, which is 0.4 gram more each year.

65. a. Put t = 2000 – 1970 = 30 into the equation $g = 140 + 0.94t$ to get
$$g = 140 + 0.94(30) = 140 + 28.2 = 168.2$$
The daily fat available for consumption in 2000 was 168.2 grams.

b. For t = 2010 – 1970 = 40,
$$g = 140 + 0.94(40) = 177.6$$
The daily fat available for consumption In 2010 was 177.6 grams.

c. In 1970, t = 0 so $g = 140 + 0.94(0) = 140$. The daily fat available for consumption In 1970 was 140 grams.

d. Since the slope 0.94 is positive, the daily fat available for consumption is increasing.

67. a. Compare $C = 2.5N + 175$ with $y = mx + b$,

where m is the slope and b the y-intercept to find that the slope is 2.5.

b. The slope represents the annual population per square mile in coastal regions increase.

c. Put $N = 55$ into the equation $C = 2.5N + 175$ to get

$$C = 2.5(55) + 175 = 137.5 + 175 = 312.5$$

There are 312.5, or 313, persons per square mile in the year 2015.

d. Yes. From the graph, the persons per square mile in 2015 is approximately 300, which is close to 313 persons per square mile.

69. Answers may vary. If (x_1, y_1) and (x_2, y_2) are any two distinct points on a horizontal line, $y_1 = y_2$ and $x_1 \neq x_2$. Thus, the slope is $m = \dfrac{y_2 - y_1}{x_2 - x_1} = \dfrac{0}{x_2 - x_1} = 0$.

71. a. $m = \dfrac{1}{1.266} \approx 0.79$

b. No. The maximum allowed is 0.22.

73. a. $m = \dfrac{0.4}{1.6} = 0.25$

b. Yes. The maximum allowed is 0.3. (With a maximum slope of 0.125 for wheelchairs, you would not want to take your wheelchair up this walking ramp!)

75. $m = \dfrac{20-16}{124-108} = \dfrac{4}{16} = 0.25$, which is not safe for parking (0.22 is the maximum allowed). The bottom of the driveway is $(x, 16)$ so $0.22 = \dfrac{20-16}{124-x}$

$$0.22(124 - x) = 4$$
$$27.28 - 0.22x = 4$$
$$-0.22x = -23.28$$
$$x \approx 105.8$$

If $x = 106$ ft, the slope would be too steep, so x should be approximately 105 ft so that the slope is the maximum allowable.

77. Let t be the total paid and h be the number of hours. Find the slope between the points to determine if it is a linear relationship. For the point (h, t), use the points (3, 110) and (1, 60) to find the slope.

$$m = \frac{110-60}{3-1} = \frac{50}{2} = 25$$

Now use the points (3, 110) and (5, 160) to find the slope.

$$m = \frac{160-110}{5-3} = \frac{50}{2} = 25$$

Finally use the points (1, 60) and (5, 160) to find the slope.

$$m = \frac{160-60}{5-1} = \frac{100}{4} = 25$$

Since the slopes are equal, we have a linear relationship. Now find a model in the form $t = mh + b$. Use the slope $m = 25$, the point (1, 60), and the point-slope form to get:

$$t - 60 = 25(h-1)$$
$$t - 60 = 25h - 25$$
$$t = 25h + 35$$

The t-intercept represents the service call with no hours, so the service call is \$35. The slope represents the hourly rate, so the hourly rate is \$25.

Skill Checker

1. $f(x) = x^2$

$$f(0) = 0^2 = 0$$
$$f(1) = 1^2 = 1$$
$$f(-1) = (-1)^2 = 1$$
$$f(2) = 2^2 = 4$$
$$f(-2) = (-2)^2 = 4$$

The corresponding ordered pairs are: (0, 0), (1, 1), (–1, 1), (2, 4) and (–2, 4). The graph of these five points is shown.

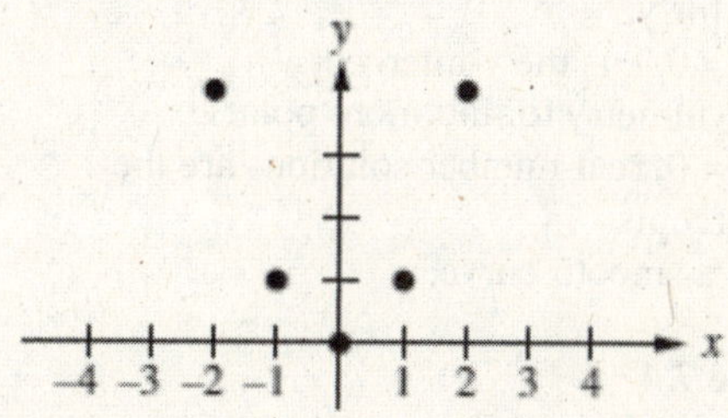

Section 7.4 Quadratic Functions and Their Graphs

Study Tips

The graph of the quadratic function, $y = f(x) = ax^2 + bx + c$, is a parabola. The parabola is a U-shaped graph. In the real world, we see parabolas all of the time: our satellite dishes!

Properties of the parabola $g(x) = ax^2$:

If $a > 0$, the parabola opens upward.
If $a < 0$, the parabola opens downward.

If $|a| > 1$, parabola narrower than $f(x) = x^2$.
If $|a| < 1$, parabola wider than $f(x) = x^2$.

To graph the function $y = ax^2 + k$:

Think: parabola, vertex at origin, *y*-axis is the line of symmetry
Note: Determine the points that satisfy the equation. To determine the points for the form $y = ax^2 + k$, you can complete this table of values:

x	−2	−1	0	1	2
y					

To graph the function $y = a(x-h)^2 + k$:

Think: parabola, vertex at (*h*, *k*), line of symmetry is $x = h$
Note: The graph is the same as that of $y = ax^2$ but moved *h* units horizontally and *k* units vertically.

To graph the function $y = ax^2 + bx + c$:

Think: parabola, follow procedure from text
Procedure summary:

1. Find vertex: let $x = -\dfrac{b}{2a}$ in the equation and solve for y.
2. Let $x = 0$, *c* is the *y*-intercept.
3. Use symmetry to find more points.
4. Let $y = 0$, real-number solutions are the *x*-intercepts.
5. Draw a smooth curve.

Exercise 7.4

1. a. $y = 2x^2$; $V(0,0)$ turns upward ($a > 0$). Select 3 values for *x* and get the corresponding *y* values:

x	0	1	−1
y	0	2	2

Graph the corresponding points and join them with a curve as shown.

b. To graph $y = 2x^2 + 2$, move the graph of $y = 2x^2$ two units up.

c. To graph $y = 2x^2 - 2$, move the graph of $y = 2x^2$ two units down.

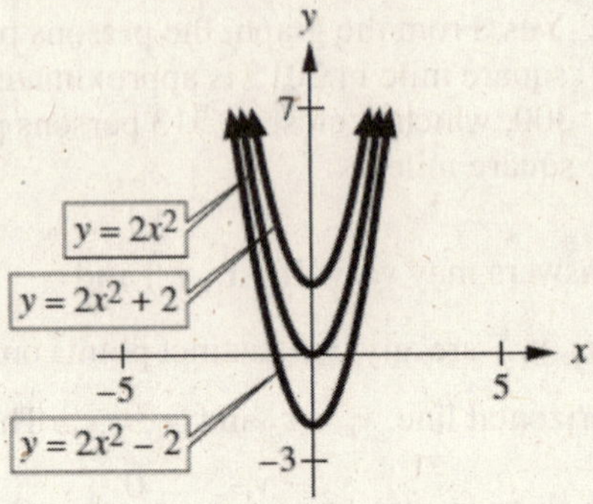

3. a. $y = -2x^2$; $V(0,0)$ turns downward ($a < 0$). Select 3 values for *x* and get the corresponding *y* values:

x	0	1	−1
y	0	−2	−2

Graph the corresponding points and join them with a curve as shown. The graph looks exactly like that of $y = 2x^2$ but "upside down," that is, it opens in the opposite direction.

b. To graph $y = -2x^2 + 1$, move the graph of $y = -2x^2$ one unit up.

c. To graph $y = -2x^2 - 1$, move the graph of $y = -2x^2$ one unit down.

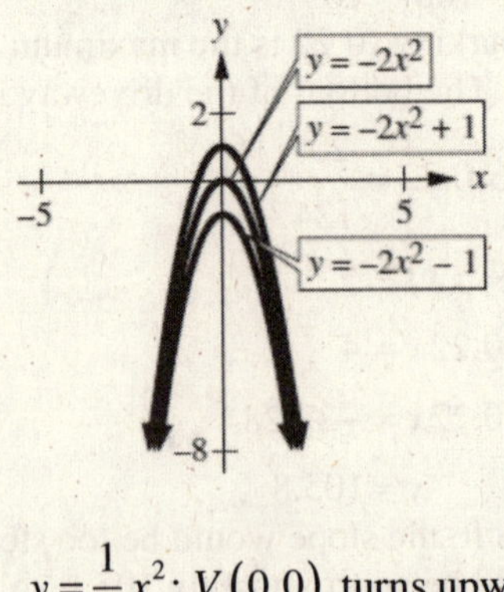

5. a. $y = \frac{1}{4}x^2$; $V(0,0)$ turns upward ($a > 0$):

x	0	2	−2
y	0	1	1

The graph has the same shape as that of $y=x^2$ but it is wider.

b. $y=-\frac{1}{4}x^2$; $V(0,0)$ looks exactly like the function in part a but opening in the opposite direction as shown in the graph:

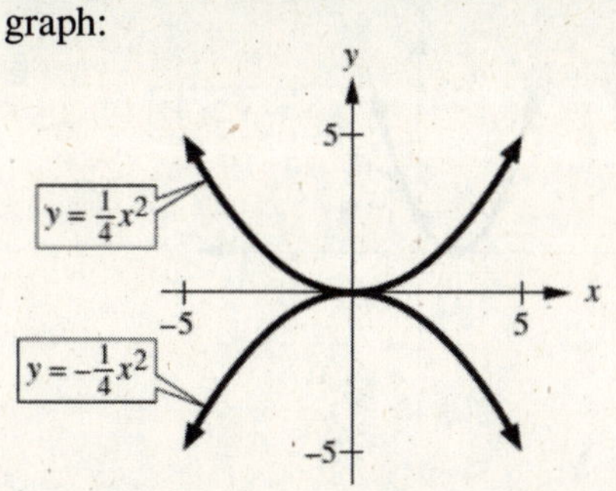

7. a. $y=\frac{1}{3}x^2+1$; $V(0,1)$ turns upward $(a>0)$:

x	0	3	−3
y	1	4	4

The graph has the same shape as that of $y=x^2$ but it is wider and moved up by one unit as shown.

b. $y=-\frac{1}{3}x^2+1$; $V(0,1)$ looks exactly like the function in part a but opening in the opposite direction:

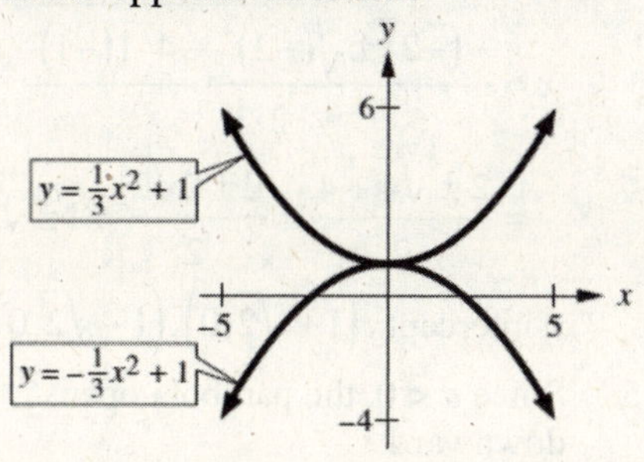

NOTE: Problems 9-15 are of the form $y=a(x-h)^2+k$ *and the vertex is at* (h, k).

9. a. $y=(x+2)^2+3$; $V(-2,3)$ turns upward $(a>0)$:

x	−2	−1	−3
y	3	4	4

The graph has the same shape as that of $y=x^2$ with the vertex moved 2 units left and 3 units up.

b. $y=(x+2)^2$; $V(-2,0)$ The graph has the same shape as that of $y=x^2$ with the vertex moved 2 units left.

c. $y=(x+2)^2-2$; $V(-2,-2)$ This time the vertex is moved 2 units left and two units down.

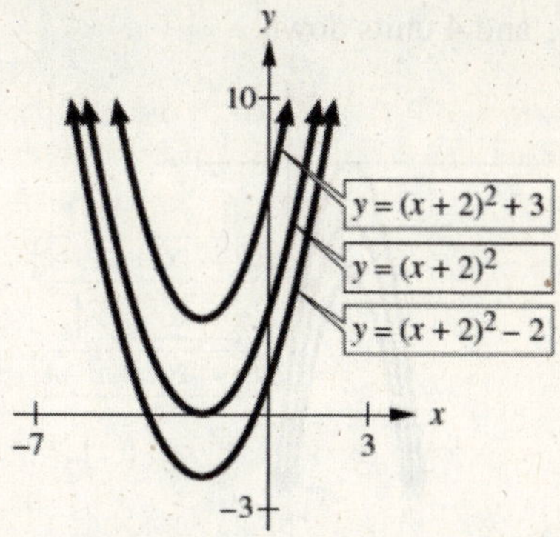

11. a. $y=-(x+2)^2-2$; $V(-2,-2)$ turns downward $(a<0)$:

x	−2	0	−4
y	−2	−6	−6

The graph has the same shape as that of $y=-x^2$ with the vertex moved 2 units left and 2 units down.

b. $y=-(x+2)^2$; $V(-2,0)$ The graph has the same shape as that of $y=-x^2$ with the vertex moved 2 units left.

c. $y=-(x+2)^2-4$; $V(-2,-4)$ This time the vertex is moved 2 units left and 4 units down.

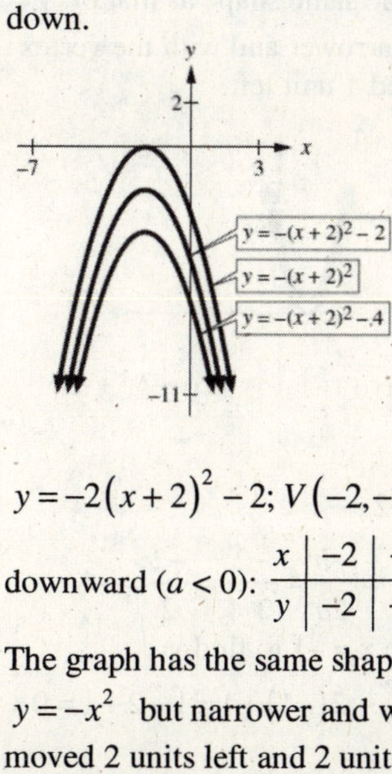

13. a. $y=-2(x+2)^2-2$; $V(-2,-2)$ turns downward $(a<0)$:

x	−2	−1	−3
y	−2	−4	−4

The graph has the same shape as that of $y=-x^2$ but narrower and with the vertex moved 2 units left and 2 units down.

b. $y=-2(x+2)^2$; $V(-2,0)$ The graph has

the same shape as that of $y=-x^2$ but narrower and with the vertex moved 2 units left.

c. $y=-2(x+2)^2-4$; $V(-2,-4)$ This time the vertex is moved 2 units left and 4 units down.

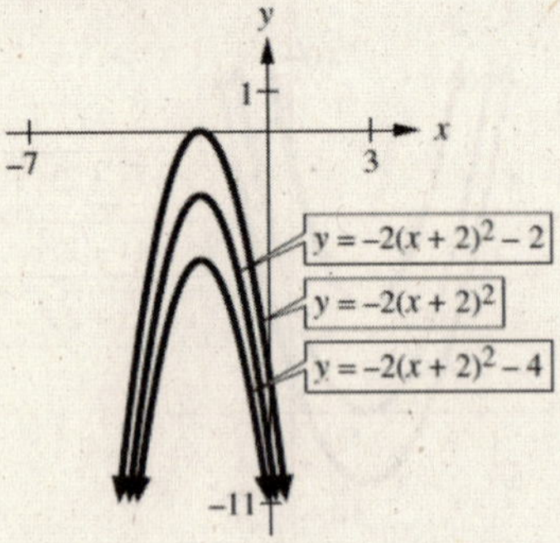

15. a. $y=2(x+1)^2+\frac{1}{2}$; $V\left(-1,\frac{1}{2}\right)$ turns upward ($a>0$):

x	−1	0	−2
y	$\frac{1}{2}$	$\frac{5}{2}$	$\frac{5}{2}$

The graph has the same shape as that of $y=x^2$ but narrower and with the vertex moved 1 unit left and $\frac{1}{2}$ unit up.

b. $y=2(x+1)^2$; $V(-1,0)$ The graph has the same shape as that of $y=x^2$ but narrower and with the vertex moved 1 unit left.

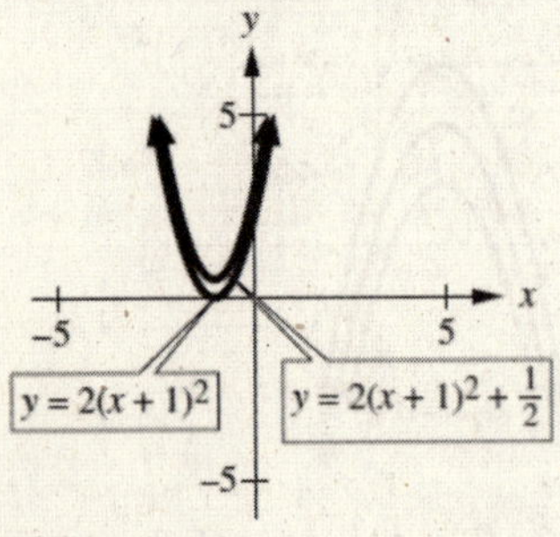

17. $y=x^2+2x+1$

1. Vertex: $x=\frac{-b}{2a}=\frac{-2}{2\cdot 1}=\frac{-2}{2}=-1$

Substitute $x=-1$ to find y

$y=(-1)^2+2(-1)+1=1-2+1=0$

$V(-1, 0)$

2. If $x=0$, $y=1$; y-intercept: (0, 1)
3. By symmetry, (–2, 1) is on the graph.
4. If $y=0$, $0=x^2+2x+1$

$0=(x+1)(x+1)$

$x=-1$ (repeated)

x-intercept: (–1, 0)

5. Since $a>0$, the parabola opens upward.

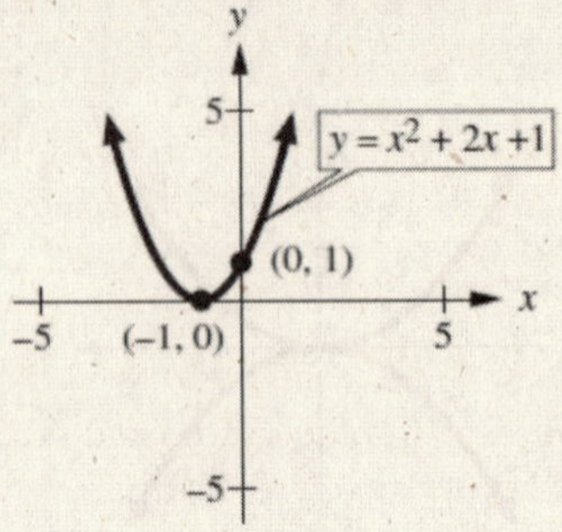

19. $y=-x^2+2x+1$

1. Vertex: $x=\frac{-b}{2a}=\frac{-2}{2(-1)}=\frac{-2}{-2}=1$

Substitute $x=1$ to find y

$y=-1^2+2(1)+1=-1+2+1=2$

$V(1, 2)$

2. If $x=0$, $y=1$; y-intercept: (0, 1)
3. By symmetry, (2, 1) is on the graph.
4. If $y=0$,

$0=-x^2+2x+1$

$0=x^2-2x-1$

$$x=\frac{-(-2)\pm\sqrt{(-2)^2-4\cdot 1(-1)}}{2\cdot 1}$$

$$=\frac{2\pm\sqrt{4+4}}{2}=\frac{2\pm 2\sqrt{2}}{2}=1\pm\sqrt{2}$$

x-intercept: $(1+\sqrt{2},0),(1-\sqrt{2},0)$

5. Since $a<0$, the parabola opens downward.

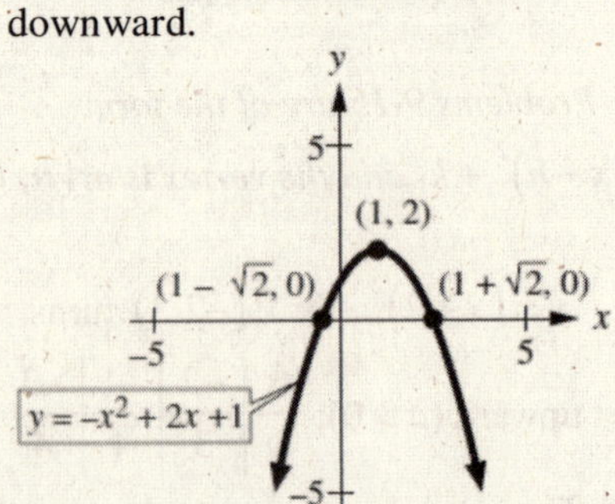

21. $y=-x^2+4x-5$

1. Vertex: $x=\dfrac{-b}{2a}=\dfrac{-4}{2(-1)}=\dfrac{-4}{-2}=2$

   Substitute $x=2$ to find y

   $y=-2^2+4(2)-5=-4+8-5=-1$

   $V(2,-1)$
2. If $x=0$, $y=-5$; y-intercept: $(0,-5)$
3. By symmetry, $(4,-5)$ is on the graph.
4. If $y=0$,

   $0=-x^2+4x-5$

   $0=x^2-4x+5$

   x^2-4x+5 is not factorable and the vertex is below the x-axis. This means that the equation has no solution and there are no x-intercepts; the graph does not cross the x-axis.
5. Since $a<0$, the parabola opens downward.

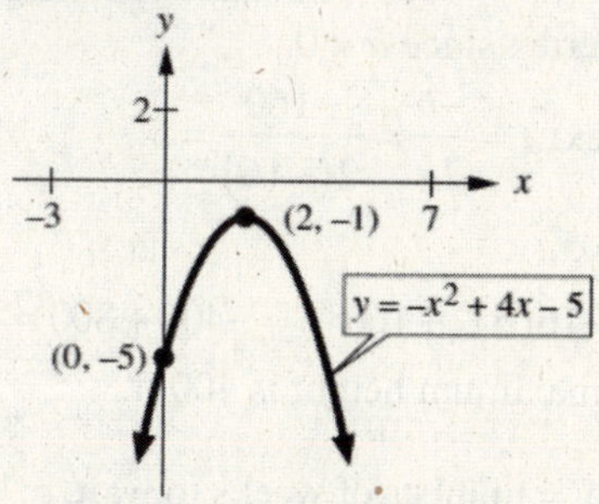

23. $y=3-5x+2x^2=2x^2-5x+3$

1. Vertex: $x=\dfrac{-b}{2a}=\dfrac{-(-5)}{2\cdot 2}=\dfrac{5}{4}$

   Substitute $x=\dfrac{5}{4}$ to find y

   $y=2\left(\dfrac{5}{4}\right)^2-5\left(\dfrac{5}{4}\right)+3=\dfrac{25}{8}-\dfrac{25}{4}+3$

   $=-\dfrac{1}{8}$

   $V\left(\dfrac{5}{4},-\dfrac{1}{8}\right)$
2. If $x=0$, $y=3$; y-intercept: $(0, 3)$
3. By symmetry, $\left(\dfrac{5}{2},3\right)$ is on the graph.
4. If $y=0$, $0=2x^2-5x+3$

   $0=(2x-3)(x-1)$

   $x=\dfrac{3}{2},\ x=1$

   x-intercepts: $\left(\dfrac{3}{2},0\right),(1,0)$
5. Since $a>0$, the parabola opens upward.

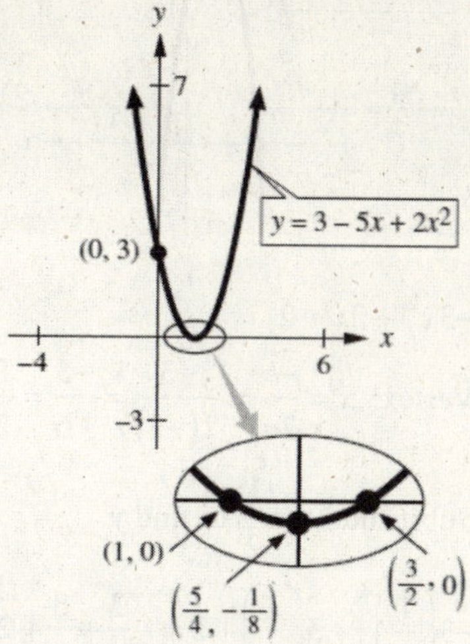

25. $y=5-4x-2x^2=-2x^2-4x+5$

1. Vertex: $x=\dfrac{-b}{2a}=\dfrac{-(-4)}{2(-2)}=\dfrac{4}{-4}=-1$

   Substitute $x=-1$ to find y

   $y=-2(-1)^2-4(-1)+5$

   $=-2+4+5=7$

   $V(-1, 7)$
2. If $x=0$, $y=5$; y-intercept: $(0, 5)$
3. By symmetry, $(-2, 5)$ is on the graph.
4. If $y=0$,

   $0=-2x^2-4x+5$

   $0=2x^2+4x-5$

   $x=\dfrac{-4\pm\sqrt{4^2-4\cdot 2(-5)}}{2\cdot 2}$

   $=\dfrac{-4\pm\sqrt{16+40}}{4}=\dfrac{-4\pm 2\sqrt{14}}{4}$

   $=\dfrac{2\pm\sqrt{14}}{2}\approx 0.9$ or -2.9

   x-intercept: $(-2.9,0),(0.9,0)$
5. Since $a<0$, the parabola opens downward.

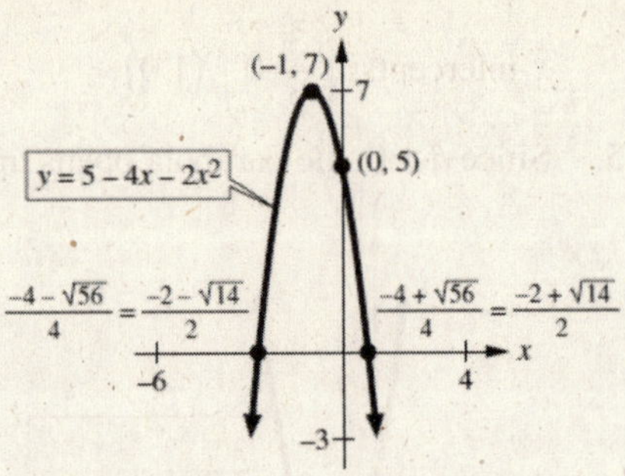

27. $y=-3x^2+3x+2$

1. Vertex: $x=\dfrac{-b}{2a}=\dfrac{-3}{2(-3)}=\dfrac{-3}{-6}=\dfrac{1}{2}$

   Substitute $x=\dfrac{1}{2}$ to find y

   $$y=-3\left(\frac{1}{2}\right)^2+3\left(\frac{1}{2}\right)+2=-\frac{3}{4}+\frac{3}{2}+2$$
   $$=\frac{11}{4}$$

   $V\left(\dfrac{1}{2},\dfrac{11}{4}\right)$
2. If $x = 0$, $y = 2$; y-intercept: (0, 2)
3. By symmetry, (1, 2) is on the graph.
4. If $y = 0$,
   $$0=-3x^2+3x+2$$
   $$x=\frac{-3\pm\sqrt{3^2-4\cdot(-3)(2)}}{2(-3)}$$
   $$=\frac{-3\pm\sqrt{9+24}}{-6}=\frac{-3\pm\sqrt{33}}{-6}$$
   $$\approx -0.5 \text{ or } 1.5$$
   x-intercept: $(-0.5,0),(1.5,0)$
5. Since $a < 0$, the parabola opens downward.

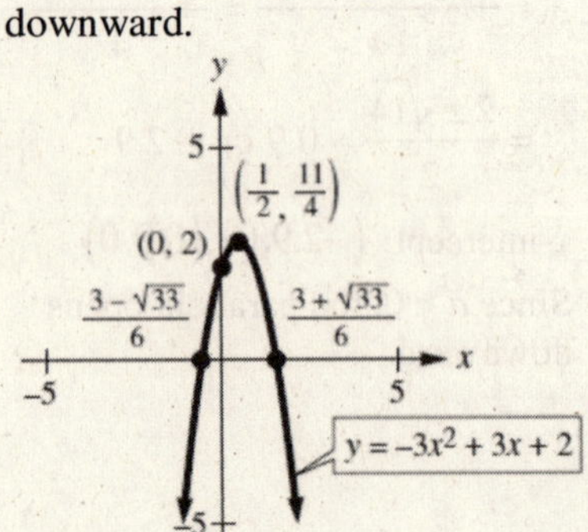

29. $P=-5000+8x-0.001x^2$ The maximum is at the vertex of the parabola since the graph curves downward ($a < 0$).

Vertex: $x=\dfrac{-b}{2a}=\dfrac{-8}{2(-0.001)}=4000$

If $x = 4000$,

$$P=-5000+8(4000)-0.001(4000)^2$$
$$=-5000+32{,}000-16{,}000=11{,}000$$

4000 items should be produced for maximum profit of \$11,000.

31. $N=50x-x^2$ The maximum is at the vertex of the parabola since the graph curves downward ($a < 0$).

Vertex: $x=\dfrac{-b}{2a}=\dfrac{-50}{2(-1)}=25$

Thus, \$25 thousand (\$25,000) should be spent to obtain maximum sales.

33. $h=-16t^2+160t$ The maximum height is at the vertex since $a < 0$.

Vertex: $t=\dfrac{-b}{2a}=\dfrac{-160}{2(-16)}=5$

If $t = 5$,

$$h=-16(5)^2+160\cdot 5=-400+800=400$$

The maximum height is 400 ft.

35. Let W = number of weeks to wait.
600 + 100W is the number of bushels dug at the end of W weeks.
1.00 – 0.10W = price per bushel at the end of W weeks.
If P is the price she'll be receiving, then

$$P=(600+100W)(1.00-0.10W)$$
$$=600W+100W-60W-10W^2$$
$$=600+40W-10W^2$$

The maximum she'll receive is at the vertex of the parabola since $a < 0$.

Vertex: $W=\dfrac{-b}{2a}=\dfrac{-40}{2(-10)}=2$. The maximum amount occurs when $W = 2$ weeks.

At $W = 2$, we have

$$P=600+40\cdot 2-10\cdot 2^2$$
$$=600+80-40=640$$

Hence, she'll receive the maximum amount, \$640, if she digs the potatoes at the end of 2 weeks.

37. a. $R = -\frac{1}{98}x^2 + \frac{6}{7}x$ The vertex is at:

$$x = \frac{-b}{2a} = \frac{-\frac{6}{7}}{2\left(-\frac{1}{98}\right)} = \frac{6}{7} \cdot \frac{49}{1} = 42 \text{ in.}$$

When $x = 42$,

$$R = -\frac{1}{98}(42)^2 + \frac{6}{7}(42) = 18$$

The vertex is (42, 18).

b. The maximum height is at the vertex of the parabola so $h = 18$ in.

c. Since the vertex is at $x = 42$, that's the midpoint of the parabola. The ending point will be 42 in. to the right of 42. Therefore the horizontal length is the distance from the beginning to the midpoint plus the distance from the midpoint to the end, that is, $42 + 42 = 84$ in.

d.

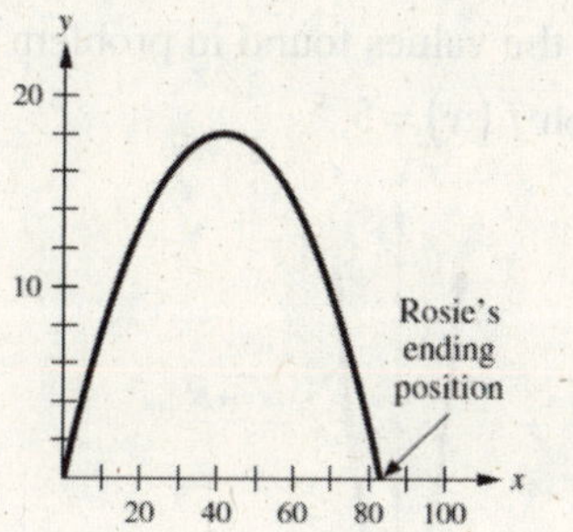

39. a. $d = -\frac{1}{400}x^2 + x$ The vertex is at:

$$x = \frac{-b}{2a} = \frac{-1}{2\left(-\frac{1}{400}\right)} = 1 \cdot \frac{200}{1} = 200 \text{ ft}$$

When $x = 200$,

$$d = -\frac{1}{400}(200)^2 + 200 = 100$$

The vertex is (200, 100).

b. The maximum height is at the vertex of the parabola so $h = 100$ ft.

c. The horizontal distance traveled is two times the x distance to the vertex or $2 \cdot 200 = 400$ ft.

d.

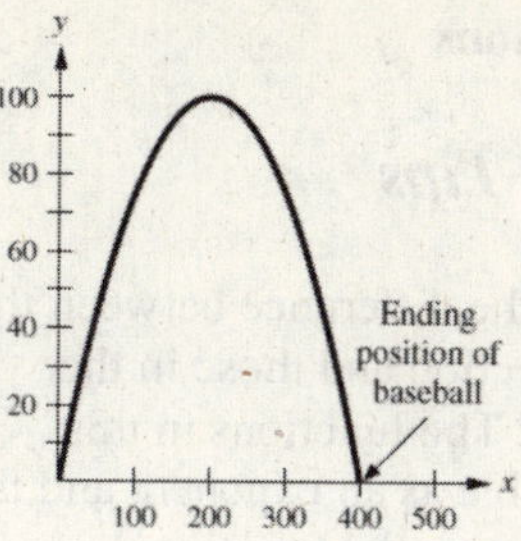

41. a. The lowest percent of low-birth-weight babies for mother who do not smoke is 6.3%, which occurs at the approximate age of 27 yr.

b. The highest percent of low-birth-weight babies for mother who do not smoke is 10.5%.

c. The approximate age of mothers who smoke that have the lowest percentage of low-birth-weight babies is 22 years.

d. The approximate age of mothers who do not smoke that have the highest percentage of low-birth-weight babies is 16 years.

43. Answers may vary. When $|a| > 1$, it is narrower. When $|a| < 1$, it is wider.

45. Answers may vary. It opens down. If it were to open up from (1, 1) it would have no x-intercept.

47. A parabola that would represent cardiovascular disease would open downward, so a is negative.

49. The coordinates for the vertex is approximately (1962, 520).

51. The profit for an item costing \$20 is expressed as $x - 20$. The profit for selling $60 - x$ items is $P = (60 - x)(x - 20) = -x^2 + 80x - 1200$.

Find the x-value of the vertex:

$$x = \frac{-b}{2a} = \frac{-80}{2(-1)} = 40$$

The price to maximize profits is \$40.

53. Let $x = 40$:

$$P = (60 - x)(x - 20)$$
$$= (60 - 40)(40 - 20)$$
$$= (20)(20) = 400$$

The maximum profit is \$400.

Section 7.5 Exponential and Logarithmic Functions

Study Tips

What is the difference between the functions in this section and those in the preceding section? The functions in this section use a variable as an exponent and thus they are called exponential functions.

To graph the function $f(x) = b^x$:
Think: curve that goes through (0, 1)
Note: When $b > 1$ the function is *increasing*, that is, the graph goes *up* from *left to right.* When $0 < b < 1$ the function is *decreasing*; the graph goes *down* from *left to right.*

Fact

$f(x) = b^x$ and $f(x) = b^{-x} = \frac{1}{b^x}$ are *symmetric* graphs, that is, reflections of each other across the *y-axis*.

Caution

In $f(x) = b^x$ the variable b must not equal 1.

To define a logarithm:
Think: a logarithm is an exponent, $b > 0$ and $b \neq 1$
Note: $y = \log_b x$ means that $b^y = x$. Note that y in the 2nd equation is an exponent!

Fact

An exponential function and a logarithmic function are *inverses* of each other.

Formulas for exponential applications:

1. Continuous compounding, growth: $A = Pe^{rt}$
2. n compoundings per year: $A = P\left(1 + \frac{r}{n}\right)^{nt}$
3. Half-life of a substance: $A(t) = A_0 e^{-kt}$
4. Rule of 72: $\frac{72}{r}$. If rate is 6%, $r = 6$.

To solve for t:
Think: take the logarithm of both sides
Note: Remember that for all real numbers k, $\ln e^k = k$. Also, $\ln e = 1$.

7.5 Exercises

1. Given 5^x:

   a. When $x = -1$, $5^{-1} = \frac{1}{5}$

   b. When $x = 0$, $5^0 = 1$

   c. When $x = 1$, $5^1 = 5$

3. Given 3^t:

   a. When $x = -2$, $3^{-2} = \frac{1}{3^2} = \frac{1}{9}$

   b. When $x = 0$, $3^0 = 1$

   c. When $x = 2$, $3^2 = 3 \cdot 3 = 9$

5. Given $10^{t/2}$:

   a. When $x = -2$, $10^{-2/2} = 10^{-1} = \frac{1}{10}$

   b. When $x = 0$, $10^{0/2} = 10^0 = 1$

   c. When $x = 2$, $10^{2/2} = 10^1 = 10$

7. a. Use the values found in problem 1 to graph $f(x) = 5^x$.

   b. Use the values found in problem 2 to graph $f(x) = 5^{-x}$.

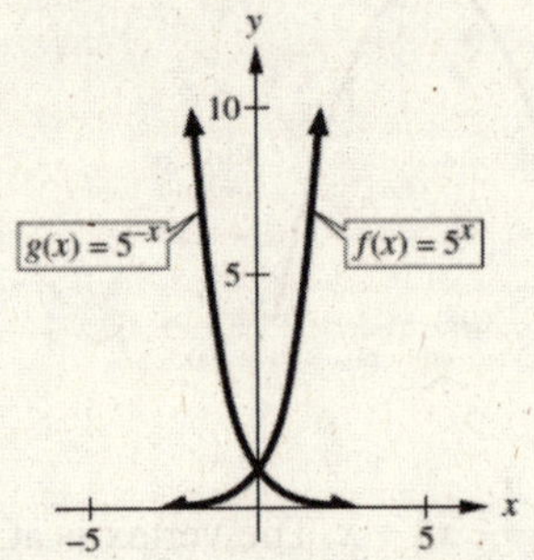

9. a. When $x = -1$, 0 and 1 the corresponding values are $10^{-1} = \frac{1}{10}$, 1 and 10 as shown in the graph.

   b. When $x = -1$, 0 and 1 the corresponding values are $10^{-(-1)} = 10$, 1 and $10^{-1} = \frac{1}{10}$ as shown in the graph.

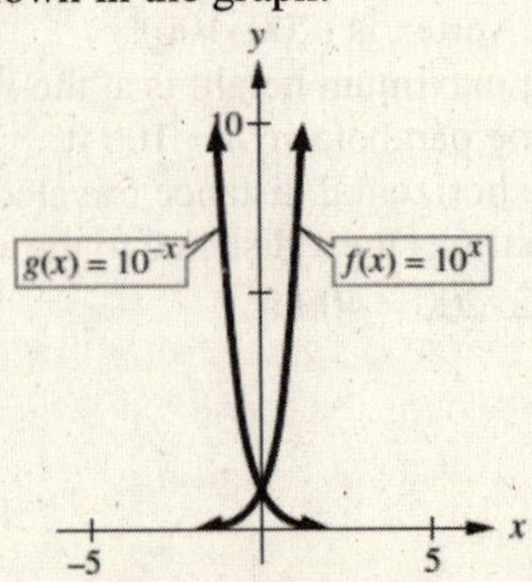

11. a. $A = Pe^{rt} = 1000e^{0.09(10)} = 1000e^{0.9}$
$= 1000(2.4596) = 2459.60$
The compounded amount is \$2459.60.

b. $A = P\left(1+\frac{r}{n}\right)^{nt} = 1000\left(1+\frac{0.09}{4}\right)^{4(10)}$
$= 1000\left(\frac{4.09}{4}\right)^{40} = 1000(1.0225)^{40}$
$= 1000(2.43519) = 2435.19$
The amount compounded quarterly is \$2435.19.

13. a. $A = Pe^{rt} = 1000e^{0.06(10)} = 1000e^{0.6}$
$= 1000(1.82212) = 1822.12$
The compounded amount is \$1822.12.

b. $A = P\left(1+\frac{r}{n}\right)^{nt} = 1000\left(1+\frac{0.06}{4}\right)^{4(10)}$
$= 1000\left(\frac{4.06}{4}\right)^{40} = 1000(1.015)^{40}$
$= 1000(1.81402) = 1814.02$
The amount compounded quarterly is \$1814.02.

15. $P = 2000(2^{0.2t})$ where t is the number of years after 2000.
a. In 2005, $t = 5$
$P = 2000(2^{0.2(5)}) = 2000(2^1) = 4000$
The population was 4000 in 2005.
b. In 2010, $t = 10$
$P = 2000(2^{0.2(10)}) = 2000(2^2) = 8000$
The population was 8000 in 2010.
c. In 2015, $t = 15$
$P = 2000(2^{0.2(15)}) = 2000(2^3) = 16{,}000$
The population was 16,000 in 2015.

17. $G = 2000e^{-1.05t}$ where t is in years.
a. At the start, $t = 0$
$G = 2000e^{-1.05(0)} = 2000e^0$
$= 2000 \cdot 1 = 2000$
2000 grams are present at the start.
b. In 1 year, $t = 1$
$G = 2000e^{-1.05(1)} = 2000e^{-1.05}$
$= 2000(0.349938) = 699.876$
699.9 grams are present in 1 year.
c. In 2 years, $t = 2$
$G = 2000e^{-1.05(2)} = 2000e^{-2.10}$
$= 2000(0.122456) = 244.912$
244.9 grams are present in 2 years.

19. $P = P_0e^{kt}$ P is the population in t years
P_0 is the initial population
k the annual growth rate $(B - D)$
t the number of years after 2000
Find k: $B - D = 0.0234 - 0.003 = 0.0204$
$P_0 = 50.5$ million, $t = 10$
$P = 50.5e^{0.0204(10)} = 50.5e^{0.204} \approx 61.93$
The number of Hispanics for the year 2020 is about 61.93 million.

21. $C = 941e^{0.0053t}$ where t is in months.
a. Total credit = C = 1000 billion
$1000 = 941e^{0.0053t}$
$\frac{1000}{941} = e^{0.0053t}$
$\ln\left(\frac{1000}{941}\right) = \ln e^{0.0053t}$
$\ln\left(\frac{1000}{941}\right) = 0.0053t$
$t = \frac{\ln\left(\frac{1000}{941}\right)}{0.0053} \approx 11.5$
The total credit will read one trillion dollars about 11.5 months later, or in December 2008.
b. December 2007 total credit doubled is $2 \cdot 941 = 1882$.
$1882 = 941e^{0.0053t}$
$2 = e^{0.0053t}$
$\ln 2 = \ln e^{0.0053t}$
$\ln 2 = 0.0053t$
$t = \frac{\ln 2}{0.0053} \approx 131$
Twice the December 2007 total credit will be in 131 months, or in 2018.

23. a. $N(t)=500{,}000\left(\frac{2}{3}\right)^t$

$N(1)=500{,}000\left(\frac{2}{3}\right)^1=333{,}333$

333,333 cans are still in use after 1 year.

b. $N(2)=500{,}000\left(\frac{2}{3}\right)^2=500{,}000\cdot\frac{4}{9}$

$=222{,}222$

After 2 years 222,222 cans are still in use.

c. $N(10)=500{,}000\left(\frac{2}{3}\right)^{10}$

$=500{,}000\left(\frac{1024}{59049}\right)\approx 8671$

8671 cans are still in use after 10 years.

25. $A(a)=14.7(10)^{-0.000018a}$, where a is in feet.

a. Let a = 29,000 ft.

$A(29{,}000)=14.7(10)^{-0.000018(29{,}000)}$

$=14.7(10)^{-0.522}\approx 4.42$

The atmospheric pressure at the top of Mt. Everest is 4.42 lb/in^2.

b. Let a = 20,000 ft.

$A(20{,}000)=14.7(10)^{-0.000018(20{,}000)}$

$=14.7(10)^{-0.36}\approx 6.42$

The atmospheric pressure at the top of Mt. McKinley is 6.42 lb/in^2.

27. The graph of $f(x)=\log_5 x$ can be obtained by reflecting the graph of $y=5^x$ along the line $y=x$ as shown below.

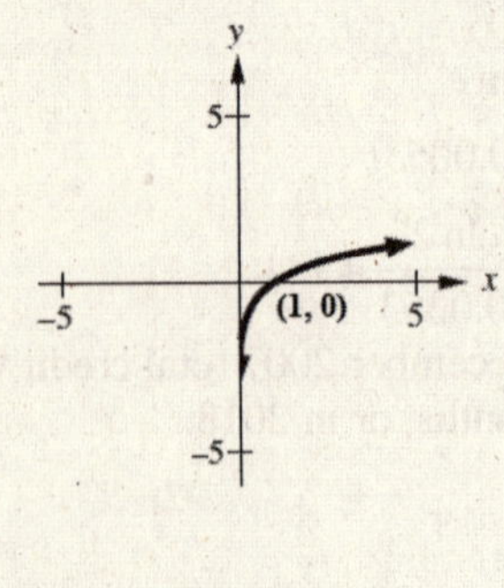

29. If A is double P and $r=5\%$, find t:

$2P=Pe^{0.05t}$

$2=e^{0.05t}$

$\ln 2=0.05t(\ln e)$

$\ln 2=0.05t$

$t=\frac{\ln 2}{0.05}=\frac{0.693}{0.05}\approx 13.863$

It takes about 13.86 years to double.

31. If A is double P and $r=6.5\%$, find t:

$2P=Pe^{0.065t}$

$2=e^{0.065t}$

$\ln 2=0.065t(\ln e)$

$\ln 2=0.065t$

$t=\frac{\ln 2}{0.065}=\frac{0.693}{0.065}\approx 10.66$

It takes about 10.66 years to double.

33. The population $P=6.1e^{rt}$, P is in billions of people, t the number of years since 2000 and r is the rate of growth. If $r=1.2\%=0.012$ and the year is 2010, $t=10$, then

$P=6.1e^{0.012(10)}=6.1e^{0.12}\approx 6.88$

The population in 2010 will be about 6.88 billion.

35. $B=1000e^{0.04t}$

$2000=1000e^{0.04t}$

$2=e^{0.04t}$

$\ln 2=0.04t(\ln e)$

$\ln 2=0.04t$

$t=\frac{\ln 2}{0.04}\approx\frac{0.693147}{0.04}\approx 17.3287$

It will take about 17.3 minutes.

37. $B=1000e^{0.04t}$

$25{,}000=1000e^{0.04t}$

$25=e^{0.04t}$

$\ln 25=0.04t(\ln e)$

$\ln 25=0.04t$

$t=\frac{\ln 25}{0.04}\approx\frac{3.218876}{0.04}\approx 80.4719$

It will take about 80.5 minutes.

39. a. $B = 100{,}000e^{-0.2t}$, when $t = 0$,

$B = 100{,}000e^{-0.2(0)} = 100{,}000e^{0}$

$= 100{,}000 \cdot 1 = 100{,}000$

b. When $t = 2$,

$B = 100{,}000e^{-0.2(2)} = 100{,}000e^{-0.4}$

$= 100{,}000(0.67032) \approx 67{,}032$

c. When $t = 10$,

$B = 100{,}000e^{-0.2(10)} = 100{,}000e^{-2}$

$= 100{,}000(0.135335) \approx 13{,}534$

d. When $t = 20$,

$B = 100{,}000e^{-0.2(20)} = 100{,}000e^{-4}$

$= 100{,}000(0.018316) \approx 1832$

41. If $A(t) = \frac{1}{2}A_0$ and $k = 0.003\% = 0.00003$, find t:

$\frac{1}{2}A_0 = A_0 e^{-0.00003t}$

$\frac{1}{2} = e^{-0.00003t}$

$\ln\left(\frac{1}{2}\right) = -0.00003t(\ln e)$

$\ln(0.5) = -0.00003t$

$t = \frac{\ln(0.5)}{-0.00003} = \frac{-0.69315}{-0.00003}$

$\approx 23{,}105$

Plutonium's half-life is 23,105 years.

43. If $A(t) = \frac{1}{2}A_0$ and $k = 5.2\% = 0.052$, find t:

$\frac{1}{2}A_0 = A_0 e^{-0.052t}$

$\frac{1}{2} = e^{-0.052t}$

$\ln\left(\frac{1}{2}\right) = -0.052t(\ln e)$

$\ln(0.5) = -0.052t$

$t = \frac{\ln(0.5)}{-0.052} = \frac{-0.69315}{-0.052} \approx 13.3$

Its half-life is 13.3 years.

45. $P = 14.7e^{-0.00005h}$, h is the altitude in feet, P is the atmospheric pressure.

a. If $h = 0$,

$P = 14.7e^{-0.00005(0)}$

$= 14.7e^{0} = 14.7 \cdot 1 = 14.7 \text{ lb/in}^2$

b. If $h = 5000$,

$P = 14.7e^{-0.00005(5000)} = 14.7e^{-0.25}$

$= 14.7(0.7788) \approx 11.4 \text{ lb/in}^2$

c. If $h = 10{,}000$,

$P = 14.7e^{-0.00005(10{,}000)} = 14.7e^{-0.5}$

$= 14.7(0.606531) \approx 8.92 \text{ lb/in}^2$

47. $S = 0.94e^{0.09h}$ million dollars, t is the number of years after 2000.

a. If $S = 3$, $3 = 0.94e^{0.09t}$

$\frac{3}{0.94} = 0.94e^{0.09t}$

$\ln\left(\frac{3}{0.94}\right) = \ln e^{0.09t}$

$\ln\left(\frac{3}{0.94}\right) = 0.09t$

$t = \frac{\ln\left(\frac{3}{0.94}\right)}{0.09} \approx \frac{1.1605}{0.09} \approx 12.89$

The players' average salaries will reach \$3 million in about 13 years, or 2013.

b. In year 1, the average salary was about 1.0 million, and in year 6 about 1.7 million. The difference is $1.7 - 1.0 = 0.7$. This is a percent increase of $\frac{1.7-1.0}{1.0} = \frac{0.7}{1.0} = .70 = 70\%$.

c. Owners revenues have a percent increase of $\frac{8.88-6.49}{6.49} = \frac{2.39}{6.49} = 0.368 = 36.8\%$.

d. The players have a higher increase but note that the years as well as the year intervals are different.

49. $A = 5000e^{-0.04t}$ where t is the number of years after 1970.

a. In 1970, $t = 0$

$A = 5000e^{-0.04(0)} = 5000e^{0} = 5000$

There were about 5000 aviation accidents in 1970.

b. In 1990, $t = 20$

$$A = 5000e^{-0.04(20)} = 5000e^{-0.8}$$
$$= 5000(0.4493) \approx 2247$$

There were about 2247 aviation accidents in 1990.

c. If $A = 1000$,

$$1000 = 5000e^{-0.04t}$$
$$0.2 = e^{-0.04t}$$
$$\ln 0.2 = -0.04t(\ln e)$$
$$-1.6094 = -0.04t$$
$$t = \frac{-1.6094}{-0.04} \approx 40$$

In 2010, 40 years after 1970, the number of aviation accidents is predicted to be 1000.

51. $B = 50{,}000e^{0.2t}$ where t is the number of hours after 12 noon.

a. At noon, $t = 0$

$$B = 50{,}000e^{0.2(0)} = 50{,}000e^{0}$$
$$= 50{,}000$$

There were 50,000 bacteria at noon.

b. At 2 p.m., $t = 2$

$$B = 50{,}000e^{0.2(2)} = 50{,}000e^{0.4}$$
$$= 50{,}000(1.4918) \approx 74{,}591$$

There were about 74,591 bacteria at 2 p.m.

c. At 6 p.m., $t = 6$

$$B = 50{,}000e^{0.2(6)} = 50{,}000e^{1.2}$$
$$= 50{,}000(3.3201) \approx 166{,}006$$

There were about 166,006 bacteria at 6 p.m.

53. $S = 1000e^{-0.1d}$ where d is the number of days after the campaign.

a. On the last day, $d = 0$

$$S = 1000e^{-0.1(0)} = 1000e^{0} = 1000$$

There were 1000 sales on the last day of the campaign.

b. 10 days after the campaign, $d = 10$

$$S = 1000e^{-0.1(10)} = 1000e^{-1}$$
$$= 1000(0.3679) \approx 368$$

There were about 368 sales 10 days after the campaign.

55. $C = 100\left(1 - e^{-0.5t}\right)$ where t is the number of hours.

a. $t = 0$

$$C = 100\left(1 - e^{-0.5(0)}\right) = 100\left(1 - e^{0}\right)$$
$$= 100(1-1) = 100 \cdot 0 = 0$$

At $t = 0$ the concentration is 0.

b. $t = 1$

$$C = 100\left(1 - e^{-0.5(1)}\right) = 100\left(1 - e^{-0.5}\right)$$
$$= 100(1 - 0.6065) = 100(0.3935) \approx 39.4$$

At $t = 1$ the concentration is 39.4.

57. $M = -2.5\log\left(\frac{B}{B_0}\right)$ where B is the brightness of the star and B_0 is the minimum of brightness.

a. $B = 2.1B_0$

$$M = -2.5\log\left(\frac{2.1B_0}{B_0}\right) = -2.5\log 2.1 \approx -0.8055$$

The stellar magnitude of the North Star is –0.8055.

b. $B = 36.2B_0$

$$M = -2.5\log\left(\frac{36.2B_0}{B_0}\right) = -2.5\log 36.2 \approx -3.8968$$

The stellar magnitude of Venus is –3.8968.

59. For monthly compounding, use $A = P\left(1 + \frac{r}{n}\right)^{nt}$, where P is the principal, r the rate of interest, n the number of periods and t the number of years.

For continuous compounding, use $A = Pe^{rt}$, where P is the principal, r the rate of interest and t the time in years.

Monthly:

$$A = 1000\left(1 + \frac{0.06}{12}\right)^{12(10)} = 1000(1.005)^{120}$$
$$= 1000(1.8194) = 1819.40$$

Compounded monthly, the amount is $1819.40.

Continuous:

$$A = 1000e^{0.06(10)} = 1000e^{0.6}$$
$$= 1000(1.82212) = 1822.12$$

Compounded continuously, the amount is

$1822.12.
In 10 years, continuous compounding yields $2.72 more than monthly compounding.

61. Answers may vary.
   a. $f(x)=1^x=1$, a horizontal line.
   b. Yes, a horizontal line is a function.

63. Answers may vary. The graphs are symmetric to the y-axis.

Skill Checker

1. To graph $x+y=4$, let $x=0$, then $y=4$. For $y=0$, $x=4$. Graph the points (0, 4) and (4, 0) and join them with a line as shown on the graph below.

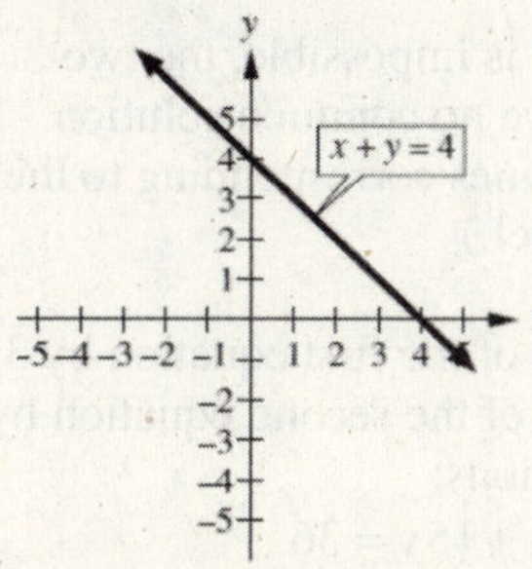

3. To graph $2x-y=5$, let $x=0$, then $y=-5$. For $y=0$, $2x=5$ which means that $x=2.5$. Graph the points (0, –5) and (2.5, 0) and join them with a line as shown on the graph below.

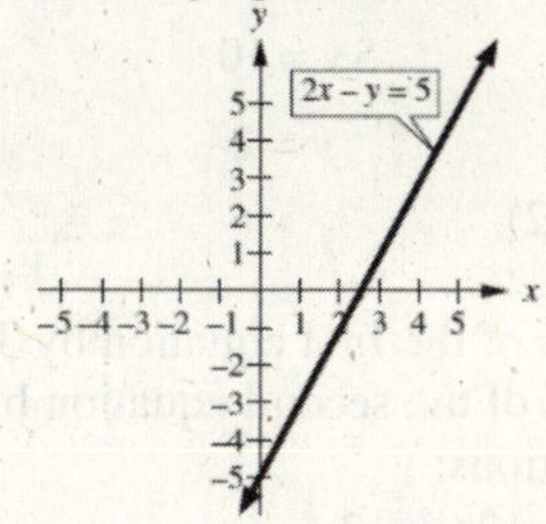

Section 7.6 Two Linear Equations in Two Variables

Study Tips

We continue to develop skills necessary to answer real-world questions by looking at systems of two linear equations.

To solve a system of two equations by graphing:
Think: graph both, where do the graphs intersect?
Note: The point of intersection satisfies both equations and thus, is a solution of both equations.

To solve a system by elimination:
Think: need same variable in both equations to be opposites, add equations, solve for remaining variable, substitute into either equation to solve for other variable
Note: Multiply one or both equations by numbers that cause the resulting coefficients for one of the variables to be of equal magnitude and opposite in sign then add the two equations to eliminate that variable.

To solve a system by substitution:
Think: in 1st equation isolate a variable on one side, substitute into 2nd equation, solve for that variable, substitute into 1st equation to solve for other variable
Note: To isolate a variable, choose the one that is the easiest algebraically to get by itself.

Types of solutions for systems of two equations:
1. *one solution*: an ordered pair of numbers satisfying both equations
2. *no solution*: a contradiction, for example $0=7$ or $-3=5$
3. *infinitely many solutions*: a true statement, for example $7=7$ or $-4=-4$

7.6 Exercises

1. Two points on the line $x+y=3$ are (0, 3) and (3, 0). Graph these points and draw a line through them. The line $2x-y=0$ passes through the origin, (0, 0), and through the point (2, 4). Graph these points and draw a line through them. The graph shows that the lines intersect at (1, 2). Check that this point satisfies both equations: $1+2=3$ and $2-2=0$. Thus, the solution is (1, 2).

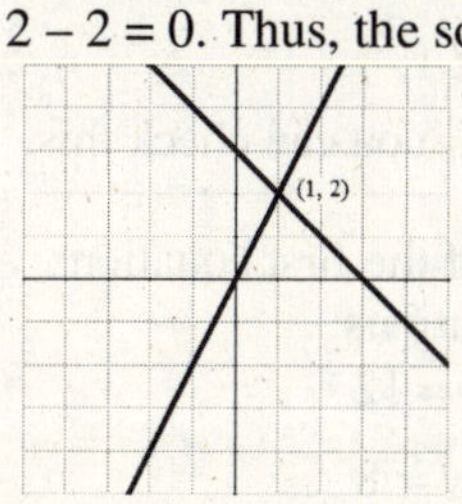

3. Two points on the line $2x-y=10$ are (0, –10) and (5, 0). Graph these points and draw a line through them. Two points on the line $3x+2y=1$ are (1, –1) and (5, –7). Graph these two points and draw a line through them. The graph shows that the lines intersect at

(3, –4). Check that this point satisfies both equations: (2)(3) – (–4) = 6 + 4 = 10 and (3)(3) + (2)(–4) = 9 – 8 = 1. The solution is (3, –4).

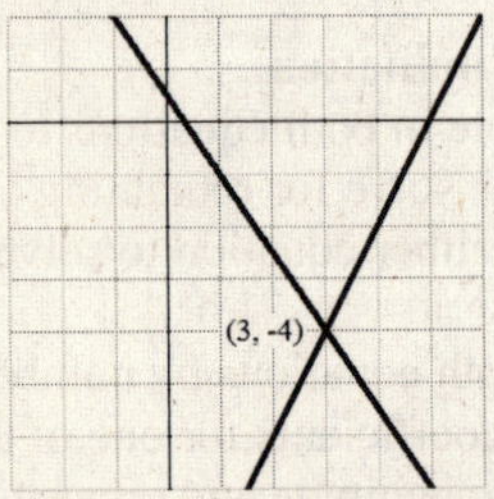

5. Two points on the line $3x + 4y = 4$ are (0, 1) and (4, –2). Graph these points and draw a line through them. Two points on the line $2x - 6y = 7$ are (7/2, 0) and (–1, –3/2). Graph these two points and draw a line through them. The graph shows that the two lines intersect at the point (2, –1/2). Check that this point satisfies both equations: (3)(2) + (4)(–1/2) = 6 – 2 = 4 and (2)(2) – (6)(–1/2) = 4 + 3 = 7. The solution is (2, –1/2).

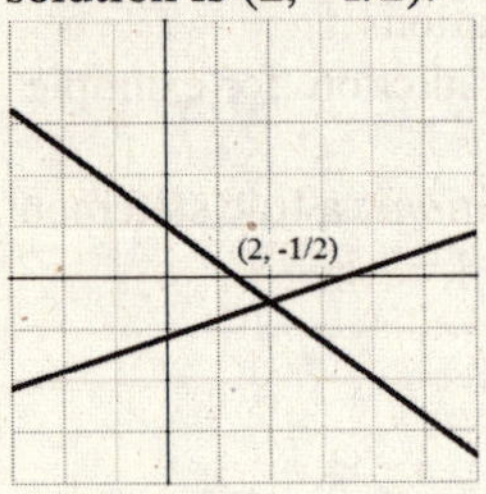

7. $x + y = 3$
$2x - y = 0$
$3x = 3$ Add the equations.
$x = 1$ Divide by 3.
$1 + y = 3$ Substitute into 1st eq.
$y = 2$ Subtract 1.
The solution is (1, 2). You can check this.

9. Multiply both sides of the first equation by 2 and add the equations:
$x + y = 6 \to 2x + 2y = 12$
$3x - 2y = 8 \to 3x - 2y = 8$
Add: $5x = 20$
Divide by 5: $x = 4$
Substitute into 1st eq: $4 + y = 6$
Subtract 4: $y = 2$
The solution is (4, 2).

11. Multiply both sides of the first equation by 2 and add the equations:
$2x - y = 10 \to 4x - 2y = 20$
$3x + 2y = 1 \to 3x + 2y = 1$
Add: $7x = 21$
Divide by 7: $x = 3$
Substitute into 1st eq: $6 - y = 10$
Solve: $y = -4$
The solution is (3, –4).

13. Multiply both sides of the first equation by –3 and add the equations:
$5x + y = 4 \to -15x - 3y = -12$
$15x + 3y = 8 \to 15x + 3y = 8$
Add: $0 = -4$
Since the last result is impossible, the two given equations have no common solution. (Note that the two lines corresponding to these equations are parallel.)

15. Multiply both sides of the first equation by 3, multiply both sides of the second equation by 5 and add the equations:
$2x + 5y = 12 \to 6x + 15y = 36$
$5x - 3y = -1 \to 25x - 15y = -5$
Add: $31x = 31$
Divide by 31: $x = 1$
Substitute into 1st eq: $2 + 5y = 12$
Subtract 2: $5y = 10$
Divide by 5: $y = 2$
The solution is (1, 2).

17. Multiply both sides of the first equation by 3, multiply both sides of the second equation by 2 and add the equations:
$3x + 4y = 4 \to 9x + 12y = 12$
$2x - 6y = 7 \to 4x - 12y = 14$
Add: $13x = 26$
Divide by 13: $x = 2$
Substitute into 1st eq: $6 + 4y = 4$
Subtract 6: $4y = -2$
Divide by 4: $y = -\frac{1}{2}$

The solution is $\left(2, -\frac{1}{2}\right)$.

19. Write the first eqution in standard form. Multiply both sides of the first equation by 2, multiply both sides of the second equation by –3 and add the equations:

$$11x+3y=-3 \rightarrow 22x+6y=-6$$
$$5x+2y=5 \rightarrow -15x-6y=-15$$

Add: $7x = -21$

Divide by 7: $x=-3$

Substitute into 1st eq: $-33+3y=-3$

Add 33: $3y=30$

Divide by 3: $y=10$

The solution is (–3, 10).

21. $x = 2y-3$

$x=-2y-1$

$2x = -4$ Add the equations.

$x=-2$ Divide by 2.

$-2=2y-3$ Substitute into 1st eq.

$1=2y$ Add 3.

$y=\frac{1}{2}$ Divide by 2.

The solution is $\left(-2, \frac{1}{2}\right)$.

23. Multiply both sides of the first equation by –2 and add the equations:

$$2x+3y+11=0 \rightarrow -4x-6y-22=0$$
$$5x+6y+20=0 \rightarrow 5x+6y+20=0$$

Add: $x-2=0$

Add 2: $x=2$

Substitute into 1st eq: $4+3y+11=0$

Subtract 15: $3y=-15$

Divide by 3: $y=-5$

The solution is (2, –5).

25. Multiply both sides of the second equation by 6 and add the equations:

$$3x-12y=-8 \rightarrow 3x-12y=-8$$
$$2x+2y=3 \rightarrow 12x+12y=18$$

Add: $15x = 10$

Divide by 15: $x=\frac{2}{3}$

Substitute into 1st eq: $2-12y=-8$

Subtract 2: $-12y=-10$

Divide by -12: $y=\frac{5}{6}$

The solution is $\left(\frac{2}{3}, \frac{5}{6}\right)$.

27. Multiply both sides of the first equation by –2 and add the equations:

$$r-4s=-10 \rightarrow -2r+8s=20$$
$$2r-8s=13 \rightarrow 2r-8s=13$$

Add: $0=33$

Since the last result is impossible, the two given equations have no common solution. (Note that the two lines corresponding to these equations are parallel.)

29. Multiply both sides of the first equation by 3, multiply both sides of the second equation by 2 and add the equations:

$$6u-2v=-27 \rightarrow 18u-6v=-81$$
$$4u+3v=8 \rightarrow 8u+6v=16$$

Add: $26u = -65$

Divide by 26: $u=-\frac{5}{2}$

Substitute into 1st eq: $-15-2v=-27$

Add 15: $-2v=-12$

Divide by -2: $v=6$

The solution is $u=-\frac{5}{2}$, $v=6$; or $\left(-\frac{5}{2}, 6\right)$.

31. a. Multiply both sides of the first equation by –1 and add the equations:

A: $y=8x+1000 \rightarrow -y=-8x-1000$

B: $y=10x+800 \rightarrow y=10x+800$

Add: $0=2x-200$

Add 200: $200=2x$

Divide by 2: $100=x$

The costs are equal for 100 guests.

b. The first equation is satisfied by (0, 1000) and (200, 2600). The line through these points is the graph of the first equation. The second equation is satisfied by (0, 800) and by (200, 2800). The line through these points is the graph of the second equation. These lines are shown below.

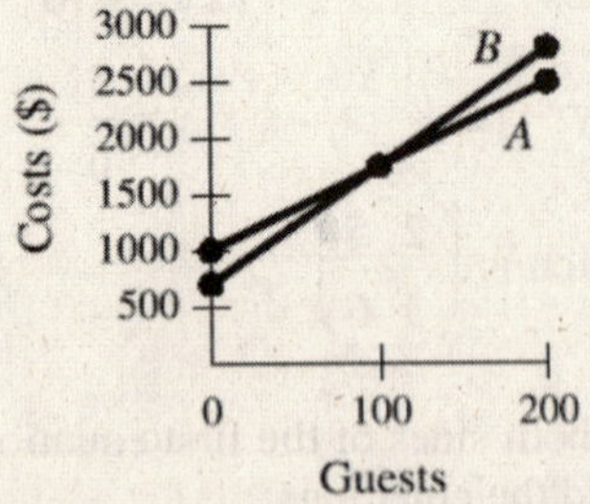

c. The graph shows that company A is cheaper for more than 100 guests.

33. a. The amount of saturated fats when you eat one McDonald's and one family-style restaurant burger is $x + y = 22$.
b. The amount of saturated fats for two Quarter Pounders is $P_Q = 2x$.
c. The amount of saturated fats for one family-style restaurant burgeris $P_F = y$.
d. The total grams consumed in two Quarter Pounders and one family-style restaurant burger is $2x + y = 30$.
e. The intersection of the two lines provides the answer. McDonald's has 8 grams saturated fat and the family-style restaurant has 14 grams.

35. a. The amount of saturated fats when you eat one McDonald's and one Wendy's Jr. Bacon Cheeseburger is $x + y = 15$.
b. The amount of saturated fats for two Quarter Pounders is $P_Q = 2x$.
c. The amount of saturated fats for one Wendy's Jr. Bacon Cheeseburger is $P_B = y$.
d. The total grams consumed in two Quarter Pounders and one Wendy's Jr. Bacon Cheeseburger is $2x + y = 23$.
e. The intersection of the two lines provides the answer. McDonald's has 8 grams saturated fat and the Wendy's Jr. Bacon Cheeseburger has 7 grams.

37. a. $C = 20 + 35m$ b.

m	6	12	18
C	230	440	650

c.

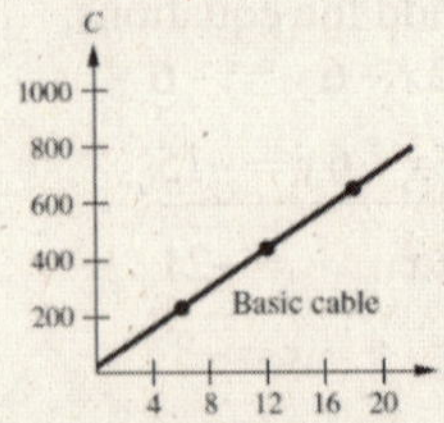

39. Use the table of values from problems 37 and 38 to plot the two graphs on the same coordinate axes.

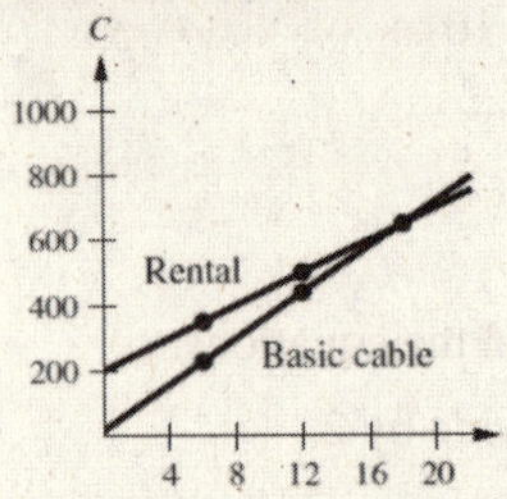

41. DVD player and rental option is cheaper if used for more than 18 months.

43. a. $W = 100 + 3t$

b.

t	5	10	15	20
W	115	130	145	160

c. Plot the four points (5, 115), (10, 130), (15, 145) and (20, 160) and join them with a line as shown on the graph:

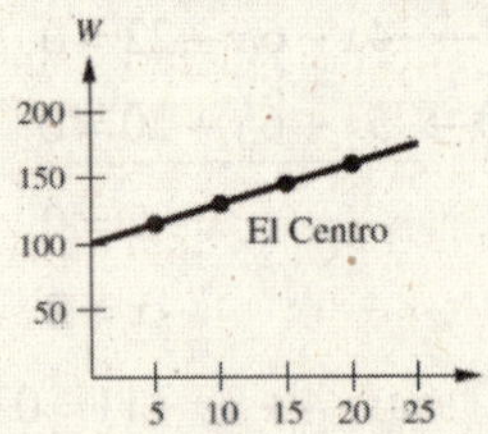

45. Cannon: $y = 100 + 80x$
Brother: $y = 130 + 70x$
Substitute $130 + 70x$ for y in the first equation and solve for x.
Substitute for y: $130 + 70x = 100 + 80x$
Simplify: $-10x = -30$
Divide by -10: $x = 3$

Substitute into 1st eq: $y = 100 + 80(3)$

Simplify: $y = 340$

The solution is $(3, 340)$.

47. Total cost for Cannon (C): $y = 100 + 80x$

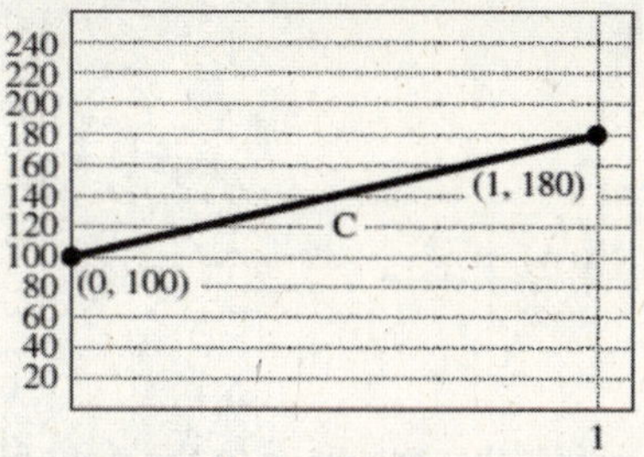

49. a. Cannon has the lower y-intercept and so has the less expensive start-up costs.
    b. The 1-year cost for Cannon is \$180, and for Epson \$220.
    c. If you use Cannon, your savings are 220 – 180 = \$40.

51. Total cost for Lexmark (L): $y = 60 + 150x$

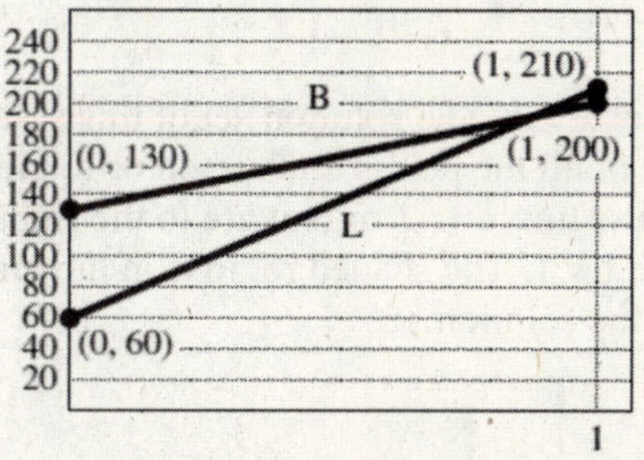

53. The Epson printer has the highest total cost for one-year. This cost is \$220.

55. Answers may vary. For a system of two linear equations, the lines will be parallel.

57. Answers may vary. The lines will be parallel. The system has no solution.

59. Answers may vary. Since the slopes are different, the lines will intersect. Thus, there will be one solution.

Section 7.7 Linear Inequalities

Study Tips

We learn about linear inequalities in preparation of using the real-world application of linear programming.

To determine a corresponding linear equation: Think: replace <, >, ≤, or ≥ by an = sign

Note: For $y < 2x + 5$, the corresponding linear equation is $y = 2x + 5$.

To graph a linear inequality: *Think*: graph corresponding linear equation, < or >: dashed line, ≤ or ≥: solid line, shade

Note: To determine which region to shade, select a test point not on the line [(0,0) is the easiest], and see if the test point satisfies the inequality. If it does, select all points on that side of the line. If it does not, select the points on the other side of the line.

Hint

Solve your inequality for *y*. If you have $y < \ldots$, then you will shade *below* the line. If you have $y > \ldots$, then you will shade *above* the line. You can always confirm by using a test point.

To solve a systems of inequalities by graphing: *Think*: graph both inequalities, solution is where there is double shading

Note: The boundary line will only be part of the solution if it is a solid line.

7.7 Exercises

1. $x + 2y \geq 2$: (0, 1) and (2, 0) are points on the line $x + 2y = 2$. Draw the line through these points. The point (0, 2) makes $x + 2y = 4 > 2$, so the required region is above the line as shown on the graph. Note that the line is drawn solid to show that all the points on it satisfy the given inequality.

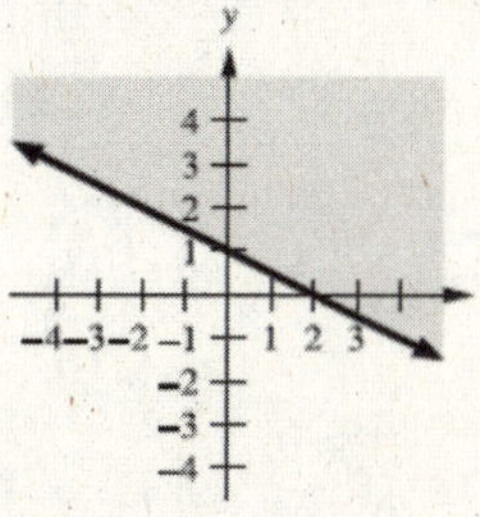

3. $x \leq 4$: Draw the line $x = 4$. Since (0, 0) satisfies the inequality, all the points in the region to the left of the line $x = 4$ also satisfy the inequality. Thus, the region to the left of

this line is shaded. The solution set consists of all the points on the line and in the shaded region.

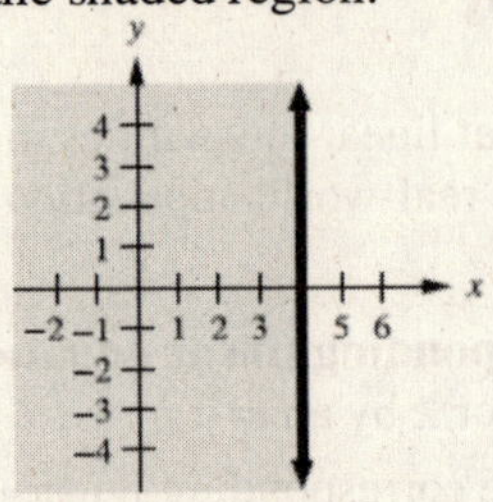

5. $3x - y < 6$: (2, 0) and (0, –6) are two points on the line $3x - y = 6$. Since the points on this line are not in the solution set of the given inequality, draw a dashed line through these points. The point (0, 0) satisfies the inequality, so all points on the same side of (above) the line as the origin form the required solution set. The shaded region in the figure is the graph of the solution set.

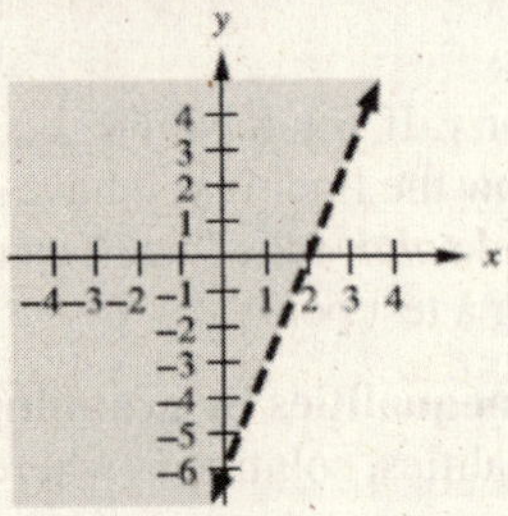

7. $2x + y \leq 4$: (2, 0) and (0, 4) are two points on the line $2x + y = 4$. Draw a solid line through these points. (All the points on this line are part of the solution set.) The point (0, 0) satisfies the inequality $2x + y < 4$, so all points on the same side of (above) the line as the origin also satisfy this inequality. So the region to the left of $2x + y \leq 4$ is shaded. The solution set consists of all the points on the line and in the shaded region.

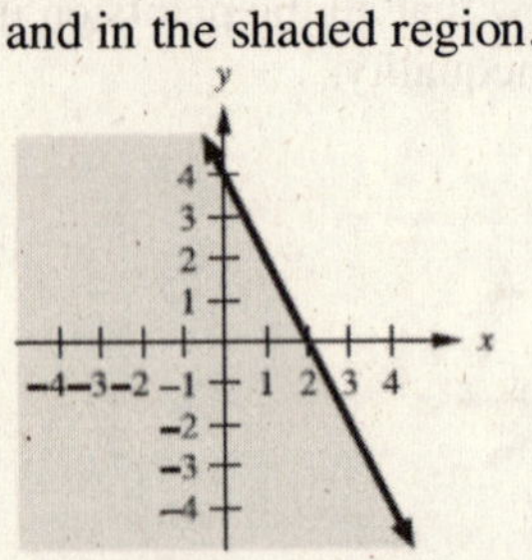

9. $4x + y > 8$: (2, 0) and (0, 8) are two points on the line $4x + y = 8$. Draw a dashed line through these two points, as the points on this line do not satisfy the given inequality. The point (0, 0) does not satisfy the inequality, so the points on the side of the line opposite the origin (right of the line) do satisfy. Shade the region to the right of the line, the required solution set as shown in the figure.

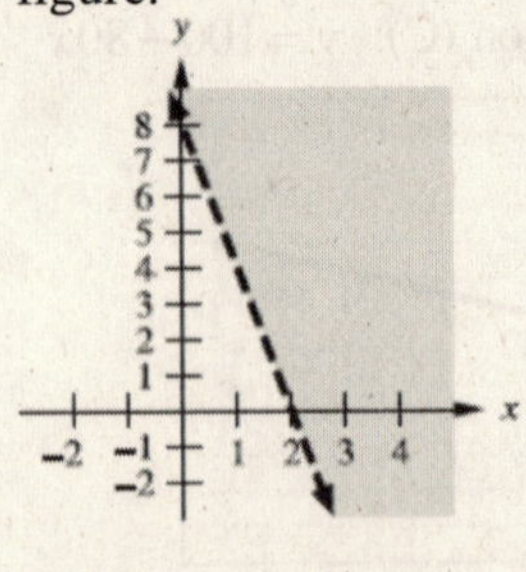

11. All the points that are on or to the right of the line $x = -4$ and also to the left of the line $x = 3$ satisfy this system. The shaded region in the graph shows the solution set.

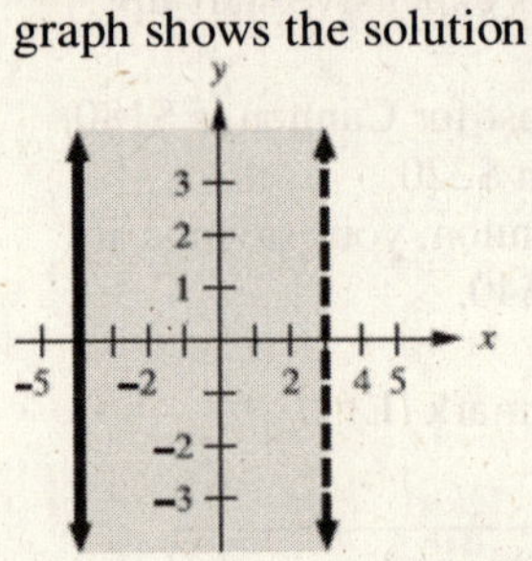

13. $x \leq -1$ or $x > 1$. The solution set of this system consist of all the points that are on or to the left of the line $x = -1$ or that are to the right of the line $x = 1$. The shaded regions in the graph shows the solution set.

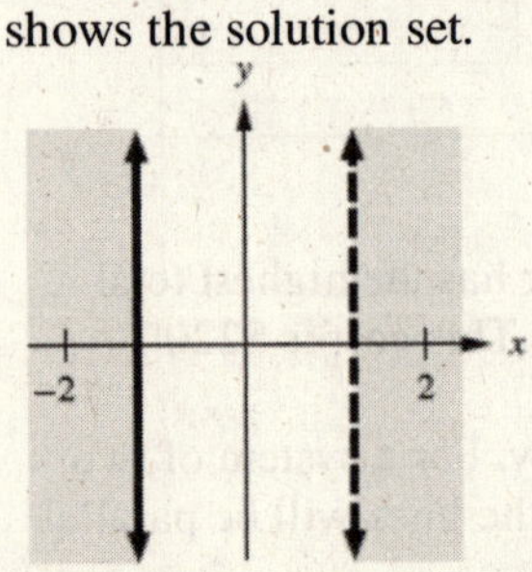

15. $x - y \geq 2,\ x + y \leq 6$. (2, 0) and (0, –2) are two points on the line $x - y = 2$. Draw a line through these two points. For (0, 0), $x - y < 2$, so all points on the side of the line opposite the origin satisfy $x - y \geq 2$. (6, 0) and (0, 6) are two points on the line $x + y = 6$. Draw a line through these two points. For (0, 0), $x + y < 6$ so all the points on the same sides of this line as the origin satisfy $x + y \leq 6$. Thus, the shaded region along with all points on the two lines below their point of intersection is the graph of the given system.

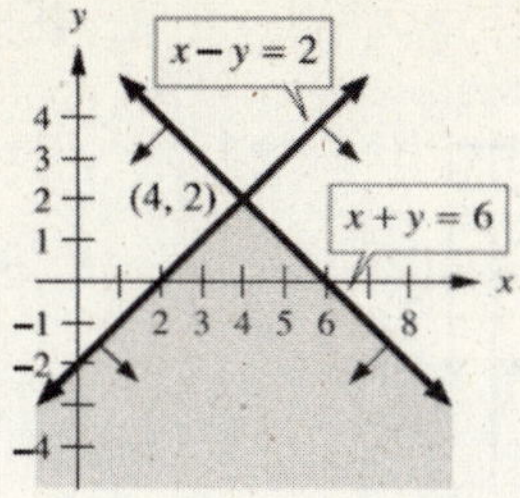

17. $2x - 3y \le 6,\ 4x - 3y \ge 12$. (3, 0) and (0, –2) are points on the line $2x - 3y = 6$. Draw a solid line through these points. Now, the point (0, 0) makes $2x - 3y = 0 < 6$, so all points on the same side of this line as the origin make $2x - 3y < 6$. (3, 0) and (0, –4) are points on the line $4x - 3y = 12$. Draw a solid line through these points. The point (0, 0) makes $4x - 3y = 0$, which is less than 12, so all points on the opposite side of this line from the origin make $4x - 3y > 12$. The shaded region, along with its boundaries, is the graph of the solution set of the given system.

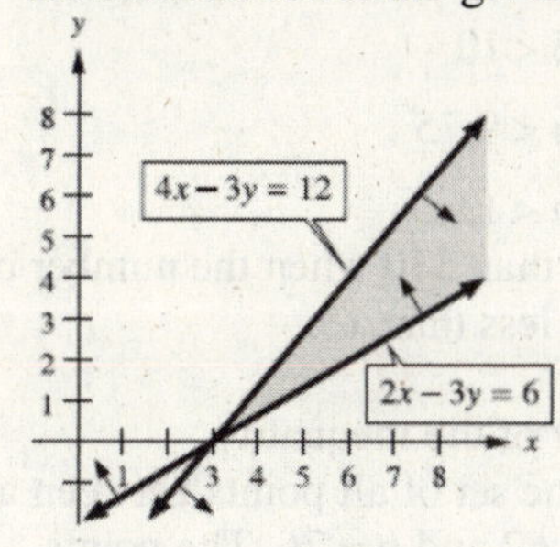

19. $2x - 3y \le 5, x \ge y, y \ge 0$. (4, 1) and (2.5, 0) are points on the line $2x - 3y = 5$. Draw a solid line through these points. (0, 0) makes $2x - 3y = 0 < 5$, so the origin side of $2x - 3y = 5$ is the region where $2x - 3y < 5$. (0, 0) and (2, 2) are points on the line $x = y$. At the point (1, 0), $x > y$; so the region below the line $x = y$ is where $x > y$. Draw a solid line a through these points. The line $y = 0$ is the x-axis. The region above the x-axis is where $y > 0$. The shaded region, along with its boundaries, is the graph of the solution set of this system.

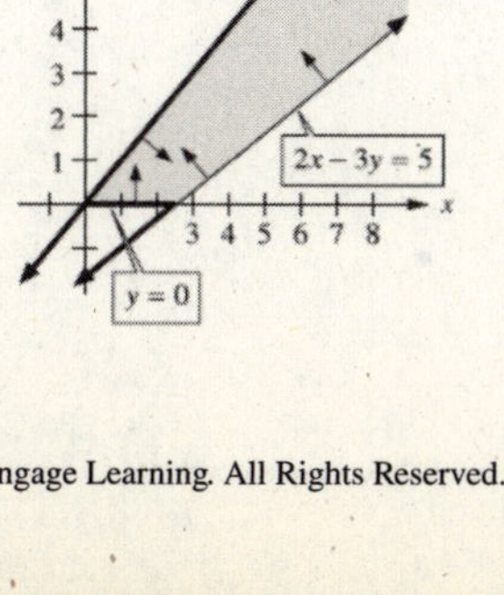

21. $x + 3y \le 6, x \ge 0, y \ge 0$. (0, 2) and (6,0) are points on the line $x + 3y = 6$. Draw a solid line through these points. (0, 0) makes $x + 3y = 0 <$ 6, so the origin side of (below) this line is where $x + 3y < 6$. The system $x \ge 0, y \ge 0$ describes the first quadrant along with its boundaries. The shaded region, along with its boundaries, is the graph of the given system of inequalities.

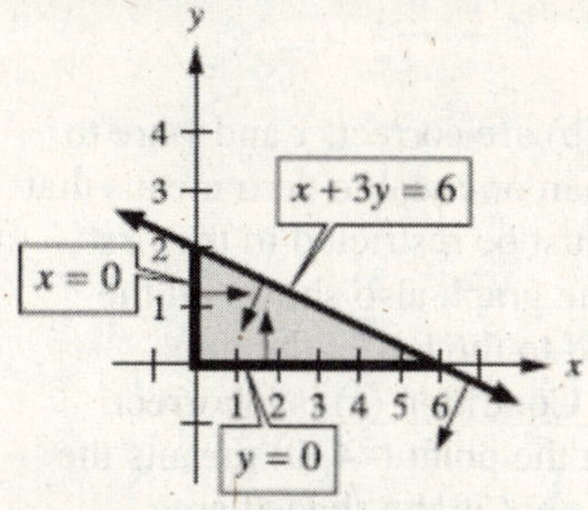

23. $x \ge 1, y \ge 1, x - y \le 1, 3y - x < 3$. The region to the right of the line $x = 1$ and above the line $y = 1$ and its boundaries is the graph of the solution set of the system $x \ge 1$ and $y \ge 1$. The points (1, 0) and (2, 1) are on the line $x - y = 1$. Draw a solid line through these points. (0, 0) makes $x - y \le 1$ true, so the same side as that point is the region where $x - y \le 1$. The points (0, 1) and (3, 2) are on the line $3y - x = 3$. Draw a dashed line through these points. (0, 0) makes $3y - x < 3$ true, so the same side as that point is the region where $3y - x < 3$.The shaded region, along with the solid line portions of the boundary, is the graph of the solution set of the given system of inequalities.

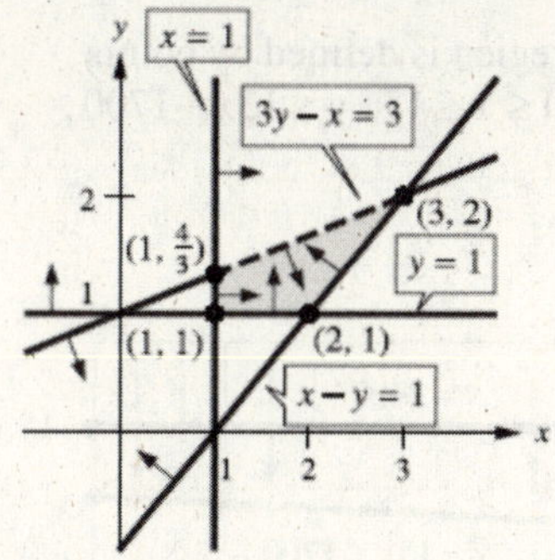

25. $x + y \ge 1, x \le 2, y \ge 0, y \le 1$. (1, 0) and (0, 1) are points on the line $x + y = 1$. (0, 0) makes $x + y = 0 < 1$, so all points in the region opposite the origin make $x + y > 1$. All points in the region to the left of $x = 2$, above the x-axis, and below the line $y = 1$ are in the solution set of the system $x \le 2, y \ge 0, y \le 1$. The shaded region, including its boundaries is the graph of the solution set of the given system of inequalities.

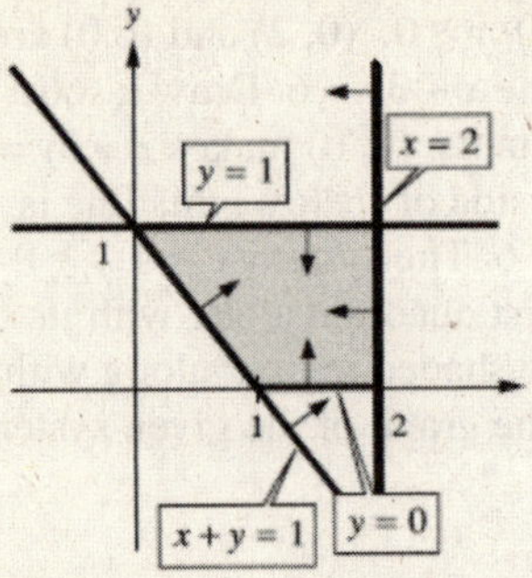

27. Conditions (b) are correct. x and y are to be greater than or equal to zero means that the points must be restricted to the first quadrant. The graph also shows all the points on and to the left of the line $3x + 2y = 6$. Condition (a) is incorrect. For example the point (–4, 0) meants the criteria but is not in the shaded area.

29. Conditions (a) are correct. The point on or to the right of $x = 2$ or on or above the line $y = 1$. Conditions (b) are incorrect. For example, the point (3, 0) is in the shaded area but does not meet the criteria.

31. The line $x + y = 1$ goes through (1, 0) and (0, 1). The condition $x + y > 1$ is true only for the points to the right of this line. The line $x - y = 1$ goes through the points (1, 0) and (0, –1). The condition $x - y < 1$ is true only for the points to the left of this line. The common region is that shown in (c).

33. The yellow region is defined by points that satisfy $0 \le x \le 15$, $y \le 13x + 1700$, and $y \ge 0$.

35. a. and b.

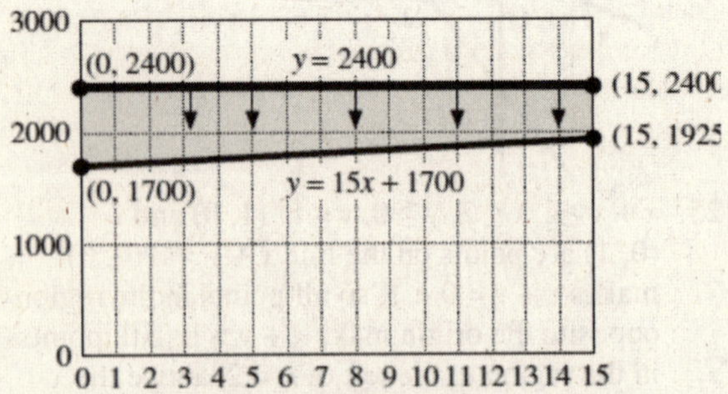

c. The region on and under the line $y = 2400$ and above and on the line $y = 15x - 1700$.

37.

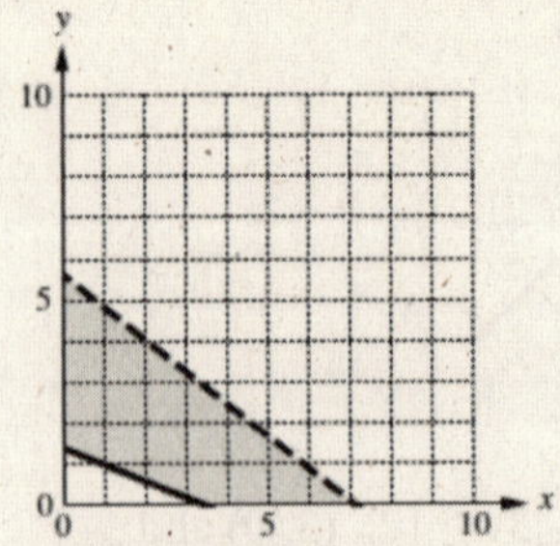

Answers may vary. Two ordered pairs that satisfy this system are (1, 3) and (2, 3).

39. $20n > 12n + 160{,}000$

$$8n > 160{,}000$$

$$n > 20{,}000$$

More than 20,000 units must be sold to make a profit.

41. $1 + 0.75(h - 1) < 10$

$$1 + 0.75h - 0.75 < 10$$

$$0.75h + 0.25 < 10$$

$$0.75h < 9.75$$

$$h < 13$$

The cost is less than \$10 when the number of hours parked is less than 13.

43. The solution set of the inequality $62 \le h \le 76$ is the set of all points between and on the lines $h = 62$ and $h = 76$. The points (62,108) and (76, 164) are on the line $w = 4h - 140$. In the figure a solid line is drawn joining these points. The points (62,134) and (76, 204) are on the line $w = 5h - 176$. A solid line is drawn joining these two points. The point (70, 150), which is inside the region enclosed by the four lines that have been drawn makes $w > 4h - 140$, since $150 > (4)(70) - 140 = 280 - 140 = 140$ and $w < 5h - 176$, since $150 < (5)(70) - 176 = 350 - 176 = 174$. This shows that the desired region is that enclosed by the four line segments drawn in the figure. Note that the boundaries are also included in the graph of the solution set of the given inequalities.

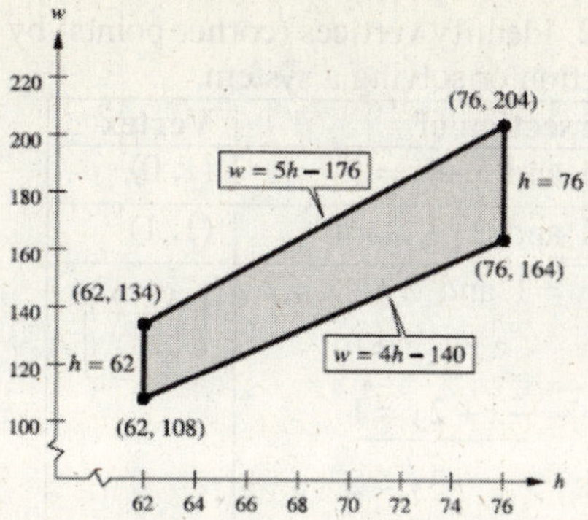

45. Answers may vary.
Suppose $c \neq 0$.
Step 1. Find the intercepts of the line $ax + by = c$.
Step 2. Draw a dashed line through these intercepts.
Step 3. Substitute (0, 0) into the equation. This gives zero for the left side.
Step 4. If $c > 0$, shade the region opposite the origin. If $c < 0$, shade the region on the origin side of the line.
Suppose $c = 0$.
Step 1. Draw the line $ax + by = 0$. This line goes through the origin and the point $(b, -a)$.
Step 2. Substitute (a, b) into the left side. This gives $a^2 + b^2 > 0$, so shade the region that is on the same side of the line as the point (a, b).

47. Answers may vary. The graph would show $x = k$ as a solid line with the region to the right of this line as the shaded region.

49.

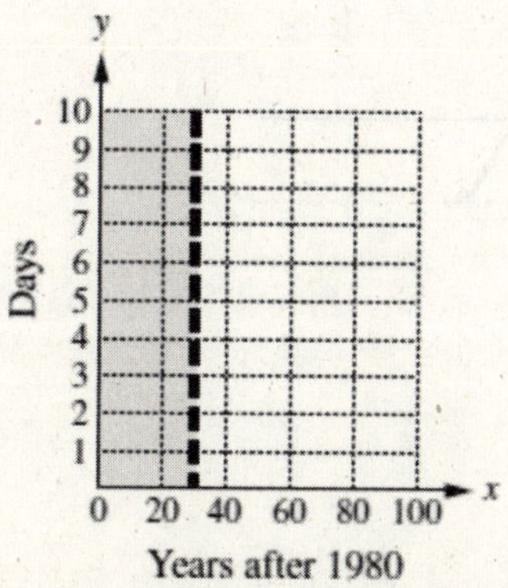

51. a. The cost for rental B is $C = 36 + 0.15m$ where m is the number of miles driven. If we want C to be \$41, we have to solve the equation

$$41 = 36 + 0.15m$$
$$5 = 0.15m$$
$$m = 33\frac{1}{3} \text{ or about 33 miles}$$

b. If you are planning to drive 100 miles, the cost for rental A is \$41. For rental B the cost is $C = 36 + 0.15(100) = \$51$, so A is cheaper.

53. Rental B is cheaper (less than \$41) when you drive less than 33 miles. Rental A is cheaper when you drive more than 33 miles. If you drive exactly $33\frac{1}{3}$ miles, the cost is the same for both, \$41.

55. Answers will vary.

57. Leave everything as in the given flowchart except for interchanging the shading instructions at the end, so that if the answer to the question, "Is c > 0 ?" is "No" The half-plane not containing the origin will be shaded, and if the answer is "Yes", then the half-plane containing the origin will be shaded. You can check this by substituting (0, 0) in the inequality.

Section 7.8 Linear Programming

Study Tips

Linear programming is used to solve the real-world problem of determining the best decision in order either maximize or minimize an outcome.

To solve a linear programming problem:
Think: make table of information, identify constraints, graph them, find corner points, evaluate objective function, make decision
Note: Only at a corner point will you be able to find the maximum and minimum of the objective function.

Become familiar with the language of linear programming:
Objective function: expression to be maximized or minimized
Constraints: system of inequalities that indicate restrictions to your decision
Feasible region: all the possible solutions that satisfy the constraints

Corner points: vertices of the feasible region that are the possible points that will either maximize or minimize the objective function

<u>Hint</u>
Making a chart of the information will help you to determine the constraint inequalities.

7.8 Exercises

1. **Step 1**: Graph the constraints and shade the feasible region.

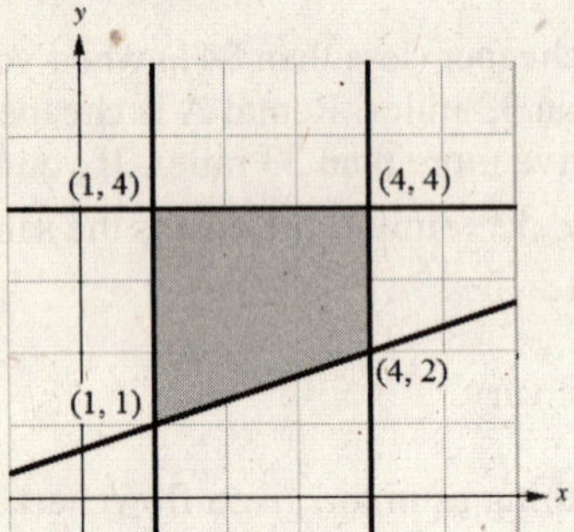

Step 2: Identify vertices (corner points) by inspection or solving a system.

Intersection of	Vertex
$x=1$ and $y=4$	(1, 4)
$x=4$ and $y=4$	(4, 4)
$x=1$ and $x-3y=-2$	(1, 1)
$x=4$ and $x-3y=-2$	(4, 2)

Step 3: Evaluate corner points in the objective function.

Vertex	$C = 2x + y$	Min/Max
(1, 4)	$2(1)+4=6$	
(4, 4)	$2(4)+4=12$	
(1, 1)	$2(1)+1=3$	Minimum
(4, 2)	$2(4)+2=10$	

The minimum value of $C = 2x + y$ is 3 at (1, 1).

3. **Step 1**: Graph the constraints and shade the feasible region.

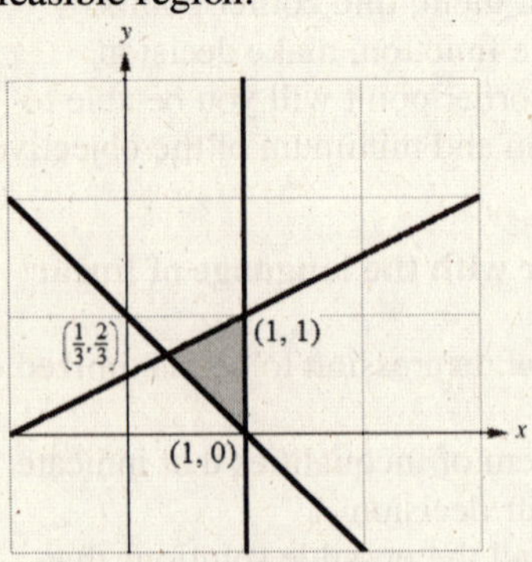

Step 2: Identify vertices (corner points) by inspection or solving a system.

Intersection of	Vertex
$x=1$ and $x+y=1$	(1, 0)
$x=1$ and $2y-x=1$	(1, 1)
$x+y=1$ and $2y-x=1$ $x+y=1$ $\underline{-x+2y=1}$ Add: $3y=2$ Divide by 3: $y=\frac{2}{3}$ Substitute into 1st eq: $x+\frac{2}{3}=1$ Subtract $\frac{2}{3}$: $x=\frac{1}{3}$	$\left(\frac{1}{3}, \frac{2}{3}\right)$

Step 3: Evaluate corner points in the objective function.

Vertex	$W = 4x + y$	Min/Max
(1, 0)	$4(1)+0=4$	
(1, 1)	$4(1)+1=5$	
$\left(\frac{1}{3}, \frac{2}{3}\right)$	$4\left(\frac{1}{3}\right)+\frac{2}{3}=2$	Minimum

The minimum value of $W = 4x + y$ is 2 at (1/3, 2/3).

5. **Step 1**: Graph the constraints and shade the feasible region.

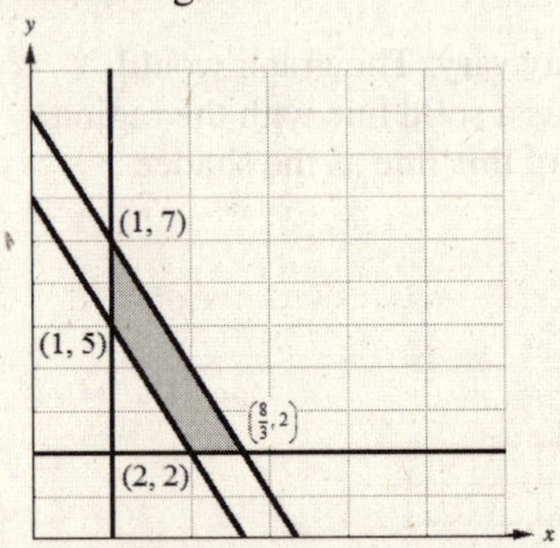

Step 2: Identify vertices (corner points) by inspection or solving a system.

Intersection of	Vertex
$3x+y=10$ and $x=1$	(1, 7)
$3x+y=8$ and $x=1$	(1, 5)
$3x+y=8$ and $y=2$	(2, 2)
$3x+y=10$ and $y=2$ Let $y=2$: $3x+2=10$ Subtract 2: $3x=8$ Divide by 3: $x=\frac{8}{3}$	$\left(\frac{8}{3}, 2\right)$

Step 3: Evaluate corner points in the objective function.

Vertex	$C = x + 2y$	Min/Max
(1, 7)	$1 + 2(7) = 15$	
(1, 5)	$1 + 2(5) = 11$	
(2, 2)	$2 + 2(2) = 6$	Minimum
$\left(\frac{8}{3}, 2\right)$	$\frac{8}{3}+2(2)=\frac{20}{3}$	

The minimum value of $C = x + 2y$ is 6 at $(2, 2)$.

7. **Step 1**: Graph the constraints and shade the feasible region.

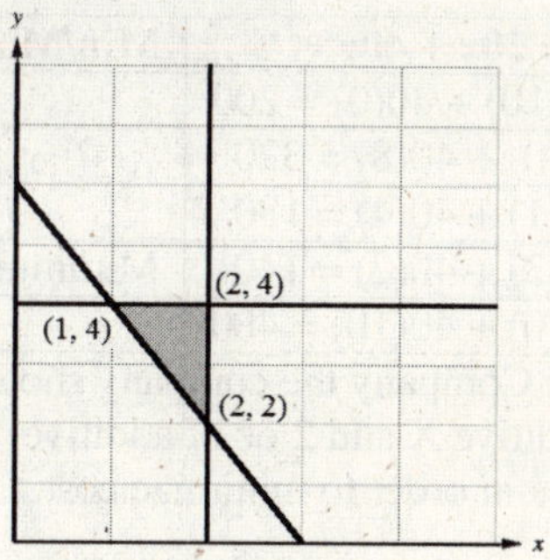

Step 2: Identify vertices (corner points) by inspection or solving a system.

Intersection of	Vertex
$2x+y=6$ and $y=4$	(1, 4)
$2x+y=6$ and $x=2$	(2, 2)
$x=2$ and $y=4$	(2, 4)

Step 3: Evaluate corner points in the objective function.

Vertex	$P = x + 2y$	Min/Max
(1, 4)	$1 + 2(4) = 9$	
(2, 2)	$2 + 2(2) = 6$	
(2, 4)	$2 + 2(4) = 10$	Maximum

The maximum value of $P = x + 2y$ is 10 at (2, 4).

9. **Step 1**: Identify the constraints.
Let x = the number of cars
y = the number of trucks

	Cars (x)	Trucks (y)
Space	$100x$	$200y$

$x + y \le 100$

$100x + 200y \le 12{,}000$ $\qquad x \ge 0,\ y \ge 0$

Step 2: Graph the constraints and shade the feasible region.

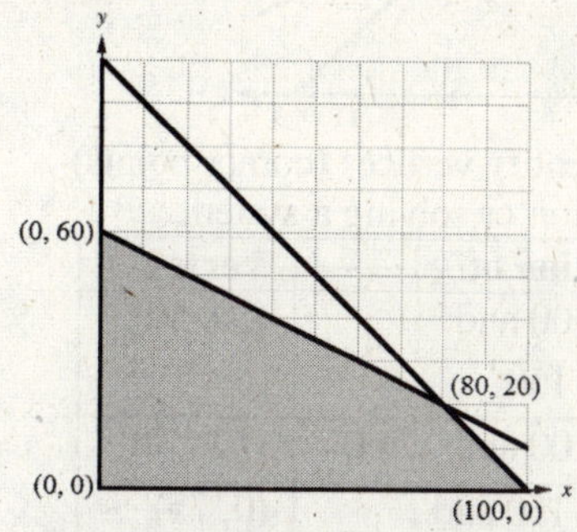

Step 3: Identify vertices (corner points) by inspection or solving a system.

Intersection of	Vertex
$x+y=100$ and $100x+200y=12{,}000$	(80, 20)
$x+y=100$ and $y=0$	(100, 0)
$100x+200y=12{,}000$ and $x=0$	(0, 60)
$x=0$ and $y=0$	(0, 0)

Step 4: Evaluate corner points in the objective function.

Vertex	$P = 20x + 35y$	Min/Max
(80, 20)	$20(80) + 35(20) = 2300$	Maximum
(100, 0)	$20(100) + 35(0) = 2000$	
(0, 60)	$20(0) + 35(60) = 2100$	
(0, 0)	$20(0) + 35(0) = 0$	

Store 80 cars and 20 trucks to provide the maximum revenue.

11. **Step 1**: Identify the constraints.
Let x = the number of the first tablets
y = the number of the second tablets

	Tablet 1 (x)	Tablet 2 (y)
B_1	$1x$	$1y$
B_2	$1x$	$2y$

$x + y \le 100$

$x + 2y \le 150$

$x \ge 0,\ y \ge 0$

Step 2: Graph the constraints and shade the feasible region.

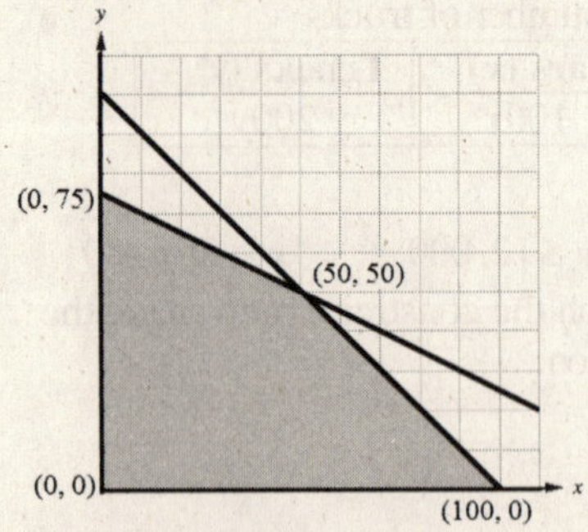

Step 3: Identify vertices (corner points) by inspection or solving a system.

Intersection of	Vertex
$x+y=100$ and $x+2y=150$	(50, 50)
$x+y=100$ and $y=0$	(100, 0)
$x+2y=150$ and $x=0$	(0, 75)
$x=0$ and $y=0$	(0, 0)

Step 4: Evaluate corner points in the objective function.

Vertex	$P = 2x + 3y$	Min/Max
(50, 50)	2(50) + 3(50) = 250	Maximum
(100, 0)	2(100) + 3(0) = 200	
(0, 75)	2(0) + 3(75) = 225	
(0, 0)	2(0) + 3(0) = 0	

The Kwik-Pep Vitamin Company should pack 50 tablets into each bottle in order to obtain the largest profit.

13. **Step 1**: Identify the constraints.
Let x = number of ounces of additive X
y = number of ounces of additive Y

	X (x)	Y (y)
A	$16x$	$4y$
B	$2x$	$2y$
C	$4x$	$14y$

$x+y\le 10$

$16x+4y\ge 32$

$2x+2y\ge 10$

$4x+14y\ge 40$

$x\ge 0,\ y\ge 0$

Step 2: Graph the constraints and shade the feasible region.

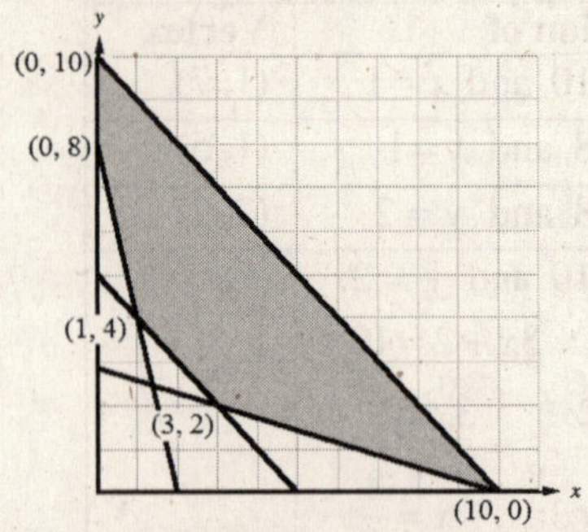

Step 3: Identify vertices (corner points) by inspection or solving a system.

Intersection of	Vertex
$x+y=10$ and $y=0$ or $4x+14y=40$ and $y=0$	(10, 0)
$16x+4y=32$ and $x=0$	(0, 8)
$16x+4y=32$ and $2x+2y=10$	(1, 4)
$2x+2y=10$ and $4x+14y=40$	(3, 2)
$x+y=10$ and $x=0$	(0, 10)

Step 4: Evaluate corner points in the objective function.

Vertex	$C = 20x + 40y$	Min/Max
(10, 0)	20(10) + 40(0) = 200	
(0, 8)	20(0) + 40(8) = 320	
(1, 4)	20(1) + 40(4) = 180	
(3, 2)	20(3) + 40(2) = 140	Minimum
(0, 10)	20(0) + 40(10) = 400	

The Jeri Tonic Company the company should put 3 oz of additive X and 2 oz of additive Y into each bottle in order to minimize cost.

15. **Step 1**: Identify the constraints.
Let x = number of orders with Refinery I
y = number of orders with Refinery II

	Refinery I (x)	Refinery II (y)
A	$1x$	$1y$
B	$3x$	$4y$
C	$1x$	$5y$

$x+y\ge 100$

$3x+4y\ge 340$

$x+5y\ge 150$

$x\le 100,\ y\le 100$

$x\ge 0,\ y\ge 0$

Step 2: Graph the constraints and shade the feasible region.

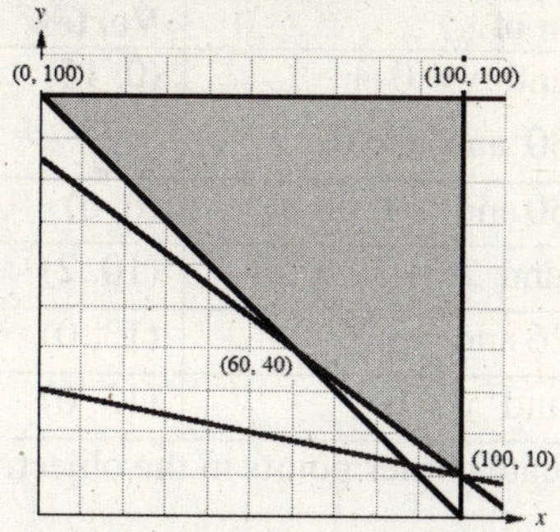

Step 3: Identify vertices (corner points) by inspection or solving a system.

Intersection of	Vertex
$x + y = 100$ and $x = 0$	(0, 100)
$x + y = 100$ and $3x + 4y = 340$	(60, 40)
$x + 5y = 150$ and $3x + 4y = 340$	(100, 10)
$x = 100$ and $y = 100$	(100, 100)

Step 4: Evaluate corner points in the objective function.

Vertex	$C = 300x + 500y$	Min/Max
(0, 100)	$300(0) + 500(100) = 50{,}000$	
(60, 40)	$300(60) + 500(40) = 38{,}000$	
(100, 10)	$300(100) + 500(0) = 35{,}000$	Minimum
(100, 100)	$300(100) + 500(100) = 80{,}000$	

The dealer should place 100 orders with Refinery I and 10 orders with Refinery II in order to minimize his cost.

17. **Step 1**: Identify the constraints.
Let x = number of boxes of oranges
y = number of boxes of grapefruit
z = number of boxes of tangerines
The total number of boxes is 800, so $x + y + z = 800$. Thus, $z = 800 - x - y$.
The constraints are

$x \geq 200$

$y \geq 100$

$z \leq 200 \rightarrow 800 - x - y \leq 200 \rightarrow x + y \geq 600$

$x \geq 0,\ y \geq 0$

Step 2: Graph the constraints and shade the feasible region.

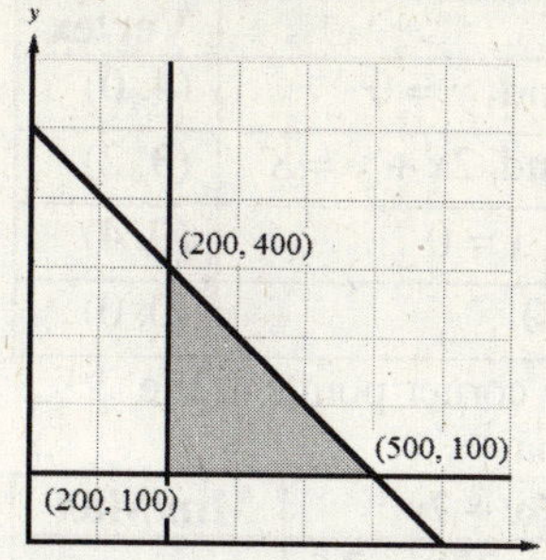

Step 3: Identify vertices (corner points) by inspection or solving a system.

Intersection of	Vertex
$x + y = 600$ and $y = 100$	(500, 100)
$x + y = 600$ and $x = 200$	(200, 400)
$x = 200$ and $y = 100$	(200, 100)

Step 4: Evaluate corner points in the objective function $P = 20x + 10y + 30z$. Use $z = 800 - x - y$ to define z at each vertex.

Vertex	$P = 20x + 10y + 30z$	Min/ Max
(500, 100) $z = 200$	$20(500) + 10(100) + 30(200) = 17{,}000$	Maxim um
(200, 400) $z = 200$	$20(200) + 10(400) + 30(200) = 14{,}000$	
(200, 100) $z = 500$	Not possible, since $z \leq 200$	

The fruit dealer should ship 500 boxes of oranges, 100 boxes of grapefruit, and 200 boxes of tangerines in order to maximize her profit.

19. **Step 1**: Identify the constraints.
Let x = number of imported fruit trees
y = number of oriental shrubs

	Fruit trees (x)	Shrubs (y)
Space	$2x$	$3y$
Prep	$2x$	$1y$

$2x + 3y \leq 12$

$2x + y \leq 8 \qquad x \geq 0,\ y \geq 0$

Step 2: Graph the constraints and shade the feasible region.

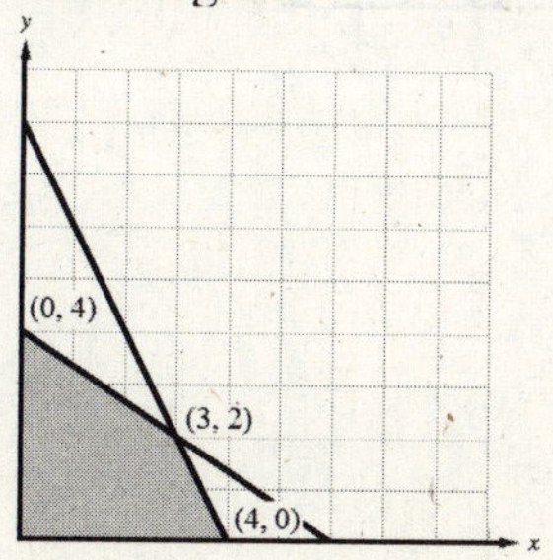

Step 3: Identify vertices (corner points) by inspection or solving a system.

Intersection of	Vertex
$2x+3y=12$ and $y=0$	(4, 0)
$2x+3y=12$ and $2x+y=8$	(3, 2)
$2x+y=8$ and $x=0$	(0, 4)
$x=0$ and $y=0$	(0, 0)

Step 4: Evaluate corner points in the objective function.

Vertex	$P = 6x + 7y$	Min/Max
(4, 0)	6(4) + 7(0) = 24	
(3, 2)	6(3) + 7(2) = 32	Maximum
(0, 4)	6(0) + 7(4) = 28	
(0, 0)	6(0) + 7(0) = 0	

Growfast should display 3 imported fruit trees and 2 oriental shrubs in order to maximize its profit.

21. **Step 1**: Identify the constraints.
Let x = number of ounces of orange juice
y = number of ounces of grapefruit juice

	Orange (x)	Grapefruit (y)
A	$2x$	$3y$
C	$3x$	$2y$
D	$0x$	$1y$

$x+y\le 15$

$2x+3y\ge 26$

$3x+2y\ge 30$

$x+y\ge 12$ $\quad x\ge 0,\ y\ge 0$

Step 2: Graph the constraints and shade the feasible region.

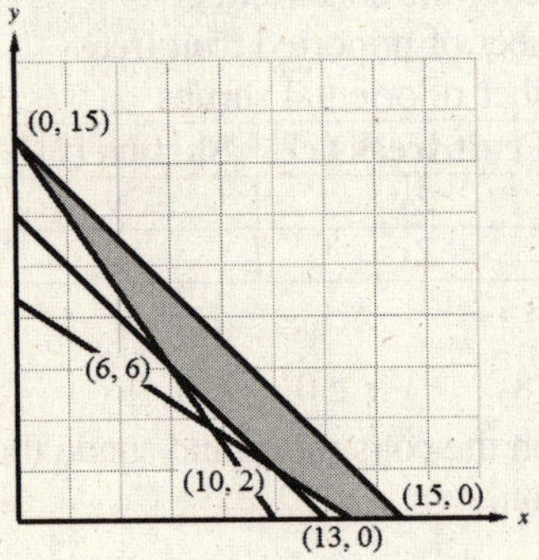

Step 3: Identify vertices (corner points) by inspection or solving a system.

Intersection of	Vertex
$x+y=15$ and $x=0$ or $3x+2y=30$ and $x=0$	(0, 15)
$3x+2y=30$ and $x+y=12$	(6, 6)
$x+y=12$ and $2x+3y=26$	(10, 2)
$2x+3y=26$ and $y=0$	(13, 0)
$x+y=15$ and $y=0$	(15, 0)

Step 4: Evaluate corner points in the objective function.

Vertex	$C = 4x + 3y$	Min/Max
(0, 15)	4(0) + 3(15) = 45	
(6, 6)	4(6) + 3(6) = 42	Minimum
(10, 2)	4(10) + 3(2) = 46	
(13, 0)	4(13) + 3(0) = 52	
(15, 0)	4(15) + 3(0) = 60	

a. The ABC Fruit Juice Company should put 6 oz of orange juice and 6 oz of grapefruit juice in each can to keep the total cost of the can as low as possible.
b. The minimum cost per can is 42¢.
c. The A vitamin content is 2(6) + 3(6) = 30 units. The C vitamin content is 3(6) + 2(6) = 30 units. The D vitamin content is 6 + 6 = 12 units.

23. **Step 1**: Identify the constraints.
Let x = number of McDonald's FOF sandwich
y = number of Burger King TGC sandwich

	FOF (x)	TGC (y)
Sugar	$5x$	$7y$
Protein	$15x$	$55y$

$5x+7y<40$

$15x+55y\ge 50$

$x\ge 0,\ y\ge 0$

Step 2: Graph the constraints and shade the feasible region.

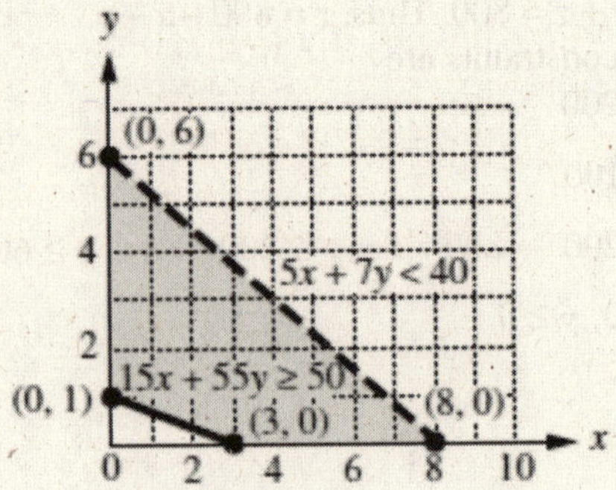

Step 3: Identify vertices (corner points) by inspection or solving a system.

Intersection of	Vertex
$5x+7y=40$ and $y=0$	(8, 0)
$5x+7y=40$ and $x=0$	(0, 6)
$15x+55y=50$ and $y=0$	(3, 0)
$15x+55y=50$ and $x=0$	(0, 1)

Step 4: Evaluate corner points in the objective function.

Vertex	*C* = 3*x* + 4*y*	Min/Max
(8, 0)	3(8) + 4(0) = 24	
(0, 6)	3(0) + 4(6) = 24	
(3, 0)	3(3) + 4(0) = 9	
(0, 1)	3(0) + 4(1) = 4	Minimum

You should each no McDonald's FOF sandwiches and only 1 Burger King TGC sandwich in order to minimize cost at $4.

25. **Step 1**: Identify the constraints.
Let x = number of McDonald's FOF
y = number of Burger King TGC

	FOF (*x*)	TGC (*y*)
Carbohydrates	$38x$	$40y$
Sugar	$5x$	$7y$

$38x+40y \le 300$

$5x+7y \ge 40$

$x \ge 0,\ y \ge 0$

Step 2: Graph the constraints and shade the feasible region.

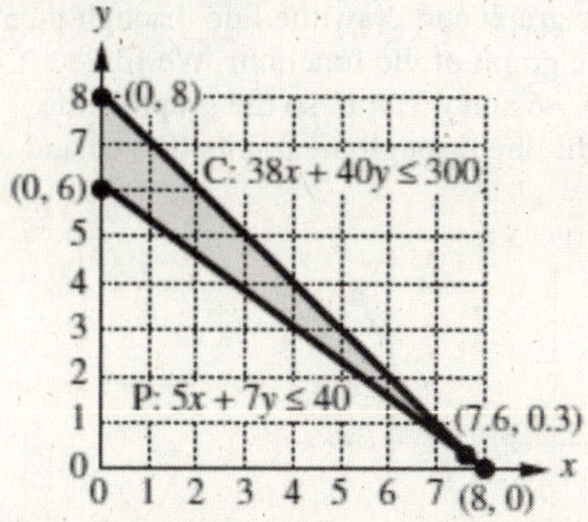

Step 3: Identify vertices (corner points) by inspection or solving a system.

Intersection of	Vertex
$38x+40y=300$ and $5x+7y=40$	(7.6, 0.3)
$38x+40y=300$ and $x=0$	(0, 8)
$5x+7y=40$ and $y=0$	(8, 0)
$5x+7y=40$ and $x=0$	(0, 6)

Step 4: Evaluate corner points in the objective function.

Vertex	*C* = 3*x* + 4*y*	Min/ Max
(7.6, 0.3)	3(7.6) + 4(0.3) = 24	
(0, 8)	3(0) + 4(8) = 32	Max
(8, 0)	3(8) + 4(0) = 24	
(0, 6)	3(0) + 4(6) = 24	

You should each no McDonald's FOF sandwiches and 8 Burger King TGC sandwiches in order to maximize cost at $32.

27. Answers may vary.
   a. The requirement that the variables of the objective function be nonnegative.
   b. A polygon such that the points of the line segment joining any two points on the boundary lie entirely inside the region.

29. c. **Step 1**: Identify the constraints.
Let x = dollars to be placed on the champion to win with A
y = dollars to be placed on the challenger to win with B

$\frac{3}{5}x - y \ge 100$

$-x+2y \ge 100$

$x \ge 0,\ y \ge 0$

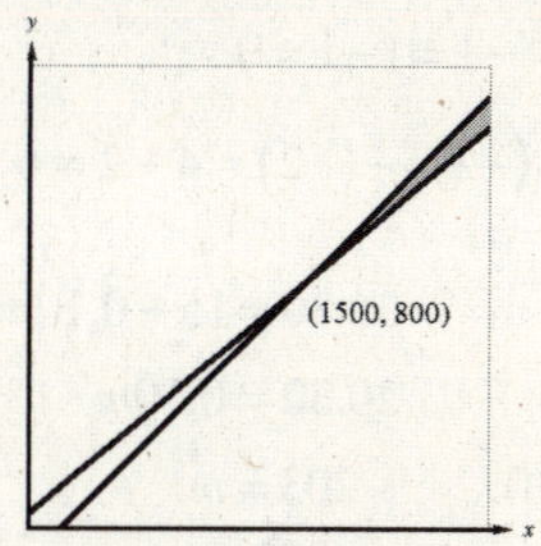

The least amount that Gary can bet for a net gain of $100 is $1500 on the champion to win and $800 on the challenger to win, for a total of $2300.

Chapter 7 Practice Test

Study Tips

Students often feel overwhelmed when there are many formulas to be remembered. An easy way to memorize these formulas is to write the formula down each and every time you need it for a problem. By the time you have written it a number of times, you will know it! As suggested earlier, make sure you write them on your test paper as soon as you get it. Then you won't worry about getting confused or forgetting these specifics.

During the test remember to read the problems very carefully. Underline important concepts as you are reading each question and see what is given. For example, if you are given a point and the slope of a line, the question is very likely to be: What is the equation of this line? The formula to use is, of course, the point-slope formula.

Here are the solutions to the practice test.

1. The domain is the set of x values: $\{0, 2, 3, 5\}$; the range is the set of y values: $\{-1, 2, 3, 4\}$.

2. The domain is the set of x values, which for this relation is the set of all real numbers; the range is also the set of all real numbers.

3. The relations in (b) and (c) are both functions. In each of these there is just one y value for each value of x in the domain.

4. Since $f(x) = x^2 - x$,

   a. $f(0) = 0^2 - 0 = 0$

   b. $f(1) = 1^2 - 1 = 1 - 1 = 0$

   c. $f(-2) = (-2)^2 - (-2) = 4 + 2 = 6$

5. Solve: $35.30 = 15 + 0.10m$

   Subtract 15: $20.30 = 0.10m$

   Divide by 0.10: $203 = m$

   Thus, the person drove 203 mi.

6. Since x is an integer between –1 and 3 inclusive, the table of values is as follows:

x	y
–1	–3
0	0
1	3
2	6
3	9

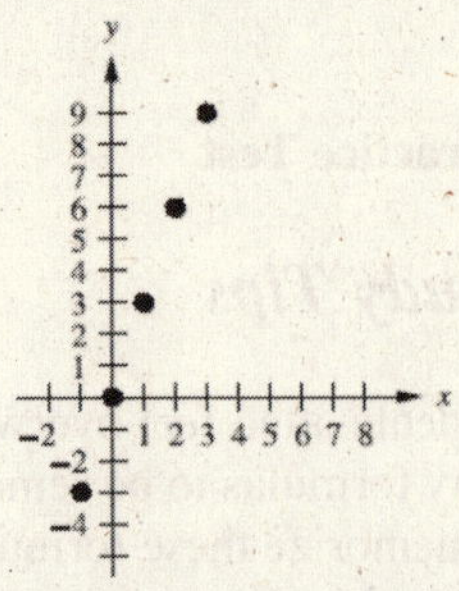

The corresponding points are the heavy dots shown in the graph.

7. Since x and y are nonnegative integers, the table of values is as follows:

x	y
0	0
0	1
0	2
1	0
1	1
2	0

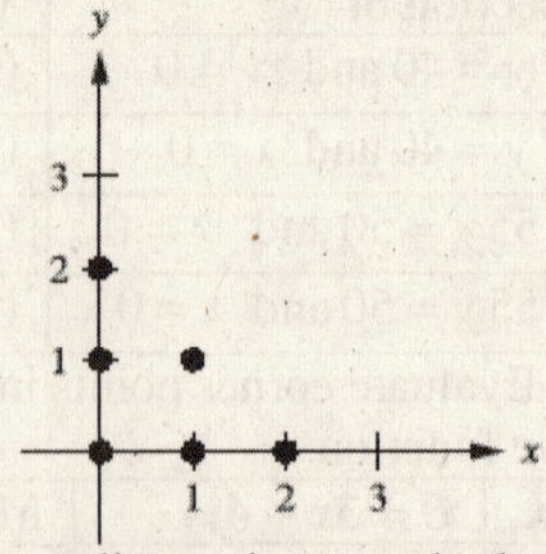

The corresponding points are the heavy dots shown in the graph.

8. x is an integer between –2 and 2, inclusive, so the table of values is as follows:

x	$g(x)$
–2	7
–1	1
0	–1
1	1
2	7

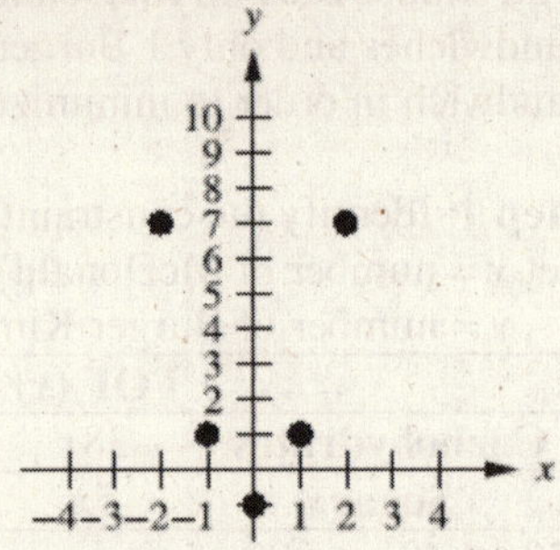

The corresponding points are the heavy dots shown in the graph.

9. The function defined by $f(x) = 2x - 6$ is a linear function, so we need to find two points on the graph and draw the line through them for the graph of the function. We find $f(0) = -6$ and $f(3) = 0$, so the graph is the straight line through the points (0, –6) and (3, 0).

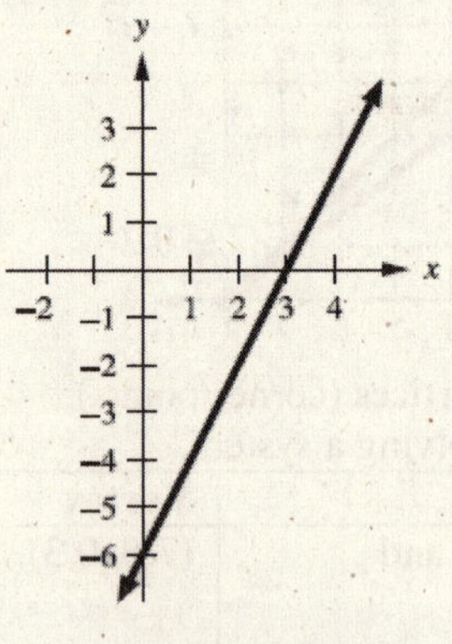

10. $3x - 2y = 5$ is the equation of a straight line. Two points that satisfy the equation are (1, –1) and (3, 2). These points are plotted on the graph and the line through them is the graph of the equation.

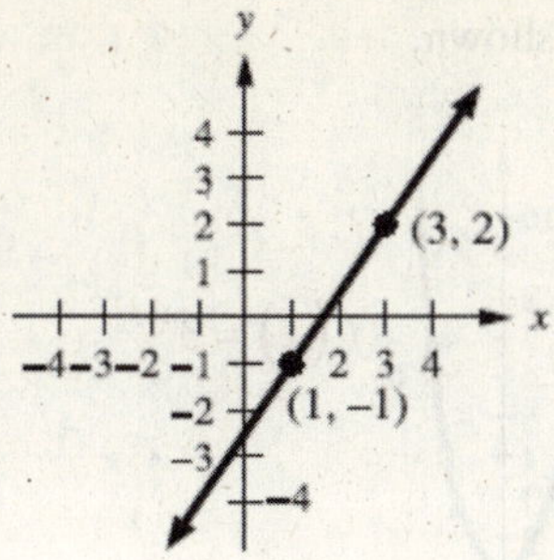

11. a. $d=\sqrt{(x_2-x_1)^2+(y_2-y_1)^2}$

$=\sqrt{(7-4)^2+(3-7)^2}$

$=\sqrt{9+16}=\sqrt{25}=5$ units

b. The two points have the same x coordinate, so are on the same vertical line. Thus,

$d=|y_2-y_1|=|-2-8|=|-10|=10$ units

12. $m=\dfrac{y_2-y_1}{x_2-x_1}=\dfrac{-2-(-3)}{9-(-1)}=\dfrac{1}{10}$

13. a. Use the point-slope form and simplify:

$$y-(-1)=-2(x-3)$$
$$y+1=-2x+6$$
$$y=-2x+5$$

b. Solve the equation $2y = 4 - 8x$ for y to get $y = -4x + 2$ as the slope-intercept form of the equation. Now, compare $y = mx + b$ with $y = -4x + 2$ and we obtain $m = -4$ and $b = 2$.

14. Use the point-slope form

$y-y_1=m(x-x_1)$ where $m=\dfrac{1}{10}$

(see problem 12) to get

$$y-(-3)=\frac{1}{10}(x-(-1))$$

Simplify: $y+3=\dfrac{1}{10}x+\dfrac{1}{10}$

Multiply by 10: $10y+30=x+1$

General form: $x-10y=29$

15. a. Solve each equation for y:

$y=-2x+1$; $y=-4x+\dfrac{4}{3}$. This shows that the slope of the first line is -2 and the slope of the second line is -4, so the lines are not parallel. By equating the two expressions for y, we get the equation:

$$-2x+1=-4x+\frac{4}{3}$$

Add $4x$ and -1: $2x=\dfrac{1}{3}$

Divide by 2: $x=\dfrac{1}{6}$

Substitute: $y=-\dfrac{1}{3}+1=\dfrac{2}{3}$

Thus, the lines intersect at $\left(\dfrac{1}{6},\dfrac{2}{3}\right)$.

b. The slope of the line $y = 2x - 5$ is $m = 2$. Solving the second equation for y, we get $y=2x-\dfrac{7}{2}$, so the slope of the second line is also $m = 2$. Since the slopes are the same and the y-intercepts are different, the two lines are parallel.

16. Since the required line is to be parallel to the line $2x - 3y = 5$, its equation must be of the form $2x - 3y = C$, where C is to be determined so that the line passes through the point $(1, -2)$. So, we substitute into the equation to get $(2)(1) - (3)(-2) = C$, which gives $C = 8$. Thus, the answer is $2x - 3y = 8$.

17. The graph is like that of $y=x^2$ but opening downward and then lowered 3 units as shown. The vertex is at $V(0, -3)$.

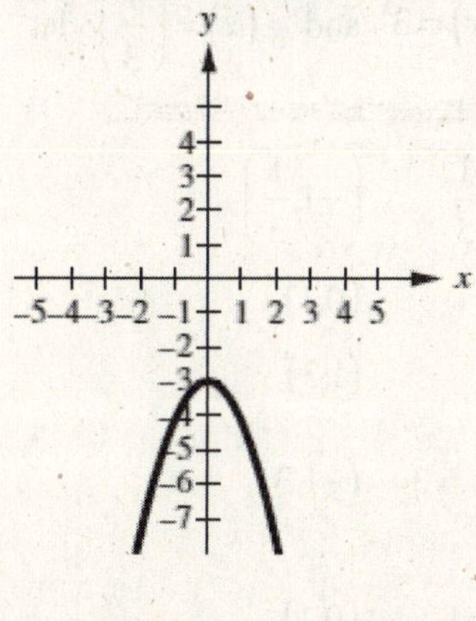

18. As you recall, the graph of $y=a(x-h)^2+k$ has its vertex at (h, k). Since $y=(x+2)^2-1$, the vertex is at $(-2, -1)$. Because $a = 1 > 0$,

the graph opens upward.

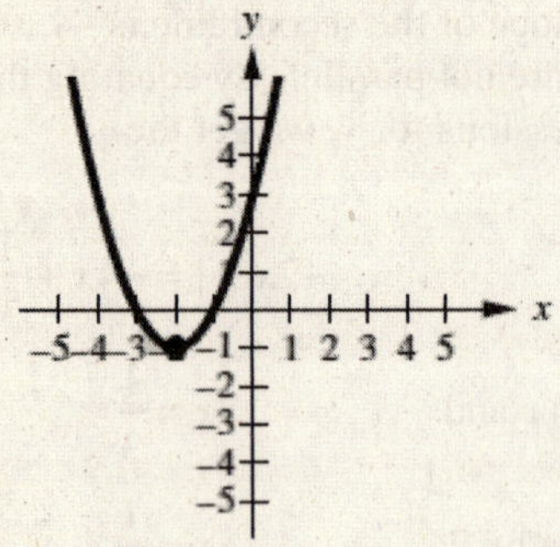

19. The graph of $y = ax^2 + bx + c$ has its vertex at $x = \dfrac{-b}{2a}$. In this case, $a = 1$ and $b = 6$, so $x = \dfrac{-6}{2 \cdot 1} = -3$. To find y substitute $x = -3$ in the original equation $y = x^2 + 6x + 8$:

$y = (-3)^2 + 6(-3) + 8 = 9 - 18 + 8 = -1$

So the vertex is at (–3, –1). Since $a > 0$, the graph opens up.

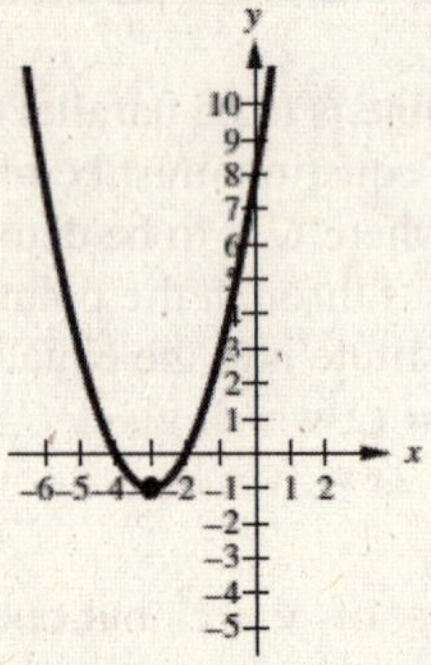

Intercepts: (–4, 0), (–2, 0), (7, 0)
Vertex: (–3, –1)

20. To graph $f(x) = 3^x$ and $g(x) = \left(\dfrac{1}{3}\right)^x$ let $x = -1$, 0, and 1.

$f(-1) = 3^{-1} = \dfrac{1}{3}$ $\quad \left(-1, \dfrac{1}{3}\right)$

$f(0) = 3^0 = 1$ $\quad (0,1)$

$f(1) = 3^1 = 3$ $\quad (1,3)$

$g(-1) = \left(\dfrac{1}{3}\right)^{-1} = 3$ $\quad (-1,3)$

$g(0) = \left(\dfrac{1}{3}\right)^0 = 1$ $\quad (0,1)$

$g(1) = \left(\dfrac{1}{3}\right)^1 = \dfrac{1}{3}$ $\quad \left(1, \dfrac{1}{3}\right)$

Both graphs are shown.

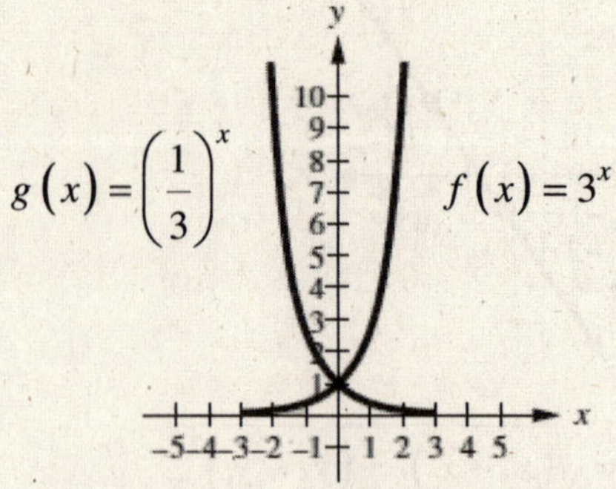

21. For convenience, use $x = -2$, 0 and 2 in $f(x) = e^{x/2}$.

$f(-2) = e^{-2/2} = e^{-1} = \dfrac{1}{e}$ (about $\frac{1}{3}$)

$f(0) = e^{0/2} = e^0 = 1$

$f(2) = e^{2/2} = e^1$ (about 2.7)

Graph the points (–2, 1/3), (0, 1) and (2, 2.7) and join them with a curve.

For $g(x) = \ln\left(\dfrac{x}{2}\right)$:

$g(x) = \ln\left(\dfrac{-2}{2}\right) = \ln(-1)$ is not defined.

$g(0) = \ln\left(\dfrac{0}{2}\right) = \ln 0$ is not defined.

$g(2) = \ln\left(\dfrac{2}{2}\right) = \ln 1 = 0$

The graph of both functions is shown.

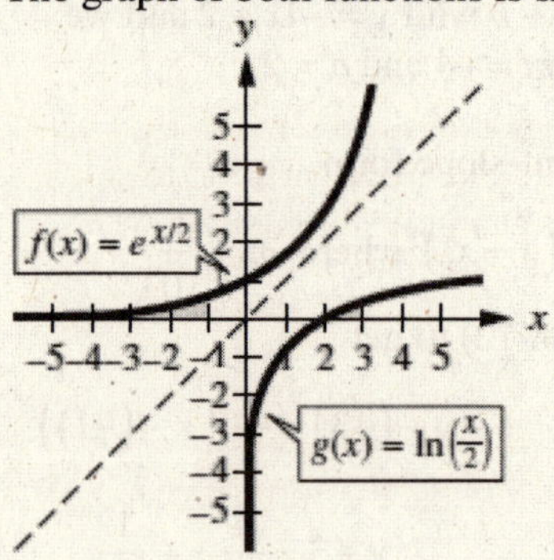

22. Set $2P = Pe^{rt}$ or $2 = e^{rt}$, then take the natural log of both sides:

$\ln 2 = \ln\left(e^{rt}\right) = rt(\ln e) = rt$

$t = \dfrac{\ln 2}{r} = \dfrac{\ln 2}{0.08} = \dfrac{0.6931471}{0.08} \approx 8.66$ years

23. Multiply both sides of the first equation by 3, multiply both sides of the second equation by 2 and add the equations:

$$3x+2y=9 \rightarrow 9x+6y=27$$
$$2x-3y=19 \rightarrow \underline{4x-6y=38}$$

Add: $13x = 65$

Divide by 13: $x=5$

Substitute into 1st eq: $15+2y=9$

Subtract 15: $2y=-6$

Divide by 2: $y=-3$

The solution is $(5,-3)$. (You should check this answer in the second equation.)

24. The first equation can be written in the form $2x - y = 3$. If both sides of this equation are multiplied by 3, the result is $6x - 3y = 9$, which is exactly the second equation. This means that the two lines corresponding to these equations are coincident, so that any solution of one equation is also a solution of the other equation. A general solution can be obtained by letting $x = a$, where a is any real number. Then, the first equation gives $y = 2a - 3$. Thus, the solution set can be written as: $\{(a, 2a - 3) \mid a \text{ is any real number}\}$.

25. To graph the solution set of $4x - 3y \le 12$, first graph the line $4x - 3y = 12$, as in the figure. The point (0, 0) makes the left side of $4x - 3y = 0$ and satisfies the inequality $4x - 3y \le 12$. Thus, all points on the origin side (above) the line are in the solution set. This region is shaded in the figure. Note that the line is drawn solid to show that it belongs in the graph of the solution set.

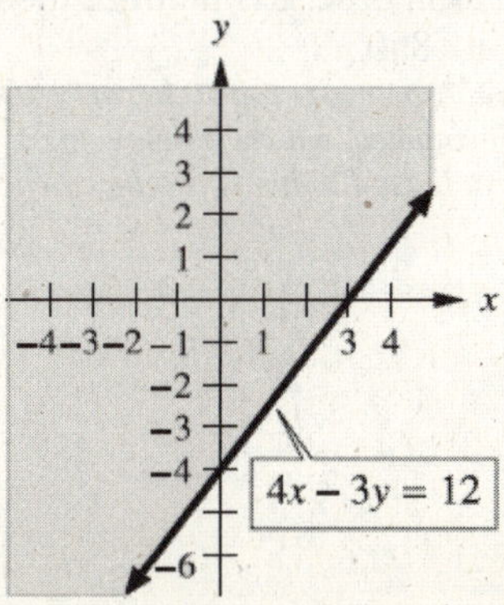

26. To graph the solution set of $x + 3y \le 6$ and $x - y \ge 2$ draw the lines $x + 3y = 6$ and $x - y = 2$, as in the figure. The point (0, 0) satisfies $x + 3y \le 6$ so the desired region is under the line $x + 3y = 6$. (0, 0) does not satisfy $x - y \ge 2$, so the desired region is under the line $x - y = 2$. Thus, the solution set is the darker region under both lines as well as the boundaries shown.

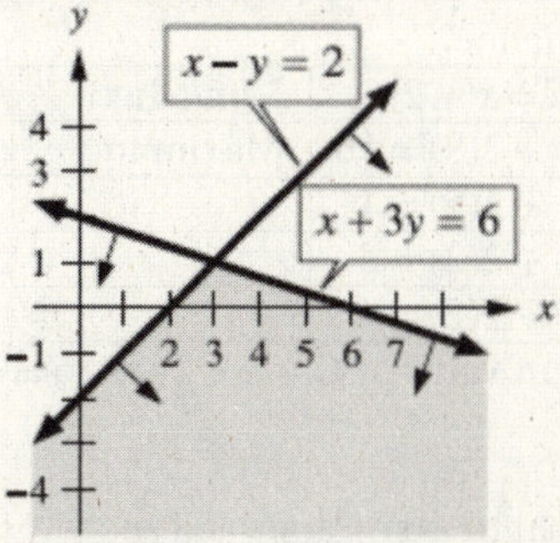

27. To graph the solution set of the system $x + 2y \le 3$, $x \le y$, and $x \ge 0$, draw the lines $x + 2y = 3$, $x = y$, and $x = 0$. The point (0, 0) satisfies $x + 2y \le 3$, so the desired region is on the origin side of (under) the line $x + 2y = 3$. The point (1, 2) satisfies $x \le y$, so the region is on the side above the line $x = y$. All points to the right of the y-axis satisfy $x \ge 0$. The shaded region along with its boundaries satisfies all inequalities and is the graph of the solution set of the given system.

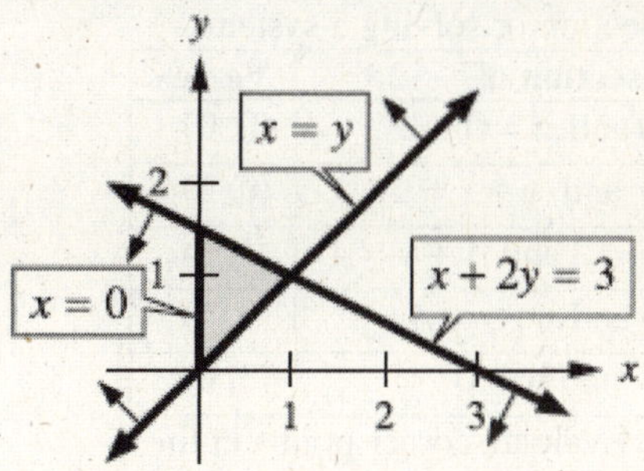

28. **Step 1**: Graph the constraints and shade the feasible region.

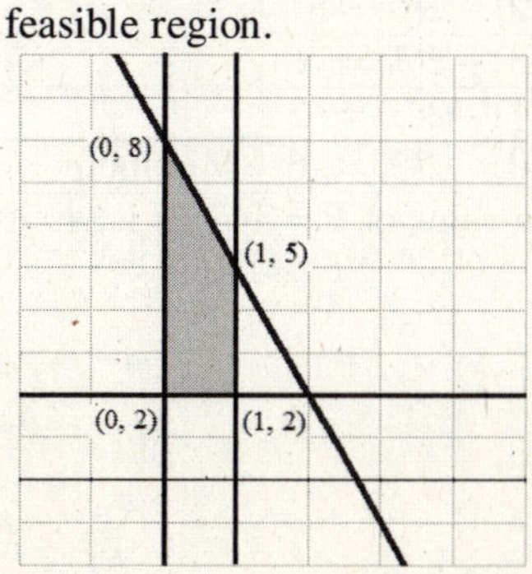

Step 2: Identify vertices (corner points) by inspection or solving a system.

Intersection of	Vertex
$3x+y=8$ and $x=0$	(0, 8)
$3x+y=8$ and $x=1$	(1, 5)
$y=2$ and $x=1$	(1, 2)
$y=2$ and $x=0$	(0, 2)

Step 3: Evaluate corner points in the objective function.

Vertex	$C=x+2y$	Min/Max
(0, 8)	0 + 2(8) = 16	Maximum
(1, 5)	1 + 2(5) = 11	
(1, 2)	1 + 2(2) = 5	
(0, 2)	0 + 2(2) = 4	

The maximum value of $C = x + 2y$ is 16 at (0, 8).

29. **Step 1**: Graph the constraints and shade the feasible region.

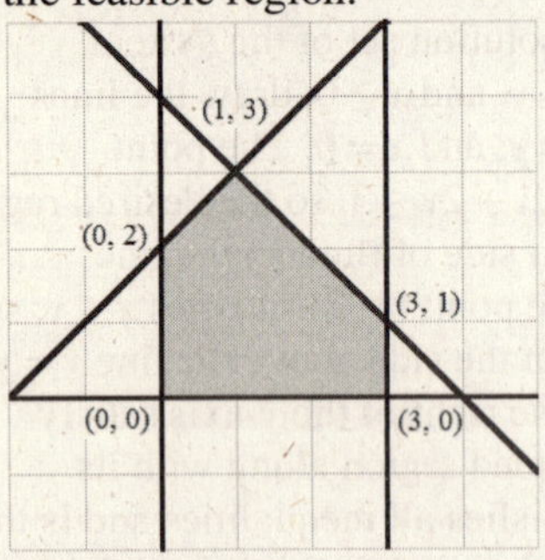

Step 2: Identify vertices (corner points) by inspection or solving a system.

Intersection of	Vertex
$y=0$ and $x=0$	(0, 0)
$x=0$ and $y-x=2$	(0, 2)
$y-x=2$ and $x+y=4$	(1, 3)
$x=3$ and $x+y=4$	(3, 1)
$x=3$ and $y=0$	(3, 0)

Step 3: Evaluate corner points in the objective function.

Vertex	$P = 3y - 2x$	Min/Max
(0, 0)	3(0) – 2(0) = 0	
(0, 2)	3(2) – 2(0) = 6	
(1, 3)	3(3) – 2(1) = 7	
(3, 1)	3(1) – 2(3) = –3	
(3, 0)	3(0) – 2(3) = –6	Minimum

The minimum value of $P = 3y - 2x$ is –6 at (3, 0).

30. **Step 1**: Identify the constraints.
Let x = the number of hours for Machine A
y = the number of hours for Machine B

Machine	Items produced per hour
A (x)	10
B (y)	12

$10x+12y \geq 420$

$x+y \leq 40$

$x \geq 0,\ y \geq 0$

Step 2: Graph the constraints and shade the feasible region.

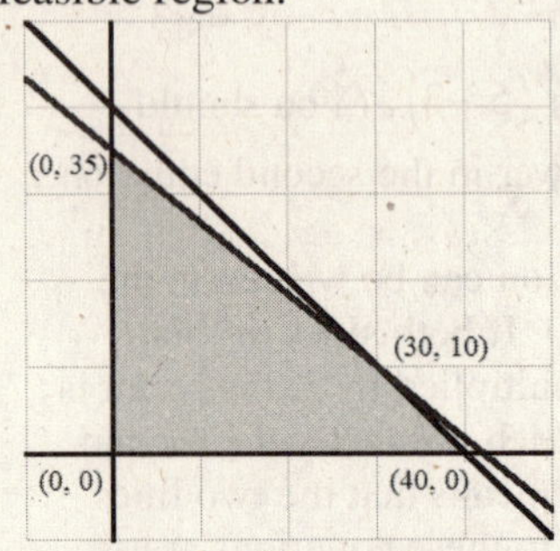

Step 3: Identify vertices (corner points) by inspection or solving a system.

Intersection of	Vertex
$x=0$ and $y=0$	(0, 0)
$x=0$ and $10x+12y=420$	(0, 35)
$10x+12y=420$ and $x+y=40$	(30, 10)
$x+y=40$ and $y=0$	(40, 0)

Step 4: Evaluate corner points in the objective function.

Vertex	$C = 20x + 25y$	Min/Max
(0, 0)	20(0) + 25(0) = 0	
(0, 35)	20(0) + 25(35) = 875	
(30, 10)	20(30) + 25(10) = 850	Minimum
(40, 0)	20(40) + 25(0) = 800	*

Machine A should run 30 hr and Machine B should run 10 hr in order to minimize the machine cost at $850.

**If only Machine A operates for 40 hr, only 400 items could be produced, which is below the requirement of at least 420 items produced each 40-hr week.*

Chapter 8 Geometry

Section 8.1 Points, Lines, Planes, and Angles

Study Tips

The study of geometry contains many undefined terms such as point, line, and plane. Learn the meaning of these concepts by looking at concrete examples and studying Figures 8.2-8.6 and 8.12-8.16, in the text.

To determine geometric figure described by union or intersection:
Think: union: *or* (all included); intersection: *and* (only where overlapping)
Note: The concept of intersection and union was studied in Chapter 2.

Fact
Lines, rays, and points can be combined to form angles. An *angle* is the figure formed by two rays with a common endpoint called the vertex.

Types of angles and measures:
Straight angle: 180°
Right angle: 90°
Acute angle: less than 90°
Obtuse angle: greater than 90° but less than 180°.
Supplementary angles: angles whose measures sum to 180°
Complementary angles: angles whose measures sum to 90°

For parallel lines cut by a transversal:
Think: of equal measure are corresponding angles, alternate interior angles, alternate exterior angles
Note:
1. All acute angles are equal.
2. All obtuse angles are equal.
3. Sum of measure of an acute angle and an obtuse angle is 180°.

Fact
The sum of the measures of the angles of any triangle is 180°.

8.1 Exercises

1. a. 

   b. 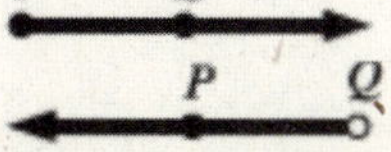

   c. 

3. The segment $\overline{BC}$

5. The ray $\overrightarrow{AD}$ (or $\overrightarrow{AC}$ or $\overrightarrow{AB}$)

7. The segment $\overline{AD}$ 9. The line $\overline{AD}$

11. $\varnothing$ 13. The point C

15. a. $\overline{AB}, \overline{AC}, \overline{AD}, \overline{BC}, \overline{BD}$, and $\overline{CD}$
    b. $\overline{AB}$ and $\overline{CD}$; $\overline{AC}$ and $\overline{BD}$; $\overline{AD}$ and $\overline{BC}$
    c. None of the edges determine parallel lines.

17. True 19. True 21. True

23. False (They can be skew lines.)

25. If a plane, say *EFG*, contains line *m*, it must contain the point where line *m* intersects plane *ABC*. So plane *EFG* must intersect plane *ABC*, and the given statement is false.

27. a. $\angle BAC$ (or $\angle CAB$) is the same as $\angle\alpha$.
    b. $\angle\beta$ (or $\angle FAE$) is the same as $\angle EAF$.

29. $\angle BAC, \angle CAD, \angle DAE, \angle EAF$

31. $\angle BAE, \angle CAE, \angle CAF$

33. a. $\angle DAE$ b. $\angle CAD$

35. a. $\angle DAF$ b. $\angle BAE$

37. $m\angle DAE = 90° - 55° = 35°$

39. $m\angle\alpha = 90° - 75° = 15°$

41. $\angle B$ is vertical to the 70° angle.

43. $m\angle A = 180° - 70° = 110°$

45. $90° - 70° = 20°$

47. $m\angle A = m\angle C = 180° - 70° = 110°$, so
    $m\angle A + m\angle C = 110° + 110° = 220°$

49. a. $m\angle A = 180° - 30° = 150°$
    b. $m\angle B = 30°$
    c. $m\angle C = m\angle A = 150°$

51. $\angle A, \angle C, \angle E, \angle G$ are all supplementary to $\angle B$.

53. a. $m\angle B = 90° - 41° = 49°$
b. $m\angle B = 180° - 41° = 139°$

55. Since the two angles are complementary, the sum of their measures is 90°. Thus,
$$(3x+15)+(2x-5)=90$$
Simplify: $5x+10=90$
Subtract 10: $5x=80$
Divide by 5: $x=16$

57. The angles are supplementary, so their measures must add to 180°. Thus, we have the equation: $(10x-5)+(2x+5)=180$
Simplify: $12x=180$
Divide by 12: $x=15$
Evaluate: $2x+5=35$
$10x-5=145$
This shows that x = 15 and one angle measures 35°; the other measures 145°.

59. The two angles are complementary, so their measures must add to 90°. Thus we have the equation:
$$(2x+10)+(5x+10)=90$$
Simplify: $7x+20=90$
Subtract 20: $7x=70$
Divide by 7: $x=10$
Thus, $2x + 10 = 30$, $5x + 10 = 60$, so that one angle measures 30° and the other measures 60°.

61. a. It moves through 1/12th of 360°, that is, through 30°.
b. It moves through 6/12ths of 360°, that is, through 180°.

63. $m\angle A + m\angle B + m\angle C = 180°$. Thus, if the number of degrees in $\angle B$ is x, then $37 + x + 53 = 180$. This gives $x = 90$, so $m\angle B = 90°$.

65. The total number of degrees in the three angles is 180°, so we have the equation:
$$(x+10)+(2x+10)+(3x+10)=180$$
Simplify: $6x+30=180$
Subtract 30: $6x=150$
Divide by 6: $x=25$

67. a. $m\angle A + m\angle B = 180°$ Angles A and B form a straight angle.
b. $m\angle C + m\angle B = 180°$ Angles C and B form a straight angle.
c. $m\angle A + m\angle B = m\angle C + m\angle B$ Both equal 180°. (Substitution)
d. $m\angle A = m\angle C$ Subtract $m\angle B$ from both sides.

69. $\angle\beta$ is an acute angle.

71. $\angle\delta$ is a right angle.

73. The angle is an obtuse angle.

75. With the phone closed, the angle would be 0°.

77. $m\angle\theta = 180° - 40° - 40° = 100°$

79. Point, line, and plane are all undefined.

81. The ray may have its endpoint only in common with the plane, or it may have some other single point in common with the plane, or it may lie entirely in the plane.

83. a. One of the usual meanings of <u>acute</u> is "sharp" or "intense." Thus "an acute pain" means "a sharp or intense pain."
b. One of the ordinary meanings of <u>obtuse</u> is "dull." Thus, "obtuse intelligence" means "dull intelligence" or "stupidity."

85. $\angle\alpha$ is an obtuse angle; $\angle\beta$ is an acute angle.

87. The triangle is isosceles and right.

89. Both angles are acute angles.

91. a. $\angle\alpha$ is an straight angle; $m\angle\alpha = 180°$
b. $\angle\beta$ is an acute angle;
$m\angle\beta = 0.22 \cdot 360° = 79.2°$
c. $\angle\delta$ is an acute angle;
$m\angle\delta = 0.19 \cdot 360° = 68.4°$
d. $\angle\gamma$ is an acute angle;
$m\angle\gamma = 0.09 \cdot 360° = 32.4°$

Section 8.2 Triangles and Other Polygons

Study Tips

Most of us are familiar with polygons, that is, triangles, rectangles, octogons, etc. However, do you know all of the terms associated with polygons?

Terms to know:
Simple path/curve: does not cross itself
Closed path/curve: starts and ends at same point
Broken lines: sequence of connected straight line segments
Polygon: path of a simple, closed broken line
Sides of polygon: line sements of the path
Vertices: endpoints of the sides
Convex polygon: no line segment joining any points on the path extends outside the polygon
Concave polygon: not convex
Included side of angle: side that lies between two angles
Included angle of sides: angle form by two sides

To name a polygon:
Think: number of sides
Note: Polygons are classified according to the number of sides they have. See Table 8.2.

To classify a triangle:
Think: by angles and by number of equal sides
Classifications Summary
Right triangle: contains right angle
Acute triangle: all angles are acute
Obtuse triangle: contains an obtuse angle
Scalene triangle: no equal sides
Isosceles triangle: two equal sides
Equilateral triangle: three sides equal
Note: An equilateral triangle is also equiangular (all angles have equal measure).

To classify a quadrilateral:
Think: by relationship between sides
Classifications Summary
Trapezoid: exactly one pair of parallel sides
Kite: two distinct pairs of consectuive equal sides
Parallelogram: opposite sides are parallel
Rectangle: parallelogram with a right angle
Rhombus: parallelogram with all equal sides
Square: rectangle will all equal sides

To determine similar triangles:
Think: equal corresponding angles and corresponding sides are proportional

To determine congruent triangles:
Think: equal corresponding angles and sides
Note: Congruent Triangle Theorems:
1. Side-Angle-Side (SAS)
2. Angle-Side-Angle (ASA)
3. Side-Side-Side (SSS)

Caution
Angle-angle-angle (AAA) cannot determine congruent triangles!

To find the sum of the measures of any convex polygon:
Think: formula $S=(n-2)180°$, n is number of sides

8.2 Exercises

1. Answers may vary.
   a.

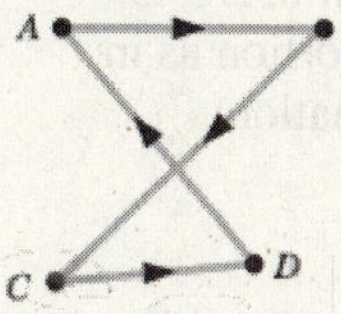

   b.

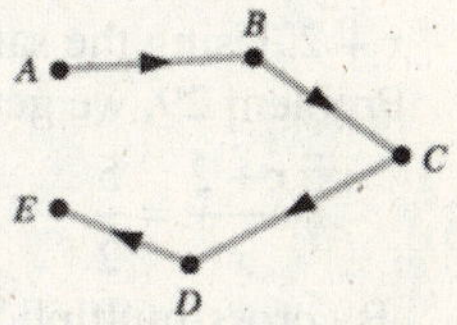

3. a. C, D, I, J, L, M, N, O, P, S, U, V, W, Z
   b. B, D, O

5. a. D, O
   b. A, E, F, G, H, K, Q, R, T, X, Y

7. Convex
9. Parallelogram

11. Rectangle
13. Trapezoid

15. Parallelogram
17. A scalene, right triangle

19. A scalene, acute triangle

21. An isosceles, acute triangle

23. A scalene, obtuse triangle

25. Triangles (a) and (c) are similar because $\frac{4/3}{4}=\frac{1}{3}=\frac{2/3}{2}$.

27. Because the triangles are similar, corresponding sides are proportional. Thus, $\frac{x}{4}=\frac{y}{5}=\frac{4}{3}$ so that $x=\frac{16}{3}=5\frac{1}{3}$ and $y=\frac{20}{3}=6\frac{2}{3}$.

29. PQ is parallel to AB, so triangle PQC is similar to triangle ABC. Thus, corresponding sides of these triangles are proportional. This gives the proportion $\frac{AC}{PC}=\frac{BC}{QC}$. Let x be the length of BC, so that $x-6$ is the length of QC. Since AP is of length 3 and PC is of length 4, AC is of length 7. With these lengths, our proportion gives the equation $\frac{7}{4}=\frac{x}{x-6}$.
By cross-multiplication, we get:
$$7x-42=4x$$
$$3x=42$$
$$x=14$$

31. Let x be the length of PC. Then, since the length of AP is 2, the length of AC is $x+2$. Using the same proportion as in Problem 29, we get the equation:
$$\frac{x+2}{x}=\frac{5}{2}$$
By cross-multiplication, we get:
$$2x+4=5x$$
$$4=3x$$
$$x=\frac{4}{3}=1\frac{1}{3}$$

33. Let x be the length of the side of the smaller triangle that corresponds to the 6 cm side of the larger triangle, and let y be the length of the side of the smaller triangle that corresponds to the 9 cm side of the larger triangle. Then, since the triangles are similar, $\frac{x}{6}=\frac{y}{9}=\frac{7}{12}$.
This proportion gives
$$x=\frac{6\cdot 7}{12}=\frac{7}{2}=3\frac{1}{2} \text{ cm and}$$
$$y=\frac{9\cdot 7}{12}=\frac{21}{4}=5\frac{1}{4} \text{ cm.}$$

35. Let the sides of the second triangle be x, y, and z in. long, respectively. Since the triangles are similar, the perimeters are also similar. Thus, $\frac{x}{2}=\frac{y}{3}=\frac{z}{4}=\frac{36}{9}=4$.
$\frac{x}{2}=4$ or $x=8$; $\frac{y}{3}=4$ or $y=12$; $\frac{z}{4}=4$. or $z=16$. The sides of the second triangle are 8 in., 12 in., and 16 in. long.
For an algebraic approach, if the sides of the second triangle are x, y and z in long respectively, we have: $x+y+z=36$.
Since the triangles are similar, $\frac{x}{2}=\frac{y}{3}=\frac{z}{4}$
This gives $x=\frac{1}{2}z$ and $y=\frac{3}{4}z$. Substituting into the first equation, we get:
$$\frac{1}{2}z+\frac{3}{4}z+z=36$$
Multiply by 4: $2z+3z+4z=144$
Add: $9z=144$
Divide by 9: $z=16$
$$x=\frac{1}{2}z=8$$
$$y=\frac{3}{4}z=12$$
The respective sides of the second triangle are 8 in., 12 in., and 16 in. long as before.

37. Let the telephone pole be x ft high. Then the similar right triangles in the figure give the proportion: $\frac{x}{5}=\frac{30}{8}$. Thus,

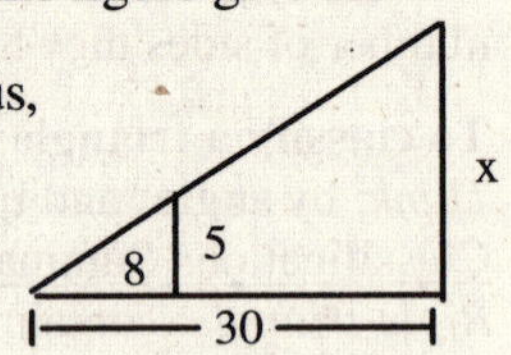

$$x=\frac{5\cdot 30}{8}=\frac{75}{4}=18\frac{3}{4},$$
so the pole is $18\frac{3}{4}$ ft high.

39. Assume that both the tree and 5-ft stake make a right angle with the ground and the angle of the tip of each shadow to the top of the respective object made by the sun's position in the sky will be the same. Therefore, the third angles will be equal, and the triangles similar. Let the height of the tree be h ft. Then the two similar right triangles give the proportion $\frac{h}{5}=\frac{40}{8}$. Thus, $h=5\cdot 5=25$, so the tree is 25 ft high.

41. The marked angles are equal and the vertical angles in the figure are equal, so the two triangles are similar. If the tunnel is t meters long, then we have the proportion:

$\frac{t}{20} = \frac{540}{18} = 30$. We have $t = 20 \cdot 30 = 600$, that is, the tunnel is 600 m long.

43. Recall that an isosceles triangle is one that has two equal sides. By the SAS Theorem, $\Delta ABD \cong \Delta CBD$. This implies that $AB \cong BC$. that is, two of the sides are equal, so ΔABC is isosceles.

45. It is given that ΔACD and ΔBCE have two pairs of sides with equal length ($AD = BC$ and $AC = BD$). They also share the side CD. Thus, by the SSS theorem $\Delta ACD \cong \Delta BCD$.

47. a. $\angle C$ is supplementary to the 100° angle, so $m\angle C = 80°$. $AC = BC$ means that $\angle A \cong \angle B$. Since $m\angle A + m\angle B = 180 - m\angle C$, we have that $2 \cdot m\angle A = 100°$. Thus, $m\angle A = m\angle B = 50°$.

    b. $\angle B$ is supplementary to the 140 ° angle, so $m\angle B = 40°$. As in part a, $\angle A \cong \angle B$. Hence, $m\angle A = 40°$. $m\angle A + m\angle B + m\angle C = 180°$. Therefore, $80° + m\angle C = 180°$. Consequently, $m\angle C = 100°$.

49. AE and BD are parallel, so $\angle ABD$ equals the 120° angle. $\angle CBD$ is supplementary to $\angle ABD$, which makes $m\angle CBD = 60°$. Because CE is perpendicular to BD, $m\angle CDB = 90°$. Since the sum of the measures of the angles of triangle BCD is 180°, $m\angle BCD = 180° - 60° - 90° = 30°$.

51. The sum of the measures of the angles of a regular n-sided polygon is $S = (n-2)180°$, where $n = 14$: $(14-2)180° = 12 \cdot 180° = 2160°$.

53. The sum of the measures of the angles of a regular pentagon is $S = (n-2)180°$, where $n = 5$: $(5-2)180° = 3 \cdot 180° = 540°$. Since there are five angles, the measure of each must be $\frac{540}{5} = 108°$.

55. The sum of the measures of the angles of a regular octagon is $S = (n-2)180°$, where $n = 8$: $(8-2)180° = 6 \cdot 180° = 1080°$. Since there are eight angles, the measure of each must be $\frac{1080}{8} = 135°$.

57. The sum of the measures of the angles of a regular decagon is $S = (n-2)180°$, where $n = 10$: $(10-2)180° = 8 \cdot 180° = 1440°$. Since there are ten angles, the measure of each must be $\frac{1440}{10} = 144°$.

59. a. $(E \cup I \cup S) = T \subset P$

    b. 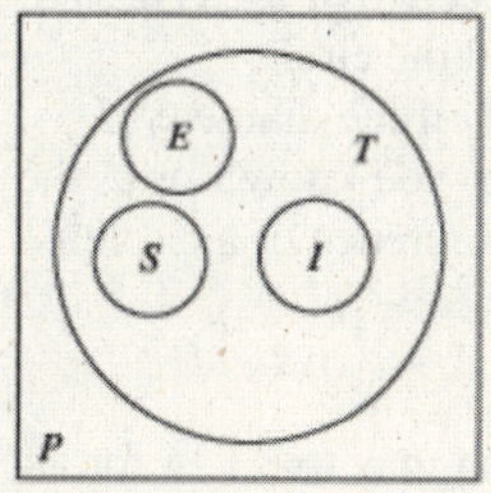

61. 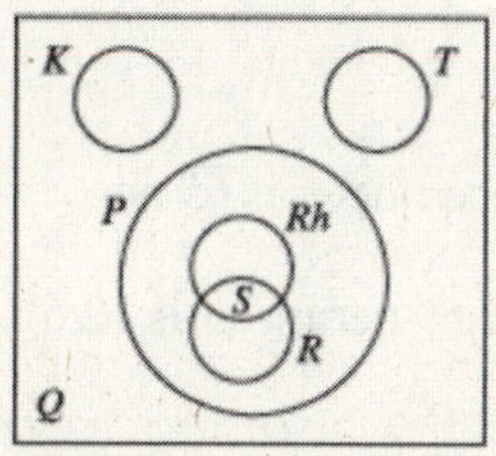

63. Answers will vary.

65. a. 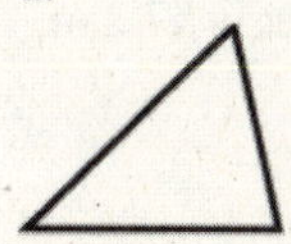 b.  c. 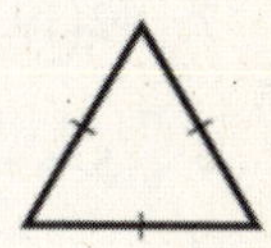

67. a.  b. 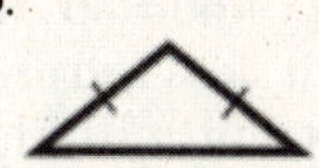 c. Impossible

Section 8.3 Perimeter and Circumference

Study Tips

The *perimeter* of a geometric figure is simply the distance around it. In the case of a circle, the perimeter is called the *circumference*.

To find the perimeter of a geometric figure:
Think: distance around the figure
Perimeter Formula Summary
Triangle: $P = s_1 + s_2 + s_3$
Trapezoid: $P = s_1 + s_2 + b_1 + b_2$
Parallelogram: $P = 2L + 2W$
Rectangle: $P = 2L + 2W$
Square: $P = 4S$

To find the circumference of a circle:
Think: distance around the circle
Note: The radius r is half the diameter d. Depending on whether you know r or d, the formula for the circumference of a circle is $C = 2\pi r$ or $C = \pi d$.

Hint
The text states that you may use 3.14 for an approximation for π.

8.3 Exercises

1. $10 + 25 + 30 = 65$; perimeter is 65 cm.

3. $2 + 5 + 1.6 + 4 = 12.6$; perimeter is 12.6 yd.

5. $2(61.2 + 31.1) = 184.6$; perimeter is 184.6 m.

7. $2(19.4 + 9.2) = 57.2$; perimeter is 57.2 m.

9. $5 \cdot 6 = 30$; perimeter is 30 cm.

11. If the width was x ft, then $2x + 2 \cdot 30 = 80$, so that $2x + 60 = 80$. By subtracting 60 from both sides and dividing both sides by 2, we obtain $x = 10$. Therefore, the width was 10 ft.

13. This distance is 1/4th of the perimeter, that is, $\frac{360}{4} = 90$ ft.

15. Let x ft be the width. Then, the length is $(x + 198)$ ft, and, in terms of x, the perimeter is $[2x + 2(x + 198)]$ ft. Thus, we have the equation:
$$2x + 2(x + 198) = 2468$$
Simplify: $4x + 396 = 2468$
Subtract 396: $4x = 2072$
Divide by 4: $x = 518$
$$x + 198 = 716$$
Thus, the dimensions are 716 ft by 518 ft.

17. $C = \pi d = \pi \cdot 7 = 7\pi \approx 7 \cdot 3.14 \approx 22.0$ m.

19. $C = 2\pi r = 2\pi \cdot 10 = 20\pi \approx 20 \cdot 3.14 \approx 62.8$ ft.

21. $C = \pi d = \pi \cdot 3 = 3\pi \approx 3 \cdot 3.14 \approx 9.42$ cm

23. $C = 2\pi r = 2\pi \cdot 4.5 = 9\pi \approx 9 \cdot 3.14 \approx 28.3$ yd

25. $C = \pi d = \pi \cdot 61 = 61\pi \approx 61 \cdot 3.14 \approx 192$
When the wheel makes one complete turn, the bicycle goes a distance of 192 cm.

27. $C = 2\pi r = 2\pi \cdot 8 = 16\pi \approx 16 \cdot 3.14 \approx 50.2$
The tip of the minute hand travels 50.2 cm in 1 hr.

29. $C = 2\pi r = 2\pi \cdot 4 = 8\pi \approx 8 \cdot 3.14 \approx 25.1$
The border should be 25.1 ft long.

31. $\pi d = 15\pi$, so $d = 15$ cm and $r = 7.5$ cm.

33. $C = \pi d = 7$, so $d = \frac{7}{\pi} \approx \frac{7}{3.14} \approx 2.23$
The diameter of the ring is 2.23 cm.

35. $\pi d = 4\frac{1}{8}$, so $d = \frac{4.125}{3.14} \approx 1.31$.
The diameter of this tiny record is 1.31 in.

37. The answer to problem 36 is approximately 776 revolutions. Thus, 20,000 mi would be approximately (20,000)(776) or 15,500,000 (rounded from 15,520,000) revolutions.

39. The diameter of the pool is $20 - 4 = 16$ yd. Thus, the circumference of the pool is $\pi d = 16\pi \approx 16 \cdot 3.14 \approx 50.2$ yd.

41. $d = 154 + 156 + 180 = 490$ mi

43. $d = 190 + 151 + 223 = 564$ mi

45. The amount of mesh needed is equal to the perimeter of the rectangular compost bin: $P = 2 \cdot 60 + 2 \cdot 70 = 260$ in. Thus, you would need 260 in. of the 48-in. high wire mesh.

47. The amount of mesh needed is equal to the circumference of the circular compost bin: $C = \pi d = 3.14 \cdot 30 = 94.20$. Thus, you would need 94.20 in. of the 36-in. high wire mesh.

49. A bicycle tire because, as a circle, it is made up of 'points' that are the same distance from the center.

51. The worn tire has a smaller circumference, so will turn more times per mile.

53. This is one fourth of the large circumference: $\frac{3}{4}$ in.

55. Point *b* travels $\frac{3}{4}$ in.

Calculator Corner

1. Problem 25:
   [61][π][ENTER] 191.6371519
   Problem 26:
   [17][π][ENTER] 53.40707511
   Problem 30:
   [2][×][6][π][ENTER] 37.69911184
   Problem 32:
   [8][÷][(][2][×][π][)][ENTER] 1.273239545
   Problem 34:
   [251.2][÷][π][ENTER] 79.95944341

Section 8.4 Area Measure and the Pythagorean Theorem

Study Tips

We have all heard of area. Do you know the formal definition? The *area* measure of a plane region is the number of unit regions contained in the given region.

To find the area of a figure:
Think: measured in square units
Area Formula Summary
Rectangle: $A = bh$
Parallelogram: $A = bh$
Triangle: $A = \frac{1}{2}bh$
Note: To find the area of polygonal regions, subdivide the region into nonoverlapping rectangular and/or triangular regions, then find the areas of those regions and add together.

Word of Advice
To memorize formulas, write all the formulas on 3 by 5 cards and practice by reading the examples and doing the problems; write down the formula each time that you use it.

To use the Pythagorean theorem:
Think: must have a right triangle, $a^2 + b^2 = c^2$
Note: If you know the length of two of the sides of a right triangle, you can always find the length of the other side by using this theorem.

Hint
Keep in mind that the longest side of a right triangle has a special name, the *hypotenuse*.

8.4 Exercises

1. This is the area of a triangle with base 6 in. and height 5 in.: $A = \frac{1}{2}bh = \frac{1}{2} \cdot 5 \cdot 6 = 15 \text{ in}^2$

3. This is the area of a parallelogram with base 5 cm and height 3 cm: $A = bh = 5 \cdot 3 = 15 \text{ cm}^2$

5. The required area can be calculated as the area of the outside rectangle minus the area of the triangle to be taken out. Thus:
   $A = 10 \cdot 3 - \frac{1}{2} \cdot 6 \cdot 2 = 30 - 6 = 24 \text{ ft}^2$

7. This area can be calculated as the sum of the area of the rectangle and the triangle on the left end. Thus:
   $A = 6 \cdot 4 + \frac{1}{2} \cdot 4 \cdot 3 = 24 + 6 = 30 \text{ ft}^2$

9. This area can be calculated as the sum of the area of the rectangle and the semicircle on the left end. Thus:

$$A = 40 \cdot 20 + \frac{1}{2}\pi \cdot 10^2 = 800 + 50\pi$$
$$\approx 800 + 50 \cdot 3.14 = 957 \text{ cm}^2$$

11. This area can be calculated as the area of the square plus the area of the four semicircles. Thus,
$$A = 8^2 + 2\pi \cdot 4^2 = 64 + 32\pi$$
$$\approx 64 + 32 \cdot 3.14 = 164 \text{ cm}^2$$

13. The shaded area is the area of the rectangle minus the area of the inscribed semicircle. Thus,
$$A = 6 \cdot 3 - \frac{1}{2}\pi \cdot 3^2 = 18 - 4.5\pi$$
$$\approx 18 - 4.5 \cdot 3.14 = 3.87 \text{ cm}^2$$

15. The shaded area is the area of the large semicircle minus the area of the two smaller semicircles. Thus,
$$A = \frac{1}{2}\pi \cdot 5^2 - 2 \cdot \frac{1}{2}\pi\left(\frac{5}{2}\right)^2 = \frac{25}{2}\pi - \frac{25}{4}\pi$$
$$= \frac{25}{4}\pi \approx 6.25(3.14) = 19.6 \text{ ft}^2$$

17. Since the displays are rectangles, the diagonal is the hypotenuse of two right triangles. Thus if x ft. is the length of the diagonal of the HD video display, then $x^2 = 72^2 + 160^2$. This equation gives $x^2 = 5184 + 25,600 = 30,784$ so that $x \approx 175$. The diagonal measurement of each video display is about 175 ft.

19. Refer to the diagram in the text, where we have a right triangle with a hypotenuse measuring 720 ft and one leg 650 ft. Thus if h ft is the height of the pole, then $h^2 + 650^2 = 720^2$. This equation gives $h^2 = 720^2 - 650^2 = 518,400 - 422,500$ $= 95,900$ so that $h \approx 310$. The height of the pole is 310 ft tall.

21. Since the DVD case is a rectangle, the diagonal is the hypotenuse of two right triangles. Thus if x cm is the length of the diagonal of the DVD case, then $x^2 = 14^2 + 11^2$. This equation gives $x^2 = 196 + 121 = 317$ so that $x \approx 17.804$. The diagonal measurement of the DVD case is about 17.804 cm.

23. Refer to the diagram in the text, where we have a right triangle with one leg measuring 45 ft and the other leg 118 ft. Thus if x ft is the length of the conveyor belt, then $x^2 = 45^2 + 118^2$. This equation gives $x^2 = 45^2 + 118^2 = 2025 + 13,924 = 15,949$ so that $x \approx 126$. The materials need to travel about 126 ft to get to the top of the conveyor belt.

25. $A = 120 \cdot 53\frac{1}{3} = 120 \cdot \frac{160}{3} = 6400$. Thus, 6400 yd^2 of turf are needed.

27. Let w ft be the width. Then, $16w = 288$, so that $w = \frac{288}{16} = 18$. The room must be 18 ft wide.

29. Let w be the width. Then, $70w = 6,720$, so that $w = \frac{6720}{70} = 96$. The screen is 96 ft wide.

31. $A = \pi r^2 = \pi \cdot 4^2 = 16\pi \text{ cm}^2$

33. Since the cost of the glass is \$4.50 at \$3/ft^2, the area of the glass is $4.50 \div 3 = 1.50 \text{ ft}^2$. Let x ft be the width of the frame, so that $1.5x$ ft is the length. The area of the glass in terms of x is $(1.5x)(x) = 1.5x^2$. Hence, we have the equation $1.5x^2 = 1.5$. Thus, $x^2 = 1$ and $x = 1$. Thus, the dimensions are 1 ft by 1.5 ft, that is, 12 in. by 18 in.

35. $A = \pi r^2 = \pi \cdot 70^2 = 4900\pi$
$$\approx 4900 \cdot 3.14 = 15,386 \text{ ft}^2$$

37. $A = \pi r^2 = \pi \cdot 16^2 = 256\pi$
$$\approx 256 \cdot 3.14 = 803.84 \text{ mi}^2$$

39. $A = \pi r^2 = \pi \cdot 5^2 = 25\pi$
$$\approx 25 \cdot 3.14 = 78.5 \text{ mi}^2$$
$$\text{Difference} = 803.84 - 78.5 = 7252.34 \text{ mi}^2$$

41. $A = \frac{1}{2}\pi r^2 = \frac{1}{2}\pi \cdot 3^2 = 4.5\pi$
$\approx 4.5(3.14) = 13.345 \text{ ft}^2$

43. Subtract the inner area from the outer area:
$A = \pi r_o^2 - \pi r_i^2 - = \pi(2.25)^2 - \pi(0.75)^2$
$= 5.0625\pi - 0.5625\pi = 4.5\pi$
$\approx 4.5(3.14) = 14.13 \text{ in}^2$

45. a. in^2 and cm^2 b. mi^2 and km^2
c. yd^2 and m^2 d. yd^2 and m^2

47. The perimeter of the first room is 20 + 24 = 44 ft, of the second room is 28 + 30 = 58 ft, and of the third room is 24 + 24 = 48 ft. Hence, the sum of the perimeters is 44 + 58 + 48 = 150 ft. The ceiling is 8 ft high, so the total wall area to be painted is $8 \cdot 150 = 1200 \text{ ft}^2$.
a. A gallon of paint covers 450 ft^2, so the number of gallons needed is
$\frac{1200}{450} = \frac{8}{3} = 2\frac{2}{3}$ gal.
b. The paint is sold only by the gallon, so 3 gallons will have to be bought at a cost of $3 \cdot 14 = \$42$.

49. The entire lot area is $100 \cdot 200 = 20{,}000 \text{ ft}^2$. The area occupied by the shed is $10 \cdot 10 = 100 \text{ ft}^2$; the area occupied by the house is $30 \cdot 50 = 1500 \text{ ft}^2$; the area occupied by the drive is $10 \cdot 50 = 500 \text{ ft}^2$. Thus, the lawn area is:
$20{,}000 - 100 - 1500 - 500 = 17{,}900 \text{ ft}^2$
a. A bag of fertilizer covers 1200 ft^2, so it will take $\frac{17{,}900}{1200} \approx 14.9^+$ bags.
Thus, 15 bags are needed.
b. The cost will be $4 \cdot 15 = \$60$.

51. a. For the 8-in. pie, $A = \pi r^2 = \pi \cdot 4^2 = 16\pi \approx 16 \cdot 3.14 \approx 50.24 \text{ in}^2$. Thus, the unit price is
$\frac{125}{50.24} \approx 2.49 \text{ cents/in}^2$.
b. For the 10-in. pie, $A = \pi r^2 = \pi \cdot 5^2 = 25\pi \approx 25 \cdot 3.14 \approx 78.5 \text{ in}^2$. Thus, the unit price is $\frac{185}{78.5} \approx 2.36 \text{ cents/in}^2$.
c. On the basis of the unit price, the 10-in. pie is the better buy.

53. The area of the rectangle taken away is Wx, and the area of the rectangle added on is hy. Since $\frac{y}{W} = \frac{x}{h}$, it follows that $hy = Wx$. Thus, the area of the new rectangle is equal to the area of the original rectangle.

Section 8.5 Volume and Surface Area

Study Tips

Three-dimensional figures called polyhedrons have volume. Some names of *polyhedrons* such as *cube* and *pyramid*, maybe even *prism*, are familiar. Not so familiar might be names such as *rectangular parallelepiped*, which we might simply call a shoebox!

To find the volume of a figure:
Think: measured in cubic units
Volume Formula Summary
Cube: $V = a^3$
Rectangular box: $V = lwh$
Cylinder: $V = \pi r^2 h$
Cone: $V = \frac{1}{3}\pi r^2 h$
Sphere: $V = \frac{4}{3}\pi r^3$
Prism: $V = Bh$
Pyramid: $V = \frac{1}{3}Bh$

To find the suface area of a figure:
Think: sum of the areas of each side
Surface Area Formula Summary
Cube: $S = 6a^2$
Rectangular box: $s = 2(lw + lh + wh)$
Cylinder: $S = 2\pi rh + 2\pi r^2$
Cone: $S = \pi r^2 + \pi rs$
Sphere: $S = 4\pi r^2$

Caution
Do not confuse *surface area* with *volume*. The *surface area* of a shoe box is, roughly, the area of the paper it will take to cover the box. The *volume*

is the amount (volume) of sand it would take to fill the box. *Surface area* is measured in *square* units, *volume* in *cubic* units.

Word of Advice

As before, memorize these formulas, practice with them, and write them each time you use them.

8.5 Exercises

1. a. *A, B, C, D, E*
   b. $\overline{AB}, \overline{AC}, \overline{AD}, \overline{AE}, \overline{BC}, \overline{BE}, \overline{CD}, \overline{CE}, \overline{DE}$

3. *ABCD*

5.

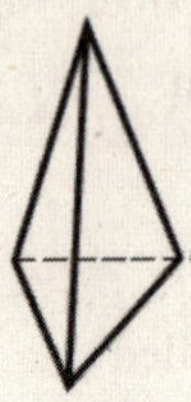

7.

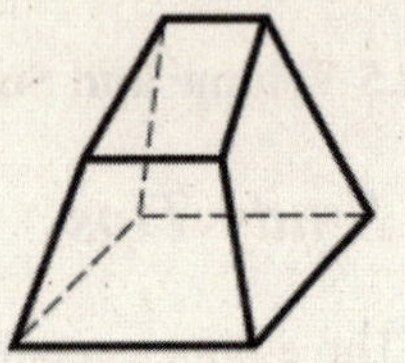

9. a. If the original edge is of length a, then $V = a^3$. The new edge is of length $2a$, so the volume is $(2a)^3 = 8a^3$. Thus, the volume is multiplied by 8.
   b. If the length of the edge is tripled, the new volume is $(3a)^3 = 27a^3$. Thus, the volume is multiplied by 27.

11. $S = 4\pi r^2 = 4\pi \cdot 24^2 = 2304\pi$
$\approx 2304 \cdot 3.14 = 7234.6 \text{ ft}^2$

13. $S = 4\pi r^2 = 4\pi \cdot 7.2^2 = 207.36\pi$
$\approx 207.36 \cdot 3.14 = 651.1 \text{ m}^2$

15. $V = Bh = 10 \cdot 5 = 50 \text{ in}^3$

17. $V = \frac{1}{3}Bh = \frac{1}{3} \cdot 492^2 \cdot 138 = 11{,}134{,}944 \text{ ft}^3$

19. The volume of the cube is $V = 10^3 = 1000 \text{ cm}^3$; the volume of the pyramid is $V_P = \frac{1}{3}Bh = \frac{1}{3} \cdot 10^2 \cdot 15$ $= 500 \text{ cm}^3$. Thus, the total volume is 1500 cm^3, or 1.5 liters.

21. a. $V = \pi r^2 h = \pi \cdot 5^2 \cdot 9 = 225\pi$
$\approx 225(3.14) \approx 707 \text{ in}^3$
$S = 2\pi rh + 2\pi r^2 = 2\pi \cdot 5 \cdot 9 + 2\pi \cdot 5^2$
$= 90\pi + 50\pi = 140\pi \approx 140(3.14) \approx 440 \text{in}^2$

    b. $V = \frac{1}{3}\pi r^2 h = \frac{1}{3}\pi \cdot 5^2 \cdot 9 = 75\pi$
$\approx 75(3.14) \approx 236 \text{ in}^3$
$S = \pi r^2 + \pi rs = \pi \cdot 5^2 + \pi \cdot 5\sqrt{106}$
$= \left(25 + 5\sqrt{106}\right)\pi \approx 76.5(3.14) \approx 240 \text{ in}^2$
(Use the Pythagorean theorem to find the slant height: $s^2 = 5^2 + 9^2$.)

23. a. $V = \pi r^2 h = \pi \cdot 3^2 \cdot 4 = 36\pi$
$\approx 36(3.14) \approx 113 \text{ ft}^3$
$S = 2\pi rh + 2\pi r^2 = 2\pi \cdot 3 \cdot 4 + 2\pi \cdot 3^2$
$= 24\pi + 18\pi = 42\pi \approx 42(3.14) \approx 132 \text{ft}^2$

    b. $V = \frac{1}{3}\pi r^2 h = \frac{1}{3}\pi \cdot 3^2 \cdot 4 = 12\pi$
$\approx 12(3.14) \approx 37.7 \text{ ft}^3$
$S = \pi r^2 + \pi rs = \pi \cdot 3^2 + \pi \cdot 3 \cdot 5$
$= (9 + 15)\pi \approx 24(3.14) \approx 75.4 \text{ ft}^2$

25. $V = \frac{4}{3}\pi r^3 = \frac{4}{3}\pi \cdot 6^3 = 288\pi$
$\approx 288(3.14) \approx 904 \text{ in}^3$

27. $V = \pi r^2 h = \pi \cdot 58^2 \cdot 754 = 2{,}536{,}456\pi$
$\approx 2{,}536{,}456(3.14) \approx 7{,}960{,}000 \text{ ft}^3$

29. $V = \frac{1}{3}\pi r^2 h = \frac{1}{3}\pi \cdot 500^2 \cdot 2000 = 166{,}666{,}666.7\pi$
$\approx 166{,}666{,}666.7(3.14) \approx 523{,}333{,}333.3 \text{ ft}^3$

31. $r = 1.5(2.54)$ cm and $h = 4(2.54)$ cm.
$V = \pi r^2 h = \pi\left[1.5(2.54)\right]^2(4)(2.54)$
$= \pi(9)(2.54)^3 \approx (3.14)(147.5) \approx 463 \text{ cm}^3$
Thus, the can will hold about 463 g.

33. a. Baseball: $r = \frac{9}{2\pi}$, so

$$S_1 = 4\pi r^2 = 4\pi\left(\frac{9}{2\pi}\right)^2 = \frac{81}{\pi}$$

$$\approx 25.8 \text{ in}^2$$

$$V_1 = \frac{4}{3}\pi r^3 = \frac{4}{3}\pi\left(\frac{9}{2\pi}\right)^3 = \frac{243}{2\pi^2}$$

$$\approx 12.3 \text{ in}^3$$

Soccer Ball: $r = \frac{27}{2\pi}$, so

$$S_2 = 4\pi r^2 = 4\pi\left(\frac{27}{2\pi}\right)^2 = \frac{729}{\pi}$$

$$\approx 232 \text{ in}^2$$

$$V_2 = \frac{4}{3}\pi r^3 = \frac{4}{3}\pi\left(\frac{27}{2\pi}\right)^3 = \frac{6561}{2\pi^2}$$

$$\approx 333 \text{ in}^3$$

Basketball: $r = \frac{30}{2\pi}$, so

$$S_3 = 4\pi r^2 = 4\pi\left(\frac{30}{2\pi}\right)^2 = \frac{900}{\pi}$$

$$\approx 287 \text{ in}^2$$

$$V_3 = \frac{4}{3}\pi r^3 = \frac{4}{3}\pi\left(\frac{30}{2\pi}\right)^3 = \frac{4500}{\pi^2}$$

$$\approx 456 \text{ in}^3$$

b. $$\frac{S_1}{S_3} = \frac{81/\pi}{900/\pi} = \frac{81}{\pi}\cdot\frac{\pi}{900} = \frac{9}{100}$$

c. $$\frac{V_1}{V_2} = \frac{243/(2\pi^2)}{6561/(2\pi^2)} = \frac{243}{2\pi^2}\cdot\frac{2\pi^2}{6561}$$

$$= \frac{1}{27}$$

35. a. $$V = \frac{4}{3}\pi r^3 = \frac{4}{3}\pi\left(\frac{27}{2\pi}\right)^3 = \frac{6561}{2\pi^2}$$

$$\approx 333 \text{ in}^3$$

b. The volumes drilled out are approximately cylinders of radii 5/8 and 1/2 in., respectively, and height 5/2 in. This volume is:

$$\pi\left(\frac{5}{8}\right)^2\cdot\frac{5}{2} + \pi\left(\frac{1}{2}\right)^2\cdot\frac{5}{2} + \pi\left(\frac{1}{2}\right)^2\cdot\frac{5}{2}$$

$$= \left(\frac{125}{128} + \frac{5}{8} + \frac{5}{8}\right)\pi = \frac{285}{128}\pi \approx 6.99 \text{ in}^3$$

Thus, the volume remaining is about $333 - 7 = 326 \text{ in}^3$ so that the final weight is about $\frac{326}{333}\cdot 16 \approx 15.66$ lb.

37. a. $V = Bh = 7\cdot 12\cdot 17 = 1428 \text{ in}^3$

b. S = area of base + area of four sides $= 7\cdot 12 + 2\cdot 7\cdot 17 + 2\cdot 12\cdot 17 = 730 \text{ in}^2$

c. Two tons of recycled paper are needed in order to save 34 trees. One ton is equal to 2000 lb, and one pound is equal to 16 oz. So, for two tons of recycled paper, we have $\frac{2\cdot 2000\cdot 16}{2} = 32{,}000$ bags have to be recycled to save 34 trees.

39. $V = lwh = 16\cdot 11\cdot 10 = 1760 \text{ in}^3$

41. $V = lwh = 16\cdot 11\cdot 13 = 2288 \text{ in}^3$

43. $V = \pi r^2 h = \pi(1.5)^2(5) = 11.25\pi$

$$\approx 11.25\cdot 3.14 \approx 35.33 \text{ in}^3$$

45. $2.5(35.33) = 88 \text{ in}^3$

47. a. $V = lwh = 22\cdot 7\cdot 8 = 1232 \text{ ft}^3$

b. $V = lwh = 3\cdot 7\cdot 3 = 63 \text{ ft}^3$

c. Truck volume: $1232 + 63 = 1295 \text{ ft}^3$
No. $1300 > 1295$.

49. $V = \pi r^2 h = \pi\left(\frac{1.5}{2}\right)^2(2.5) = 1.41\pi$

$$\approx 1.41\cdot 3.14 \approx 4.43 \text{ ft}^3$$

51. $V_{Bottom} = \pi r^2 h = \pi\left(\frac{1.5}{2}\right)^2(2.5) = 1.41\pi$

$$\approx 1.41\cdot 3.14 \approx 4.43 \text{ ft}^3$$

$$V_{Top} = \frac{1}{2}\left(\frac{4}{3}\pi r^3\right) = \frac{1}{2}\cdot\frac{4}{3}\pi\left(\frac{1.5}{2}\right)^3$$

$$\approx 0.88 \text{ ft}^3$$

$$V_{Total} = 4.43 + 0.88 \approx 5.31 \text{ ft}^3$$

53. $\dfrac{500\text{ ft}^3}{5.31\text{ ft}^3} \approx 94.16 \approx 94$ cans of trash

55. Answers will vary.

57. $C_M = 2\pi r = 2\pi \cdot 1 = 2\pi$

$C_B = 2\pi r = 2\pi \cdot (1.75) = 3.50\pi$

The ratio of the circumferences is $\dfrac{3.50\pi}{2\pi} = 1.75$, so the Burger King burger is 75% larger than the McDonald burger.

59. Answers will vary. Length is one dimensional, area is two dimensional, and volume is three dimensional.

61. No. This is true only if the radius is more than three units long. (Compare $\frac{4}{3}\pi r^3$ and $4\pi r^2$.)

63. Since the sum of the face angles at a vertex must be less than 360°, three, four or five equilateral triangles, three squares or three pentagons can be put together at any vertex. There are no other possibilities, so only the five regular polyhedrons listed are possible.

65.

Figure	*F*	*V*	*E*	*F* + *V*
8.68B	6	8	12	$14 = E + 2$
8.69C	6	5	9	$11 = E + 2$
8.69D	7	10	15	$17 = E + 2$
8.70	5	6	9	$11 = E + 2$
8.71	6	6	10	$12 = E + 2$

Thus, Euler's formula is F + V = E + 2.

Calculator Corner

1. [4] [×] [π] [×] [6] [y^x] [3] [÷] [3] [ENTER]

$904.78 \approx 905$

Section 8.6 Networks, Non-Euclidean Geometry, and Topology

Study Tips

Did you know that some figures are impossible to trace, a right angle is not always 90°, and a straight line and the letter C can be considered equivalent? Traversability, non-Euclidean geometry, and topology provide these interesting results!

To determine if a network is traversable:
Think: exactly two odd vertices or no odd vertices
Note: So, if you have just *one odd* vertex or *more than two odd* vertices, the network is *not* traversable.

Hint
To determine traversability, the key is to know whether the vertices (the endpoints of the arcs, or edges, forming the figure) are odd or even.

To determine the angle measure of a quadrilateral:
Think: what geometry is being used?
Euclidean plane: 90°
hyperbolic: less than 90°
elliptic: more than 90°
Note: Hyperbolic geometry is based on a pseudosphere, while elliptic geometry is based on a sphere.

Fact
Non-Euclidean geometry is based on the rejection of Euclid's Fifth Postulate: Given a line and a point not on that line, there is one and only one line through that point parallel to the original line.

To determine topologically equivalent figures:
Think: how many holes do you see? equivalent if have same number
Note: The term *genus* describes how many holes in a figure. If a figure has three holes, the genus is 3.

8.6 Exercises

1. a. 3 b. 0
   c. This network is traversable; all three vertices are possible starting points.

3. a. 3 (*B*, *D*, and *E*) b. 2 (*A* and *C*)
   c. This network is traversable; start at either *A* or *C*.

5. a. 1 (*A* only) b. 4 (*B*, *C*, *D*, *E*)
   c. This network is not traversable; it has more than 2 odd vertices.

7. a. 5 (*A*, *C*, *D*, *E*, *G*) b. 2 (*B*, *F*)
   c. This network is traversable; start at *B* or *F*.

9. a. 1 (The vertex of the pyramid)
   b. 4 (The vertices of the base)
   c. This network is not traversable; it has more than 2 odd vertices.

11. Think of each region as a vertex with the individual line segments in its boundary as the number of paths to the vertex. The boundary of region *A* has four segments, so the corresponding network point would be even. The boundary of region *B* has five segments, so the corresponding network point would be odd. The boundary of region *C* has four segments, so the corresponding network point would be even. The boundary of region *D* has five segments, so the corresponding network point would be odd. The boundary of region *E* has 10 segments, so the corresponding network point would be even. Thus, the network would have two odd vertices (*B* and *D*). By starting in region *B* or *D*, it is possible to draw a simple connected broken line that crosses each line segment exactly once.

13. Region *A* has three doorways, so the corresponding vertex would be odd. Regions *B*, *C*, *D* and *E* each have two doorways, so the corresponding vertices would be even. Region *F* has three doorways, so the corresponding vertex would be odd. (Two doorways lead to the outside region, so the outside region is also an even vertex.) Thus, the network would have two odd vertices, *A* and *F*. By starting in *A* and ending in *F* (or vice versa), it is possible to take a walk and pass through each doorway exactly once. It is not possible to start and end outside, because the walk must start in one of the regions corresponding to one odd vertex and end in the region corresponding to the other odd vertex.

15. *A* and *D* each have three doorways and all the other rooms have an even number of doorways. Thus, the corresponding network would have two odd vertices, *A* and *D*. By starting in either *A* or *D* and ending in the other, it is possible for a walk to pass through each doorway exactly once. It is not possible to start and end outside.

17. All the rooms and the outside have an even number of doorways, so the corresponding network would have no odd vertices. The walk can start in any room or outside and end in the same place and pass through each doorway exactly once.

19. Rooms *B* and *D* each have three doorways and all the other rooms have an even number of doorways (two doorways lead to the outside region, so the outside region is also an even vertex), so the corresponding network would have two odd vertices. The walk can start in either *B* or *D* and end in the other, passing through each doorway exactly once. It is not possible to start and end outside.

21. Room *A* and the outside *D* each have three doorways and *B* and *C* each have an even number of doorways, so the corresponding network would have two odd vertices. By starting in *A* and ending in *D* (or vice versa), a walk could pass through each doorway exactly once. It is not possible to start and end outside.

23. Given a line and any point not on that line, there is one and only one line through that point and parallel to the given line.

25. Given a line and any point not on that line, there is no line through that point that is parallel to the given line.

27. In hyperbolic geometry

29. The surface of a rectangular box

31. The surface of a sphere

33. (a) - (e) are topologically equivalent

35. (a) - (e) are of genus 2. (f) is of genus 1.

37. Topo is correct. If you cut through a loop of the left-hand figure, you can unwind it into a single strip as you can with the circular cord.

39. a. 6 points are marked on the network.
    b. Each point has 2 edges.
    c. Yes, because there are no odd vertices.
    d. 10,471 km
    e. 108 hr
    f. $\frac{10,471\text{ km}}{108\text{ hr}} \approx 97\text{ km/hr}$

    $\frac{97\text{ km}}{\text{hr}} \cdot \frac{1\text{ mi}}{1.609\text{ km}} \approx 60\text{ mi/hr}$

41. a. Yes. Removing any bridge will reduce by one the edges at two vertices (making

them even) leaving only two odd vertices. So, for instance, close bridge *f*.

b. You have to start at one of the odd vertices.

c. You will end at the other odd vertex.

43. Since each arc has two endpoints, the total number of endpoints must be even. An odd vertex accounts for an odd number of endpoints, while an even vertex accounts for an even number of endpoints. Thus, there must be an even number of odd vertices.

Exercises 45 – 51, refer to the table.
You can check that the results in problems

Figure	Vertices (V)	Regions (R)	Arcs (A)
45	3	2	3
47	5	2	5
49	2	3	3
51	6	3	7

45 – 51 all fit the formula $V + R = A + 2$.

Section 8.7 Right Triangle Trigonometry

Study Tips

How can you measure a distance without actually measuring? This is what you will encounter in this section on right triangle trigonometry.

To find the trigonmetric ratios:
Think: adjacent, opposite, hypotenuse, then these relationships:
cos A = adjacent/hypotenuse
sin A = opposite/hypotenuse
tan A = opposite/adjacent
Note: The concepts presented here will only apply to right triangles.

To solve a triangle:
Think: measures of angles, length of sides
Note: Often the values of sine, cosine, and tangent are rounded, so there may be a slight variation in results.

To solve an application problem:
Think: draw diagram, use trigonometry ratios
Note: Apply the problem solving procedure RSTUV.

Word of Advice
When working application problems, it is cruicial that you draw a *diagram* of the situation. Many mistakes are avoided by a simple drawing!

Word of Advice
Memorizing the definitions of *cosine*, *sine*, and *tangent* will be very helpful. Using a mnemonic such as **CAH SOH TOA** will help with the memorization. A silly question is "What do you do when you stub your toe?" You go(cah) soak(soh) your toe(toa).

8.7 Exercises

1. Substitute the known values into $c^2 = a^2 + b^2$.

$c^2 = 3^2 + 4^2 = 9 + 16 = 25$

$c = 5$

3. Substitute the known values into $c^2 = a^2 + b^2$.

$c^2 = 8^2 + 15^2 = 64 + 225 = 289$

$c = 17$

5. Substitute the known values into $b^2 = c^2 - a^2$.

$b^2 = 15^2 - 9^2 = 225 - 81 = 144$

$b = 12$

7. Substitute the known values into $a^2 = c^2 - b^2$.

$a^2 = 26^2 - 24^2 = 676 - 576 = 100$

$a = 10$

9. a. $\cos A = \frac{b}{c} = \frac{3}{5}$ b. $\sin A = \frac{a}{c} = \frac{4}{5}$

c. $\tan A = \frac{a}{b} = \frac{4}{3}$

11. a. $\cos A = \frac{b}{c} = \frac{16}{34} = \frac{8}{17}$

b. $\sin A = \frac{a}{c} = \frac{30}{34} = \frac{15}{17}$

c. $\tan A = \frac{a}{b} = \frac{30}{16} = \frac{15}{8}$

13. a. Substitute the known values into $c^2 = a^2 + b^2$.

$c^2 = 9^2 + 12^2 = 81 + 144 = 225$

$c = 15$

b. $\cos A = \dfrac{9}{15} = \dfrac{3}{5}$; $\sin A = \dfrac{12}{15} = \dfrac{4}{5}$

$\tan A = \dfrac{12}{9} = \dfrac{4}{3}$

15. a. Substitute the known values into $c^2 = a^2 + b^2$.

$c^2 = 12^2 + 35^2 = 144 + 1225 = 1369$

$c = 37$

b. $\cos A = \dfrac{12}{37}$; $\sin A = \dfrac{35}{37}$

$\tan A = \dfrac{35}{12}$

17. a. $\tan A = \dfrac{a}{16} = 1.43$, then solve for *a*.

$a = 16 \cdot 1.43 \approx 23$ in.

b. Substitute the known values into $c^2 = a^2 + b^2$.

$c^2 = 23^2 + 16^2 = 529 + 256 = 785$

$c \approx 28$ in.

19. a. $\tan A = \dfrac{24}{b} = 0.36$, then solve for *b*.

$b = \dfrac{24}{0.36} \approx 67$ in.

b. Substitute the known values into $c^2 = a^2 + b^2$.

$c^2 = 24^2 + 67^2 = 576 + 4489 = 5065$

$c \approx 71$ in.

21. Let *h* be the height of the tree.

$\tan 72° = \dfrac{\text{opposite}}{\text{adjacent}} = \dfrac{h}{32}$

$3 = \dfrac{h}{32}$

$h = 3 \cdot 32 = 96$

The tree is 96 ft high.

23. Let *a* be altitude of the plane. The trajectory of the plane is $d = r \cdot t = 5 \cdot 5 = 25$ mi.

$\sin 30° = \dfrac{\text{opposite}}{\text{hypotenuse}} = \dfrac{a}{25}$

$0.5 = \dfrac{a}{25}$

$a = 0.5 \cdot 25 \approx 13$

The altitude of the plane is 13 mi.

25. Let *h* be the height of the bird. The trajectory of the bird is $d = r \cdot t = 200 \cdot 7 = 1400$ ft.

$\sin 12° = \dfrac{\text{opposite}}{\text{hypotenuse}} = \dfrac{a}{1400}$

$0.21 = \dfrac{a}{1400}$

$a = 0.21 \cdot 1400 \approx 294$

The height of the bird is 294 ft.

27. Let *x* be the length of the runway. The altitude of the plane is 40 ft.

$\tan 10° = \dfrac{\text{opposite}}{\text{adjacent}} = \dfrac{40}{x}$

$0.18 = \dfrac{40}{x}$

$x = \dfrac{40}{0.18} \approx 222$

The length of the runway is 222 ft.

29. Let *x* be the distance the materials travel. The height of the belt at the end is 40 ft.

$\sin 20° = \dfrac{\text{opposite}}{\text{hypotenuse}} = \dfrac{40}{x}$

$0.34 = \dfrac{40}{x}$

$x = \dfrac{40}{0.34} \approx 118$

The distance the materials travel is 118 ft. The length of the conveyor belt is $118 \cdot 2 = 236$ ft.

31. Answers will vary.

33. Answers may vary. The tangent of the angle gets infinitely large.

35. If $c = 2$, then $\cos 30° = \frac{a}{2}$. Solve for a:

$$\frac{\sqrt{3}}{2} = \frac{a}{2};\ a = 2 \cdot \frac{\sqrt{3}}{2} = \sqrt{3}$$

If $c = 2$, then $\sin 30° = \frac{b}{2}$. Solve for b:

$$\frac{1}{2} = \frac{b}{2};\ b = 2 \cdot \frac{1}{2} = 1$$

Alternatively, since c and a are known, we can use $b^2 = c^2 - a^2$ to find b.

$$b^2 = 2^2 - (\sqrt{3})^2 = 4 - 3 = 1$$

$$b = 1$$

37. If $a = \sqrt{3}$, then $\sin 60° = \frac{\sqrt{3}}{c}$.

Solve for c:

$$\frac{\sqrt{3}}{2} = \frac{\sqrt{3}}{c};\ c = \frac{2}{\sqrt{3}} \cdot \sqrt{3} = 2$$

If $a = \sqrt{3}$, then $\cos 60° = \frac{b}{2}$. Solve for b:

$$\frac{1}{2} = \frac{b}{2};\ b = 2 \cdot \frac{1}{2} = 1$$

Alternatively, since c and a are known, we can use $b^2 = c^2 - a^2$ to find b.

$$b^2 = 2^2 - (\sqrt{3})^2 = 4 - 3 = 1$$

$$b = 1$$

39. $\sin 30° = \frac{b}{c}$ so that $b = c \sin 30°$ and

$\cos 30° = \frac{a}{c}$ so that $a = c \cos 30°$

$$\tan 30° = \frac{b}{a} = \frac{c \sin 30°}{c \cos 30°} = \frac{\sin 30°}{\cos 30°}$$

$$= \frac{\frac{1}{2}}{\frac{\sqrt{3}}{2}} = \frac{1}{2} \cdot \frac{2}{\sqrt{3}} = \frac{1}{\sqrt{3}} \cdot \frac{\sqrt{3}}{\sqrt{3}} = \frac{\sqrt{3}}{3}$$

41. Given $(\sin A)^2 + (\cos A)^2 = 1$, then $(\sin 30°)^2 + (\cos 30°)^2 = 1$ and with the Pythagorean theorem $a^2 + b^2 = c^2$, we have $a = \sin 30°$ and $b = \cos 30°$. We also know that $\tan A = \frac{\text{opposite}}{\text{hypotenuse}} = \frac{a}{b}$, so

$\tan 30° = \frac{\sin 30°}{\cos 30°}$. Substitute the values for $\sin 30°$ and $\cos 30°$:

$$\tan 30° = \frac{\frac{1}{2}}{\frac{\sqrt{3}}{2}} = \frac{1}{2} \cdot \frac{2}{\sqrt{3}} = \frac{1}{\sqrt{3}} \cdot \frac{\sqrt{3}}{\sqrt{3}} = \frac{\sqrt{3}}{3}$$

This answer is identical to the answer for problem 39.

43. Given $(\sin A)^2 + (\cos A)^2 = 1$, then $(\sin 15°)^2 + (\cos 15°)^2 = 1$. Substitute cos 15° value and solve for sin 15°.

$$(\sin 15°)^2 + (0.97)^2 = 1$$

$$(\sin 15°)^2 = 1 - (0.97)^2 = 1 - 0.9409 = 0.0591$$

$$\sin 15° = \sqrt{0.0591} \approx 0.24$$

$\tan 15° = \frac{\sin 15°}{\cos 15°}$. Substitute the values for sin 15° and cos 15°.

$$\tan 15° = \frac{0.24}{0.97} \approx 0.25$$

Section 8.8 Chaos and Fractals

Study Tips

If you have watched a movie, undoubably you have heard of computer-generated scenes. We can thank the concepts of *chaos* and *fractals* for those scenes.

To find an attractor:
Think: repeated calculation, always use previous answer in each next calculation
Note: Use the ANS key on your calculator. For instance, enter the value of x, say $x = 0.2$, and press ENTER. Then enter the equation using ANS instead of x. This will allow you to easily take an answer given and use it in your next calculation.

Hint
To repeat a calculation press ENTRY.

To construct a fractal:
Think: repeatedly remove a portion

Note: The Koch snowflake has *infinite perimeter* but the area is just 1.6 times the area of the original starting triangle.

Word of Advice
With fractal geometry, be patient and use a pencil, not a pen, to draw the requested drawings.

8.8 Exercises

Note: Answers may vary because of rounding.

1.

x	0.200	0.862	0.850	0.865	0.846
y	0.560	0.416	0.446	0.409	0.456

x	0.868	0.841	0.871	0.835	0.874
y	0.402	0.468	0.393	0.482	0.385

x	0.829	**0.875**	**0.827**	**0.875**	**0.827**
y	0.496	**0.383**	**0.501**	**0.383**	**0.501**

There are four attractors: **0.383, 0.501, 0.827**, and **0.875**.

3.

x	0.400	0.605	0.609	0.613	0.615
y	0.720	0.717	0.714	0.712	0.710

x	0.618	0.620	0.621	0.623	0.624
y	0.708	0.707	0.706	0.705	0.704

x	0.625	0.626	0.628	0.629	0.630
y	0.703	0.702	0.701	0.700	**0.699**

x	**0.631**	**0.631**
y	**0.699**	**0.699**

There are two attractors: **0.631** and **0.699**.

5. Take s as the length of one side of the Original triangle, thus the original perimeter is $3s$. For each iteration, the length of each piece of the broken line will decrease by $\frac{1}{3}$, and the number of pieces in the broken line will increase by a factor of 4.
   a. The perimeter of the third step (Step C in Figure 8.94) would be
   $$3\left(\frac{1}{3^1}s\cdot 4^1\right)=4s$$
   b. The perimeter of the fourth step (Step D in Figure 8.94) would be
   $$3\left(\frac{1}{3^2}s\cdot 4^2\right)=\frac{16}{3}s$$

7. On completing the second step, the fractal appears like this.

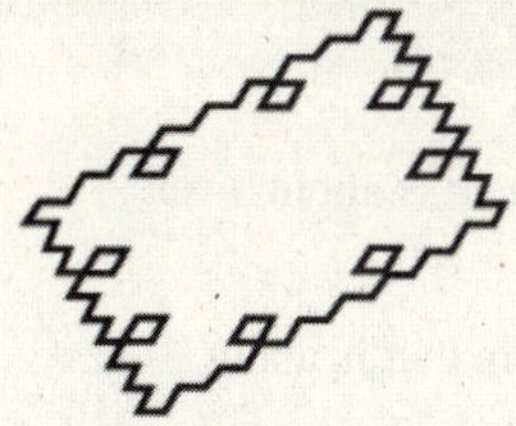

9. For each step, each small square is divided into fourths. Thus, the area will decrease by $\frac{1}{4}$. So for the first step the area of first small square is $\frac{1}{4}\text{ in}^2$, the second step $\frac{1}{16}\text{ in}^2$, the third step $\frac{1}{64}\text{ in}^2$, and the fourth step $\frac{1}{256}\text{ in}^2$. Then for the fifth step the area of the small square will be $\frac{1}{1024}\text{ in}^2$.

11. On completing the second step, the fractal appears like this.

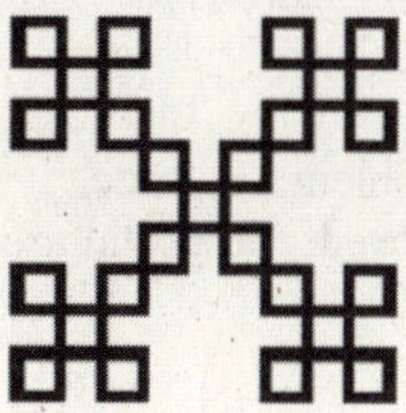

13. After the second step, the area will decrease by a factor of $\frac{5}{9}$ (see problem 12). Thus, the area after the second step would be $\frac{5}{9}\cdot\frac{5}{9}s^2=\frac{25}{81}s^2$.

15. The remaining area approaches zero.

17. a.

b. The initial segment is about $\frac{1}{2}$ in.,

the segment from that is about $\frac{1}{4}$ in.

Their ratio is $\frac{\frac{1}{2}}{\frac{1}{4}} = \frac{1}{2} \cdot 4 = 2$.

c. The length of the next segment would be $\frac{1}{8}$ in. This ratio is

$\frac{\frac{1}{4}}{\frac{1}{8}} = \frac{1}{4} \cdot 8 = 2$. Yes, the ratio is maintained.

d. The measure of the angle is about 165°.

e. Yes, the angle measure is maintained.

19. In the first step of problem 18, three triangles will be removed, for a total area of

$$3\left(\frac{1}{2} \cdot \frac{1}{3} s \cdot \frac{\sqrt{3}}{6} s\right) = 3\left(\frac{\sqrt{3}}{36} s^2\right) = \frac{\sqrt{3}}{12} s^2.$$

Subtract from the original area to obtain the new area $\frac{\sqrt{3}}{4} s^2 - \frac{\sqrt{3}}{12} s^2 = \frac{\sqrt{3}}{6} s^2$.

21. The perimeter will increase without limit.

Chapter 8 Practice Test

Study Tips

Hopefully, you have been taking notes both during your individual study and in class. This chapter contains so many concepts that we shall offer a slightly different test-preparation technique.

Get a ruled writing tablet or pad with a one inch margin (usually it has two red, parallel, vertical lines for margin. By the way, see how geometry terms come up in everyday life?)

Now, divide the writing area into two columns.

a. To the left of the red line, write the concept, key idea or word you are studying.

b. In the middle column, write the definition, theorem, procedure or formula that corresponds to the idea.

c. In the far right column, if it is a...
definition, think of examples that illustrate it and write them
theorem, do the same as for a definition
procedure or a *formula*, write what the procedure or formula does. What kind of problems use this procedure or formula? Write examples.

To get some practice, do this with the concepts of line, angle, the Pythagorean Theorem. Try reading the chapter again and jotting down the important concepts using this three column studying technique. Here are the solutions to the practice test.

1. a. $\overline{XY}$ b. Point Y c. The ray $\overrightarrow{WZ}$

2.

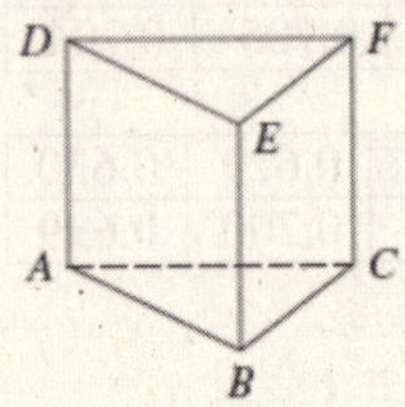

a. $\overline{AD}$, $\overline{BE}$, and $\overline{CF}$; $\overline{BC}$ and $\overline{EF}$; $\overline{DE}$ and $\overline{AB}$; and $\overline{DF}$ and $\overline{AC}$

b. $\overline{AB}$ and $\overline{CF}$; $\overline{AB}$ and $\overline{DF}$; $\overline{AB}$ and $\overline{EF}$; $\overline{BC}$ and $\overline{AD}$; $\overline{BC}$ and $\overline{DE}$; $\overline{BC}$ and $\overline{DF}$; $\overline{AC}$ and $\overline{BE}$; $\overline{AC}$ and $\overline{DE}$; $\overline{AC}$ and $\overline{EF}$; $\overline{DE}$ and $\overline{CF}$; $\overline{EF}$ and $\overline{AD}$; $\overline{DF}$ and $\overline{BE}$

c. $\overline{AB}$, $\overline{AC}$, and $\overline{AD}$; $\overline{AD}$, $\overline{DE}$, and $\overline{DF}$; $\overline{BE}$, $\overline{AB}$, and $\overline{BC}$; $\overline{BE}$, $\overline{EF}$, and $\overline{ED}$; $\overline{CF}$, $\overline{AC}$, and $\overline{BC}$; $\overline{CF}$, $\overline{EF}$, and $\overline{DF}$

3. a. $\frac{4}{12} \cdot 360° = 120°$ b. $\frac{2}{12} \cdot 360° = 60°$

4. Let $m\angle B = x°$. Then $m\angle A = 3x°$. Since the two angles are supplementary,

$$x + 3x = 180$$
$$4x = 180$$
$$x = 45$$

Therefore, $m\angle B = 45°$ and $m\angle A = 135°$.

5. a. $\angle C$ and the 50° angle are supplementary, so $m\angle C = 130°$.
   b. $\angle E$ and the 50° angle are equal, so $m\angle E = 50°$.
   c. $\angle D$ and $\angle C$ are equal, so $m\angle D = 130°$.

6. a. $m\angle C = 180° - 38° - 43° = 99°$
   b. Let $m\angle C = x°$, so that $m\angle A = m\angle B = 2x°$. Then,

$$x + 2x + 2x = 180$$
$$5x = 180$$
$$x = 36$$

Thus, $m\angle C = 36°$ and $m\angle A = m\angle B = 72°$.

7. a.

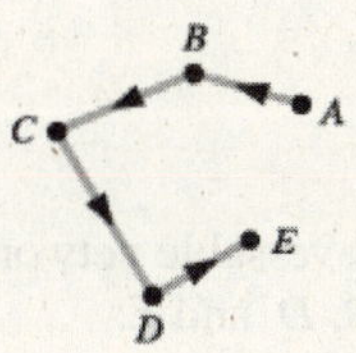

b.

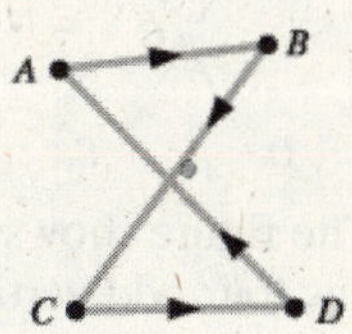

8. Since the triangles are similar, the corresponding sides are proportional. Thus, we have (see the figure)

$$\frac{x}{3} = \frac{y}{4} = \frac{3}{2}$$

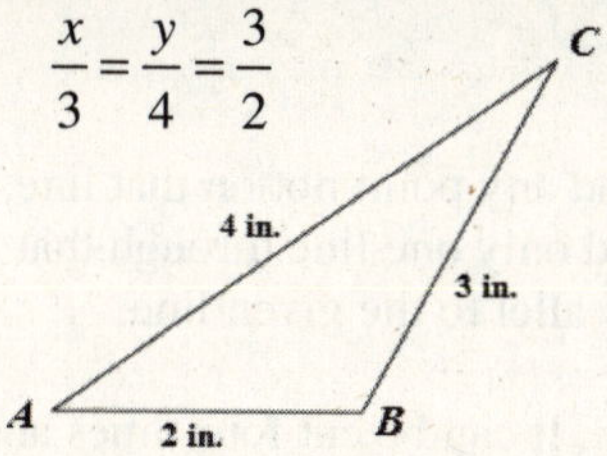

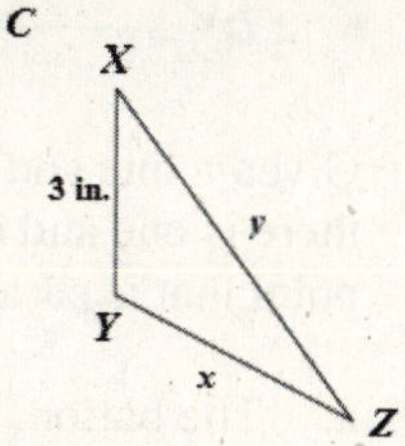

Therefore, $x = \frac{9}{2} = 4\frac{1}{2}$ in. and

$y = \frac{12}{2} = 6$ in.

9. a. The sum of the measures of the 9 angles is $(9-2)180°$, so the measure of one angle is $\frac{(9-2)180°}{9} = 140°$.
   b. The sum of the measures of the 10 angles is $(10-2)180°$, so the measure of one angle is $\frac{(10-2)180°}{10} = 144°$.

10. If w is the width, then the length is $2w$. Hence, the perimeter is $6w$, and this is to be 120 yd. Thus, $6w = 120$, so that $w = 20$. This means that the width is 20 yd and the length is 40 yd.

11. The figure shows that d is the hypotenuse of a right triangle whose legs are each 2 cm long.

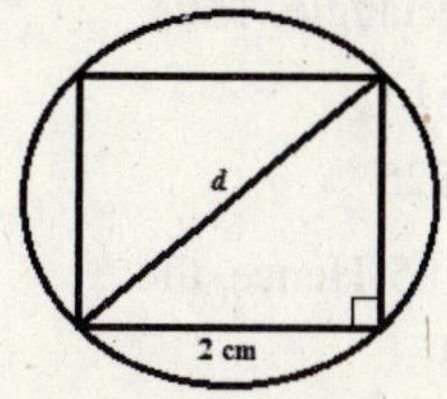

Thus, $d^2 = 2^2 + 2^2 = 8$, so $d = \sqrt{8} = 2\sqrt{2}$ cm, and the circumference of the circle is $C = \pi d = 2\sqrt{2}\pi$ cm.

12. The circumference of the cylinder would be 11 in. Hence, if the diameter is d, then $\pi d = 11$, so that $d = \frac{11}{\pi}$ in.

13. The area that is inside the circle and outside the square is $\pi r^2 - s^2$, where r is the radius of the circle and s is the side of the square. Thus, this area is $\pi\left(\sqrt{2}\right)^2 - 2^2 = (2\pi - 4)$ cm^2.

14. As the figure shows, we need to find the area of a rectangle of base 3 ft and height 4 ft, and the area of a semicircle of radius 3/2 ft.

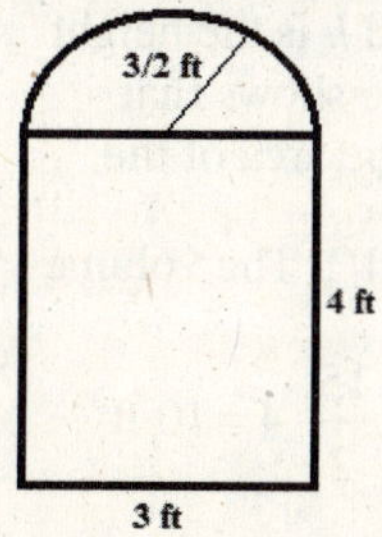

Thus, $A = bh + \frac{1}{2}\pi r^2 = 3 \cdot 4 + \frac{1}{2}\pi\left(\frac{3}{2}\right)^2$

$$= 12 + \frac{9}{8}\pi \approx 12 + 1.125(3.14)$$

$$\approx 15.5 \text{ ft}^2$$

15. The area that is outside the circle and inside the square is $s^2 - \pi r^2$, where r is the radius of the circle and s is the side of the square. Thus, this area is

$$2^2 - \pi \cdot 1^2 = (4 - \pi)$$

$$\approx 4 - 3.14 \approx 0.86 \text{ in}^2$$

16. The diagonal of a rectangle is the hypotenuse of a right triangle whose legs are the sides of the rectangle. Thus,

$$d^2 = b^2 + h^2 = 84^2 + 13^2$$

$$= 7056 + 169 = 7225$$

so that $d = \sqrt{7225} = 85$. Hence, the diagonal is 85 ft long.

17. $S = 2(lw + lh + wh)$

$$= 2(5 \cdot 3 + 5 \cdot 2 + 3 \cdot 2) = 62 \text{ ft}^2$$

18. The volume of a pyramid is given by the formula $V = \frac{1}{3}BH$ where B is the area of the base and H is the height of the pyramid.

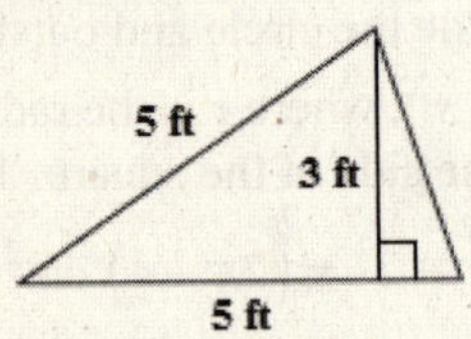

The given triangle is the base of the pyramid, and its area is $\frac{1}{2}bh$, where b is the length of the base and h is the height of the triangle. The figure shows that $b = 5$ ft and $h = 3$ ft, so the area of the triangle is $\frac{1}{2} \cdot 5 \cdot 3 = \frac{15}{2}$ ft^2. The volume of the pyramid is $V = \frac{1}{3} \cdot \frac{15}{2} \cdot 4 = 10$ ft^3.

19. The volume of the hemisphere is

$$\frac{1}{2}\left(\frac{4}{3}\pi r^3\right) = \frac{2}{3}\pi \cdot 2^3 = \frac{16}{3}\pi \text{ in}^3.$$

The volume of the cone is

$\frac{1}{3}\pi r^2 h = \frac{1}{3}\pi \cdot 2^2 \cdot h = \frac{4}{3}\pi h$ in^3, where h is the number of inches in the height of the cone. Since the volumes are equal, we have the equation $\frac{4}{3}\pi h = \frac{16}{3}\pi$. Dividing both sides by $\frac{4}{3}\pi$, we get $h = 4$. Thus, the cone is 4 in. high.

20. The surface area of the sphere is $4\pi r^2$ and of the cylinder is $2\pi rh + 2\pi r^2$. Since these areas are to be equal, we have

$$2\pi rh + 2\pi r^2 = 4\pi r^2$$

Subtract $2\pi r^2$: $\quad 2\pi rh = 2\pi r^2$

Divide by $2\pi r$: $\quad h = r$

The height of the cylinder is equal to the given radius, 10 in.

21. The figure shows a traversable network; it has no odd vertices.

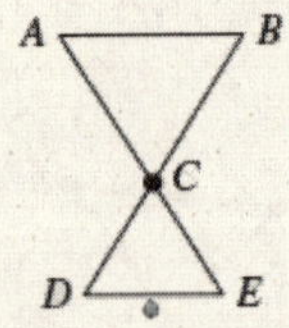

The figure shows a non-traversable network; it has four odd vertices: *A*, *B*, *D*, and *E*.

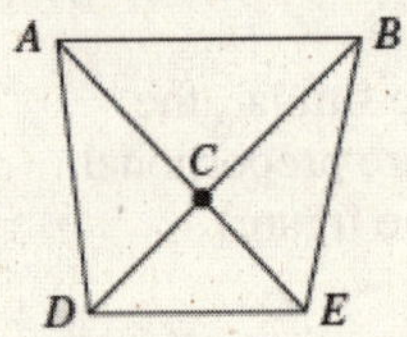

22. Given a line and any point not on that line, there is one and only one line through that point that is parallel to the given line.

23. a. The button. It can be cut four times and still be in one piece. This cannot be done with any of the other objects.
    b. Zero (No holes)
    c. Four (Four holes)
    d. Zero

24. a. Substitute the known values into $c^2 = a^2 + b^2$.

$$c^2 = 20^2 + 21^2 = 400 + 441 = 841$$

$$c = 29$$

b. $\cos A = \dfrac{20}{29}$; $\sin A = \dfrac{21}{29}$

$\tan A = \dfrac{21}{20}$

25. $\tan 40° = \dfrac{26}{b} = 0.84$, then solve for b.

$b = \dfrac{26}{0.84} \approx 31 \text{ in.}$

26. $\cos 15° = \dfrac{16}{c} = 0.97$, then solve for c.

$c = \dfrac{16}{0.97} \approx 16 \text{ in.}$

27. Let x be the height of the Statue of Liberty. The length of the cable is 337 ft.

$\sin 65° = \dfrac{\text{opposite}}{\text{hypotenuse}} = \dfrac{x}{337}$

$0.9063 = \dfrac{x}{337}$

$x = 0.9063 \cdot 337 \approx 305 \text{ ft}$

The height of the Statue of Liberty is 305 ft.

28. Let a be altitude of the plane. The trajectory of the plane is $d = r \cdot t = 3 \cdot 5 = 15 \text{ mi}$.

$\sin 15° = \dfrac{\text{opposite}}{\text{hypotenuse}} = \dfrac{a}{15}$

$0.21 = \dfrac{a}{15}$

$a = 0.21 \cdot 15 \approx 3 \text{ mi}$

The altitude of the plane is 3 mi.

29. $y_4 = 1.455$

x	0.100	0.4248	1.455
y	–0.18	–0.4887	

30.

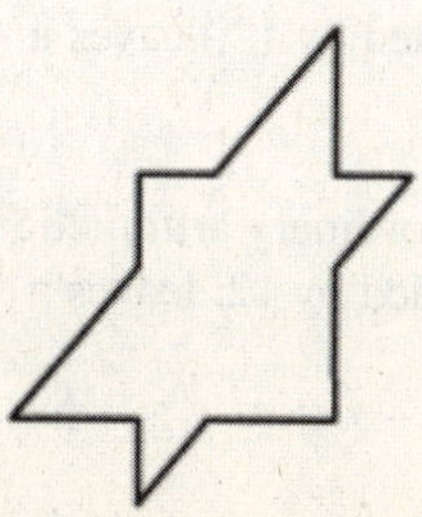

Chapter 9 Mathematical Systems

Section 9.1 Clock and Modular Arithmetic

Study Tips

Whether you realize it, you probably already know how to do clock arithmetic. For instance, if it is currently 10:00 and you have a meeting in 3 hours, you would not say that the meeting is at 13:00; you would say 1:00. (Unless you are on military time, which uses a 24-hour clock!)

The fundamental operations defined for 12-hours clock are generalized to a mathematical system called *modular arithmetic*. Be sure to note the similarities and differences.

An interesting topic will be check digits. An understanding of modular arithmetic will help you to determine whether a credit card is valid.

To add or subtract in clock arithmetic:
Think: move the hour hand on a clock
Note: Figure 9.3 – Figure 9.5 illustrate the basic operations. Table 9.1 gives you all the facts for addition in clock arithmetic.

To solve an equation in clock arithmetic:
Think: by inspection of Table 9.1,
Note: If you are subtracting, change to addition and express number as $12 - n$. Example 2(b) shows how Table 9.5 can be used for subtraction problems.

To multiply and divide in clock arithmetic:
Think: answer is the remainder when product is divided by 12
Note: Multiplication: $a \otimes b = \underbrace{b \oplus b \oplus b \oplus \cdots \oplus b}_{a\text{ times}}$

Division: $\frac{a}{b} = n$ if and only if $a = b \otimes n$.

Caution
Division in clock arithmetic is not always possible just like in our number system division by **0** is *not* defined.

To perform operations in modulo *m*:
Think: answer is the remainder when sum or product is divided by m
Note: Use Tables 9.2 and 9.3 for addition and multiplication of mod 5.

Hint
Rewrite subtraction in terms of addition, and division in terms of multiplication. You may also use the distributive property.

To solve an equation in modulo *m*:
Think: if $m = 5$, by inspection of Tables 9.2 (or 9.4) and 9.3
Note: If in the equation, n is allowed to be any integer that makes the given congruence true, then each answer can be modified by adding or subtracting any desired *multiple* of m.

Hint
One can also list the numbers 0 1 2 3 4 and then start on a given number and move to the right or left the number indicated.

9.1 Exercises

1. $9 \oplus 7$ means 7 hours after 9 o'clock, so the answer is 4.

3. $8 \oplus 3$ means 3 hours after 8 o'clock, so the answer is 11.

5. $7 \oplus 8$ means 8 hours after 7 o'clock, so the answer is 3.

7. $8 \oplus 11$ means 11 hours after 8 o'clock so the answer is 7.

9. $8 \ominus 3$ means 3 hours before 8 o'clock, so the answer is 5.

11. $9 \ominus 12$ means 12 hours before 9 o'clock, so the answer is 9.

13. $8 \ominus 7$ means 7 hours before 8 o'clock, so the answer is 1.

15. $n \oplus 7 = 9$ is true if $n = 9 \ominus 7 = 2$.

17. $2 \oplus n = 1$ is true if $n = 1 \ominus 2 = 11$.

19. $3 \ominus 5 = n$ means 5 hours before 3 o'clock, so the answer is 10.

21. $1 \ominus n = 12$ is true if $1 = n \oplus 12$, so $n = 1$.

23. $4 \otimes 3 = 3 \oplus 3 \oplus 3 \oplus 3 = 12$

25. $9 \otimes 2 = 6$ because in ordinary arithmetic $9 \cdot 2 = 18$, which divided by 12, leaves a remainder of 6.

27. $2 \otimes 8 = 4$ because in ordinary arithmetic $2 \cdot 8 = 16$, which divided by 12, leaves a remainder of 4.

29.

⊗	1	2	3	4	5	6	7	8	9	10	11	12
1	1	2	3	4	5	6	7	8	9	10	11	12
2	2	4	6	8	10	12	2	4	6	8	10	12
3	3	6	9	12	3	6	9	12	3	6	9	12
4	4	8	12	4	8	12	4	8	12	4	8	12
5	5	10	3	8	1	6	11	4	9	2	7	12
6	6	12	6	12	6	12	6	12	6	12	6	12
7	7	2	9	4	11	6	1	8	3	10	5	12
8	8	4	12	8	4	12	8	4	12	8	4	12
9	9	6	3	12	9	6	3	12	9	6	3	12
10	10	8	6	4	2	12	10	8	6	4	2	12
11	11	10	9	8	7	6	5	4	3	2	1	12
12	12	12	12	12	12	12	12	12	12	12	12	12

31. $\frac{3}{5} = n$ only if $3 = 5 \otimes n$. The table above shows that $5 \otimes 3 = 3$, so $n = 3$.

33. $\frac{1}{11} = n$ only if $1 = 11 \otimes n$. The table above shows that $11 \otimes 11 = 1$, so $n = 11$.

35. $\frac{n}{5} = 8$ is true if $n = 5 \otimes 8 = 4$. (See problem 29 table.)

37. $\frac{n}{2} \oplus 4 = 8$ is true if $\frac{n}{2} = 8 \ominus 4 = 4$, and $\frac{\text{n}}{2} = 4$ if $n = 2 \otimes 4 = 8$. (See problem 29 table.)

39. $\frac{2}{n} = 3$ is true only if $2 = 3 \otimes n$. The table shows no entry of 2 in the row for 3. Therefore, there is no solution.

41. False; $4 - 2 = 2$ is not a multiple of 3.

43. False; $7 - 6 = 1$ is not a multiple of 5.

45. False; $9 - 8 = 1$ is not a multiple of 10.

47. $3 + 4 = 7 \equiv 2 \pmod 5$

49. $3 + 1 \equiv 4 \pmod 5$

51. $4 \times 2 = 8 \equiv 3 \pmod 5$

53. $2 \times 3 = 6 \equiv 1 \pmod 5$

55. $2 - 4 = -2 = 5 - 2 \equiv 3 \pmod 5$

57. $1 - 3 = -2 = 5 - 2 \equiv 3 \pmod 5$

59. By the distributive property,
$4 \times (3 + 0) \equiv [(4 \times 3) + (4 \times 0)] \pmod 5$.
Thus, $n = 0$ will make the sentence
$4 \times (3 + 0) \equiv [(4 \times 3) + (4 \times n)] \pmod 5$ true.

61. By the distributive property,
$2 \times (0 + 3) \equiv [(2 \times 0) + (2 \times 3)] \pmod 5$.
Thus, $n = 0$ will make the sentence
$2 \times (0 + 3) \equiv [(2 \times n) + (2 \times 3)] \pmod 5$ true.

63. By inspection of Table 9.2 (or 9.4), we see that $2 + 1 \equiv 3 \pmod 5$. Thus $n = 1$ will make the sentence $2 + n \equiv 3 \pmod 5$ true.

65. By inspection of Table 9.3, we see that $2 \times 2 \equiv 4 \pmod 5$. Thus $n = 2$ will make the sentence $2 \times n \equiv 4 \pmod 5$ true.

67. $n - 3 \equiv 4 \pmod 5$ if and only if $n \equiv (3 + 4) \pmod 5$. From Table 9.2 (or 9.4), we see that $3 + 4 \equiv 2 \pmod 5$. Thus $2 - 3 \equiv 4 \pmod 5$, so $n = 2$.

69. $3 \equiv (n - 4) \pmod 5$ if and only if $3 + 4 \equiv n \pmod 5$ From Table 9.2 (or 9.4), we see that $3 + 4 \equiv 2 \pmod 5$. Thus $3 \equiv (2 - 4) \pmod 5$, so $n = 2$.

71. $\frac{n}{2} \equiv 4 \pmod 5$ is true if and only if $n \equiv (2 \times 4) \pmod 5$. From Table 9.3, we see that $3 \equiv (2 \times 4) \pmod 5$. Thus $\frac{3}{2} \equiv 4 \pmod 5$, so $n = 3$.

73. $\frac{3}{4} \equiv n \pmod 5$ is true if and only if $3 \equiv (4 \times n) \pmod 5$ From Table 9.3, we see that $3 \equiv (4 \times 2) \pmod 5$. Thus

$\frac{3}{4} \equiv 2(\text{mod}\,5)$, so $n = 2$.

75. Yes, all the entries in the table are elements of S.

77. Yes. The identity is 1 because, if a is an element of S, then $1 \times a = a \times 1 \equiv a(\text{mod}\,5)$.

79. $\frac{1985}{12} = 165$ with a remainder of 5, which corresponds to the Ox.

81. No. There are many numbers that can be divided by 12 and leave a remainder of 5. For example 1949 and 1961.
Yes. If the year is, say 2009, they were born in the 80's. $\frac{1980}{12} = 164$ with a no remainder, so 1980 + 5 = 1985 will leave a remainder of 5 when divided by 12.

83. What decade they were born in or if they are in their 20's, 30's, and so on.

85. The year of the Rat occurs when the remainder is 4. Multiply a sufficiently large number by 12, say 165, and add 4. We obtain $165 \cdot 12 + 4 = 1984$. So 1984 and all other years obtained by subtracting 12 or adding 12 successively will be years of the Rat. Thus, 1984 – 12 = 1972 and 1984 + 12 = 1996 are years of the Rat. Similarly, 1948, 1960, 1972, 1984, 1996, 2008, and so on are years of the Rat.

87. A military clock has a 24-hour cycle. One of the advantages is that there is no confusion over a.m. (day) and p.m. (night) time.

89. Answers will vary.

91. Write the digits in the ISBN under the numbers 10, 9, 8, . . ., 2 as shown in the text. Multiply the two numbers in a vertical line and add the results as below:

10	9	8	7	6	5	4	3	2
0	0	6	0	4	0	6	1	3

0 + 0 + 48 + 0 + 24 + 0 + 24 + 3 + 6 = 105
Then solve the congruence
$105 + c \equiv 0(\text{mod}\,11)$.
Since 110 is the multiple of 11 that is just greater than 105, and 105 + 5 = 110, we have $c = 5$. You can also divide 105 by 11 and get a remainder of 6. Since the check number c is the solution of $c + 6 \equiv 0(\text{mod}\,11)$, we still get $c = 5$.

93. Follow the same procedure as in problem 91:

10	9	8	7	6	5	4	3	2
0	3	1	2	8	7	8	6	7

0 + 27 + 8 + 14 + 48 + 35 + 32 + 18 + 14 = 196
Now solve the congruence
$196 + c \equiv 0(\text{mod}\,11)$.
Since 198 is the multiple of 11 that is just greater then 196, and 196 + 2 = 198, we have $c = 2$. You can also divide 196 by 11 obtaining a remainder of 9 and then solving $c + 9 \equiv 0(\text{mod}\,11)$, which yields $c = 2$.

95. Let x be the blurred digit and proceed as in problems 91 and 93:

10	9	8	7	6
5	4	3	2	
0	0	3	0	5
8	9	x	4	

$0 + 0 + 24 + 0 + 30 + 40 + 36 + 3x + 8$
$= 138 + 3x$
Since the check number is $c = 2$, we must have $138 + 3x + 2 \equiv 0(\text{mod}\,11)$ or $140 + 3x \equiv 0(\text{mod}\,11)$. Now, 143 is the multiple of 11 that is just greater than 140, and 143 = 140 + 3. Therefore, we must have $3x = 3$, so $x = 1$.

97. Following the procedure of Example 10, we have

$$3n_1 + n_2 + 3n_3 + n_4 + 3n_5 + n_6 + 3n_7 + n_8 + 3n_9 + n_{10} + 3n_{11} + c \equiv 0(\text{mod}10)$$

$$3 \cdot 0 + 4 + 3 \cdot 3 + 1 + 3 \cdot 6 + 8 + 3 \cdot 3 + 8 + 3 \cdot 5 + 4 + 3 \cdot 0 + c \equiv 0(\text{mod}10)$$

$$76 + c \equiv 0(\text{mod}10)$$

Since $76 + c$ must be a multiple of 10, we make $c = 4$.

99. Following the procedure of Example 10, we have

$$3 \cdot 0 + 2 + 3 \cdot 4 + 0 + 3 \cdot 0 + 0 + 3 \cdot 0 + 2 + 3 \cdot 1 + 4 + 3 \cdot 0 + c \equiv 0(\text{mod}10)$$

$$23 + c \equiv 0(\text{mod}10)$$

Since $23 + c$ must be a multiple of 10, we make $c = 7$.

101. Following the procedure of Example 10, we

have

$3 \cdot 0+7+3 \cdot 4+1+3 \cdot 0+1+3 \cdot 7+4$
$+3 \cdot 0+1+3 \cdot 0+c \equiv 0(\bmod 10)$

$47+c \equiv 0(\bmod 10)$

Since 47 + c must be a multiple of 10, we make $c = 3$.

103. Let x be the missing digit.

$3 \cdot 0+4+3 \cdot 1+3+3 \cdot 3+x+3 \cdot 0+3$
$+3 \cdot 1+8+3 \cdot 3+7 \equiv 0(\bmod 10)$

$49+x \equiv 0(\bmod 10)$

Since 49 + x must be a multiple of 10, we make $x = 1$.

105. Let x be the missing digit.

$3 \cdot 0+x+3 \cdot 0+4+3 \cdot 0+4+3 \cdot 0+0$
$+3 \cdot 1+0+3 \cdot 0+2 \equiv 0(\bmod 10)$

$13+x \equiv 0(\bmod 10)$

Since 13 + x must be a multiple of 10, we make $x = 7$.

107. $1+0+0+0+2+5+0+0=8$

$5+0+1+0+4+9+0+3=22$

$8+22=30$

30 is a multiple of 10 so the card is valid.

109. $6+3+6+4+5+0+8+5=37$

$4+5+4+1+9+2+6+9=40$

$40+37=77$

77 is not a multiple of 10 so the card is not valid. The check digit should be 2.

111. $6+0+8+1+4+8+8=35$

$0+2+4+6+9+0+1=22$

$35+22=57$

57 is not a multiple of 10 so the card is not valid. The check digit should be 4.

113. $6+3+9+4+6+6+8=42$

$6+3+0+3+9+8+9=38$

$42+38=80$

80 is a multiple of 10 so the card is valid.

115. $3+2+0+6+0+0+2+6=19$

$0+1+0+0+5+7+0+1=14$

$19+14=33$

33 is not a multiple of 10 so the card is not valid. The check digit should be 8.

117. $7 \cdot 0+3 \cdot 3+9 \cdot 1+7 \cdot 3+3 \cdot 1+9 \cdot 3+7 \cdot 5$
$+3 \cdot 6+9 \cdot 2=140 \equiv 0(\bmod 10)$

The routing number is valid.

119. $7 \cdot 0+3 \cdot 6+9 \cdot 3+7 \cdot 1+3 \cdot 0+9 \cdot 7$
$+7 \cdot 5+3 \cdot 1+9 \cdot 4=189 \equiv 9(\bmod 10) \neq 0(\bmod 10)$

There is an error in the routing number.

To read a project concerning checking numbers under different schemes, go to:
http://www.missioncollege.org/depts/math/sanitate/lynn.htm

Section 9.2 Abstract Mathematical Systems: Groups and Fields

Study Tips

This section is concerned with the rules and laws that are obeyed by an abstract mathematical system. The text will often use *tables* to represent these systems.

Know the definitions of properties:
Closure: for a and b in A, $a * b$ is also in A
Associative operation: for a, b, and c in A, $(a*b)*c = a*(b*c)$; can change grouping
Commutative operation: for a and b in A, $a*b = b*a$; can change order
Identity: for element e in A, $x*e = x = e*x$; any operation with element e will not change values
Inverse: for a and b in A, $a*b = e = b*a$; so a is the inverse of b
Distributive property: for a, b, and c in A, $a*(b\#c) = a\#b*a\#c$; an operation will distribute over another operation

Table features of properties:
Closure: all the elements in the table come from the original set of elements
Commutative operation: the table is symmetric with respect to a diagonal drawn from top left to bottom right
Identity: the column under the identity is the same as the left-most column
Inverse: the identity element will be within the table

To determine a group:
Think: must have four specific properties
Note: The properties are closure, associative, identity, inverse.

Fact
If the system also has the commutative property, then the group is called a commutative (or *abelian*) group.

To determine a field:
Think: must be a group AND distributive property.
Note: If the system is *not* a group, it *cannot* be a field.

9.2 Exercises

1. a. $a@b=a$ b. $b@c=c$
   c. $c@a=b$

3. a. $b@(a@b)=b@a=a$
   b. $(b@a)@b=a@b=a$ c. Yes

5. a. $b@c=c$ b. $c@b=c$
   c. Yes

7. Yes; if $x\in S$ and $y\in S$, then $x@y\in S$, that is, all the entries in the table are elements of *S*.

9. a. Yes; if *a* and *b* are natural numbers, then *a* F *b* = *a*, which is a natural number.
   b. Yes. *a* F (*b* F *c*) = *a* F *b* = *a*, and (*a* F *b*) F *c* = *a* F *c* = *a*. Thus, *a* F (*b* F *c*) = (*a* F *b*) F *c*.
   c. No. *a* F *b* = *a*, and *b* F *a* = *b*, so if *a* ≠ *b*, then *a* F *b* ≠ *b* F *a*.

11. a. No. For example, 1 + 3 = 4, which is not an odd number.
    b. Yes; the product of two odd numbers is an odd number.
    c. Yes; the sum of two even numbers is an even number.
    d. Yes; the product of two even numbers is an even number.

13.

$\cap$	$\varnothing$	$\{a\}$	$\{b\}$	$\{a,b\}$
$\varnothing$	$\varnothing$	$\varnothing$	$\varnothing$	$\varnothing$
$\{a\}$	$\varnothing$	$\{a\}$	$\varnothing$	$\{a\}$
$\{b\}$	$\varnothing$	$\varnothing$	$\{b\}$	$\{b\}$
$\{a,b\}$	$\varnothing$	$\{a\}$	$\{b\}$	$\{a,b\}$

15. a. $(\{b\}\cap\{a,b\})\cap\{a\}=\{b\}\cap\{a\}=\varnothing$
    b. $\{b\}\cap(\{a,b\}\cap\{a\})=\{b\}\cap\{a\}=\varnothing$
    c. Yes

17. Yes; all the entries in the table from problem 13 are elements of *S*.

19. a.

L	1	2	3	4
1	1	2	3	4
2	2	2	3	4
3	3	3	3	4
4	4	4	4	4

    b. The first row is the same as the L row, and the first column is the same as the L column, so the identity element is 1.

21. The table given for this problem shows that 4 is the identity element. This information is needed here.
    a. There is no 4 in the first row of the table, so 1 has no inverse.
    b. There is no 4 in the second row of the table, so 2 has no inverse.
    c. There is no 4 in the third row of the table, so 3 has no inverse.
    d. Since 4 S 4 = 4, the element 4 is its own inverse.

23. Yes; $1\times a=a\times 1=a$ for each element *a* of the set *S*, so 1 is the identity element.

25. a. *A* is the identity element because if *B* is any subset of *A*, then $A\cap B=B\cap A=B$.
    b. No; there is only the one identity element.

27. Yes. The 0 row of the table is the same as the @ row, and the 0 column is the same as the @ column, so 0 is the identity element.

29. a. 3 F (4 L 5) = 3 F 5 = 3, because 3 is the first of the numbers 3 and (4 L 5).
    b. 4 F (5 L 6) = 4 F 6 = 4, because 4 is the first of the numbers 4 and (5 L 6).

31. Suppose *a*, *b*, *c* are real numbers. Then *a* F (*b* L *c*) = *a* and (*a* F *b*) L (*a* F *c*) = *a* L *a* = *a*. This shows that the distributive property of F over L holds for all real numbers. (Keep in mind that F means to select the first of the numbers in the expression *a* F *b*.)

33. Yes. If a, b, c are real numbers, we may replace $b - c$ by $b + (-c)$. Then, since the distributive property of multiplication over addition holds, we have $a(b-c)$ $= a[b+(-c)] = ab + a(-c) = ab - ac$. This shows that multiplication is distributive over subtraction.

35. Yes; fractions are real numbers and multiplication is distributive over addition for the set of real numbers.

37. Yes. We check the four requirements as follows:
    1. The set S is closed under the operation $*$ because every entry in the table is an element of S.
    2. For all x, y, z that are elements of S: $x*(y*z) = (x*y)*z$. This can be checked for all possibilities. For instance, $a*(b*c) = a*b = c$ and $(a*b)*c = c*c = c$. Thus, the set S has the associative property under the operation $*$.
    3. The bottom row is the same as the $*$ row, and the last column is the same as the $*$ column, so c is the identity element.
    4. The identity element c occurs once in each row and column of the table, so every element has an inverse. For instance, $a*b = b*a = c$, so a and b are inverses of each other.

    This shows that the set S under the operation $*$ meets the four requirements and is thus a group. (Note that it is a commutative group.)

39. No. The set of odd integers under multiplication is not a group because, except for the number 1, no number in this set has an inverse that is an element of the set. (The fractions are not in this set.)

41. No. The set of even integers under multiplication is not a group because it has no identity element. (The number 1 is not in this set.)

43. No. The set of positive integers under multiplication is not a group because, except for the number 1, no number in this set has an inverse that is an element of the set. (The fractions are not in this set.)

45. No. The set of integers under multiplication is not a group because, except for the numbers 1 and –1, no number in this set has a multiplicative inverse that is an element of the set. (The fractions are not in this set.)

47. Yes. We check the four requirements as follows:
    1. If a and b are real numbers, then $a + b$ is a real number, so the set of real numbers is closed under addition.
    2. If a, b, c are real numbers then $(a+b)+c = a+(b+c)$, so the set of real numbers has the associative property under addition.
    3. The number 0 is a real number and $0 + a = a + 0 = a$ for every real number a. Thus, 0 is the identity element for the set of real numbers under addition.
    4. If a is a real number, then $-a$ is also a real number and $a+(-a) = (-a)+a = 0$. Thus, every real number has an inverse under addition.

    Since the set of real numbers under addition meets all four requirements, the system is a group. (Note that it is a commutative group because $a + b = b + a$.)

49. No. The set {–1, 0, 1} under multiplication is not a group because the number 0 has no multiplicative inverse under multiplication.

51. No. The set {a, b, c, d, e} under the operation # defined by the table is not a group because there is no identity element. (There is no column of the table that reads a, b, c, d, e in that order.)

53. No. The set of positive even integers has no multiplicative inverses that are elements of the set, so this system is not a field.

55. No. The set of integral multiples of 2 is the set of even integers. This set has no multiplicative inverses that are elements of the set, so this system is not a field.

57. Check to see that a ◆ b is always an element of S, that is, every entry in the table is a member of S.

59. Check the table to see if there is an element

e in *S* such that the column under *e* is identical to the column at the far left and the row opposite *e* is identical to the top row. If there is, then *e* is the identity element. If there is no such element, then there is no identity element.

61. You have to check that $x*(y*z)=(x*y)*z$ for all possible values of *x, y, z* from the set $\{a, b, c\}$. If you had to check all possible cases, there would be 27 of these because each of the 3 places has three possible values. However, since the operation has the commutative property, the number of cases to be checked is greatly reduced. Think about it.

63.

$\oplus$	1	2	3	4	5	6	7
1	1	1	1	1	1	1	1
2	1	2	2	2	2	2	2
3	1	2	3	3	3	3	3
4	1	2	3	4	4	4	4
5	1	2	3	4	5	5	5
6	1	2	3	4	5	6	6
7	1	2	3	4	5	6	7

a. Yes, all elements in the table are elements of $\{1, 2, 3, 4, 5, 6, 7\}$.

b. Yes, it is commutative because $a \oplus b = b \oplus a$, where *a* and *b* are in the set $\{1, 2, 3, 4, 5, 6, 7\}$.

c. Yes, it is associative because $(a \oplus b) \oplus c = a \oplus (b \oplus c)$, where *a, b* and *c* are in the set $\{1, 2, 3, 4, 5, 6, 7\}$.

d. The identity element is 7 because $7 \oplus b = b \oplus 7 = b$, where *b* are in the set $\{1, 2, 3, 4, 5, 6, 7\}$.

65. Yes, $3 \otimes (4 \oplus 5) = 3 \otimes 4 = 3 + 4 = 7$ and $(3 \otimes 4) \oplus (3 \otimes 5) = 7 \oplus 3 = 7$. Thus, $3 \otimes (4 \oplus 5) = (3 \otimes 4) \oplus (3 \otimes 5)$.

Note: You are comparing $a + \min\{b, c\}$ to $\min\{a + b, a + c\}$.

67. $6 \times 9999 = 6(10{,}000 - 1) = 60{,}000 - 6$
$= 59{,}994$

69. $7 \times 59 = 7(60 - 1) = 420 - 7 = 413$

71. $4 \times 9995 = 4(10{,}000 - 5) = 40{,}000 - 20$
$= 39{,}980$

73. Let *x* be the number you select, then the following steps show why the puzzle works:

Think of a number:	x
Add 3 to it:	$x + 3$
Triple the result:	$3x + 9$
Subtract 9:	$3x$
Divide by the number with which you started:	$\frac{3x}{x} = 3$

It does not matter what value of *x* is used, the left side will always be 3.

Section 9.3 Game Theory

Study Tips

Game theory is a very interesting application of matrices. This theory has important everyday applications in business, economics, and science.

To find the value of a strictly determined game defined by a matrix:
Think: optimal strategy, saddle point is the value of the game
Note: A saddle point is *both* the smallest element in its row and the largest element in its column.

To determine optimal pure strategy in a strictly determined game:
Row player: Circle smallest element in each row, choose row with largest of circled values
Column player: Box largest element in each column, choose column with smallest of boxed number

To determine expected payoff and proportions:
Think: payoff matrix, *p* and $1 - p$,
Note: Set expected values equal to each other to find the value of *p*.

Caution
If a saddle point does not exist, the game is not strictly determined.

9.3 Exercises

1. The 4 in the first row is the smallest element

in its row and the largest in its column. The game is strictly determined. Optimum strategy: Row player should play row 1; column player should play column 1. Value of the game is 4.

3. There is no element that is the smallest in its row and the largest in its column. Hence, the game is not strictly determined.

5. The 4 in the first row is the smallest element in its row and the largest in its column. The game is strictly determined. Optimum strategy: Row player should play row 1; column player should play column 3. Value of the game is 4.

7. Either 4 in the third row is the smallest element in its row and the largest in its column. The game is strictly determined. Optimum strategy: Row player should play row 3; column player should play column 3 (or column 1). The value of the game is 4.

9. a. There is no element that is the smallest in its row and the largest in its column, so there is no saddle point.

   b. Row player's expected value is given by $E_R =$

$$p \cdot 2 + (1-p)(-1) = p \cdot 1 + (1-p) \cdot 4$$
$$2p - 1 + p = p + 4 - 4p$$
$$3p - 1 = 4 - 3p$$
$$6p = 5$$
$$p = \frac{5}{6}$$

This means that the row player should play row 1 five-sixths of the time and row 2 one-sixth of the time. (The row player's expected value is

$$E_R = 3\left(\frac{5}{6}\right) - 1 = \frac{5}{2} - 1 = \frac{3}{2} = 1.5.)$$

11. a. There is no element that is the smallest in its row and the largest in its column, so there is no saddle point.

   b. Row player's expected value is given by $E_R =$

$$p \cdot 3 + (1-p)(-1) = p(-2) + (1-p) \cdot 0$$
$$3p - 1 + p = -2p$$
$$4p - 1 = -2p$$
$$6p = 1$$
$$p = \frac{1}{6}$$

This means that the row player should play row 1 one-sixth of the time and row 2 five-sixths of the time. The row player's expected value is:

$$E_R = 4\left(\frac{1}{6}\right) - 1 = \frac{2}{3} - 1 = -\frac{1}{3}.$$

NOTE: In Exercises 13 – 23 and 29, there is no saddle point.

13. Since row 2 dominates row 3, row 3 may be discarded and we may use rows 1 and 2 for the payoff matrix. The row player's expected value is given by

$$E_R = p \cdot 6 + (1-p) \cdot 2 = p \cdot 0 + (1-p) \cdot 4$$
$$6p + 2 - 2p = 4 - 4p$$
$$4p + 2 = 4 - 4p$$
$$8p = 2$$
$$p = \frac{1}{4}$$

This means that the row player should play row 1 one-fourth of the time and row 2 three-fourths of the time.
The row player's expected value is

$$E_R = 4\left(\frac{1}{4}\right) + 2 = 1 + 2 = 3.$$

15. The expected row value is $E_R =$

$$p \cdot 70 + (1-p) \cdot 85 = p \cdot 80 + (1-p) \cdot 75$$
$$70p + 85 - 85p = 80p + 75 - 75p$$
$$85 - 15p = 5p + 75$$
$$10 = 20p$$
$$p = \frac{1}{2}$$

This means that you should study 2 hr half the time and 4 hr the other half the time. (Your expected score is

$$E_R = 85 - 15\left(\frac{1}{2}\right) = 85 - 7.5 = 77.5.)$$

17. Since the first row is dominated by both the other rows, Ann should not use the first row. This leaves a 2-by-2 matrix for which Ann's expected value is

$$E_R = p \cdot 10 + (1-p) \cdot 15 = p \cdot 12 + (1-p) \cdot 10$$
$$10p + 15 - 15p = 12p + 10 - 10p$$
$$15 - 5p = 2p + 10$$
$$5 = 7p$$
$$p = \frac{5}{7}$$

This means that Ann should buy no bonds, buy stocks with five-sevenths and money market funds with two-sevenths of her investment. Ann's expected return is found from $E_R = 15 - 5\left(\frac{5}{7}\right) = 15 - \frac{25}{7} = \frac{80}{7} = 11\frac{3}{7}$ that is, her expected return would be $11\frac{3}{7}\%$.

19. The expected row value is $E_R =$
$$p \cdot 60 + (1-p) \cdot 30 = p \cdot 50 + (1-p) \cdot 70$$
$$60p + 30 - 30p = 50p + 70 - 70p$$
$$30p + 30 = 70 - 20p$$
$$50p = 40$$
$$p = \frac{4}{5}$$
This means that station R should price its gasoline at \$1.00 four-fifths of the time and at \$1.10 the other fifth of the time. (The expected row value would be $E_R = 30\left(\frac{4}{5}\right) + 30 = 24 + 30 = 54$, that is, with this strategy, station R would expect to get 54% of the business.)

21. a. The payoff matrix can be written as follows:

	Young	Old
Performance	70%	20%
Safety	40%	80%

b. The expected row value is $E_R =$
$$p \cdot 70 + (1-p) \cdot 40 = p \cdot 20 + (1-p) \cdot 80$$
$$70p + 40 - 40p = 20p + 80 - 80p$$
$$30p + 40 = 80 - 60p$$
$$90p = 40$$
$$p = \frac{4}{9}$$
This means that 4/9 of the ads should be based on performance and 5/9 should be based on safety. (With this strategy, the expected row value would be $E_R = 30\left(\frac{4}{9}\right) + 40 = \frac{40}{3} + 40 = \frac{160}{3}$, that is, the ads would be effective on $53\frac{1}{3}\%$ of the buyers.)

23. With the numbers in dollars, the required matrix can be written as

		Freeze? Yes	Freeze? No
Water?	Yes	6000	–400
	No	–4000	4000

The expected row value is $E_R =$
$$p \cdot 6000 + (1-p)(-4000) = p(-400) + (1-p) \cdot 4000$$
$$6000p - 4000 + 4000p = -400p + 4000 - 4000p$$
$$10{,}000p - 4000 = 4000 - 4400p$$
$$14{,}400p = 8000$$
$$p = \frac{5}{9}$$
This means that the farmer's optimal strategy would be to water five-ninths of the time and not to water four-ninths of the time. The corresponding expected payoff is
$$E_R = 10{,}000\left(\frac{5}{9}\right) - 4000 = \frac{50{,}000 - 36{,}000}{9} = \frac{14{,}000}{9}$$
So the farmer's expected payoff is $\frac{14{,}000}{9}$ or about \$1556.

25. If row i dominates row j, this means that, in the long run, playing row i is more profitable than playing row j. Thus, row j may be eliminated from the row player's options.

27. Both entries in row 1 are less than the corresponding entries in row 2. Therefore, row 1 is dominated by row 2.

29. With row 1 eliminated, the payoff matrix is:

	Allergic	On a Diet
Poems	2	5
Candy	3	−2

The expected row value is

$$E_R = p\cdot 2+(1-p)\cdot 3 = p\cdot 5+(1-p)(-2)$$
$$2p+3-3p = 5p-2+2p$$
$$3-p = 7p-2$$
$$5 = 8p$$
$$p = \frac{5}{8}$$

Hence, the student's optimal strategy is to send poems five-eighths of the time and candies three-eighths of the time. He should not send flowers.

Chapter 9 Practice Test

Study Tips

Using your time wisely is extremely important when taking the exam. Here are some suggestions:

1. Answer the easy questions first.
2. If you are unsure of how to solve a problem, skip it and come back to it later. If time permits, check over your work and answers.
3. If you are unsure of the directions, ask the instructor to clarify.
4. A test is not a race. You have the entire period, so do not become frustrated when other students begin leaving the classroom.

Try these ideas as you take this practice test.

Here are the solutions to the practice test.

1. a. $3\oplus 11 = 2$ (11 hours after 3 o'clock is 2 o'clock.)
   b. $8\oplus 9 = 5$ (9 hours after 8 o'clock is 5 o'clock.)
   c. $3\ominus 9 = 6$ (9 hours before 3 o'clock is 6 o'clock.)
   d. $5\ominus 12 = 5$ (12 hours before 5 o'clock is 5 o'clock.)

2. a. $3\otimes 5 = 3$ ($3\cdot 5 = 15$ and $15-12 = 3$)
   b. $6\otimes 8 = 12$ ($6\cdot 8 = 48$, a multiple of 12)
   c. Look at the clock arithmetic multiplication table and go along the row for 11 until you come to 3. The number in the top line is 9. Thus, in clock arithmetic, $\frac{3}{11} = 9$. You can check this by noting that $11\cdot 9 = 99$, which leaves a remainder of 3 when divided by 12.
   d. There is no solution. If you look at the multiplication table and go along the row for 6, you will see that there is no 5 in this row.

3. a. True. $5-2$ is a multiple of 3.
   b. True. $9-5$ is a multiple of 4.
   c. False. $9-2 = 7$ is not a multiple of 6.

4. a. By inspection of Table 9.2, we see that $3+2\equiv 0 \pmod 5$. Thus $n = 0$ will make the sentence $3+2\equiv n \pmod 5$ true.
   b. By inspection of Table 9.3, we see that $4\times 3\equiv 2 \pmod 5$. Thus $n = 2$ will make the sentence $4\times 3\equiv n \pmod 5$ true.
   c. $2-4\equiv n \pmod 5$ if and only if $2\equiv 4+n \pmod 5$. From Table 9.2 (or 9.4), we see that $2\equiv 4+3 \pmod 5$. Thus $2-4\equiv 3 \pmod 5$, so $n = 3$.
   d. $\frac{3}{2}\equiv n \pmod 5$ is true if and only if $3\equiv 2n \pmod 5$. From Table 9.3, we see that $3\equiv 2\times 4 \pmod 5$. Thus $\frac{3}{2}\equiv 4 \pmod 5$, so $n = 4$.

5. a. $6+n\equiv 1 \pmod 7$ is true if $6+n = 8$ that is, if $n = 2$. Since any multiple of 7 may be added to this solution, the general solution is $n = 2+7k$, k any integer.
   b. $3-n\equiv 4 \pmod 7$ is true if $3\equiv n+4 \pmod 7$, which is true if $n+4 = 10$, that is, if $n = 6$. Since any multiple of 7 may be added to this solution, the general solution is $n = 6+7k$, k any integer.

6. a. $2n\equiv 1 \pmod 3$ is true if $2n = 4+3k$, k any integer. Thus, the general solution is $n = 2+\frac{3k}{2}$, k any integer.

b. $\frac{n}{2} \equiv 2(\text{mod}3)$ is true if $n \equiv 4(\text{mod}3)$ $= 1(\text{mod}3)$, so the general solution is $n = 1 + 3k$, k any integer.

7. a. ($ ∗ ¢) ∗ # = # ∗ # = #
   b. $ ∗ (¢ ∗ #) = $ ∗ ¢ = #
   c. ($ ∗ #) ∗ (% ∗ ¢) = $ ∗ $ = %

8. Yes. The set *S* is closed with respect to the operation * because all the entries in the table are elements of *S*.

9. Yes. The operation is commutative because the table is symmetric with respect to the main diagonal from upper left to lower right. This means that if *a* and *b* are any two elements of *S*, then $a * b = b * a$.

10. The identity element is # because the # column is the same as the column under ∗ and the # row is the same as the top row of the table. This means that if *a* is any element of *S*, then a ∗ # = # ∗ $a = a$.

11. a. The inverse of # is itself because # is the identity element.
    b. Because $ ∗ ¢ = #, the inverse of $ is ¢.
    c. Because % ∗ % = #, the inverse of % is %.
    d. Because ¢ ∗ $ = #, the inverse of ¢ is $.

12. a. There is no identity element because no column reads exactly the same as the column under @.
    b. Since there is no identity element, no element has an inverse.
    c. No. The table is not symmetric to the diagonal from upper left to lower right.

13. a. Yes. Since the operation S selects the second of the two numbers in *a* S *b*, *a* S (*b* L *c*) = *b* L *c*. Also, *a* S *b* = *b* and *a* S *c* = *c*, so that (*a* S *b*) L (*a* S *c*) = *b* L *c*. Thus, it is true that *a* S (*b* L *c*) = (*a* S *b*) L (*a* S *c*). Therefore, S is distributive over L.
    b. Yes. Since *b* S *c* = *c*, it follows that *a* L (*b* S *c*) = *a* L *c*. Also, it is true that (*a* L *b*) S (*a* L *c*) = *a* L *c*, because *a* L *c* is the second number. Thus, *a* L (*b* S *c*) = (*a* L *b*) S (*a* L *c*), so L is distributive over S.

14. Yes. With the operation of addition mod 3, the set $S = \{0, 1, 2\}$ has the five properties required to make it a commutative group
    1. *S* has the closure property because the sum (mod 3) of any two elements of *S* is an element of *S*.
    2. *S* has the associative property because ordinary addition does.
    3. 0 is the identity element.
    4. Every element of *S* has an inverse. The inverse of 0 is 0 and the elements 1 and 2 are inverses of each other. [$1 + 2 = 2 + 1 = 3$ and $3 \equiv 0(\text{mod}3)$.]
    5. *S* has the commutative property because ordinary addition does.

15. Yes. With the operation of multiplication mod 3, the set $T = \{1, 2\}$ has the five properties required to make it a commutative group.
    1. *T* has the closure property because the product of any two elements (mod 3) is an element of *T*.
    2. *T* has the associative property because ordinary multiplication does.
    3. The identity element is 1.
    4. Since $1 \cdot 1 \equiv 1(\text{mod}3)$ and $2 \cdot 2 \equiv 1(\text{mod}3)$, the elements 1 and 2 are their own inverses.
    5. *T* has the commutative property because ordinary multiplication does.

16. Yes. The system has the six required properties (closure, associative, commutative, identity, distributive of multiplication over addition, and inverses, except there is no inverse for zero with respect to multiplication), so the system is a field.

17. Yes, all the requirements of the definition of a group are satisfied. Let $S = \{0, 1\}$. We check the requirements for *S* to be a group under the operation ⊕.
    1. *S* has the closure property because all the entries in the table are elements of *S*.
    2. You can check that the operation ⊕ corresponds to addition (mod 2), so *S* has the associative property under this operation.
    3. The identity element is 0 because $0 \oplus 0 = 0$, and $0 \oplus 1 = 1 \oplus 0 = 1$.
    4. Each row and column of the table has an identity element, so each element of

S has an inverse. The table shows that 0 and 1 are both their own inverses. Therefore, the set *S* with the operation ⊕ is a group.

18. No. You can check as in problem 17 that the set *S* under the operation ⊗ has the closure and associative properties. The identity element is 1. However, the element 0 has no inverse, so the system is not a group.

19. Yes. In problems 17 and 18, we have checked all the requirements for a field except the distributive property of ⊗ over ⊕ . Since the operation ⊕ corresponds to addition (mod 2) and the operation ⊗ corresponds to multiplication (mod 2), the system does have this distributive property. Thus, the system is a field.

20. a. There is no element that is both the least in its row and the greatest in its column, so there is no saddle point and the game is not strictly determined.
    b. The 1 in the third row, third column is the least in its row and the greatest in its column, so this is a saddle point. The game is strictly determined and its value is 1.
    c. The 2's in the bottom row are the least in the row and the greatest in their respective columns, so these are saddle points. The game is strictly determined and its value is 2.

21. The –6 in the first row is the saddle point. The row player should play row 1 and the column player should play column 2. The payoff for the row player is –6.

22. a. With the given percents written as decimals, the payoff matrix is:

$$\begin{array}{r c} & \begin{matrix} \text{H} & \text{C} \end{matrix} \\ \begin{matrix} \text{This Year} \\ \text{Next Year} \end{matrix} & \begin{bmatrix} 0.6 & 0.4 \\ 0.3 & 0.7 \end{bmatrix} \end{array}$$

    b. There is no saddle point, so this is not a strictly determined game. If E_R is the expected value for HART and p is the probability that the first row is what will happen, then $E_R =$

$$p \cdot 0.6 + (1-p) \cdot 0.3 = p \cdot 0.4 + (1-p) \cdot 0.7$$
$$0.6p + 0.3 - 0.3p = 0.4p + 0.7 - 0.7p$$
$$0.3p + 0.3 = 0.7 - 0.3p$$
$$0.6p = 0.4$$
$$p = \frac{2}{3}$$

Hence,

$E_R = 0.3\left(\frac{2}{3}\right) + 0.3 = 0.2 + 0.3 = 0.5$, so that the expected percentage of persons using HART next year is 50%.

Chapter 10 Counting Techniques

Section 10.1 The Sequential Counting Principle (SCP): A Problem-Solving Tool

Study Tips

To study the ideas of probability in the next chapter, we need to learn how to *count* the number of ways in which an event can occur.

To use a tree diagram to count the number of ways:
Think: first make a branch for each possible first choice, from each of those branches make a branch for each possible second choice, and so on; follow the branches horizontally to see possible outcomes
Note: Tree diagrams are used when there are not very many outcomes.

To use SCP to count the number of ways:
Think: make a blank for each event, write the number of possible choices in each blank, multiply
Note: SCP stands for Sequential Counting Principle.

Word of Advice
Study *Problem-Solving: The Sequential Counting Principle* and the illustrative examples to understand how the SCP is used.

10.1 Exercises

1. 8 different outfits

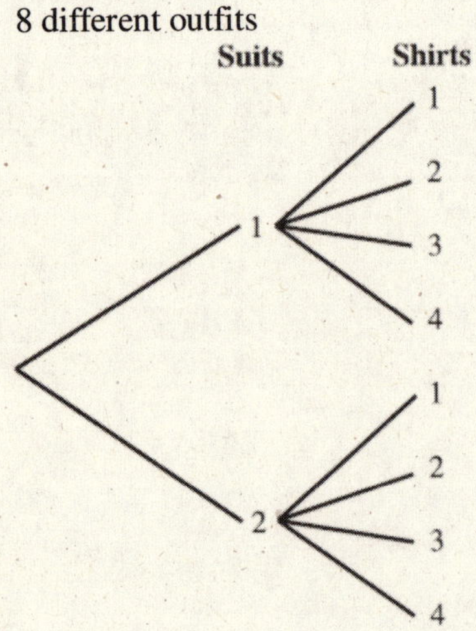

3. $4 \times 3 = 12$ outcomes

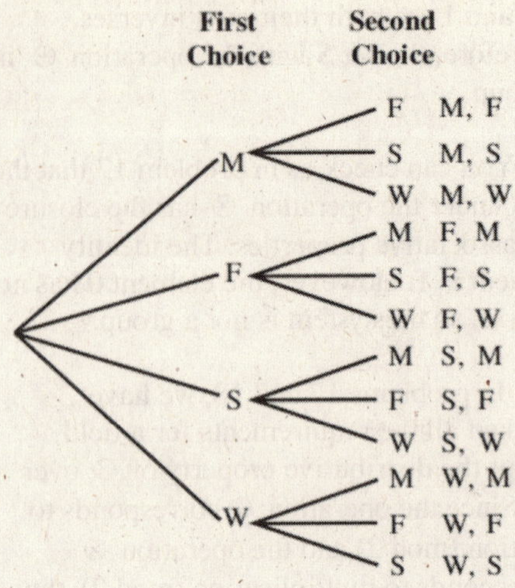

5. a. $1 \cdot 3 \cdot 3 = 9$ ways to get 3 bells.
   b. $3 \cdot 6 \cdot 7 = 126$ ways to get 3 oranges.
   c. $5 \cdot 1 \cdot 5 = 25$ ways to get 3 plums.

7. a. $4 \cdot 3 = 12$ ways to get a pair of aces.
   b. $4 \cdot 3 \cdot 4 \cdot 3 = 144$ ways to get a pair of aces, then a pair of eights.

9. $5 \cdot 6 = 30$ ways to select 1 man and 1 woman from 5 men and 6 women.

11. $26 \cdot 26 \cdot 26 = 26^3 = 17{,}576$ different sets of 3 initials from the English alphabet.

13. $7 \cdot 6 \cdot 9 = 378$ meal choices.

15. a. $6 \cdot 5 \cdot 4 \cdot 3 \cdot 2 \cdot 1 = 720$ ways to list the possible treatments.
    b. $4 \cdot 3 \cdot 2 \cdot 1 = 24$ ways to rank the four top-ranked treatements.

17. $9 \cdot 10 = 90$ 2-digit numbers in the set of natural numbers.

19. a. $9 \cdot 10^8 = 900{,}000{,}000$ possible Social Security numbers.
    b. $10^9 = 1{,}000{,}000{,}000$ possible Social Security numbers.

21. a. $40 \cdot 39 \cdot 38 = 59{,}280$ combinations if no number is used twice.
    b. $40^3 = 64{,}000$ combinations if repetition of numbers is allowed.

23. a. Billy has 10 choices from group A.

b. Billy has 2 choices for soup.
c. Billy has 4 choices for dessert.
d. Billy has $10 \cdot 2 \cdot 4 = 80$ choices for an item from group A, soup, and dessert.

25. a. Pedro has 19 choices from group A or group B.
b. Pedro has $19 \cdot 2 \cdot 4 = 152$ choices for an item from group A or B, soup, and dessert.

27. Sam and Sally have $10 \cdot 9 \cdot 2 \cdot 4 = 720$ choices.

29. Sam and Sally have $10 \cdot 6 \cdot 2 \cdot 4 = 480$ choices.

31. (*a*, @), (*a*, &), (*a*, %), (*b*, @), (*b*, &). (*b*, %), (*c*, @), (*c*, &), (*c*, %), (@, *a*), (&, *a*), (%, *a*), (@, *b*), (&, *b*), (%, *b*), (@, *c*), (&, *c*), (%, *c*)

33. $2 \cdot 2 = 4$ choices as to when and what to buy.

35. $3 \cdot 2 = 6$ investment possibilities.

37. The first step indicates that the product can be developed (D) or abandoned (A). Second, if the product is developed (D) they can decide to get a patent (P) or not (N). Third, if they get a patent (P) they can license the product (Li) or sell it (S). Finally, if they decide to sell it (S) the sales can be high (H), low (L) or medium (M).

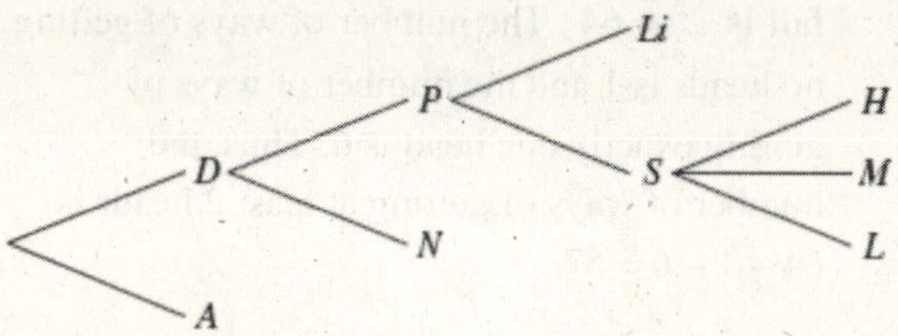

39. The first decision is to treat (T) or not to treat (~T). If the person is treated, there may be some side effect (S) or not (~S). If the person is not treated (~T) the person may get HIV (H) or not (~H). Regardless of side effects, the person may get HIV (H) or not (~H).
Note: The tilde (~) is used to mean NOT.

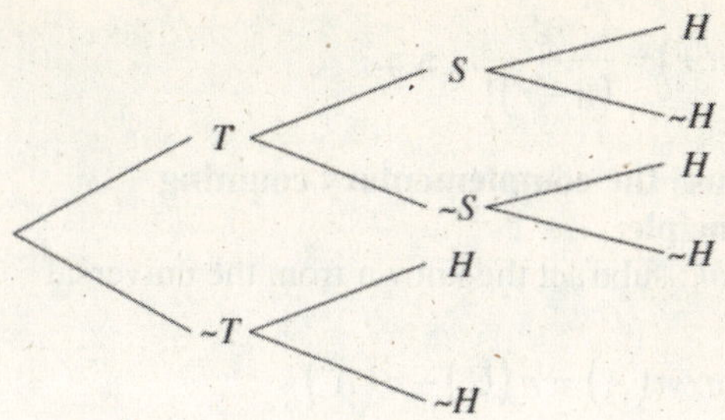

41. If a single event can occur in *m* ways or in *n* ways, then the total number of ways in which the event can occur is $m + n$. (This assumes that the *m* ways and the *n* ways are all distinct.)

43. $4 \cdot 3 \cdot 2 \cdot 1 = 24$

45. Yes. Only 17,576 different sets of initials are possible. Thus, if there are 27,000 people, at least two people must have the same set of initials.

47. $12 \cdot 12 = 144$ ways in which the front and rear air valves could come to rest.

49. It will point to 3 if no slippage occurs.

Section 10.2 Permutations

Study Tips

An *ordered* arrangement of *r* objects selected from a set of *n* distinct objects without repetition is called a *permutation* and is denoted by $P(n, r)$.

To compute the factorial *n*!:
Think: multiply descending numbers starting with *n*
Note: $4!$ means to find the product $4 \cdot 3 \cdot 2 \cdot 1$.
An important definition is that $0! = 1$.

When to use the permutation formula:
Think: order must be important
Note: The notation ${}_nP_r$ is sometimes used instead of $P(n,r)$. Also remember that $P(n,n) = n!$ and another way to calculate a permutation is $P(n,r) = n(n-1)(n-2)\ldots(n-r+1)$.

How to use the permutation formula:
Think: identify the number of objects *n* and how many to take at a time *r*
Note: The permutation formula is

$$P(n,r)=\frac{n!}{(n-r)!}, r\le n.$$

To use the complementary counting principle:
Think: subtract the known from the universal set
Note: $n(A)=n(U)-n(A')$

Hint
You will use this principle when it is easier to count the set that you don't want, called the complementary set, in order to find the number in the set that you do want.

To use the additive counting principle:
Think: union of two sets that intersect
Note: $n(A\cup B)=n(A)+n(B)-n(A\cap B)$

10.2 Exercises

1. $4\cdot 3\cdot 2\cdot 1=24$ different orders can be written with the letters in the set $\{a, b, c, d\}$.

3. $6\cdot 5\cdot 4\cdot 3\cdot 2\cdot 1=720$ different ways that the horses can finish the race.

5. $5\cdot 4\cdot 3\cdot 2\cdot 1=120$ different ways to list the 5 top reasons people hate scanning their own groceries.

7. $8!=8\cdot 7\cdot 6\cdot 5\cdot 4\cdot 3\cdot 2\cdot 1=40{,}320$

9. $9!=9\cdot 8\cdot 7\cdot 6\cdot 5\cdot 4\cdot 3\cdot 2\cdot 1=362{,}880$

11. $\frac{11!}{8!}=\frac{11\cdot 10\cdot 9\cdot 8!}{8!}=990$

13. $\frac{9!}{5!4!}=\frac{9\cdot 8\cdot 7\cdot 6\cdot 5!}{5!\cdot 4\cdot 3\cdot 2\cdot 1}=9\cdot 7\cdot 2=126$

15. $P(10,2)=10\cdot 9=90$

17. $\frac{P(5,2)}{2!}=\frac{5\cdot 4}{2\cdot 1}=10$

19. $3\cdot P(8,5)=3\cdot 8\cdot 7\cdot 6\cdot 5\cdot 4=20{,}160$

21. $5!=5\cdot 4\cdot 3\cdot 2\cdot 1=120$ ways to stack the 5 books she must carry.

23. a. $5!=5\cdot 4\cdot 3\cdot 2\cdot 1=120$ permutations of the 5 qualities.
 b. $P(5,3)=5\cdot 4\cdot 3=60$ different buildings.
 c. Yes, 7,601,445 is divisible by 3:
 $7{,}601{,}445\div 3=2{,}533{,}815$
 Yes, 7,601,445 is divisible by 5:
 $7{,}601{,}445\div 5=1{,}520{,}289$

25. $P(13,3)=13\cdot 12\cdot 11=1716$ different ways to draw 3 hearts without replacement.

27. $P(26,2)=26\cdot 25=650$ different ways to draw 2 red cards without replacement.

29. $P(5,3)=5\cdot 4\cdot 3=60$ different 3-digit numbers with no repetitions from the digits, 1, 3, 5, 7, and 9.

31. $3\cdot 3=9$ different ways for both dice to come up even numbers.

33. There is just one way to have no boys. Therefore, there are $2^5-1=32-1=31$ ways to have at least one boy.

35. There is just one way to have no tails. All the coins must come up heads.

37. The number of ways in which the six coins can fall is $2^6=64$. The number of ways of getting no heads is 1 and the number of ways of getting exactly one head is 6. Thus, the number of ways of getting at least 2 heads is $64-1-6=57$.

39. There are 50 multiples of 2; there are 20 multiples of 5; and there are 10 multiples of $2\cdot 5$. Use the equation $n(A\cup B)=n(A)+n(B)-n(A\cap B)$ with $n(A)=50, n(B)=20$, and $n(A\cap B)=10$. Then, $n(A\cup B)=50+20-10=60$.

41. a. $10^6=1{,}000{,}000$
 b. $P(10,6)=10\cdot 9\cdot 8\cdot 7\cdot 6\cdot 5=151{,}200$

c. There will only be 151,200 license plate numbers available, which is not enough for a population of 608,827.

43. There are two events involved: pick the first letter (2 choices, K or W) and then pick 3 other letters with no repetitions $P(25, 3)$, since there are 25 letters left in the alphabet to choose the 3 letters from. The total number of possibilities is $2 \cdot P(25,3) = 2 \cdot 25 \cdot 24 \cdot 23 = 27{,}600$.

45. (1) $n! = n(n-1)!$ for $n > 1$. This formula holds for $n = 1$ only if 0! is defined to be 1.

(2) The formula $P(n,r) = \dfrac{n!}{(n-r)!}$ holds for $n = r$ only if 0! is defined to be 1.

47. The number of elements in the union of two sets is found by finding the sum of the number of elements in each of the sets, then subtracting the number of elements that have been counted twice (that is, the number of elements common to the two sets).

49. 5! = 120 different ways that the race could end.

51. 3. *A* 4th, *B* 5th; *B* 4th, *A*5th and a tie for 4th place.

53. If one of the persons is seated, then the others can be seated in $(n-1)!$ ways. Except for rotation around the table, this is the total number of ways in which the n persons could be seated.

55. After *A* is seated, there are two ways to seat *B* next to *A* (so that *B* does not face *A*). Then there are two ways of seating the remaining two persons. Thus, there are $2 \cdot 2 = 4$ ways of seating the four persons.

Calculator Corner

1. Problem 7:

[8][2nd][!][=]40,320

Problem 14:

[9][$_nP_r$][4][=]3024

Section 10.3 Combinations

Study Tips

When order is *not important*, a set of r objects selected from a set of n objects, is called a *combination* and is denoted by $C(n, r)$.

When to use the combination formula:
Think: order is not important

Note: The notations ${}_nC_r, C_{n,r}, C_r^n$, and $\binom{n}{r}$ are also used to denote combinations. Remember that *combinations* are used when the order is *not important.*

How to use the combination formula:
Think: identify the number of objects n and how many to take at a time r

Note: The formula to use is $C(n,r) = \dfrac{P(n,r)}{r!}$, which is usually written as $\dfrac{n!}{r!(n-r)!}$.

<u>Word of Advice</u>
When you work with these formulas, use "cancellation" wisely. Thus, to simplify $\dfrac{7!}{5!2!}$, write the 7! as $7 \cdot 6 \cdot 5!$ and "cancel" the 5!. Do not multiply it out.

10.3 Exercises

1. $C(5,2) = \dfrac{5 \cdot 4}{2 \cdot 1} = 10;\ P(5,2) = 5 \cdot 4 = 20$

3. $C(7,3) = \dfrac{7 \cdot 6 \cdot 5}{3 \cdot 2 \cdot 1} = 35; P(7,3) = 7 \cdot 6 \cdot 5 = 210$

5. 5. $C(9,6) = \dfrac{9 \cdot 8 \cdot 7}{3 \cdot 2 \cdot 1} = 84;$

$P(9,6) = 9 \cdot 8 \cdot 7 \cdot 6 \cdot 5 \cdot 4 = 60{,}480$

7. $C(5,4) = \dfrac{5 \cdot 4!}{4! \cdot 1!} = 5$ combinations with 4 of the 5 tips given in the list.

9. $C(5,3) = \dfrac{5!}{3!2!} = 10$ combinations if the 5 tips in the list are taken 3 at a time.

11. $C(8,2)=\dfrac{8\cdot7\cdot6!}{2\cdot1\cdot6!}=\dfrac{56}{2}=28$ subsets of 2 elements from a set of 8 elements.

13. $C(12,8)=\dfrac{12\cdot11\cdot10\cdot9\cdot8!}{4\cdot3\cdot2\cdot1\cdot8!}=495$ different 8-element subsets from a set of 12 elements.

15. a. $C(12,3)=\dfrac{12\cdot11\cdot10\cdot9!}{3!9!}=220$ subsets of 3 elements.

b. $C(12,0)+C(12,1)+C(12,2)$

$=1+12+\dfrac{12\cdot11\cdot10!}{2!10!}=13+66=79$ subsets of less than 3 elements.

c. $C(12,0)=\dfrac{12!}{0!12!}=1$ subset with no elements.

d. $C(12,10)+C(12,11)+C(12,12)$

$=\dfrac{12\cdot11\cdot10!}{10!2!}+\dfrac{12\cdot11!}{11!1!}+\dfrac{12!}{12!0!}$

$=66+12+1=79$ subsets of more than 9 elements.

17. $C(5,4)=\dfrac{5!}{4!1!}=5$ different sums of money if 4 coins are used.

19. a. $C(5,3)=C(5,2)=\dfrac{5\cdot4}{2\cdot1}=10$ subsets of 3 elements.

b. $C(5,0)+C(5,1)+C(5,2)+C(5,3)$

$=1+5+10+10=26$ subsets that have no more than 3 elements.

21. $C(24,3)=\dfrac{24\cdot23\cdot22\cdot21!}{3\cdot2\cdot1\cdot21!}=2024$ ways to select 3 different Greek letters when order does not matter.

23. $C(10,3)=\dfrac{10\cdot9\cdot8\cdot7!}{3!7!}=120$ different committees.

25. $C(52,5)=\dfrac{52\cdot51\cdot50\cdot49\cdot48\cdot47!}{5\cdot4\cdot3\cdot2\cdot1\cdot47!}$

$=2{,}598{,}960$ different 5-card poker hands.

27. $C(100,5)=\dfrac{100\cdot99\cdot98\cdot97\cdot96\cdot95!}{5\cdot4\cdot3\cdot2\cdot1\cdot95!}$

$=75{,}287{,}520$ different 5-member committees.

29. $C(4,1)+C(4,2)+C(4,3)+C(4,4)$

$=2^4-1=16-1=15$ different sums of money.

31. $C(6,1)+C(6,2)+C(6,3)+C(6,4)$

$+C(6,5)+C(6,6)$

$=2^6-1=64-1=63$ different sums of money.

33. This is the number of ways that 4 of the 8 people can be selected without regard to order:

$C(8,4)=\dfrac{8\cdot7\cdot6\cdot5\cdot4!}{4\cdot3\cdot2\cdot1\cdot4!}=70$

35. Answers will vary. The important difference is that permutations take account of order and combinations do not.

37. The order of the numbers is important, so a permutation is being used.

39.

$$\begin{array}{ccccccc} & 1+5 & 5+10 & 10+10 & 10+5 & 5+1 & \\ & \downarrow & \downarrow & \downarrow & \downarrow & \downarrow & \\ n=6 \quad 1 & 6 & 15 & 20 & 15 & 6 & 1 \end{array}$$

$$\begin{array}{cccccccc} & 1+6 & 6+15 & 15+20 & 20+15 & 15+6 & 6+1 & \\ & \downarrow & \downarrow & \downarrow & \downarrow & \downarrow & \downarrow & \\ n=7 \quad 1 & 7 & 21 & 35 & 35 & 21 & 7 & 1 \end{array}$$

41. a. $(a+b)^4$

$=a^4+4a^3b+6a^2b^2+4ab^3+b^4$

b. $(a+b)^5=a^5+5a^4b+10a^3b^2$

$+10a^2b^3+5ab^4+b^5$

43. $C(5,0)=\dfrac{5!}{0!5!}=1$ different way for the outcome to be 0 heads and 5 tails.

45. $C(5,3)=\dfrac{5\cdot4\cdot3!}{2\cdot1\cdot3!}=10$ different ways for the outcome to be 31 head and 2 tails.

47. $C(5,5) = \frac{5!}{5!0!} = 1$ different way for the outcome to be 5 heads and 0 tails.

49. The left side is the sum of the number of ways in which there could be 0 heads and n tails, 1 head and $n-1$ tails, 2 heads and $n-2$ tails, and so on, to n heads and 0 tails. The right side is exactly the number of ways in which n coins can fall either heads or tails. Thus,

$$C(n,0)+C(n,1)+C(n,2)+\ldots+C(n,n) = 2^n$$

Calculator Corner

1. Example 6:
   a. [26] [$_nC_r$] [3] [ENTER] Answer: 2600
   b. [8] [$_nC_r$] [2] [ENTER] Answer: 28

   Example 7:
   [52] [$_nP_r$] [2] [÷] [2!] [ENTER] Answer: 1326

   Example 8:
   a. [6] [$_nC_r$] [1] [ENTER] Answer: 6
   b. [6] [$_nP_r$] [2] [÷] [2!] [ENTER] Answer: 15
   c. [6] [$_nP_r$] [3] [÷] [3!] [ENTER] Answer: 20
   d. [6] [$_nC_r$] [4] [ENTER] Answer: 15
   [6] [$_nC_r$] [2] [ENTER] Answer: 15

Section 10.4 Miscellaneous Counting Methods

Study Tips

To decide when to use permutations and when to use combinations, let this figure help you.

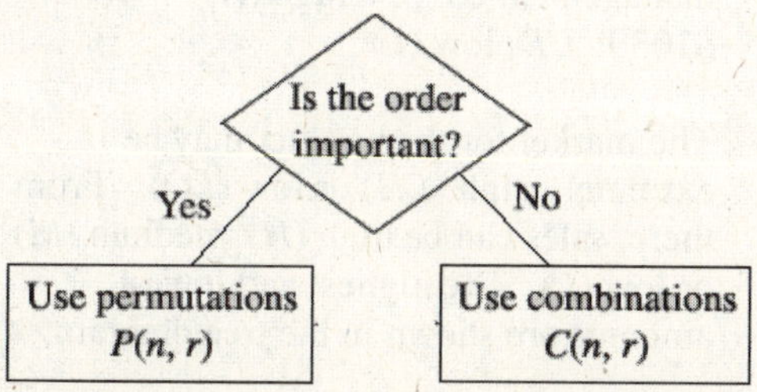

Hint
Use **permutations** when you see the words such as 'different ways for items,' 'how many permutations,' or problems involving placing books on a shelf, people in a line, electing a president, vice president, etc., placement in a race, items drawn without replacement.

Hint
Use **combinations** when you see the words such as 'different sums/subsets,' 'how many combinations,' or problems involving committees, forming different groups of objects, where order does not matter.

Caution
Permutations and combinations can easily get confused. So remember that with n objects taken r at a time, there will *always* be more permutations than combinations.

To find permutations of nondistinct objects:
Think: count the number of like object(s), divide by the factorial of that number(s)
Note: The total number of *distinct* arrangements of n objects is $\frac{n!}{n_1!n_2!\cdots n_r!}$.

10.4 Exercises

1. a. Order important:
   $P(52,3) = 52\cdot 51\cdot 50 = 132{,}600$ different orders of 3 cards dealt.
   b. Order not important:
   $C(52,3) = \frac{52\cdot 51\cdot 50\cdot 49!}{3\cdot 2\cdot 1\cdot 49!} = 22{,}100$
   different 3-card hands.

3. a. $50\cdot 50\cdot 50 = 125{,}000$ different ways to select 3 plays for the next 3 downs. (The plays do not have to be all different.)
   b. Order not important :
   $C(50,3) = \frac{50\cdot 49\cdot 48\cdot 47!}{3\cdot 2\cdot 1\cdot 47!} = 19{,}600$
   different ways to select a set of 3 plays to study. (The plays would be different.)

5. $C(5,3) = \frac{5\cdot 4\cdot 3!}{2\cdot 1\cdot 3!} = 10$ different ways to take 3 different courses on Mondays.

7. a. Order not important:
$C(7,3) = \dfrac{7 \cdot 6 \cdot 5 \cdot 4!}{3 \cdot 2 \cdot 1 \cdot 4!} = 35$ ways to select 3 electives from a group of 7 courses.
b. Peter can select the 3 hours in $P(7,4) = 840$ ways.

9. a. $P(14,2) \cdot P(10,2) = 14 \cdot 13 \cdot 10 \cdot 9$ $= 16,380$ possibilities.
b. $P(12,2) \cdot P(10,2) = 12 \cdot 11 \cdot 10 \cdot 9 =$ $11,880$ possibilities if 2 of the boys refuse to participate.

11. a. $C(7,3) \cdot C(8,1) = \dfrac{7 \cdot 6 \cdot 5 \cdot 4!}{3 \cdot 2 \cdot 1 \cdot 4!} \cdot \dfrac{8}{1} = 280$ ways to fill 4 vacancies with 1 woman and 3 men if 7 men and 8 women are available to serve.
b. $C(5,3) \cdot C(2,1) = \dfrac{5 \cdot 4 \cdot 3!}{2 \cdot 1 \cdot 3!} \cdot \dfrac{2}{1} = 20$ ways to fill 4 vacancies with 1 woman and 3 men if 5 men and 2 women are available to serve.

13. There are 11 letters in all: 1 T, 3 A's, 2 L's, 1 H, 2 S's, and 2 E's, so the answer is
$$\frac{11!}{1!3!2!1!2!2!} = 831,600$$

15. There are 9 letters in all: 2 R's, 2E's, 2 D's. 2 I's, and 1 V, so the answer is
$$\frac{9!}{2!2!2!2!1!} = 22,680$$

17. The contractor can select 2 components from the first subcontractor in C(7, 2) ways, 3 components from the second subcontractor in C(5, 3) ways, and 2 components from the third subcontractor in C(2, 2) ways. Thus, he can select his 7 components in $C(7,2) \cdot C(5,3) \cdot C(2,2)$
$$= \frac{7 \cdot 6}{2 \cdot 1} \cdot \frac{5 \cdot 4 \cdot 3}{3 \cdot 2 \cdot 1} \cdot \frac{2 \cdot 1}{2 \cdot 1} = 21 \cdot 10 \cdot 1 = 210$$
ways.

19. We have to choose the least value of *n* for which *C*(*n*, 5) ≥ 21, that is, for which $\dfrac{n(n-1)(n-2)(n-3)(n-4)}{5!} \geq 21$. To simplify the calculation, multiply both sides by 5! = 120:
$n(n-1)(n-2)(n-3)(n-4) \geq 2520$. Try $n = 6$. The left side becomes $6 \cdot 5 \cdot 4 \cdot 3 \cdot 2 = 720$. Too small! Try $n = 7$. The left side becomes $7 \cdot 6 \cdot 5 \cdot 4 \cdot 3 = 2520$. Thus, to carry out the program, the network must have at least **7** different movies.

21. She can select 2 of the 3 courses and the order in which they come in the morning. Then she can select 2 of the 4 morning hours for these 2 courses. That leaves her the choice of 1 of the 2 afternoon hours for the third course. Thus, she has
$$P(3,2) \cdot C(4,2) \cdot C(2,1) = 3 \cdot 2 \cdot \frac{4 \cdot 3}{2 \cdot 1} \cdot \frac{2}{1}$$
$= 6 \cdot 6 \cdot 2 = 72$ choices.

23. a. The first decision is to select a mutual fund (*M*), a management company (*C*) or a *CD*. The mutual fund can have high (10%) or low (6%) yields, the management company can have high (10%) or low yields (5%) and the *CD* can have high (4%) or low (3%) yield. The highest (↑) and lowest (↓) investments are shown in the tree diagram:

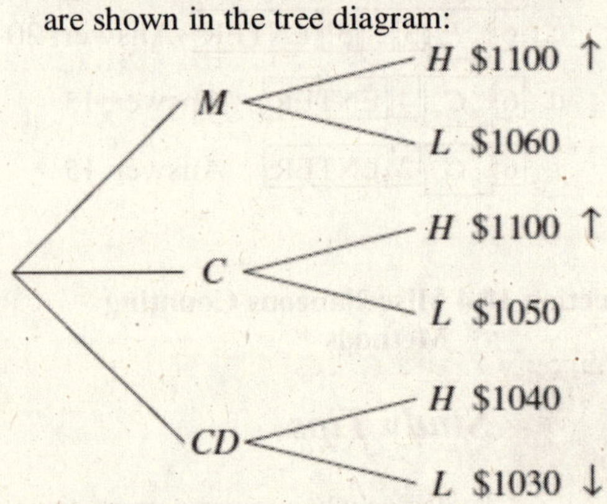

b. $1100. Mutual Fund (*M*), high (*H*) or management co. (*C*), high (*H*)
c. $1030. *CD*, low (*L*)

25. a. The market for the product may be extremely high (*EH*) or low (*LO*). From there, sales can be high (*H*), medium (*M*) or low (*L*) The highest and lowest amounts are shown in the tree diagram:

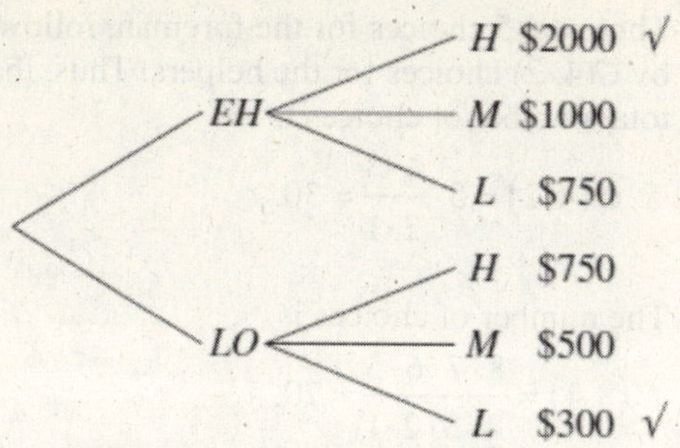

b. $2000. Market size extremely high (*EH*), sales high (*H*).

c. $300. Market size low (*LO*), sales low (*L*)

27. Answers will vary. Read the text to see what each of the SCP, *P*(*n*, *k*) and *C*(*n*, *k*) formulas calculates.

29. $144 = 2^4 \cdot 3^2$. The factor 2 can be used 0, 1, 2, 3, or 4 times, and the factor 3 can be used 0, 1, or 2 times. So the number of exact divisors is $(4+1)(2+1) = 5 \cdot 3 = 15$.

31. The 2 can be a factor 0, 1, 2, . . ., or *a* times; (*a* + 1) choices.
The 3 can be a factor 0, 1, 2, . . ., or *b* times; (*b* + 1) choices.
The 5 can be a factor 0, 1, 2, . . ., or *c* times; (*c* + 1) choices.
The 7 can be a factor 0, 1, 2, . . ., or *d* times; (*d* +1) choices.
Thus, the number of exact divisors is $(a+1)(b+1)(c+1)(d+1)$.

Chapter 10 Practice Test

Study Tips

Your main task is to know *when* to use the formulas given in the text. Once again, remember: permutations are used when the order is important and combinations when the order is not important.

As you take your test read each question carefully to make sure you are giving the answer that is requested. It is helpful to underline important concepts or words as you are reading each question. Small words can make a huge difference in the answer.

Here are the solutions to the practice test.

1.

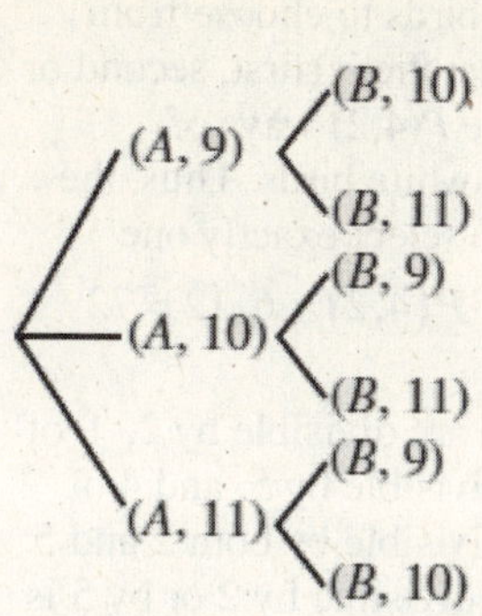

2. $2 \cdot 3 \cdot 5 = 30$ different meals are possible.

3. a. Since each die can fall in 6 different ways, $6 \cdot 6 = 36$ different results are possible.
   b. To get a sum of 5, one die must come up 1 and the other 4, or one must come up 2 and the other 3. Since each of these events can occur in 2 ways, the total number of ways is $2 \cdot 2 = 4$.

4. There are 2 black jacks and 26 red cards, so the total number of ways to get a black jack followed by a red card is $2 \cdot 26 = 52$.

5. You have a choice of 3 flights followed by a choice of 5 flights, so the answer is $3 \cdot 5 = 15$.

6. a. $7! = 7 \cdot 6 \cdot 5 \cdot 4 \cdot 3 \cdot 2 \cdot 1 = 5040$
   b. $\frac{7!}{4!} = \frac{7 \cdot 6 \cdot 5 \cdot 4 \cdot 3 \cdot 2 \cdot 1}{4 \cdot 3 \cdot 2 \cdot 1} = 210$

7. a. $3! \cdot 4! = 3 \cdot 2 \cdot 1 \cdot 4 \cdot 3 \cdot 2 \cdot 1 = 144$
   b. $3! + 4! = 3 \cdot 2 \cdot 1 + 4 \cdot 3 \cdot 2 \cdot 1$
   $= 6 + 24 = 30$

8. a. $P(5,5) = 5! = 5 \cdot 4 \cdot 3 \cdot 2 \cdot 1 = 120$
   b. $P(6,6) = 6! = 6 \cdot 5 \cdot 4 \cdot 3 \cdot 2 \cdot 1 = 720$

9. a. $P(8,2) = 8 \cdot 7 = 56$
   b. $P(7,3) = 7 \cdot 6 \cdot 5 = 210$

10. They can be arranged in $P(4,4) = 4! = 4 \cdot 3 \cdot 2 \cdot 1 = 24$ ways.

11. Each couple can be seated in 2 ways, and the 3 couples can be arranged in *P*(3, 3) ways, so the answer is $2^3 \cdot 3! = 8 \cdot 6 = 48$.

12. There are 2 white birds to choose from and 3 ways to select them (first, second or third) and there are $P(4,2)$ ways of selecting two non-white birds. Thus, the number of ways to select exactly one white bird is $2 \cdot 3 \cdot P(4,2) = 6 \cdot 12 = 72$.

13. 24 of the numbers are divisible by 2, 9 of the numbers are divisible by 5, and 4 of the numbers are divisible by both 2 and 5. Thus, the number divisible by 2 or by 5 is $24 + 9 - 4 = 29$.

14. The number of different sums is
$$C(4,2) = \frac{4 \cdot 3}{2 \cdot 1} = 6.$$

15. The number of subsets is
$$C(6,3) = \frac{6 \cdot 5 \cdot 4}{3 \cdot 2 \cdot 1} = 20.$$

16. a. $C(5,2) = \frac{5 \cdot 4}{2 \cdot 1} = 10$

b. $C(6,4) = C(6,2) = \frac{6 \cdot 5}{2 \cdot 1} = 15$

17. a. $C(6, 0) = 1$

b. $C(5,4) = C(5,1) = 5$

$$C(5,3) = C(5,2) = \frac{5 \cdot 4}{2 \cdot 1} = 10$$

$$\frac{C(5,4)}{C(5,3)} = \frac{5}{10} = \frac{1}{2}$$

18. The number of different sets is
$$C(52,2) = \frac{52 \cdot 51}{2 \cdot 1} = 1326.$$

19. The number of different schedules is $P(4,4) = 4! = 4 \cdot 3 \cdot 2 \cdot 1 = 24$.

20. The number of different sums is
$$C(5,1) + C(5,2) + C(5,3)$$
$$= \frac{5}{1} + \frac{5 \cdot 4}{2 \cdot 1} + \frac{5 \cdot 4 \cdot 3}{3 \cdot 2 \cdot 1} = 5 + 10 + 10 = 25.$$

21. There are 5 choices for the foreman, followed by $C(4, 2)$ choices for the helpers. Thus, the total number of choices is
$$5 \cdot C(4,2) = 5 \cdot \frac{4 \cdot 3}{2 \cdot 1} = 30.$$

22. The number of choices is
$$C(8,4) = \frac{8 \cdot 7 \cdot 6 \cdot 5}{4 \cdot 3 \cdot 2 \cdot 1} = 70.$$

23. There are 8 letters: 2 B's, 4 O's, 1 G and 1 A, so the number of distinct arrangements is
$$\frac{8!}{2!4!1!1!} = \frac{8 \cdot 7 \cdot 6 \cdot 5 \cdot 4 \cdot 3 \cdot 2 \cdot 1}{2 \cdot 1 \cdot 4 \cdot 3 \cdot 2 \cdot 1 \cdot 1 \cdot 1} = 840.$$

24. There are 11 letters: 4 M's, 4 A's, 2 D's and 1 I, so the number of distinct arrangements is
$$\frac{11!}{4!4!2!1!} = \frac{11 \cdot 10 \cdot 9 \cdot 8 \cdot 7 \cdot 6 \cdot 5 \cdot 4 \cdot 3 \cdot 2 \cdot 1}{4 \cdot 3 \cdot 2 \cdot 1 \cdot 4 \cdot 3 \cdot 2 \cdot 1 \cdot 2 \cdot 1 \cdot 1}$$
$$= 34,650.$$

25. There are 1977 marks in all: 1189 a's, 460 d's, and 328 n's, so the number of distinct arrangements is $\frac{1977!}{1189!460!328!}$.

Chapter 11 Probability

Section 11.1 Sample Spaces and Probability

Study Tips

You may be very familiar with probabilities, for instance, in news articles, gambling, or even weather reports, yet not fully understand probability. *Probability* is the science of determining the likelihood or chance that an event will occur.

To find the probability of an event:
Think: divide the number of ways the event can occur by the sample space
Note: In the study of probability, tossing a coin and drawing a card from a deck are activities called *experiments* and the set of all possible outcomes is the *sample space*.

Fact
The probability of an event is always a number between **0** and **1**. If the event cannot occur its probability is **0** and if the event is certain to occur its probability is **1**.

To find the empirical probability of an event:
Think: experimental probability
Note: $P(E) = \dfrac{\text{number of times } E \text{ has occurred}}{\text{total number of times the experiment is performed}}$

Caution
Be aware of these differences: *theoretical* probability is based on the theory that the outcomes are equally likely, while *empirical* probability shows what really happened in the experiment.

Hint
Make sure you remember the sample space in Example 6, it will be used again!

11.1 Exercises

1. There are 6 possible outcomes of which exactly one is favorable, so $P(E)=\frac{1}{6}$.

3. There are 2 favorable outcomes and 6 possible outcomes, so $P(E)=\frac{2}{6}=\frac{1}{3}$.

5. There are 10 possible outcomes of which exactly one is favorable, so $P(E)=\frac{1}{10}$.

7. There are 9 favorable outcomes and 10 possible outcomes, so $P(E)=\frac{9}{10}$.

9. There are no favorable outcomes possible, so $P(E)=0$.

11. There are 4 aces and 52 cards in all, so $P(\text{ace})=\frac{4}{52}=\frac{1}{13}$.

13. There are 13 spades out of 52 cards, so $P(\text{spade})=\frac{13}{52}=\frac{1}{4}$.

15. There are 13 spades and 9 face cards that are not spades, so there are $13+9=22$ favorable outcomes out of 52 possible outcomes. Therefore, $P(E)=\frac{22}{52}=\frac{11}{26}$.

17. a. There is 1 favorable outcome out of 5 possible outcomes, so $P(E)=\frac{1}{5}$.
    b. There are 3 favorable outcomes out of 5 possible outcomes so $P(E)=\frac{3}{5}$.
    c. There are 4 favorable outcomes out of 5 possible outcomes, so $P(E)=\frac{4}{5}$.

19. a. There is 1 favorable outcome out of 3 possible outcomes, so $P(E)=\frac{1}{3}$.
    b. There is 1 favorable outcome out of 3 possible outcomes, so $P(E)=\frac{1}{3}$.

21. There is 1 favorable outcome and 8 possible outcomes, so $P(E)=\frac{1}{8}$.

23. There are 3 favorable outcomes and 8 possible outcomes, so $P(E)=\frac{3}{8}$.

25. A sum of 7 is either a 1 and 6, 2 and 5, 3 and 4, 4 and 3, 5 and 2, or 6 and 1. There are 6 favorable outcomes and 36 possible outcomes, so $P(E)=\frac{6}{36}=\frac{1}{6}$.

27. There are 6 possible ways to have the same number on both dice and there are 36 possible outcomes. So, $P(E)=\frac{6}{36}=\frac{1}{6}$.

29. There are 18 possible ways to obtain an even sum on the dice and there are 36 possible outcomes. So, $P(E)=\frac{18}{36}=\frac{1}{2}$.

31. a. $P(B)=\frac{25}{100}=\frac{1}{4}$

b. $P(R)=\frac{28}{100}=\frac{7}{25}$

c. $P(G)=\frac{24}{100}=\frac{6}{25}$

d. $P(Y)=\frac{23}{100}$

e. Theoretical probabilities: $\frac{1}{4}$ for each

f. $P(B)$

33.

Outcome	Empirical Probability	Theoretical Probability
a. Red	$\frac{24}{100}=\frac{6}{25}$	$\frac{1}{6}$
b. Blue	$\frac{9}{100}$	$\frac{1}{9}$
c. Green	$\frac{29}{100}$	$\frac{5}{18}$
d. White	$\frac{20}{100}=\frac{1}{5}$	$\frac{2}{9}$
e. Yellow	$\frac{18}{100}=\frac{9}{50}$	$\frac{2}{9}$

35. a. $P(11)=\frac{6}{50}=\frac{3}{25}$

b. $P(7)=\frac{9}{50}$

c. $P(O)$

$=P(3)+P(5)+P(7)+P(9)+P(11)$

$=\frac{4}{50}+\frac{4}{50}+\frac{9}{50}+\frac{3}{50}+\frac{6}{50}=\frac{26}{50}=\frac{13}{25}$

d. $P(\sim O)=1-P(O)=1-\frac{13}{25}=\frac{12}{25}$

e. Getting a sum of 8.

f. Getting a sum of 2 or 4 or 12.

g. Getting a sum of 2 or 4 or 12.

h. No

37. a. $P(\text{None})=\frac{172}{502}=\frac{86}{251}$

b. $P(1)=\frac{73}{502}$

c. Having exactly 4 credit cards.

d. Having no credit cards.

e. $P(\text{at least } 1)=1-P(\text{None})$

$=1-\frac{86}{251}=\frac{165}{251}$

39. a. $P(\text{Victim})=\frac{75}{500}=\frac{3}{20}$

b. $P(\text{Not a victim})=\frac{370}{500}=\frac{37}{50}$

c. $P(\text{Not sure})=\frac{55}{500}=\frac{11}{100}$

41. a. $P(\text{Reading})=\frac{89}{239}$

b. $P(\text{Watch movie})=\frac{33}{239}$

c. $P(\text{Not watch movie})$

$=1-P(\text{Watch movie})$

$=1-\frac{33}{239}=\frac{206}{239}$

d. $P(\text{Sleeping})=\frac{24}{239}$

43. a. $P(\text{Very often})=\frac{140}{5543}$

b. $P(\text{Never}) = \dfrac{2804}{5543}$

c. $P(\text{Regift}) = 1 - P(\text{Never})$

$= 1 - \dfrac{2804}{5543} = \dfrac{2739}{5543}$

45. a. There are 650 favorable outcomes and 850 possible outcomes, so $P(E) = \dfrac{650}{850} = \dfrac{13}{17}$.

b. There are 200 favorable outcomes and 850 possible outcomes, so $P(E') = \dfrac{200}{850} = \dfrac{4}{17}$

This problem can also be done by using the result of part (a) and the equation $P(E') = 1 - P(E)$ to get the same answer:

$P(E') = 1 - \dfrac{13}{17} = \dfrac{4}{17}$.

47. a. $P(\text{NL player home run})$

$= \dfrac{150}{5493} = \dfrac{50}{1831}$

b. $P(\text{AL player home run}) = \dfrac{158}{5533}$

49. AL players; their probability of hitting a home run $\dfrac{158}{5533} = 0.029$ is higher than that of NL players $\dfrac{150}{5493} = 0.027$.

51. Answers will vary. Experimental (empirical) probablity is based on actual results from an experiment, whereas, theoretical probablity is based on logical reasoning.

53. The probability formula does not apply if the events are not all equally likely to occur. For example, if a die is weighted so that a 6 is twice as likely to come up as any other number, then to calculate the probability that an even number comes up, it would be wrong to use the fact that three of the six faces are even so that the probability is 1/2. Instead, the 6 face must be given a weight of 2 and the other faces weights of 1. Then the weight of the even faces is $1 + 1 + 2 = 4$ and the weight of all the faces is $1 + 1 + 1 + 1 + 1 + 2 = 7$. This means that the probability that an even number comes up is $\dfrac{4}{7}$.

55. a. Choosing a number at random from 1 to 7 has an equally likely outcome.

b. Tossing a fair coin has an equally likely outcome.

c. Choosing a letter at random from MISSISSIPPI does not have equally likely outcomes, because each letter does not occur the same number of times.

d. Choosing a number at random from 1 to 10 has an equally likely outcome.

57. There were 465,300 successes and 472,000 operations, so the probability of a success is $P(S) = \dfrac{465300}{472000} = \dfrac{4653}{4720} \approx 0.986$.

59. There were 506,000 successes and 508,000 operations, so the probability of a success is $P(S) = \dfrac{506000}{508000} = \dfrac{253}{254} \approx 0.996$.

Section 11.2 Counting Techniques and Probability

Study Tips

We will now utilize the counting techniques from Chapter 10 in order to determine probability.

To use a tree diagram to find the probability of an event:
Think: draw a tree diagram, all outcomes are found by following the branches
Note: For the probability of a certain event, count the number of times it shows up and divide by the total number of outcomes.

To use permutations and combinations to find the probability of an event:
Think: permutations if order is important, combinations if order is not important
Note: For the probability of a certain event, divide number of ways for the specific event (draw 2 kings from a standard deck) to occur by the total number of ways of the general event (draw 2 cards from a standard deck).

Word of Advice
Some problems can be worked using permutations, combinations or the SCP. Review these techniques from Chapter 10, if necessary.

11.2 Exercises

1. The total number of possibilities is $3\cdot 2\cdot 3=18$ and the number favorable is 1. Therefore, the probability is $\frac{1}{18}$.

3. In 1 throw, *P*(even number) = $\frac{1}{2}$, so in 3 throws, $P(\text{Even})=\left(\frac{1}{2}\right)^3=\frac{1}{8}$.

5. The total number of possible choices for the committee of four is *C*(25, 4). The number of choices to select both Helen and Patty are *C*(2, 2), and the total number of choices of the remaining committee members is *C*(23, 2). So the probability that both Helen and Patty will be selected is

$$P(E)=\frac{C(2,2)C(23,2)}{C(25,4)}$$
$$=\frac{1\cdot 23\cdot 22}{2\cdot 1}\cdot\frac{4\cdot 3\cdot 2\cdot 1}{25\cdot 24\cdot 23\cdot 22}=\frac{1}{50}.$$

7. a. Suppose that Mr. C. Nile chooses a restaurant. The probability that Mr. D. Mented goes to that restaurant is $\frac{1}{5}$

   b. The probability that they miss each other is $1-\frac{1}{5}=\frac{4}{5}$.

9. a. To get 35¢, a person must get first a dime and then a quarter or first a quarter and then a dime. Since there are 2 dimes and 2 quarters, there are $2\cdot(2\cdot 2)=8$ ways of getting 35¢. There are 7 coins, so there are $7\cdot 6=42$ ways of getting 2 coins. Thus, $P(35¢)=\frac{8}{42}=\frac{4}{21}$.

   b. To get 50¢, a person must get the 2 quarters in either order, and there are 2 ways of doing this. Hence, $P(50¢)=\frac{2}{42}=\frac{1}{21}$.

11. $P(2\text{ kings})=\frac{P(4,2)}{P(52,2)}=\frac{C(4,2)}{C(52,2)}=\frac{1}{221}$

You can also use the SCP. There are 4 possibilities for selecting the first king and 3 for selecting the second king. Thus, there are $4\cdot 3$ favorable ways of selecting the two kings. Since there are $52\cdot 51$ ways of selecting two cards from a deck, the answer is $\frac{4\cdot 3}{52\cdot 51}=\frac{1}{221}$.

13. There are 13 spades and 3 kings (other than the king of spades), so there are $13\cdot 3$ ways of getting a spade and one of the 3 kings in that order. The number of ways of drawing 2 cards is $52\cdot 51$. Thus, the probability of drawing a spade and one of the other 3 kings, in that order, is $\frac{13\cdot 3}{52\cdot 51}=\frac{1}{68}$.

15. Since there are 26 red cards and 52 cards in all, the probability of drawing 2 red cards in 2 draws is

$$P(2\text{ red cards})=\frac{P(26,2)}{P(52,2)}=\frac{C(26,2)}{C(52,2)}=\frac{25}{102}.$$

17. a. There are 5 \$50 bills and 3 \$10 bills, so there are $5\cdot 3=15$ ways of getting a \$50 bill and a \$10 bill in that order.

   b. There are 5 + 4 + 3 + 2 + 1 = 15 bills in all, so there are $15\cdot 14$ ways of selecting 2 bills. Thus, the probability of getting a \$5 bill and a \$10 bill, in that order, is $\frac{5\cdot 3}{15\cdot 14}=\frac{1}{14}$.

   c. Since there are 2 ways to get a specific \$50 bill and a \$10 bill (without regard to order), the probability is twice that found in part b: $2\cdot\frac{1}{14}=\frac{1}{7}$.

19. The probability that at least one of the two tires is defective is

$$1-P(\text{neither defective})=1-(0.98)^2$$
$$=1-0.9604=0.0396$$

21. The probability is

$$\frac{6}{100}\cdot\frac{8}{10{,}000}=\frac{3}{62{,}500}=0.000048.$$

23. a. Since the order does not matter, we can use combinations rather than permutations. There are 4 kings and 4 aces and 44 other cards in the deck of 52 cards. Therefore, the probability of getting 2 kings, 2 aces, and one other card is $\dfrac{C(4,2)C(4,2)C(44,1)}{C(52,5)}$.

b. The probability of getting 3 kings and 2 aces is $\dfrac{C(4,3)C(4,2)}{C(52,5)}$.

25. The number of different combinations of 2 of the 4 crew members is $C(4, 2) = 6$. There are $C(3, 1) = 3$ ways of selecting 1 of the 3 men and $C(4, 2) = 6$ ways of selecting 1 woman. Thus, the probability that 1 man and the 1 woman are selected for the space walk is

$$\frac{C(3,1)\cdot C(1,1)}{C(4,2)}=\frac{3\cdot 1}{6}=\frac{3}{6}=\frac{1}{2}$$

27. There are 4 possible royal flushes, so the probability of getting a royal flush is

$$P(\text{royal flush})=\frac{4}{C(52,5)}=\frac{1}{649{,}740}$$
$$\approx 0.0000015$$

29. There are 13 sets of four of a kind and if one of these occurs, the fifth card can be any one of the remaining 48 cards. Thus, the probability of getting four of a kind is

$$P(4\text{ of a kind})=\frac{13\cdot 48}{C(52,5)}$$
$$=\frac{1}{4165}\approx 0.00024$$

31. The number of combinations of 5 cards in a given suit is C(13, 5) and 10 of these are straights (including a royal flush). Thus, the number of flushes that are not straights is $C(13, 5) - 10$. The probability of getting such a flush is

$$P(E)=\frac{4\left[C(13,5)-10\right]}{C(52,5)}=\frac{5148-40}{C(52,5)}$$
$$=\frac{5108}{C(52,5)}=\frac{1277}{649{,}740}\approx 0.0020$$

33. The total probability for a low-producing well (See Example 9) is
$0.6(0.25)+0.4(0.125)=0.2$

35. Dome structure, dry with probability
$0.6(0.6)=0.36$.

37.

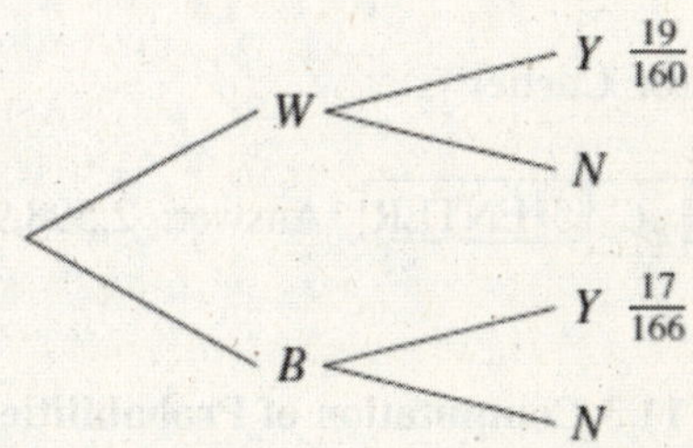

a. The probability that a white convict received the death penalty is $\dfrac{19}{160}$.

b. The probability that a black convict received the death penalty is $\dfrac{17}{166}$.

39. The number of convicts that did not received the death penalty is 141 + 149. The total number of convicts is 326. Thus, the probability that a convict of either race did not receive the death penalty is

$$\frac{141+149}{326}=\frac{290}{326}=\frac{145}{163}.$$

41. No. The coin is probably weighted to come up heads. Bet on heads.

43. No. The coin is probably weighted to come up heads as the odds of obtaining this result from a 'fair' coin is $\dfrac{1}{2^{10}}$. Bet on heads.

45. The number of combinations of 150 women that could be selected is $C(40{,}000, 150)$ and

the number of combinations of 150 men is $C(36{,}000, 150)$. The number of combinations of 300 persons that could be selected is $C(76{,}000, 300)$. Hence, the probability that 150 men and 150 women will be selected is

$$\frac{C(40{,}000,150)C(36{,}000,150)}{C(76{,}000,300)}.$$

47. The number of possible combinations of six women is $C(16, 6)$ and of six men is $C(14, 6)$. The number of combinations of 12 people selected from the 30 is $C(30, 12)$. Thus, the probability that the final jury will consist of six men and six women is $\dfrac{C(16,6)C(14,6)}{C(30,12)}$.

Calculator Corner

52 $_nC_r$ 5 ENTER Answer: 2,598,960

Section 11.3 Computation of Probabilities

Study Tips

This section offers additional work with the computation of probabilities.

Properties to help solve probability problems...

Property 1: impossible event, the $P(E)=0$

Property 2: range of probability is from 0 to 1

Property 3: two events overlapping, use $P(A\cup B)=P(A)+P(B)-P(A\cap B)$

Property 4: mutually exclusive (do not overlap), use $P(A\cup B)=P(A)+P(B)$

Note

To learn how to use Venn diagrams to solve probability problems, look at the *Discovery* part of Exercise 11.3.

11.3 Exercises

1. $E=\varnothing$, so $P(E)=0$; *Property 1*

3. Since the sum is always between 0 and 13, $P(E)=1$; *Property 2*

5. Let A be the event that the number is even and B the event that the number is greater than 7. Then $P(A)=\frac{5}{10}, P(B)=\frac{3}{10}$, $P(A\cap B)=\frac{2}{10}$. Now, we use *Property 3*, $P(A\cup B)=P(A)+P(B)-P(A\cap B)$, to get $P(A\cup B)=\frac{5}{10}+\frac{3}{10}-\frac{2}{10}=\frac{6}{10}=\frac{3}{5}$.

7. Since E is certain to occur, $P(E)=1$.

9. There are 14 favorable outcomes and 52 possible outcomes, so $P(E)=\frac{14}{52}=\frac{7}{26}$.

11. There are 13 diamonds and 52 cards in all, so $P(E)=\frac{13}{52}=\frac{1}{4}$.

13. There are 12 picture cards and 52 cards in all, so $P(E)=\frac{12}{52}=\frac{3}{13}$.

15. Let N be the event that the stock remained unchanged. Since 9 stocks remained unchanged, $P(N)=\frac{9}{50}$. The event that the stock did not change is N', so the probability that a randomly selected stock did change is $P(N')=1-P(N)=1-\frac{9}{50}=\frac{41}{50}$.

17. Let E_4 be the event that more than 3 persons are in line. From the table, we read that $P(E_4)=0.20$.

19. The table gives 0.10 as the probability that 0 persons are in line, and 0.15 that 1 person is in line, so the probability that less than 2 persons are in line is $0.10 + 0.15 = 0.25$. Since the event E_4 that more than 3 persons are in line and the event, say E_2, that less than 2 persons are in line are mutually exclusive, $E_2\cap E_4=\varnothing$. Therefore, the probability that more than 3 or less than 2 persons are in line is

$$P(E_2\cup E_4)=P(E_2)+P(E_4)$$
$$=0.25+0.20=0.45$$

21. a. Using Table 11.4, we find the probability that a male who is alive at 20 is alive at age 60 is

$P(E)=\frac{85,227}{98,541}$. Hence, the probability that a male who is alive at age 20 is not alive at age 60 is

$$P(E')=1-P(E)=1-\frac{85,227}{98,541}$$

$$=\frac{13,314}{98,541}=\frac{4438}{32,847}$$

b. Using Table 11.4, we find the probability that a female who is alive at 20 is alive at age 60 is

$P(E)=\frac{91,220}{98,983}$. Hence, the probability that a female who is alive at age 20 is not alive at age 60 is

$$P(E')=1-P(E)=1-\frac{91,220}{98,983}$$

$$=\frac{7763}{98,983}$$

23. a. Using Table 11.4, we find the probability that a male who is alive at 30 is alive at age 70 is

$P(E)=\frac{72,066}{97,147}$. Hence, the probability that a male who is alive at age 30 is not alive at age 70 is

$$P(E')=1-P(E)=1-\frac{72,066}{97,147}=\frac{25,081}{97,147}.$$

b. Using Table 11.4, we find the probability that a female who is alive at 30 is alive at age 70 is

$P(E)=\frac{81,944}{98,466}$. Hence, the probability that a female who is alive at age 30 is not alive at age 70 is

$$P(E')=1-P(E)=1-\frac{81,944}{98,466}$$

$$=\frac{16,522}{98,466}=\frac{8261}{49,233}$$

25. a. Since the assumption is that no one attained 130 years of age, then the probability that a male will live less than 80 more years after attaining age 50 is $P(E) = 1$.

b. Since the assumption is that no one attained 130 years of age, then the probability that a female will live less than 80 more years after attaining age 50 is $P(E) = 1$.

27. The table gives the number of correct forms with no itemized deductions as $15 + 40 = 55$. Since the total number of forms is 100, the probability that a form had no itemized deductions and was correctly filled out is

$$P(E)=\frac{55}{100}=\frac{11}{20}.$$

29. The table gives the number of short forms as 20, so the number of long forms was 80. Hence, the probability that a form was not a short form is $P(E)=\frac{80}{100}=\frac{4}{5}$.

31. The green and yellow lights are on for a total of 65 seconds and the entire cycle takes 85 seconds. Thus, the probability of finding the light green or yellow is $P(E)=\frac{65}{85}=\frac{13}{17}$.

33. The probability of an event being 0, means that the event cannot occur.

35. If A and B have no elements in common, so that $A\cap B=\varnothing$, then

$$P(A\cup B)=P(A)+P(B).$$

37. We can tabulate John Dough's points as follows:

Item	Points
Age 27	5
Time at Address, 3 yr.	5
Age of Auto, 2 yr.	16
Monthly Auto Payment \$200	0
Housing Cost \$130/month	10
Checking Account Only	2
No Finance Co. Reference	15
Major Credit Cards 1	5
Debt to Income Ratio 12%	20
Total Score	78

From the table showing the probability of repayment, we estimate that John's probability is between 0.81 and 0.84, say about 0.83.

39. Answer will depend on the number of points you score but if you want to see a real way of scoring your credit history go to **http://www.myfico.com/.** Generally speaking, five factors are taken into account: Payment history, how much you owe, the length of time of your credit history, how much new credit you have established and the type of credit you have (credit cards, mortgages, finance companies, etc.)

The Venn diagram shows the data given for problems 41.

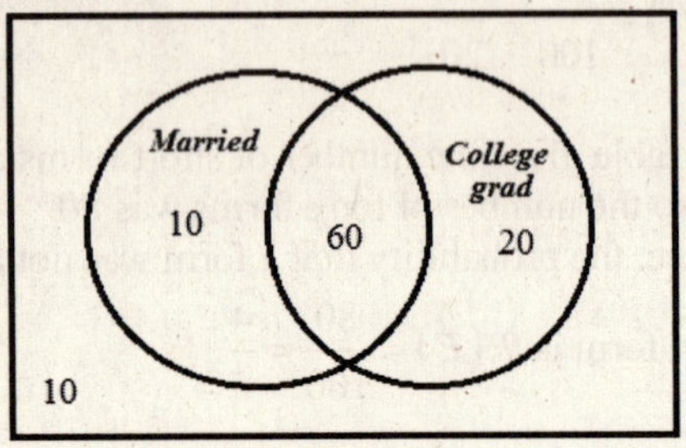

41. Refer to the Venn diagram. The number of persons in $M \cup C$ is $10 + 60 + 20 = 90$, so the probability that a person selected at random is married or a college graduate is $P(M \cup C) = \frac{90}{100} = \frac{9}{10}$.

The Venn diagram shows how the voters voted for problems 43 – 47.

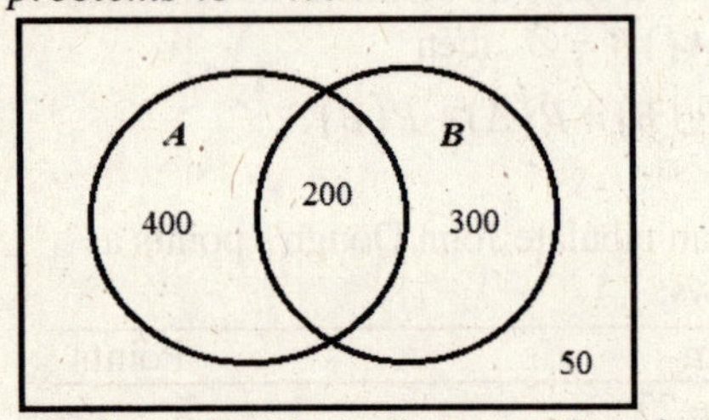

43. Refer to the Venn diagram. Note that the numbers add up to 950, so 50 persons did not vote at all. The diagram shows that 400 voted for *A* and not *B*. Thus, the probability that an eligible voter selected at random voted for *A* and not for *B* is $P(E) = \frac{400}{1000} = \frac{2}{5}$.

45. Refer to the Venn diagram. Since 200 voted for both *A* and *B*, the probability that an eligible voter selected at random voted for both *A* and *B* is $P(E) = \frac{200}{1000} = \frac{1}{5}$.

47. Refer to the Venn diagram. Since 50 did not vote at all, the probability that an eligible voter selected at random did not vote at all is $P(E) = \frac{50}{1000} = \frac{1}{20}$.

Section 11.4 Conditional Probability

Study Tips

When we use a subset of the sample space to compute probabilities, we are finding a conditional probability.

To determine a conditional probability:
Think: what is the condition, or given?, find $P(A \cap B)$ and $P(B)$

Note: The conditional probability formula is

$$P(A \mid B) = \frac{P(A \cap B)}{P(B)}.$$

Word of Advice
Do not mix up what is the condition (given) with the event for which you are finding the probability. In the notation $P(A \mid B)$, the *given* is *B*, and the event for which you are finding the probability is *A*.

11.4 Exercises

1. Let *F* be the event that a 5 came up and let *E* be the event that an even number came up. Then $F \cap E = \varnothing$, so $P(F \cap E) = 0$ and $P(F|E) = 0$.

3. a. Let A be the event that the numbers coming up were identical, and let *B* be the event that the sum was 8. There is only one way for the sum to be 8 and the numbers to be identical: that is for both numbers to be 4. Since the probability that both numbers were 4's is 1/36, and the probability that the numbers were identical is 1/6, we have

$$P(B|A)=\frac{P(A\cap B)}{P(A)}=\frac{\frac{1}{36}}{\frac{1}{6}}=\frac{1}{6}.$$

b. It is impossible to get a sum of 9 if the two numbers coming up are identical, so $P(9|\text{numbers identical})=0$.

c. If the two numbers coming up are identical, then the sum is even, so $P(\text{even sum}|\text{numbers identical})=1$.

d. It is impossible to get an odd sum if the two numbers coming up are identical, so $P(\text{odd sum}|\text{numbers identical})=0$.

5. $P(BB|BB \text{ or BG})$

$$=\frac{P(BB \text{ and } (BB \text{ or BG}))}{P(BB \text{ or BG})}=\frac{\frac{1}{4}}{\frac{1}{2}}=\frac{1}{2}.$$

7. If the first child is a girl and there are exactly two girls, then we must have GGB or GBG. Since there are $2^3=8$ possibilities, $P(GGB \text{ or } GBG)=\frac{1}{4}$. The probability that the first child is a girl is 1/2. Therefore, the probability that there are exactly 2 girls given that the first child is a girl is

$P(\text{exactly 2 } G|1^{\text{st}} \text{ child is } G)$

$$=\frac{P(GGB \text{ or } GBG)}{P(|1^{\text{st}} \text{ child is } G)}=\frac{\frac{1}{4}}{\frac{1}{2}}=\frac{1}{2}.$$

9. a. From the table, we have the probability that an employee missed 6–10 days, given that the employee is a woman, $P=\frac{20\%}{100\%}=\frac{1}{5}$.

b. Suppose there were 100 men and 100 women employees. The table shows that 40 men and 20 women missed 6-10 days. Thus, the probability that the employee is a woman, given that the employee missed 6-10 days is

$$P=\frac{20}{60}=\frac{1}{3}.$$

11. Use the following table.

Reason to Work	AGE 16–64	65+
Need Money (*N*)	49	17
Wants to (*W*)	20	54
Both (*B*)	31	27

a. The table shows that 49 + 17 = 66 people work because they need the money (*N*). 17 of those are 65+ in age. Thus, the probability that a randomly selected person is 65+ given that they work because they need money is

$$P(65+|N)=\frac{P(65+\cap N)}{P(N)}=\frac{17}{66}.$$

b. The table shows that 49 + 17 = 66 people work because they need the money (*N*). 49 of those are 16-64 in age. Thus, the probability that a randomly selected person is 16-64 given that they work because they need money is

$$P(16-64|N)=\frac{P(16\text{-}64\cap N)}{P(N)}=\frac{49}{66}.$$

c. The table shows that 17 of the 98 people 65+ in age are those that work because they need money (*N*). Thus, the probability that a randomly selected person works because they need money given that they are 65 + is $\frac{17}{98}=17.3\%$.

13. If the first card was a king, there are 3 kings left in the remaining 51 cards, so the probability that the second card is a king is

$$P=\frac{3}{51}=\frac{1}{17}.$$

15. There are 25 low-risk stocks, of which 5 are computer stocks, so the probability that the person selected a computer stock, given that the person selected a low-risk stock is

$$P=\frac{5}{25}=\frac{1}{5}.$$

17. $P(B|A)=\frac{P(A\cap B)}{P(A)}=\frac{0.4}{0.6}=\frac{2}{3}$

19. a. $P(\text{Good}|\text{Tenured})$

$$= \frac{P(\text{Good} \cap \text{Tenured})}{P(\text{Tenured})}$$

$$= \frac{72}{72+168} = \frac{72}{240} = \frac{3}{10}$$

b. $P(\text{Good}) = \dfrac{72+84}{72+84+168+76}$

$$= \frac{156}{400} = \frac{39}{100}$$

21. a. $P(\text{None}|65+) = \dfrac{P(\text{None} \cap 65+)}{P(65+)}$

$$= \frac{280}{70+287+280} = \frac{280}{637} \approx 0.44 = 44\%$$

b. $P(\text{None}|18\text{-}24) = \dfrac{P(\text{None} \cap 18\text{-}24)}{P(18\text{-}24)}$

$$= \frac{11}{108+101+11} = \frac{11}{220} = 0.05 = 5\%$$

23. a. The table shows that 135 of the 200 patients have improved. Thus, the probability that a patient chosen at random has improved is

$$P(I) = \frac{70+65}{200} = \frac{135}{200} = \frac{27}{40}.$$

b. The table shows that 70 of the 100 patients taking the experimental drug have improved, so the probability that one of these patients chosen at random has improved is

$$P(I|E) = \frac{70}{100} = \frac{7}{10}.$$

25. $P(D|W)$ is the probability that the person gets the death penalty (D) given that the person is white (W). In the first column of the table, there are 160 whites of which 19 got the death penalty, thus

$$P(D|W) = \frac{19}{160} \approx 0.119.$$

27. For these 326 convicted murderers,

$P(D|W) = \dfrac{19}{160} \approx 0.119$ and

$P(D|B) = \dfrac{17}{166} \approx 0.102$. There is not much difference $(0.119 - 0.102 = 0.017)$ between outcomes for whites and blacks.

29. $P(D|BW)$ means the probability that the person gets the death penalty (D) given that the defendant is black and the victim is white (BW). From the first table, where the race of the victim is white, and the defendant is black (last column), we see that there are 11 blacks that got the death penalty out of the 63 in the column. Thus, $P(D|BW) = \dfrac{11}{63} \approx 0.175$.

31. $P(D|BB)$ means the probability that the person gets the death penalty (D) given that the defendant is black and the victim is black (BB). From the second table, where the race of the victim is black, and the defendant is black, (last column), we see that there are 6 blacks that got the death penalty out of the 103 in the column. Thus, $P(D|BB) = \dfrac{6}{103} \approx 0.058$.

33. Answers will vary.

35. The total number of crimes listed in the table is 128.

a. The probability that the victim of one of these crimes was a male is

$$P(\text{male}) = \frac{5+18+52}{128} = \frac{75}{128}.$$

b. The probability that the victim was a female is $P(\text{female}) = \dfrac{2+9+42}{128} = \dfrac{53}{128}$

c. Male; [P(male) > P(female)]

37. $P(\text{assault}|\text{female}) = \dfrac{P(\text{assault} \cap \text{female})}{P(\text{female})}$

$$= \frac{9}{2+9+42} = \frac{9}{53}$$

Section 11.5 Independent Events

Study Tips

To determine whether two events are *independent*, that is, whether the occurrence of one of them affects the probability of the other, we have to study the definitions.

To determine independence of events:
Think: conditional probability of event equals probability of the event
Note: Events are independent if $P(A \mid B) = P(A)$. Another way to express this is $P(A \cap B) = P(A) \cdot P(B)$.

Multiplication Rule for Independent Events:

$$P(E_1 \cap E_2 \cap E_3 \cap \cdots \cap E_n) = P(E_1) \cdot P(E_2) \cdot P(E_3) \cdot \cdots \cdot P(E_n)$$

To use the stochastic process:
Think: tree diagram, multiply probabilities by following the branches to find terminal probabilities
Note: Stochastic process is a sequence of experiments in which the outcome of each experiment depends on chance. The relationship is $P(H \cap T) = P(H \mid T) \cdot P(T)$.

Hint
When specified events are independent, the probability of the event, that all of them occur, is just the product of their separate probabilities.

Note
For experiments involving repeated trials of the same experiment with two outcomes we need the *binomial* probability formula, studied in the *Using Your Knowledge* part of Exercise 11.5.

11.5 Exercises

1. Yes. $P(E_1 \mid E_2) = P(E_1)$, so E_1 and E_2 are independent.

3. a. Since the three events are independent
$P(\text{bad modem, bad CPU, bad drive})$
$$= \frac{1}{4} \cdot \frac{1}{8} \cdot \frac{1}{3} = \frac{1}{96}.$$
b. We need
$P(\text{bad modem, bad CPU, good drive})$
$$= \frac{1}{4} \cdot \frac{1}{8} \cdot \frac{2}{3} = \frac{1}{48}.$$
c. Here we need
$P(\text{gd modem, gd CPU, gd drive})$
$$= \frac{3}{4} \cdot \frac{7}{8} \cdot \frac{2}{3} = \frac{7}{16}.$$

5. a. $P(HTH) = \frac{1}{2} \cdot \frac{1}{2} \cdot \frac{1}{2} = \frac{1}{8}$
b. If a coin is tossed three times the possible outcomes are: {*HHH, HHT, HTH, THH, TTH, THT, HTT, TTT*}
The favorable outcomes are {*HHH, HHT, HTH, THH*}. Since the events are independent
P(*HHH*) + *P*(*HHT*) + *P*(*HTH*) + *P*(*THH*)
$= \frac{4}{8} = \frac{1}{2}$. Thus, the probability of at least 2 heads is $\frac{1}{2}$.
c. $P(HHH) = \frac{1}{8}$. Therefore the probability of at most two heads is
$1 - P(HHH) = 1 - \frac{1}{8} = \frac{7}{8}$.

7. a. There are 13 spades and 52 cards in all, so the probability that the first card is a spade is $\frac{13}{52} = \frac{1}{4}$.
b. Since the first card is returned to the deck, the probability that the second card is a spade is also $\frac{1}{4}$.
c. The probability that both cards are spades is $\frac{1}{4} \cdot \frac{1}{4} = \frac{1}{16}$.
d. The probability that the first card is not a spade is $1 - \frac{1}{4} = \frac{3}{4}$. This is also the probability that the second card is not a spade. Thus, the probability that neither card is a spade is $\frac{3}{4} \cdot \frac{3}{4} = \frac{9}{16}$.

9. a. At most one girl means either no girls or exactly one girl, that is, *BBB, GBB, BGB, BBG*. Thus, $P(M) = \frac{1}{8} + \frac{3}{8} = \frac{1}{2}$. (Note that the probability that the first child is a girl and the other two are boys is

$\frac{1}{2}\cdot\frac{1}{2}\cdot\frac{1}{2}=\frac{1}{8}$. This is also the probability that the second child is a girl and the other two are boys and also is the probability that the third child is a girl and the other two are boys. Hence, the probability that there is exactly one girl is $\frac{3}{8}$.) (You can verify this by making a tree diagram.)

b. $P(B)=1-P(\text{no boys})-P(\text{no girls})$
$=1-\frac{1}{8}-\frac{1}{8}=\frac{3}{4}$

c. M is the event that the family has at most 1 girl, and B is the event that the family has children of both sexes, so $M\cap B$ is the event that the family has exactly 1 girl. From part a, $P(1\text{ girl})=\frac{3}{8}$, so $P(M\cap B)=\frac{3}{8}$.

d. In part a, we found $P(M)=\frac{1}{2}$ and from part b, $P(B)=\frac{3}{4}$. Since $\frac{3}{8}=\frac{1}{2}\cdot\frac{3}{4}, P(M\cap B)=P(M)P(B)$ which shows that M and B are independent.

11. $P(GR)=P(G)\cdot P(R)=\frac{5}{18}\cdot\frac{1}{6}=\frac{5}{108}$

13. $P(GG)=P(G)\cdot P(G)=\frac{5}{18}\cdot\frac{5}{18}=\frac{25}{324}$

15. $P(R'R')=(1-P(R))^2=\left(1-\frac{1}{6}\right)^2$
$=\left(\frac{5}{6}\right)^2=\frac{25}{36}$

17. There are 5 red marbles out of 20, so $P(R)=\frac{5}{20}$. There are 9 blue marbles out of the remaining 19 marbles, so $P(B)=\frac{9}{19}$. Thus, $P(RB)=\frac{5}{20}\cdot\frac{9}{19}=\frac{9}{76}$.

19. There are 3 green marbles out of 20, so $P(G)=\frac{3}{20}$. There are 3 yellow marbles out of the remaining 19 marbles, so $P(Y)=\frac{3}{19}$. Thus, $P(GY)=\frac{3}{20}\cdot\frac{3}{19}=\frac{9}{380}$.

21. There are 7 red marbles out of 25, so $P(R)=\frac{7}{25}$. Since the marble is replaced, there are 4 yellow marbles out of 25 marbles, so $P(Y)=\frac{4}{25}$. Thus, $P(RY)=\frac{7}{25}\cdot\frac{4}{25}=\frac{28}{625}$.

23. Since no blue marble is chosen, one of the remaining 17 marbles is chosen. So, $P(B')=\frac{17}{25}$. Thus, $P(B'B')=\frac{17}{25}\cdot\frac{17}{25}=\frac{289}{625}$.

25. $P(\text{at least 1 red})=1-P(\text{no reds})$
$=1-P(R'R')$
$=1-\frac{18}{25}\cdot\frac{18}{25}\ =1-\frac{324}{625}=\frac{301}{625}$.

27. $P(\text{none healthcare})$
$=(0.59)(0.59)(0.59)=0.205379$

29. $P(\text{none direct product})$
$=0.86(0.86)(0.86)=0.636056$

31. a. $P(3\text{ successful})=\frac{1}{4}\cdot\frac{2}{3}\cdot\frac{1}{2}=\frac{1}{12}$

b. $P(0\text{ successful})$
$=\left(1-\frac{1}{4}\right)\left(1-\frac{2}{3}\right)\left(1-\frac{1}{2}\right)$
$=\frac{3}{4}\cdot\frac{1}{3}\cdot\frac{1}{2}=\frac{1}{8}$

33. a. $P(2\text{ tails})=\frac{1}{2}\cdot\frac{1}{2}=\frac{1}{4}$

b. $P(H\text{ and }6)=\frac{1}{2}\cdot\frac{1}{6}=\frac{1}{12}$

c. $P(H \text{ and even no.}) = \frac{1}{2} \cdot \frac{3}{6} = \frac{1}{4}$

35. $\frac{91,220}{98,466} \cdot \frac{91,220}{97,586} \approx 0.86579$

37. $(0.90)^5 \approx 0.59$

39. a. $P(\text{all 3 fail}) = \left(\frac{1}{20}\right)^3 = \frac{1}{8000}$

b. $P(\text{exactly 2 fail})$

$$= C(3,1) \cdot \frac{1}{20} \cdot \frac{1}{20} \cdot \frac{19}{20}$$

$$= 3 \cdot \frac{19}{8000} = \frac{57}{8000}$$

41. a. National League:

$P(\text{hit}) = \frac{54}{254} \approx 0.2126$

American League:

$P(\text{hit}) = \frac{66}{274} \approx 0.2409$

b. The probabily of getting a hit in the postseason is higher in the American League.

c. National League:

$P(4 \text{ hits}) = (0.2126)^4 \approx 0.00204$

American League:

$P(4 \text{ hits}) = (0.2409)^4 \approx 0.00337$

43.

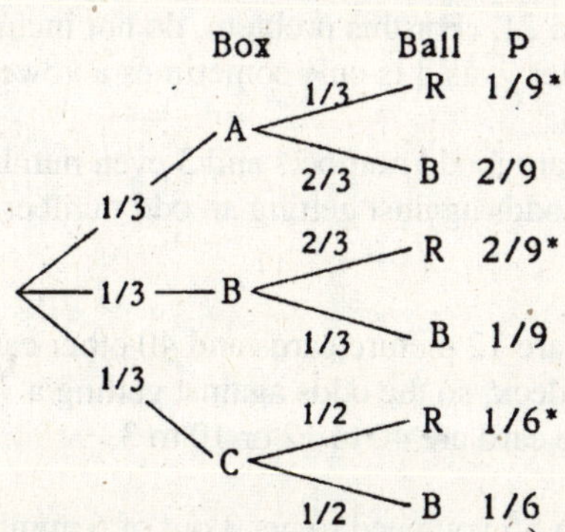

Add the starred items to get:

$$P = \frac{1}{9} + \frac{2}{9} + \frac{1}{6} = \frac{2}{18} + \frac{4}{18} + \frac{3}{18} = \frac{9}{18} = \frac{1}{2}$$

45. Draw a tree diagram, letting F be the fair coin and U the unbalanced coin:

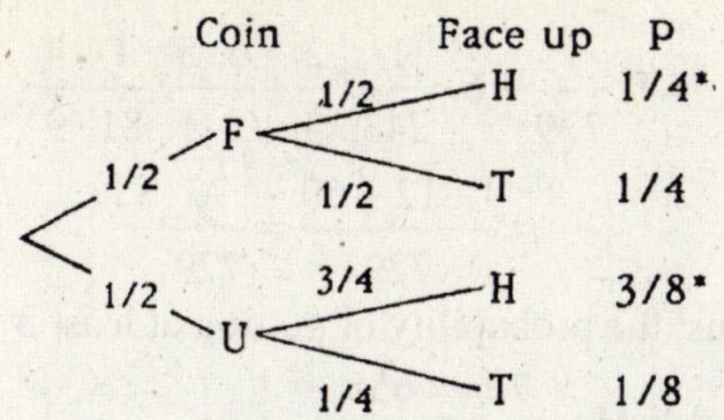

Use the starred numbers as weights to get:

$$P(U|H) = \frac{P(U \cap H)}{P(H)} = \frac{\frac{1}{2} \cdot \frac{3}{4}}{\frac{3}{8} + \frac{1}{4}} = \frac{\frac{3}{8}}{\frac{5}{8}} = \frac{3}{5}$$

47. a. The probability of either event does not depend on the probability of the other event. The occurrence of either event does not depend on the occurrence or non-occurrence of the other event.

b. If A and B are independent events, then $P(A \cap B) = P(A) \cdot P(B)$.

49. $C(50, 25)$ is the number of ways to get exactly 25 heads and 2^{50} is the total number of ways in which the coin can fall. Thus,

$P(\text{exactly 25 heads}) = \frac{C(50,25)}{2^{50}}$. We can arrive at the same result by using the formula.

Letting $n = 50$, $x = 25$, $p = \frac{1}{2}$, and $q = \frac{1}{2}$, we have

$$C(50,25)\left(\frac{1}{2}\right)^{25}\left(\frac{1}{2}\right)^{50-25}$$

$$= C(50,25)\left(\frac{1}{2}\right)^{25}\left(\frac{1}{2}\right)^{25}$$

$$= C(50,25)\left(\frac{1}{2}\right)^{50} = \frac{C(50,25)}{2^{50}}$$

51. The numbers of ways of getting 0, 1, or 2 heads (without regard to order) out of the 6 tosses are $C(6, 0)$, $C(6, 1)$, and $C(6, 2)$, respectively. Thus, for the biased coin, the probability of getting less than 3 heads is

$$C(6,0)\left(\frac{1}{3}\right)^6 + C(6,1)\left(\frac{1}{3}\right)^5\left(\frac{2}{3}\right)$$

$$+ C(6,2)\left(\frac{1}{3}\right)^4\left(\frac{2}{3}\right)^2$$

$$=1\cdot\frac{1}{729}+6\cdot\frac{1}{243}\cdot\frac{2}{3}+\frac{6\cdot 5}{2\cdot 1}\cdot\frac{1}{81}\cdot\frac{4}{9}$$
$$=\frac{1+12+60}{729}=\frac{73}{729}$$

Thus, the probability of getting at least 3 heads is: $1-\frac{73}{729}=\frac{656}{729}$.

53. The number of ways (without regard to order) of getting exactly two 3's out of 5 tosses is $C(5, 2)$, so the probability is:
$$C(5,2)\cdot\left(\frac{1}{6}\right)^2\left(\frac{5}{6}\right)^3=\frac{5\cdot 4}{2\cdot 1}\cdot\frac{1}{36}\cdot\frac{125}{216}$$
$$=\frac{625}{3888}$$

55. The probability that a person fools the first machine is $1 - 0.98 = 0.02$.

57. The probability that a person fools all three machines is $(0.02)(0.02)(0.02)=0.000008$.

Section 11.6 Odds and Mathematical Expectation

Study Tips

Most students are extremely interested in this section because they want to know the *odds* for many events.

To determine odds "for" or in favor of an event:
Think: ratio of favorable f to unfavorable u occurrences for an event

Note: $\frac{f}{u}$; f to u; $f:u$

To determine odds "against" an event:
Think: ratio of unfavorable u to favorable f occurrences for an event

Note: $\frac{u}{f}$; u to f; $u:f$

Hint
The sum of the favorable f and unfavorable u occurrences will equal the total outcomes $(f + u)$ for the event.

Fact
In terms of probabilities, an equivalent definition is as follows: The odds in favor of an event E are given by $P(E)$ to $P(\text{not } E)$.

To determine an expected value:
Think: probability of each outcome, value of each outcome, multiply, then add the products
Note: Let a = value, p = probability, then
$E=a_1p_1+a_2p_2+a_3p_3+\cdots++a_kp_k$.

What is a fair bet?
Think: the odds of f to u in favor of E occurring

Note: $P(E)=\frac{f}{n}$ and $P(\text{not } E)=\frac{u}{n}$, n is total number of possible outcomes.

What is a fair game?
Think: expected value is 0
Note: You would have an equal chance of winning versus losing.

11.6 Exercises

1. There is 1 favorable outcome and there are 5 unfavorable outcomes, so the odds are 1 to 5 in favor of getting a 2.

3. There are 4 aces and 48 other cards, so the odds are 4 to 48 or 1 to 12 in favor of getting an ace.

5. You can get *TT*, *TH*, *HT*, *HH*, so there is 1 favorable outcome and 3 unfavorable outcomes. Thus, the odds in favor of getting 2 tails are 1 to 3.

7. There are 5 vowels {a, e, i, o, u} and 21 other letters, so the odds in favor of getting a vowel are 5 to 21. (For this problem, do not include the letter y, as it is only sometimes a vowel.)

9. There are 3 odd numbers and 3 even numbers, so the odds against getting an odd number are 1 to 1.

11. There are 12 picture cards and 40 other cards in the deck, so the odds against getting a picture card are 40 to 12 or 10 to 3.

13. To win $5 you need to get 3 out of 6 numbers and the probability of that is:
$$\frac{C(6,3)\cdot C(47,3)}{C(53,6)}=\frac{324{,}300}{22{,}957{,}480}=\frac{32{,}430}{2{,}295{,}748}$$
The odds in favor are 32,430 to 2,263,318.

15. The probability that he wins is $\frac{3}{3+2} = \frac{3}{5}$.

17. Probability is $\frac{630}{100{,}000} = \frac{63}{10{,}000}$; the odds in favor of being a victim are 63 to 9937.

19. For the probability $\frac{1}{2} = \frac{1}{1+1}$, the odds in favor of having complications are 1 to 1.

21. For the odds 10 to 4867, the probability of publishing 1 of the 10 best-selling novels of the year is $\frac{10}{4867+10} = \frac{10}{4877}$.

23. For the odds 3 to 21, the probability of being a top executeive is $\frac{3}{21+3} = \frac{3}{24} = \frac{1}{8}$.

25. For the probability $0.33 = \frac{33}{100} = \frac{19}{33+67}$, the odds in favor of growing up incompetent in math are 33 to 67.

27. The probability that the sum is even is $\frac{1}{2}$, and this is also the probability that the sum is not even, so the expected value is $E = \$10 \cdot \frac{1}{2} - \$20 \cdot \frac{1}{2} = \$5 - \$10 = -\$5$.

29. a. The expected value of the policy is $\$300{,}000(1-0.998349)$ $= \$300{,}000(0.001651) = \$495.30$
    b. The fair price for the annual premium is \$495.30.
    c. The monthly payment is $\$495.30 \div 12 = \$41.28$.

31. Expected value is $3 \cdot 0.25 + 4 \cdot 0.375 + 5 \cdot 0.375 = 4.125$ games.

33. For the first location,
$E_1 = \$100{,}000 \cdot \frac{2}{3} - \$50{,}000 \cdot \frac{1}{3} = \$50{,}000$.
For the second location,
$E_2 = \$150{,}000 \cdot \frac{2}{5} - \$80{,}000 \cdot \frac{3}{5} = \$12{,}000$.
Since $E_1 > E_2$, the first location should be chosen.

35. a. $E_1 = \$20{,}000 \cdot \frac{1}{5} - \$10{,}000 \cdot \frac{4}{5}$ $= -\$4000$
    b. $E_2 = \$30{,}000 \cdot \frac{1}{5} - \$5000 \cdot \frac{4}{5} = \$2000$
    c. Since continuing the campaign would result in an expected loss of \$4000, and discontinuing the campaign would result in an expected gain of \$2,000, the campaign should be discontinued.

37.

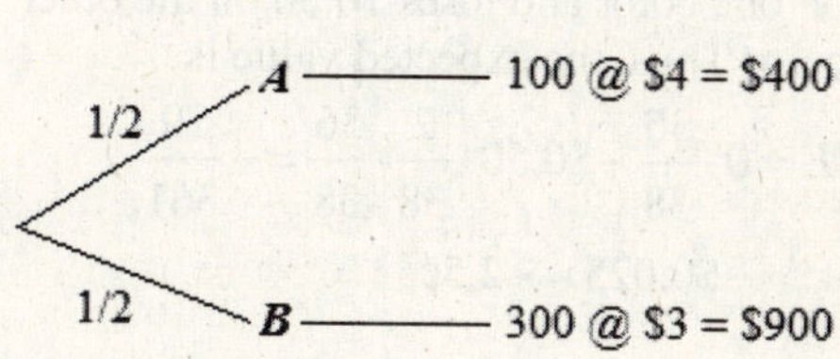

a. $E_A = \$400 \cdot \frac{1}{2} = \$200$
b. $E_B = \$900 \cdot \frac{1}{2} = \$450$
c. Location B has the better expected value.

39. The mathematical expectation is that you lose \$0.58 per \$2 bet on the average, so it is not a fair bet. (For a fair bet, the mathematical expectation is 0.)

41. a. The curve shows that the probability of winning is about 0.81.
    b. The probability of losing is $1 - 0.81 = 0.19$.

43. Odds against you are 9 to 1, so put up \$1 for each \$9. Thus, you should bet $\left(\frac{1}{9}\right) \cdot \$10$, or about \$1.11, to make the bet fair.
Alternative method:
The curve shows that the probability of winning is 0.9, so the probability of losing is $1 - 0.9 = 0.1$. If $\$x$ is put up against the man's \$10, the expected value is $E = \$x(0.9) - \$10(0.1)$. For the bet to be fair,

E must be 0, so we have

$$0.9x - 1 = 0$$
$$9x - 10 = 0$$
$$9x = 10$$
$$x = \frac{10}{9} = 1.111...$$

and (to the nearest cent) \$1.11 should be put up against the man's \$10.

45. The probability of a black or a red number is $\frac{36}{38}$ and if one of these colors comes up, player wins \$0.50 and loses \$0.50, so he breaks even. The probability of 0 or 00 is $\frac{2}{38}$ and if 0 or 00 turns up, then the player gets his money back (but gains nothing) on one color and loses \$0.50 on the other color Thus, the expected value is

$$E = 0 \cdot \frac{36}{38} - \$0.50 \cdot \frac{2}{38} \cdot \frac{36}{38} = -\frac{\$9}{361}$$
$$\approx -\$0.025 \approx -2.5¢$$

Chapter 11 Practice Test

Study Tips

How can you "ACE" the test you are about to take? Here is a study tip you can use when taking the practice tests in the book.

Answer all the problems in the practice test without notes or references.

Correct any wrong answers and work that particular problem again.

Examine your understanding of each problem by analyzing the objective of each question and avoiding the common errors associated with that type of question.

Here are the solutions to the practice test.

1. a. This event is certain, so $P(E) = 1$.
   b. 4 of the 6 numbers are greater than 2, so the probability is $\frac{4}{6} = \frac{2}{3}$.

2. a. Of the 5 balls 2 are even numbered, so the probability is $\frac{2}{5}$.
   b. There is just 1 ball numbered 2, so the probability is $\frac{1}{5}$.
   c. The probability is $1 - \frac{1}{5} = \frac{4}{5}$.

3. a. There are 5 possible outcomes on the first draw and 4 possible outcomes on the second draw, so the sample space has $5 \cdot 4 = 20$ elements.
   b. There are 5 possible outcomes on each of the two draws, so the sample space has $5 \cdot 5 = 25$ elements.

4. a. The probability that the first card is red is $\frac{26}{52} = \frac{1}{2}$. If the first card is red, there are 25 red cards out of the 51 cards left, so the probability that the second card is red is $\frac{25}{51}$. Thus, the probability that both cards are red is $\frac{26}{52} \cdot \frac{25}{51} = \frac{1}{2} \cdot \frac{25}{51} = \frac{25}{102}$.
   b. The probability that the first card is not an ace is $\frac{48}{52} = \frac{12}{13}$. The probability that the second card is not an ace is then $\frac{47}{51}$. Thus, the probability that neither card is an ace is $\frac{48}{52} \cdot \frac{47}{51} = \frac{12}{13} \cdot \frac{47}{51} = \frac{188}{221}$.

5. a. 26 of the 52 cards are red, so the probability that both cards will be red is $\frac{26}{52} \cdot \frac{26}{52} = \frac{1}{2} \cdot \frac{1}{2} = \frac{1}{4}$.
   b. If neither card is red then both cards must be black. The probability that both cards are black is the same as the probability that both cards are red, that is $\frac{1}{4}$.

6. $P(\text{no heads}) = \left(\frac{1}{2}\right)^5 = \frac{1}{32}$. Therefore,
   $P(\text{at least one head}) = 1 - \frac{1}{32} = \frac{31}{32}$.

7. a. There are 5 white and 3 black balls out of a total of 10 balls, so $P(\text{white or black}) = \frac{8}{10} = \frac{4}{5}$.

b. This is the same as part (a), so the probability is $\frac{4}{5}$.

8. There are 12 face cards and a total of 52 cards in the deck, so

$$P(3 \text{ face cards}) = \frac{12}{52} \cdot \frac{11}{51} \cdot \frac{10}{50} = \frac{3}{13} \cdot \frac{11}{51} \cdot \frac{1}{5} = \frac{11}{1105} \approx 0.01$$

9. Let E stand for passing English and M stand for passing math. Then, $P(E \cup M) = 0.9, P(E) = 0.8$, and $P(E \cap M) = 0.6$ are all given. Use the equation $P(E \cup M) = P(E) + P(M) - P(E \cap M)$ to get $0.9 = 0.8 + P(M) - 0.6$, which gives $P(M) = 0.7$.

10. This is just the probability that the second die comes up 6, that is, $\frac{1}{6}$.

11. In order to get a sum of 11 and have the second die come up an even number, this die must come up 6. Since there are 3 even numbers, the probability that the die comes up 6, given that it is an even number is 1/3. To get a sum of 11, the first die must come up 5. The probability of this is $\frac{1}{6}$. Thus, the probability that the sum is eleven given that the second die comes up an even number is $\frac{1}{6} \cdot \frac{1}{3} = \frac{1}{18}$.

12. a. There are 3 ways to get a sum of 10: 1st die 4, 2nd die 6; 1st die 5, 2nd die 5; 1st die 6, 2nd die 4. Since the dice can come up in 36 ways, the probability of getting a sum of 10 is $\frac{3}{36} = \frac{1}{12}$.

b. 3 of the 6 numbers are odd, so the probability of getting an odd number is $\frac{3}{6} = \frac{1}{2}$.

c. No. Let A be the event that the sum is 10 and B the event that the first die comes up an odd number. If the first number is odd, then the sum of 10 can be obtained only if both numbers are 5's. From part a, we see that the probability of this is $\frac{1}{36}$. Thus, we have $P(A \cap B) = \frac{1}{36}$ and $P(A)P(B) = \frac{1}{12} \cdot \frac{1}{2} = \frac{1}{24}$. Since these two results are not equal, A and B are not independent.

13. a. $P = (0.04)^3 = 0.000064$

b. $P = (1 - 0.04)^3 = (0.96)^3 = 0.884736$

14. There are 2 courses and 3 possible hours. This makes $2 \cdot 3 = 6$ possible arrangements. Thus, the probability that Roland will have English at 8 A.M. and history at 3 P.M. is $\frac{1}{6}$.

15. $P = (1 - 0.97)^2 = (0.03)^2 = 0.0009$

16. a. There are 4 kings and 48 other cards, so the odds in favor of selecting a king are 4 to 48 or (dividing by 4) 1 to 12.

b. The odds against selecting a king are just reversed from the answer in part a. The answer is 12 to 1.

17. a. Since $P(E) = \frac{3}{7}, P(E') = 1 - \frac{3}{7} = \frac{4}{7}$. Thus, the odds in favor of E are 3 to 4.

b. The odds against E are 4 to 3 (just the reverse of the answer in part a).

18. a. If the odds in favor are 3 to 7, the odds against the event are 7 to 3.

b. The probability that the event will not occur is $\frac{7}{7+3} = \frac{7}{10}$.

19. The expected value of this game is

$$E = \$5 \cdot \frac{1}{2} + \$5 \cdot \frac{1}{4} = \$2.50 + \$1.25 = \$3.75$$

so we should be willing to pay \$3.75 to play this game.

20. From Example 4, you pay \$1 to play the Florida Lottery. Thus, the mathematical expectation of being an instant winner is

$$E = \$2 \cdot \frac{1}{10} + \$5 \cdot \frac{1}{50} + \$25 \cdot \frac{1}{600} + \$50 \cdot \frac{1}{1200} - \$1$$

$$= \$0.20 + \$0.10 + \$\frac{1}{24} + \$\frac{1}{24} - \$1$$

$$= \$0.30 + \$\frac{1}{12} - \$1 = \$0.38\frac{1}{3} - \$1$$

$$= -\$0.61\frac{2}{3}, \text{ or } -61\frac{2}{3} \text{ cents}$$

Chapter 12 Statistics

Section 12.1 Sampling, Frequency Distributions, and Graphs

Study Tips

Statistics are often used to sway opinion, so you should find this to be a very interesting topic. Statistics studies different ways of organizing and reporting data.

To determine a….

Target population: all the people or items that belong to a specified group

Sample: portion that is representative of the population

Simple random sample: selected at random and same chance of being chosen

Comment

The simplest way to organize a list of numbers is to make a frequency distribution. If the number of items is too large, we can group the data. In addition, for better understanding we can make a picture of the data using a histogram and a frequency polygon.

Summary for a frequency distribution:

1. Decide on number of classes.

a. Class width

$$\approx \frac{\text{largest data value} - \text{smallest data value}}{\text{desired number of classes}}$$

b. *Lower class limit* and *upper class limit*: lowest and highest values in each class

c. *Midpoint*: center of the class

d. *Class boundaries*: use halfway points between upper limit of one class and lower limit of another or use upper class limit of first class as the lower class limit of the second class

2. Sort or tally data into classes.

3. Count the number of items in each class.

4. Display the results in a table.

5. Can make either a histogram and/or frequency polygon.

Hint

If a data value falls on the upper limit it should be included in the next higher class.

Summary for a histogram/frequency polygon:

1. Make a frequency distribution.

2. Label y-axis to represent the frequencies and x-axis the scores

3. Draw the histogram or frequency polygon.

Note

Example 2, the explanation that follows it and the *Problem-Solving* show you how to construct histograms and frequency polygons.

12.1 Exercises

1. Descriptive statistics is the science of collecting, organizing and summarizing data.

3. A sample usually consists of just a part of the target population.

5. a. The implied population is all households in the U. S.
   b. The sample is the 1006 households surveyed.

7. a. No, this is not a random sample.
   b. The sample includes only those viewers who are willing to pay for the call.

9. a. Make a card for each student, number the cards and mix them up. Then draw ten cards at random and select the corresponding students.
   b. Make a card for each student, number the cards and:
      i. Pick only even numbered cards.
      ii. Pick only odd numbered cards.
      iii. Don't mix the cards and pick the first 10.

11. a. The target population is all TV watchers.
    b. The sample population is people who visit the Web site and are willing to participate in the survey.
    c. No. Respondents do not necessarily represent the target population (all TV watchers). Those in the target population may not have internet access, may not use that provider, may not hit that pop-up, and so on.

13. Although both surveys ultimately had 50 responses, the second survey should produce more accurate results because $\frac{50}{75} \approx 67\%$ of the people responded compared with only $\frac{50}{100} = 50\%$ for the first survey.

15. a. Attended a football game last year?

Y	720
N	160
NR	120
Total	1000

b. So the total will add to 1000.

17. a.

Number of Hours	*Tally Marks*	*Frequency*
0	|	1
1	|	1
2	|||	3
3	|||	3
4	|||	3
5	|||	3
6	||||	4
7	||	2

b. The most frequent number is 6 hours.

c. 6 people spend more than the 5.18 hr daily average; $\frac{6}{20} = 0.3 = 30\%$

d. 14 people spend few than the 5.18 hr daily average; $\frac{14}{20} = 0.7 = 70\%$

19. a.

Outcome	*Tally Marks*	*Frequency*
1	卌 卌 |	11
2	||	2
3	卌 |	6
4	||	2
5	卌 |||	8
6	|	1

b. The most frequent outcome is the 1. It occurred 11 times.

c. The least frequent outcome is the 6. It occurred 1 time.

21. a.

Number of Hours	*Tally Marks*	*Frequency*
0	||	2
1	|||	3
2	||	2
3	|||	3
4	|||	3
5	||	2
6	||	2
7	||	2
8	||||	4
9	|	1
10	||	2
11		0
12	||	2
13		0
14	|	1
15	|	1

b. The most frequent number of hours watched is 8.

c. $0 + 2 + 0 + 1 + 1 = 4$ students watched television more than 10 hr.

d. $2 + 3 + 2 + 3 + 3 + 2 = 15$ students watched television 5 hr or less.

e. $4 + 1 + 2 + 0 + 2 + 0 + 1 + 1 = 11$ students watched television more than 7 hr, which is $\frac{11}{30} \approx 0.367 = 36.7\%$.

23. a.

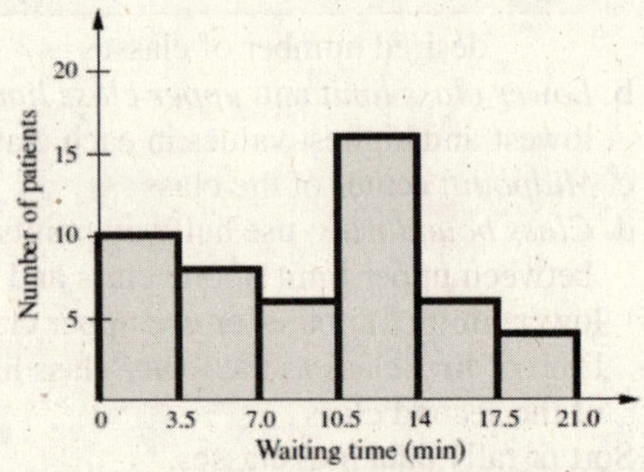

b. There were 18 out of 50 patients, that is, $\frac{18}{50} = 0.36 = 36\%$, who waited 7.0 min. or less.

c. There were 26 out of 50 patients, that is, $\frac{26}{50} = 0.52 = 52\%$ who had to wait more than 10.5 min.

25. a.

Age	Tally Marks	Frequency
6	I	1
7	I	1
8	III	3
9	III	3
10	II	2

b.

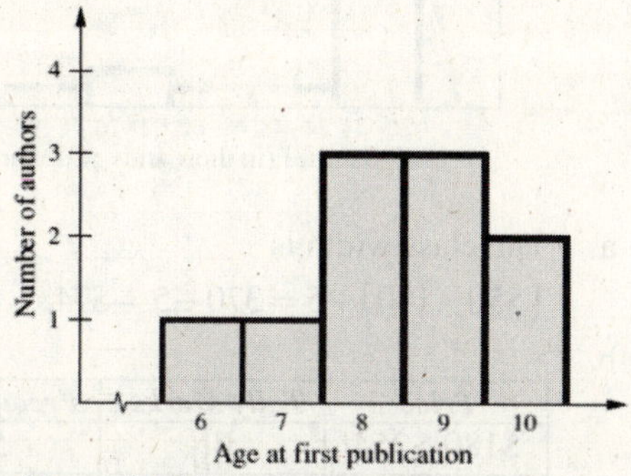

c. From the histogram, we see that $\frac{2}{10} = 0.20 = 20\%$ authors were less than 8 years old when they published their first book.

27. a.

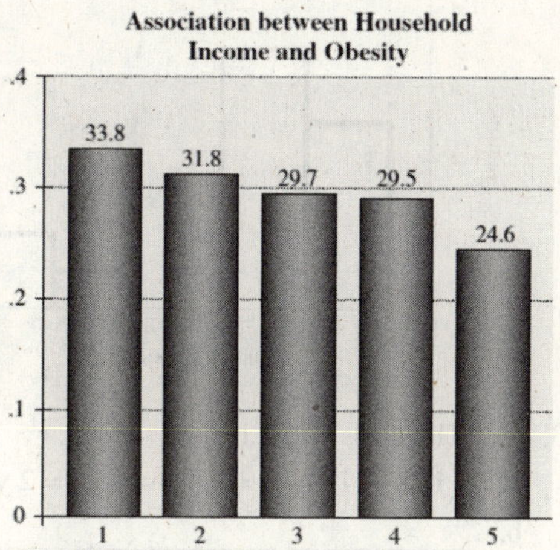

b. The highest percentage of obese adults is category 1: Less than \$15,000.

c. The obesity rate for adults making \$50,000 or more (Category 5) is 24.6%.

d. As household income *increases*, the percent of obese people *decreases*.

29.

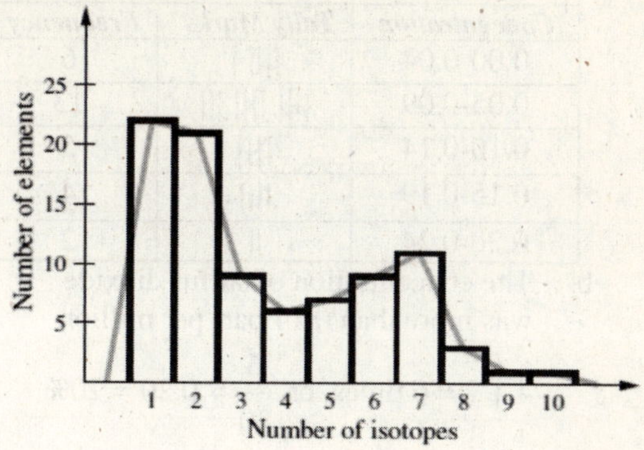

31. a.

Letter	Frequency	Letter	Frequency
a	10	n	13
b	4	o	14
c	3	p	2
d	4	q	0
e	18	r	7
f	2	s	10
g	3	t	17
h	9	u	5
i	12	v	2
j	0	w	5
k	1	x	0
l	4	y	2
m	4	z	0

b. The letter e occurs most frequently.

c. The vowels are a, e, i, o, and u, of which there are $10 + 18 + 12 + 14 + 5 = 59$. Thus, $\frac{59}{151} \approx 0.391 = 39.1\%$ of the letters are vowels.

33. a. and b.

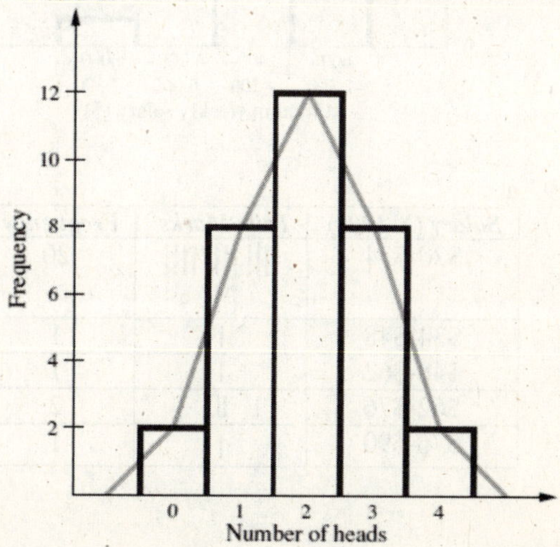

35. a.

Concentration	*Tally Marks*	*Frequency*
0.00-0.04	𝍸 I	6
0.05-0.09	𝍸 𝍸 III	13
0.10-0.14	𝍸	5
0.15-0.19	IIII	4
0.20-0.24	II	2

b. The concentration of sulfur dioxide was more than 0.14 part per million $4 + 2 = 6$ times, or $\frac{6}{30} = 0.20 = 20\%$ of the time.

c. The frequency distribution groups data so use the raw data for the exact count. The concentration of sulfur dioxide was more than 0.075 part per million 17 times, or $\frac{17}{30} \approx 0.567 = 56.7\%$ of the time.

37. a.

Weekly Salary	*Tally Marks*	*Frequency*
\$600-\$2400	𝍸 IIII	9
\$2400-\$4200		0
\$4200-\$6000	II	2
\$6000-\$7800	I	1

b.

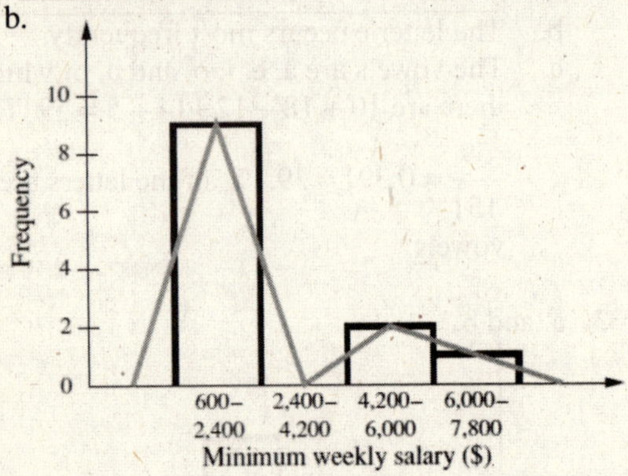

39. a.

Salary (X 1000)	*Tally Marks*	*Frequency*
\$20-\$34	𝍸 𝍸 𝍸 𝍸	20
\$34-\$48	I	1
\$48-\$62	I	1
\$62-\$76	II	2
\$76-\$90	I	1

b.

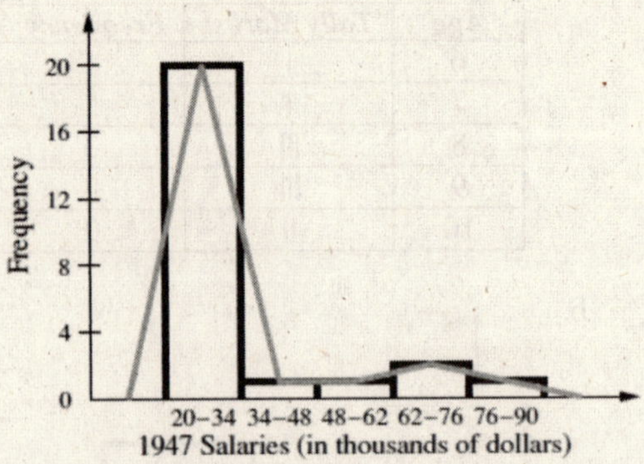

41. a. The class width is $(550 - 180) \div 5 = 370 \div 5 = \$74$.

b.

Price	*Tally Marks*	*Frequency*
\$180-\$254	𝍸	5
\$254-\$328	𝍸 II	7
\$328-\$402		0
\$402-\$476	I	1
\$476-\$550	II	2

c.

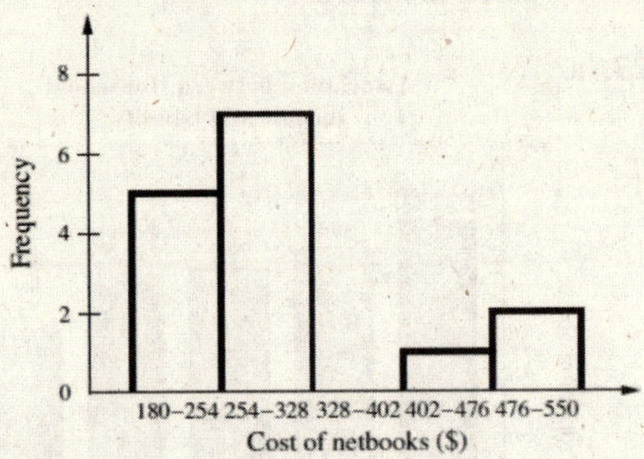

43. a. The class width is $(122 - 115) \div 4 = 7 \div 4 = 1.75 \approx 2$ years.

b.

Age	*Tally Marks*	*Frequency*
115–117	𝍸 𝍸 𝍸 IIII	19
117–119	IIII	4
119–121	I	1
121–123	I	1

c.

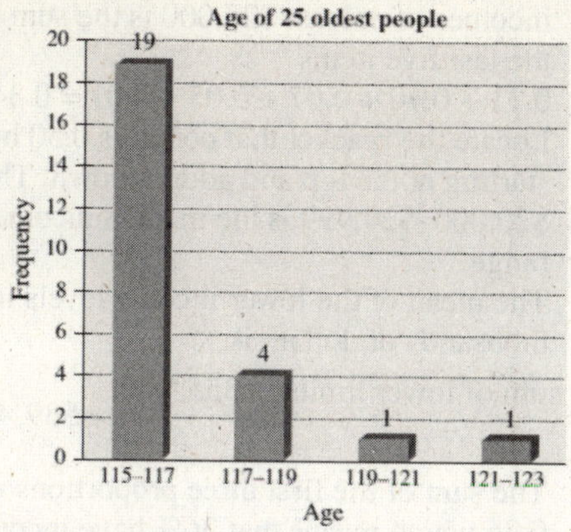

45. Tract 1304 has the fewest renter-occupied homes.

47. Chose tract 1305 as it has the most renter-occupied homes.

49. No. The women were not selected at random. Only women who thought that they might be pregnant would chose to buy the test.

51. No. Not all members of the population had the same chance of being chosen.

53. The lower and upper class limits, respectively, are the least and the greatest values in that class. Each class boundary is the midpoint between the upper limit of the respective class and the lower limit of the next class.

55.

Digit	*Frequency*
0	1
1	4
2	5
3	6
4	4
5	4
6	3
7	3
8	5
9	5

57. Yes, the jump in sales from first 9 months to first quarter is 3,205,000 – 3,104,729 = 100,271.

59. Yes. In each case, most of the bar is omitted.

Section 12.2 Measures of Central Tendency: The Mean, Median, and Mode

Study Tips

If we want to describe an entire sample or a population by a number or an average, we can use three measures of central tendency: the mean, the mode, and the median.

To find the mean of a data set:
Think: "average"
Note: This measure of central tendency is probably the most familiar.

For a set of n numbers, the mean is $\overline{x} = \dfrac{\text{sum}}{n}$.

To find the mode of a data set:
Think: occurs most often
Note: The mode is the number of the set that occurs most often. It is possible to have more than one mode, or to have no mode.

To find the median of a data set:
Think: middle number
Note: Arrange the numbers from smallest to greatest. If no single middle number, the median is the mean (average) of the two middle numbers.

Word of Advice
Study the examples carefully and make sure you learn the differences between the mean, median and mode by reading the comparison of these measurements given at the end of the section.

12.2 Exercises

1. a. Mean $= \dfrac{1+5+9+13+17}{5} = \dfrac{45}{5} = 9$ and median is the 5th term: 9

   b. Mean $= \dfrac{1+5+9+27+81}{5} = \dfrac{121}{5} = 24.2$ and median is the 5th term: 9

   c. Mean $= \dfrac{1+4+9+16+25}{5} = \dfrac{55}{5} = 11$ and median is the 5th term: 9

   d. The mean and medium are the same for part a only. The median is 9 for all three. None has a mode.

3. Mean $= \dfrac{121}{20} = 6.05$, the median is the average of the 10th and 11th terms:

$\frac{6+7}{2}=\frac{13}{2}=6.5$, and the mode is the most frequent score, 8. The mode is the least representative.

5. Betty is correct. She took account of the number of students making each score, and Agnes did not.

7. The frequency distribution is

Number of Letters	*Frequency*
2	1
3	5
4	4
5	5
6	0
7	3
8	2
9	0
10	0
11	1

a. There are two modes: 3 and 5.
b. The sum of the frequencies is 21, so add the numbers in the frequency column until we first get a sum of 11 or more. This occurs in the row where the number of letters is 5, so the median is 5.
c. The mean is

$$\frac{2\cdot 1+3\cdot 5+\cdots+10\cdot 0+11\cdot 1}{21}=\frac{106}{21}\approx 5.05$$

d. No. There is too much repetition for this to be representative of ordinary English writing.

9. b. – c. Answers will vary, but should be approximately 8.

11. If the mean in 81, then

$$\frac{1560+x}{20}=81$$
$$1560+x=1620$$
$$x=60$$

The grade on the lost paper is 60.

13. Company A: $20\cdot \$90=\$1800$
Company B: $30\cdot \$80=\$2400$
The new mean salary is

$$\frac{1800+2400}{50}=\frac{1200}{50}=\$84 \text{ per week}$$

15. a. The proportion of families that have incomes of at least \$25,000 is the sum of the last five items
$0.11+0.10+0.07+0.05+0.01=0.34$
b. Locate the bracket that contains 0.50 by starting at the top and adding down. Thus \$20,000–\$24,999 is the median income range.
c. The mean of the lower income levels in thousands of dollars is

$$\frac{\text{sum of lower limits}}{9}=\frac{355,000}{9}\approx \$39,444$$

d. The sum of the first three proportions is 0.36 which means that 36% have incomes less than \$20,000.

17. a. $\bar{x}=\frac{0.28+0.74+0.75+0.77+0.82}{5}$

$$=\frac{3.36}{5}\approx \$0.67$$

Median is 3rd term: \$0.75
No mode.
b. No. The mean, median and mode are not close.
c. The median is most representative.
d. Without Caracas:

$$\bar{x}=\frac{0.74+0.75+0.77+0.82}{4}$$
$$=\frac{3.08}{4}=\$0.77$$

Median is average of the 2nd and 3th terms: $\frac{0.75+0.77}{2}=\$0.76$
No mode.
The mean and median are most representative.

19. a. $\bar{x}=\frac{1553+1449+1411+1311}{5}$

$$=\frac{7199}{5}\approx \$1439.80$$

Arrange the weekly salaries in order of magnitude: \$1311, \$1411, \$1449, \$1475, \$1553. The median is 3rd term: \$1449
No mode.
b. The median is higher.

21. $\bar{x}=\frac{247+481+817+764}{4}=\frac{2309}{4}=577.25$

Arrange the number of houses in order of

magnitude: 247, 481, 764, 817. The median is average between the 2nd and 3rd terms: $\frac{481+764}{2}=\frac{1245}{2}=622.5$

23. The mean is

$$\bar{x}=\frac{-18+8+5+46+62}{5}=\frac{103}{5}=20.6$$

25. The mean number of spams in July is

$$\bar{x}=\frac{14+28+18+40+2+30+16+28+24}{9}$$
$$=\frac{200}{9}\approx 22.2$$

27. The mode is 28, since it occurs twice.

29. The category that increased the most from July to August was Internet.

31. Disneyland:

$$\bar{x}=\frac{10.3+14.1+15.0+14.3}{4}$$
$$=\frac{53.7}{4}=13.425 \text{ million}$$

Arrange the attendance in order of magnitude: 10.3, 14.1, 14.3, 15.0. The median is average between the 2nd and 3rd terms: $\frac{14.1+14.3}{2}=\frac{28.4}{2}=14.2$ million.
The current attendance is greater than the averages, so attendance has increased.

33. Disney-MGM:

$$\bar{x}=\frac{8.0+9.5+10.0+10.5}{4}$$
$$=\frac{38}{4}=9.5 \text{ million}$$

Arrange the attendance in order of magnitude: 8.0, 9.5, 10.0, 10.5. The median is average between the 2nd and 3rd terms: $\frac{9.5+10.0}{2}=\frac{19.5}{2}=9.75$ million.
The current attendance is greater than the mean, so attendance has increased.

35. The average number of visitors to the top 10 Web sites overall is

$$\bar{x}=\frac{\text{sum of unique visitors}}{10}$$
$$=\frac{4750}{10}=475 \text{ million}$$

37. a. $\bar{x}=$

$$\frac{39,468+39,276+40,827+46,778+49,733}{5}$$
$$=\frac{216,082}{5}=\$43,216.40$$

Arrange the salaries in order of magnitude: \$39,276, \$39,468, \$40,827, \$46,778, \$49,733. The median is the 3rd term: \$40,827. There is no mode.

b. The median is most representative of the person's salary.

39. The median of a set of scores is the middle number (if there is one) when the scores are arranged in order of magnitude. If there is no middle number, the median is the average of the two middle numbers. The median is not a good measure of a set of scores as it gives no indication of how the scores are spread. (The median could be considered a good measure of a set of scores if one wants to negate the effect of an outlier, that is, a score that is either very high or low as compared to the other scores.)

41. a. The average (mean) number of people in line is $\bar{x}=\frac{2+3+2+3+2}{5}=\frac{12}{5}=2.4$.

b. The mode is 2 because there are 2 people in line in three of the five minutes.

43. This "average" is the mean in dollars as shown.

$\bar{x}=$

$$\frac{100,000+50,000+25,000+2\cdot 10,000+6\cdot 6000}{11}$$
$$=\frac{231,000}{11}=\$21,000$$

Scrooge is using the mean of all the salaries.

45. B. Crooked is using the mean of a secretary's and a worker's salary as his "average" as shown.

$$\bar{x}=\frac{10,000+6000}{2}=\frac{16,000}{2}=\$8000$$

Section 12.3 Measures of Dispersion: The Range and Standard Deviation

Study Tips

After learning about individual measurements, and averages, two measures of *dispersion*, the *range* and the *standard deviation* tell us how *far* from these averages you may be. For instance, standard deviation is used in many everyday applications such as polls, test scores and so on.

To find the range:
Think: largest minus smallest
Note: The range is easy to find, since it is simply the difference between the least and the greatest of the numbers in a given set.

Standard Deviation Summary:
1. Find the mean, $\bar{x}$
2. Find the difference (deviation) between each number and the mean $\bar{x}$, $x-\bar{x}$
3. Square the deviations, $(x-\bar{x})^2$
4. Sum all the deviations, then divide by $n-1$
5. Take the square root of this quotient

Standard Deviation of a Sample Formula

$$s=\sqrt{\frac{(x_1-\bar{x})+(x_2-\bar{x})+(x_3-\bar{x})^2}{n-1}}$$

Word of Advice
Complete this table to help organize your calculations for finding the standard deviation.

x	$x-\bar{x}$	$(x-\bar{x})^2$
$\Sigma x=$		$\Sigma(x-\bar{x})^2=$
$\bar{x}=\frac{\text{sum}}{n}=$		$\frac{\Sigma(x-\bar{x})^2}{n}=$
Σ means to add up the numbers in the column.		$s=\sqrt{\frac{\Sigma(x-\bar{x})^2}{n}}=$

Fact
If the standard deviation is to be calculated for an entire population, then divide by N, the size of the entire population.

Hint
If you are allowed to use a calculator, with $\boxed{\Sigma+}$ and $\boxed{\bar{x}}$ keys, the calculator will do it for you. See information on sum at end of Section 12.2 Exercises.

12.3 Exercises

1. a. Range = 21 − 3 = 18
   b.

x	$x-\bar{x}$	$(x-\bar{x})^2$
3	−7	49
5	−5	25
8	−2	4
13	3	9
21	11	121
50		**208**
$\bar{x}=\frac{50}{5}=10$		$\frac{208}{4}=52$ $s=\sqrt{52}\approx 7.21$

3. a. Range = 25 − 5 = 20
   b.

x	$x-\bar{x}$	$(x-\bar{x})^2$
5	−10	100
10	−5	25
15	0	0
20	5	25
25	10	100
75		**250**
$\bar{x}=\frac{75}{5}=15$		$\frac{250}{4}=62.5$ $s=\sqrt{62.5}\approx 7.91$

5. a. Range = 9 − 5 = 4
   b.

x	$x-\bar{x}$	$(x-\bar{x})^2$
5	−2	4
6	−1	1
7	0	0
8	1	1
9	2	4
35		**10**
$\bar{x}=\frac{35}{5}=7$		$\frac{10}{4}=2.5$ $s=\sqrt{2.5}\approx 1.58$

7. a. Range = 9 − 1 = 8

b.

x	$x-\bar{x}$	$(x-\bar{x})^2$
5	0	0
9	4	16
1	−4	16
3	−2	4
8	3	9
7	2	4
2	−3	9
35		**58**
$\bar{x}=\frac{35}{7}=5$		$\frac{58}{6}\approx 9.67$ $s=\sqrt{9.67}\approx 3.11$

9. a. Range = 3 − (−3) = 6

b.

x	$x-\bar{x}$	$(x-\bar{x})^2$
−3	−3	9
−2	−2	4
−1	−1	1
0	0	0
1	1	1
2	2	4
3	3	9
0		**28**
$\bar{x}=\frac{0}{7}=0$		$\frac{28}{6}\approx 4.67$ $s=\sqrt{4.67}\approx 2.16$

11. a. The most frequent number is the mode, 8.

b. The number of students is even, so the median is the average of the 10th and 11th terms: $\frac{6+7}{2}=\frac{13}{2}=6.5$

c. The mean is $\bar{x}=\frac{\text{sum}}{20}=\frac{120}{20}=6.$

d.

x	$x-\bar{x}$	$(x-\bar{x})^2$
0	−6	36
0	−6	36
1	−5	25
2	−4	16
4	−2	4
4	−2	4
5	−1	1
6	0	0
6	0	0
6	0	0
7	1	1
8	2	4
8	2	4
8	2	4
8	2	4
9	3	9
9	3	9
9	3	9
10	4	16
10	4	16
10	4	16
120		**198**
$\bar{x}=\frac{120}{20}=6$		$\frac{198}{19}\approx 10.42$ $s=\sqrt{10.42}\approx 3.23$

e. The table shows that 14 of the 20 scores, or $\frac{14}{20}=0.7=70\%$ of the scores, lie within 1 standard deviation of the mean.

f. The table shows that 20 of the 20 scores, or $\frac{20}{20}=100\%$ of the scores, lie within 2 standard deviations of the mean.

13. a. The most frequent number is the mode, 110.

b. Arrange the hours in order of magnitude: 102, 103, 105, 106, 110, 110, 110, 110, 111, 113. The number of terms is even, so the median is the average of the 5th and 6th terms: $\frac{110+110}{2}=\frac{220}{2}=110$

c. The mean is $\bar{x}=\frac{\text{sum}}{10}=\frac{1080}{10}=108.$

d.

x	$x-\bar{x}$	$(x-\bar{x})^2$
103	–5	25
110	2	4
113	5	25
102	–6	36
105	–3	36
110	2	4
111	3	9
110	2	4
106	–2	4
110	2	4
1080		**124**
$\bar{x}=\frac{1080}{10}=108$		$\frac{124}{9}\approx 13.78$
		$s=\sqrt{13.78}\approx 3.71$

e. The values more than 1 standard deviation of the mean are 102, 103, and 113. Thus, $\frac{3}{10}=0.30=30\%$ of the values are more than 1 standard deviation of the mean.

15. Range = 17 – 2 = 15
Mean =
$\bar{x}=\frac{6+2+17+3+5+9}{6}=\frac{42}{6}=7$

x	$x-\bar{x}$	$(x-\bar{x})^2$
6	–1	1
2	–5	25
17	10	100
3	–4	16
5	–2	4
9	2	4
42		**150**
$\bar{x}=\frac{42}{6}=7$		$\frac{150}{5}=30$
		$s=\sqrt{30}\approx 5.48$

17. Range = 12 – 3 = 9
Mean = $\bar{x}=\frac{6+7+9+12+3+5}{6}=\frac{42}{6}=7$

x	$x-\bar{x}$	$(x-\bar{x})^2$
6	–1	1
7	0	0
9	2	4
12	5	25
3	–4	16
5	–2	4
42		**50**
$\bar{x}=\frac{42}{6}=7$		$\frac{50}{5}=10$
		$s=\sqrt{10}\approx 3.16$

19. Coral Reef: Range = \$150 – \$70 = \$80
Waikiki: Range = \$495 – \$105 = \$390
Radisson: Range = \$215 – \$215 = \$0
Waikiki Parkside: Range = \$148 – \$59 = \$89
Pagoda: Range = \$145 – \$68= \$77

21. The average of the price averages is
$\bar{x}=\frac{110+300+215+103.5+106.5}{5}=\frac{835}{5}=\$167$

23. Renting for 1 day:
Range = \$37.99 – \$23.99 = \$14

25.

x	$x-\bar{x}$	$(x-\bar{x})^2$
23.99	–8.2	67.24
26.99	–5.2	27.04
34.98	2.79	7.7841
37.99	5.8	33.64
36.99	4.8	23.04
160.94		**158.7441**
$\bar{x}=\frac{160.94}{5}=32.19$		$\frac{158.7441}{4}\approx 39.67$
		$s=\sqrt{39.67}\approx 6.30$

The standard deviation of the prices for renting a compact car for 1 day is \$6.30.

27. The numbers are all the same. If zero is the standard deviation, then $(x-\bar{x})^2=0$ for all x in the set.

29. Choose the numbers 1, 1, 5, and 5 (remember, repetitions are allowed). The mean of these

numbers is 3 and 1, 1, 5, and 5 are as far from the mean 3 (2 units) as possible.

31. $s=\sqrt{180\cdot\frac{1}{6}\cdot\frac{5}{6}}=\sqrt{\frac{900}{36}}=\sqrt{25}=5$

33. $s=\sqrt{400\cdot\frac{1}{4}\cdot\frac{3}{4}}=\sqrt{\frac{1200}{16}}=\sqrt{75}\approx 8.66$

35. If you multiply each data point by 1.1 (100% + 10% = 1.1), it increases the spread of the points by a factor of 1.1. For example, let the mean be \$64,000, and one standard deviation below the mean is at \$44,000. We see that one standard deviation is \$20,000. So if two employees making \$64,000 and \$44,000 are given a 10% raise, they are
$1.1\times 64{,}000-1.1\times 44{,}000$
$=1.1\times(64{,}000-44{,}000)$, or \$22,000, apart. This is an increase of \$2000, or
$\frac{2000}{20{,}000}=0.1=10\%$. Thus, the standard deviation increases by 10%.

Section 12.4 The Normal Distribution: A Problem-Solving Tool

Study Tips

Have you ever heard the comment, "I hope the instructor grades on a curve"? This curve is the *normal curve* and describes data that have a very large number of values distributed among the population in a *bell shape*. The normal curve shows how the data in a normal distribution are spread out.

Properties of a normal curve:
1. Smooth and symmetric
2. High point over the mean of the population
3. Levels out but never touches *x*-axis
4. Total area is 1, proportion of data values between 1, 2, and 3 standard deviations to either side of the mean

To calculate a *z*-score:
Think: mean and standard deviation in the formula $z=\frac{x-\mu}{\sigma}$
Note: The *z*-scores are used to compare scores in different subjects.

To relate distribution of z-scores and probability:
Think: find *z*-score, reference normal curve for percent or Table II
Note: The percent references the percent of the area under the curve; this total area is 1.

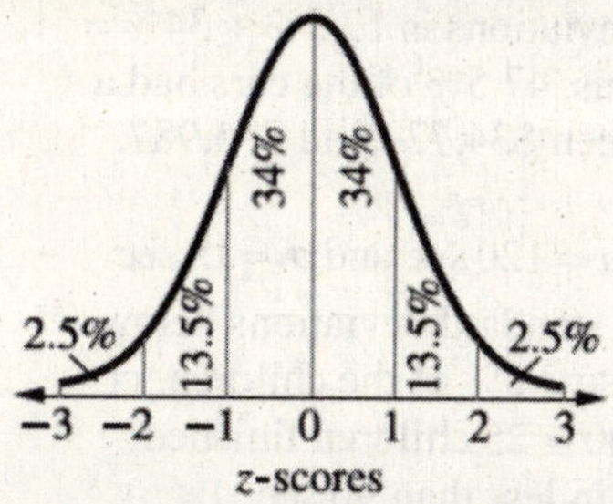

When determining probability, remember....
a. The percent above the mean is 50% and below the mean is 50%.
b. Table II shows the probability (area) within various standard deviations away from the mean.
c. The probability of all data values with a *z*-score below, say 1, would not be just 0.34 but would be 0.34 + 0.50 = 0.84, or 84%.
d. The probability of all data values with a *z*-score below −1 would be 0.025 + 0.135 = 0.16, or 16%.
e. The probability between two *z*-scores will be the area between those scores.

Hint
Draw the normal curve when working a problem and mark the *z*-score(s). This will help you to visualize the area (probability) that you want to find.

Word of Advice
Memorize the percents given on the normal curve as shown above. Write these percents and formulas on a note card for easy reference.

Note
The *Using Your Knowledge* in the exercises discusses the subject of percentiles and *Discovery* discusses Chebyshev's Theorem, $\left(1-\frac{1}{h^2}\right)N$.

12.4 Exercises

1. a. The mean is under the highest point on the graph, \$35,987.
   b. From the graph, it appears that the value \$34,772 is approximately 2 standard deviations from the mean, \$35,987.

Thus, the standard deviation is

$$\frac{\$35,987-\$34,772}{2}=\frac{\$1215}{2}=\$607.50.$$

c. There are two standard deviations between \$34,772 and \$35,987. Refer to the Figure 12.10 to see that the area under a normal curve for two standard deviations is 13.5% + 34% = 47.5%. Thus, 47.5% of the cars had a price between \$34,772 and \$35,987.

3. We are given $\mu = 120$ sec and $\sigma = 15$ sec.

a. 90 sec is 2 standard deviations below the mean, so 2.5% of the children, or $0.025 \cdot 1000 = 25$ children finished the puzzle in less than 90 sec.

b. 150 sec is 2 standard deviations above the mean, so 2.5% of the children, or $0.025 \cdot 1000 = 25$ children finished the puzzle in more than 150 sec.

c. 68% of the children, or $0.68 \cdot 1000 = 680$, are within 1 standard deviation of the mean.

5. We are given $\mu = 50$ and $\sigma = 10$.
A: 2.5% get this grade. 2.5% of 500 = 12.5, so 12 or 13 get A.
B: 13.5% get this grade. 13.5% of 500 = 67.5, so 67 or 68 get B.
C: 68% get this grade. 68% of 500 = 340, so 340 get C.
D: 13.5% get this grade. 13.5% of 500 = 67.5, so 67 or 68 get D.
F: 2.5% get this grade. 2.5% of 500 = 12.5, so 12 or 13 get F.
(Of course, the teacher's choices must add up to 500.)

7. Given $\mu = 26,000$ mi and $\sigma = 2500$ mi. 21,000 mi is

$$\frac{26,000-21,000}{2500}=\frac{5000}{2500}=2 \text{ standard}$$

deviations below the mean. Thus, $0.025 \cdot 200 = 5$ tires gave out before 26,000 mi.

9. Since $\mu = 20$ ft and $\sigma = 0.5$ in., almost all the measurements fall between $20\text{ft} - 3(0.5\text{in.})$ and $20\text{ft} + 3(0.5\text{in.})$, that is, between 19 ft, 10.5 in. and 20 ft, 1.5 in.

11. The purchasing director will decide to buy. If the lifetimes are normally distributed, 2.5% will last for 30 + 2(5) = 40 or more days. Thus, of the 8000 ball bearings, about $8000 \cdot 0.025 = 200$ would last for 40 or more days, so the purchasing director would decide to buy.

13. For the S.A.T., the z-score is

$z = \dfrac{635-514}{117} \approx 1.03$. For the A.C.T., the z-score is $z = \dfrac{26.2-21}{5.2} = 1$. Thus, the student who scored 635 on the S.A.T. has the higher score relative to the test.

15. a. $\sigma = 10$ b. $\sigma = 5$

c. $\mu = 40$ d. $\mu = 65$

e. A z-score of +2 corresponds to

$$\frac{x-40}{10}=2$$
$$x-40=20$$
$$x=60$$

A z-score of −1 corresponds to

$$\frac{x-40}{10}=-1$$
$$x-40=-10$$
$$x=30$$

f. A z-score of +2 corresponds to

$$\frac{x-65}{5}=2$$
$$x-65=10$$
$$x=75$$

A z-score of −1 corresponds to

$$\frac{x-65}{5}=-1$$
$$x-65=-5$$
$$x=60$$

17. Given $\mu = 16.15$ oz and $\sigma = 0.05$ oz, a z-score of 0 corresponds to $\dfrac{x-16.15}{0.05}=0$

$$x-16.15=0$$
$$x=16.15$$

The weight of a box with a z-score 0 is 16.15 oz.

19. Given $\sigma = 10$, $\mu = 70$, and z-score is +2, we have $\frac{x-70}{10} = 2$

$$x - 70 = 20$$
$$x = 90$$

Pedro's exam score is 90.

21. Given $\mu = 5$ and $\sigma = 1.25$.

a. If $x = 6$, then $z = \frac{6-5}{1.25} = \frac{1}{1.25} = 0.8$

b. If $x = 7$, then $z = \frac{7-5}{1.25} = \frac{2}{1.25} = 1.6$

c. If $x = 7.5$, then $z = \frac{7.5-5}{1.25} = \frac{2.5}{1.25} = 2$

23. For the German test, the z-score is $z = \frac{85-75}{20} = 0.5$. For the English test, the z-score is $z = \frac{85-80}{15} = \frac{5}{15} \approx 0.33$.

Thus, the German test score is the better score relative to the test.

25. 20 minutes is 2 standard deviations from the mean of 30 minutes. Thus, it has a z-score of –2. The probability of arriving at school in 20 to 30 minutes is 47.7%, using the values of 34.1% and 13.6% instead of the approximate 34% and for 13.5%, respectively.

27. Find the z-score for values of $\mu = 4.5$ min, $\sigma = 1.5$min, and $x = 5$ min.

$$z = \frac{5-4.5}{1.5} = \frac{0.5}{1.5} \approx 0.33$$

From Table II the probability associated with 0.33 is 0.129. So the probability of being served within 5 minutes is $0.50 + 0.129 = 0.629 = 62.9\%$.

29. Given $\mu = 100$ and $\sigma = 15$.

a. The z-score for $x = 110$ is

$$z = \frac{110-100}{15} = \frac{10}{15} \approx 0.667$$

From Table II, we find the probability that a randomly selected score is between the mean and 0.667 standard deviation above the mean to be 0.249.

b. The z-score for $x = 130$ is

$$z = \frac{130-100}{15} = \frac{30}{15} = 2$$

Again, from Table II, we find the probability that a randomly selected score is between the mean and 2 standard deviations above the mean to be 0.477.

31. a. Since $z = \frac{55-100}{15} = -\frac{45}{15} = -3$ and $z = \frac{145-100}{15} = \frac{45}{15} = 3$, 55 and 145 are each 3 standard deviations from the mean. From Table II, we find the probability that a randomly selected score falls between the mean and 3 standard deviations above the mean to be 0.499. This means that the probability that the score is between 100 and 145 is 0.499. Because of the symmetry of the normal curve, the probability that the score is between 55 and 100 is also 0.499. Therefore, the probability that the score is between 55 and 145 is $0.499 + 0.499 = 0.998$.

b. $z = \frac{60-100}{15} = -\frac{40}{15} = -2.67$. So 60 is 2.67 standard deviations "below" the mean. Table II gives 0.496 as the probability that a randomly selected score is between the mean and 2.67 standard deviations from the mean. Since the probability that such a score is below the mean is 0.5, the probability that the score is below 60 is $0.5 - 0.496 = 0.004$.

33. In problem 29, we found 0.249 to be the probability that a randomly selected score is between 100 and 110, and 0.477 to be the corresponding probability for the score to be between 100 and 130. Consequently, the probability that the score is between 110 and 130 is $0.477 - 0.249 = 0.228$.

35. This student is to be shortest of the tallest 20%. From Table II, the area from the mean to the top 20% is found by finding the z-score corresponding to $0.5 + 0.3 = 0.8$. Thus, the z-score of his height must be greater than 0.8. The height x in. of the shortest student in this group must be such that

$$\frac{x - 5\text{ ft }7\text{in.}}{3\text{ in.}} = \frac{x - 67\text{ in.}}{3\text{ in.}} = 0.8$$

$$x - 67\text{ in.} = 2.4\text{ in.}$$

$$x = 69.4\text{ in. or }5\text{ ft }9\frac{1}{2}\text{ in.}$$

Thus, the shortest student in the tallest group is about 5 ft 9 ½ in. tall.

37. $z = \frac{4400 - 4200}{200} = \frac{200}{200} = 1.$
A z-value of 1 indicates that the value $4400 is one standard deviation above the mean of $4200. (Someone with this income is earning more than 84% of other budget analysts.)

39. We can be certain that the ice cream will be within 3 standard deviations of the mean. Since the mean is 8 oz and the standard deviation is 0.25 oz, the weight of the ice cream must be between 8 – 3(0.25) = 7.25 oz and 8 + 3(0.25) = 8.75 oz.

41. Find the z-scores for the values of $x = 8$ and $x = 8.4$ given that $\mu = 8\text{hr}$ and $\sigma = 0.25\text{ hr}$.

$$z = \frac{8 - 8}{0.25} = \frac{0}{0.25} = 0$$

$$z = \frac{8.4 - 8}{0.25} = \frac{.4}{0.25} = 1.6$$

From Table II the probability associated with 1.6 is 0.445. So the probability of Caruca will get her restorative sleep is 0.445 or 44.5%.

43. Since 68% of the values on the normal curve are within one standard deviation (10 gal) of the mean (110 gal), the daily water usage lies between 110 – 10 = 100 gal and 110 + 10 = 120 gal.

45. Since 110 gal is the mean and 50% of all values in the normal curve are to the left of the mean, the probability that a person selected at random will use less than 110 gal per day is 50%.

47. Since the mean is 110, and the standard deviation is 10, the value 100 is one standard deviation below the mean. Thus, 34% of the people use between 100 gal and 110 gal.

49. She has provided poor service if her tips are less than $70. The z-value for $70 is

$$z = \frac{70 - 100}{20} = -\frac{30}{20} = -1.5.$$

From Table II, the value for 1.5 is 0.433, so the probability that the waitress provided poor service is 0.5 – 0.433 or 0.067, that is, 6.7%. (Remember, you are looking for the area under the normal curve that is to the left of the z-score corresponding to –1.5).

51. Yes. See Figure 12.14 on page 708. Here both curves have the mean 0. The standard deviation for A is $\frac{1}{3}$ unit while that for B is 1 unit.

53. No. If curve A in Figure 12.14 on page 708 is moved 2 units to the right, this would be an example. The standard deviation for A would be $\frac{1}{3}$ unit and that for B would be 1 unit, while the mean for A would be 2 and that for B would be 0.

55. $\text{Percentile} = \frac{40}{50} \cdot 100 = 0.8 \cdot 100 = 80.$ Her percentile is 80th.

57. a. There are 2 scores less than 90, so the percentile is $\frac{2}{10} \cdot 100 = 0.2 \cdot 100 = 20^{\text{th}}$.

b. There are 6 scores less than 97, so the percentile is $\frac{6}{10} \cdot 100 = 0.6 \cdot 100 = 60^{\text{th}}$.

c. There are 9 scores less than 100, so the percentile is $\frac{9}{10} \cdot 100 = 0.9 \cdot 100 = 90^{\text{th}}$.

d. There are 0 scores less than 83, so the percentile is $\frac{0}{10} \cdot 100 = 0 \cdot 100 = 0^{\text{th}}$.

59. a. 40 and 60 are each 2 standard deviations from the mean, so we use Chebyshev's theorem with $h = 2$ and $N = 100$:

$$\left(1 - \frac{1}{2^2}\right) \cdot 100 = \left(1 - \frac{1}{4}\right) \cdot 100$$

$$= \left(\frac{3}{4}\right) \cdot 100 = 75$$

Thus, at least 75 of the measurements must be between 40 and 60.

b. 35 and 65 are each 3 standard deviations from the mean, so we use Chebyshev's theorem with $h = 3$ and $N = 100$:

$$\left(1 - \frac{1}{3^2}\right) \cdot 100 = \left(1 - \frac{1}{9}\right) \cdot 100$$

$$= \left(\frac{8}{9}\right) \cdot 100 \approx 89$$

Thus, at least 89 of the measurements must be between 35 and 65.

c. 43 and 57 are each $\frac{7}{5}$ standard deviations from the mean, so we use Chebyshev's theorem with $h = \frac{7}{5}$ and $N = 100$:

$$\left(1 - \frac{1}{\left(\frac{7}{5}\right)^2}\right) \cdot 100 = \left(1 - \frac{25}{49}\right) \cdot 100$$

$$= \left(\frac{24}{49}\right) \cdot 100 \approx 49$$

Thus, at least 49 of the measurements must be between 43 and 57.

61.

x	$x - \bar{x}$	$(x - \bar{x})^2$
1	–5	25
1	–5	25
1	–5	25
2	–4	16
6	0	0
10	4	16
11	5	25
11	5	25
11	5	25
54		**182**
$\bar{x} = \frac{54}{9} = 6$		$\frac{182}{8} = 22.75$ $s = \sqrt{22.75} \approx 4.77$

Three of the numbers (2, 6, and 10) lie within 1 standard deviation from the mean, and all 9 of the numbers lie within 2 standard deviations from the mean. The theorem makes no prediction for 1 standard deviation; it predicts at least 75% for 2 standard deviations.

63. No; the theorem predicts the least number that must lie within 2 standard deviations from the mean, and it predicts the least number that must lie within 3 standard deviations from the mean, but it does not predict how many must lie between 2 and 3 standard deviations from the mean. However, since at least 75% of the numbers must lie within 2 standard deviations from the mean, the number of numbers between 2 and 3 standard deviations from the mean cannot be more than 25% of all the numbers.

Section 12.5 Statistical Graphs: A Problem-Solving Tool

Study Tips

Have you seen all those graphs in newspapers and magazines? Graphs display data so that trends and relationships can be easily seen.

To make and interpret a linear graph:
Think: label *x*-axis and *y*-axis, graph line by plotting points, connect graph
Note: A linear graph is very effective in showing trends over time.

To make and interpret a bar graph:
Think: label *x*-axis and *y*-axis, use horizontal or vertical bars to represent amounts
Note: A bar graph is very effective in showing comparisons between groups or items.

To make and interpret a circle graph:
Think: sections proportional to actual percentage, use protractor and simple compass
Note: A circle graph is very effective in showing what percent of the whole (budget) is used for a particular item (food, housing, clothing, etc.).

Caution
These graphs can also be *distorted* to present a wrong impression of the data. Learn how that is done in the *Discovery* in the exercises.

12.5 Exercises

1.

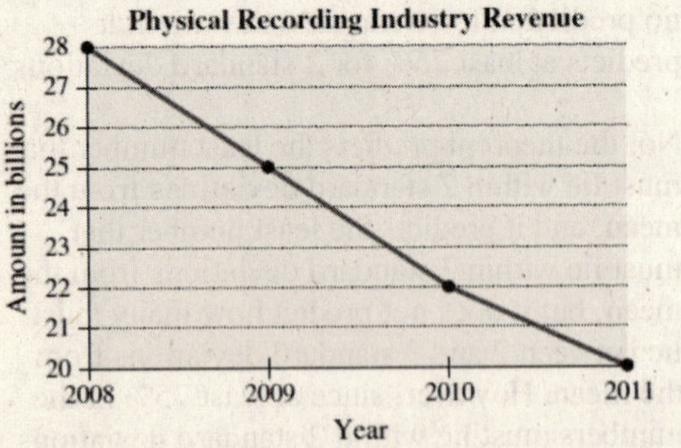

This is just a portion of the complete graph.

3.

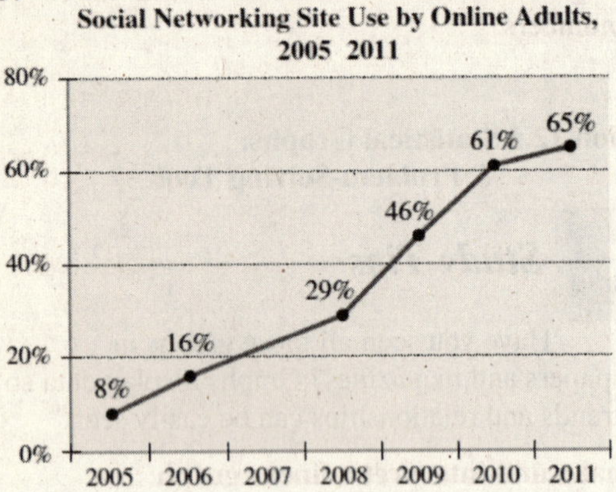

5.

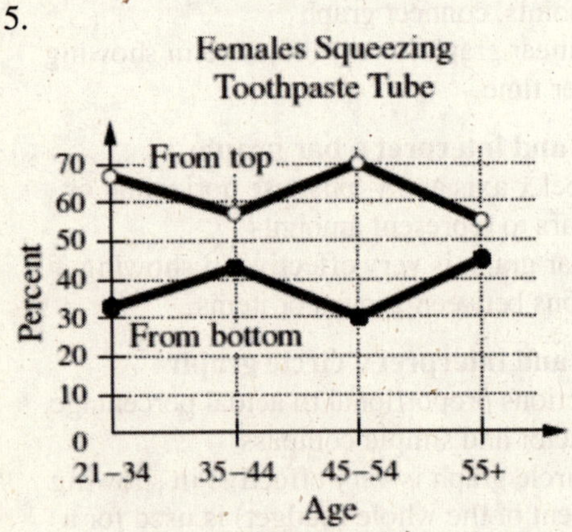

7.

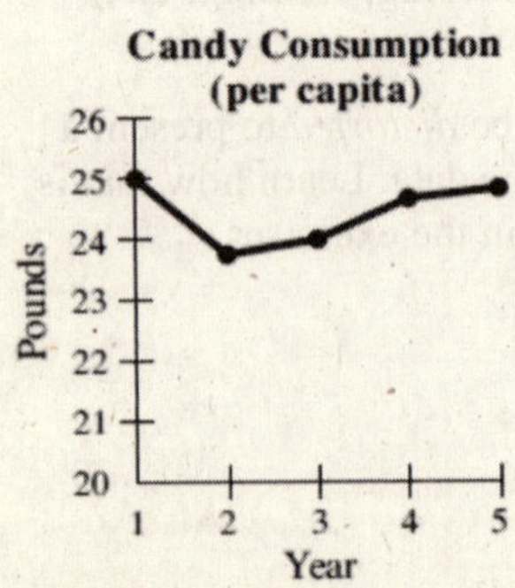

9.

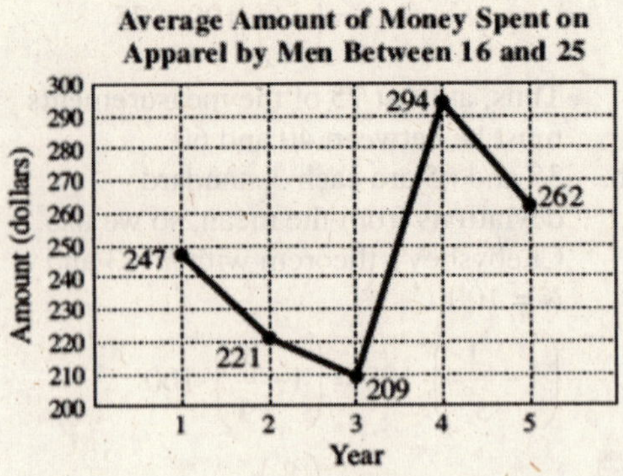

11.

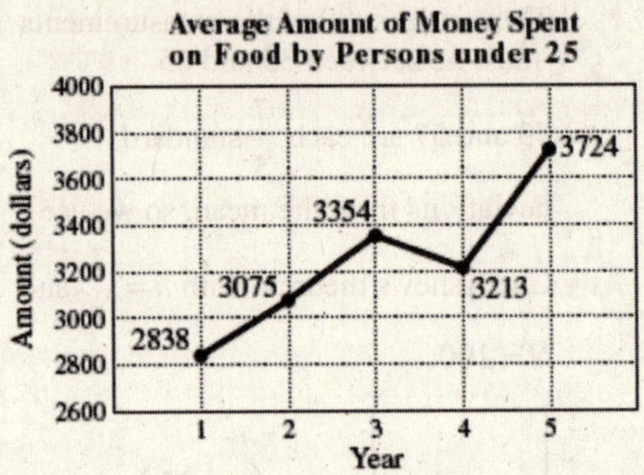

13.

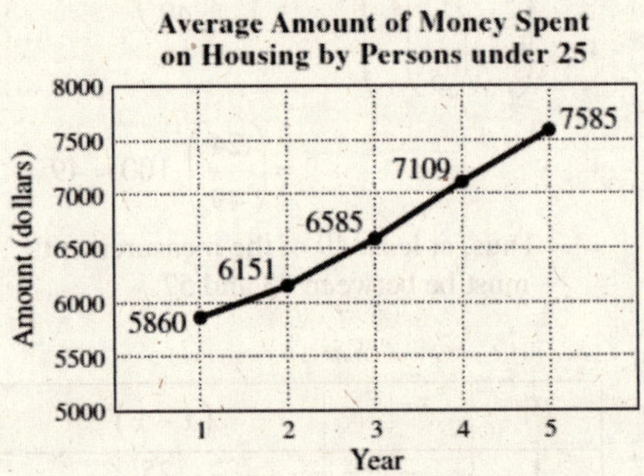

15.

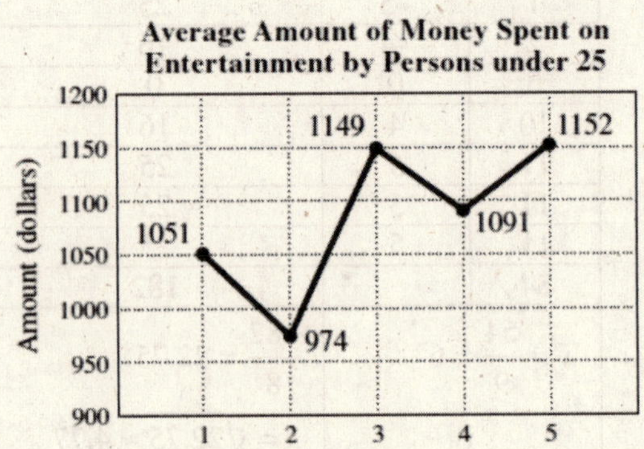

17.

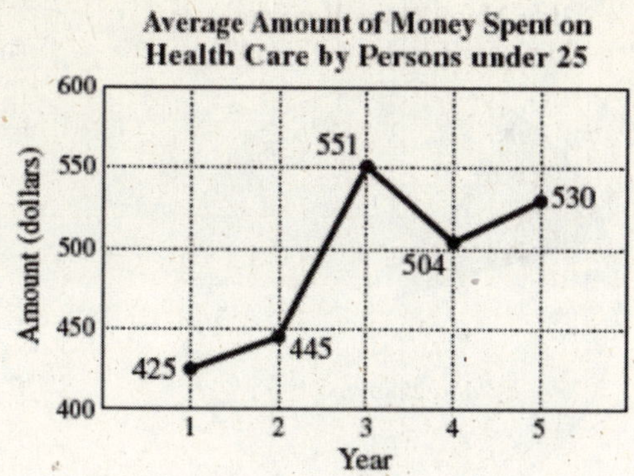

19.

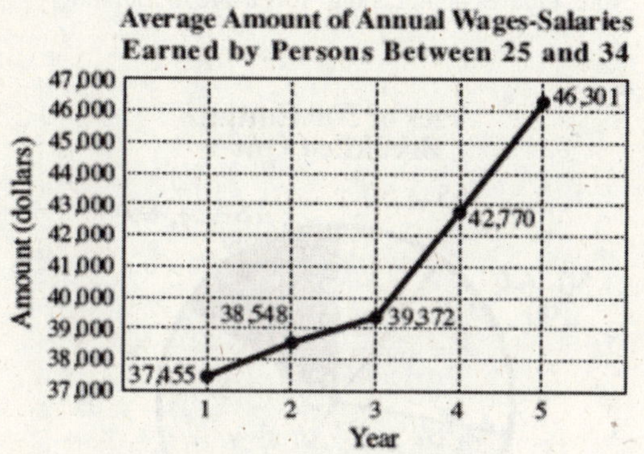

21.

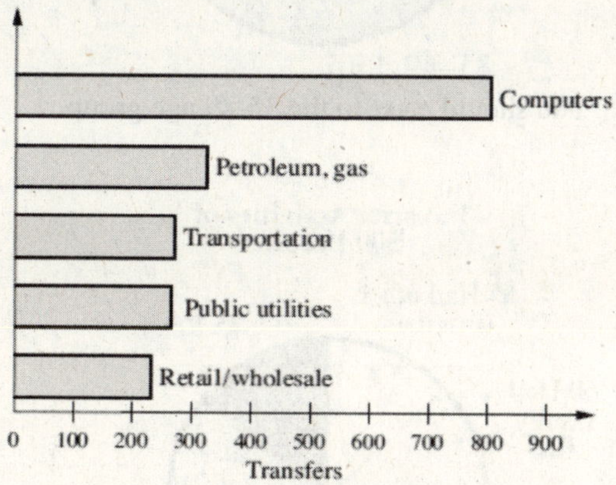

23. a.

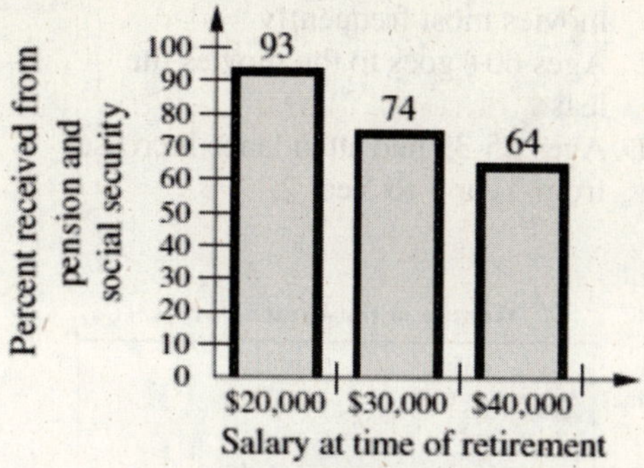

b. You should expect $0.93 \cdot \$20,000 = \$18,600$ from your pension and Social Security.

25. a.

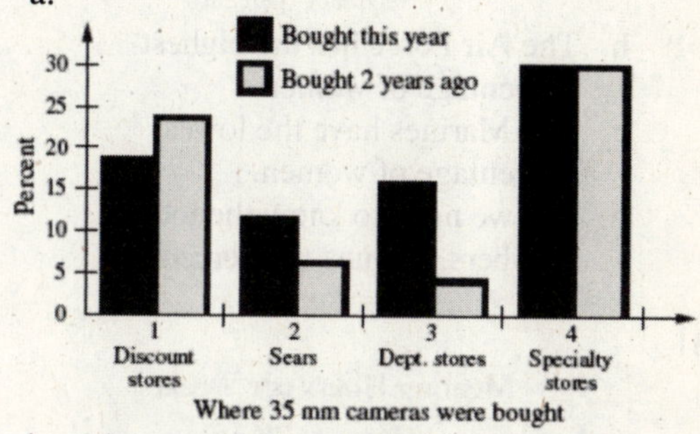

b. Most people buy their 35-mm cameras at specialty stores.

c. Department stores lost the most in the 2-year interval.

d. Specialty stores had the most consistent sales.

27. a.

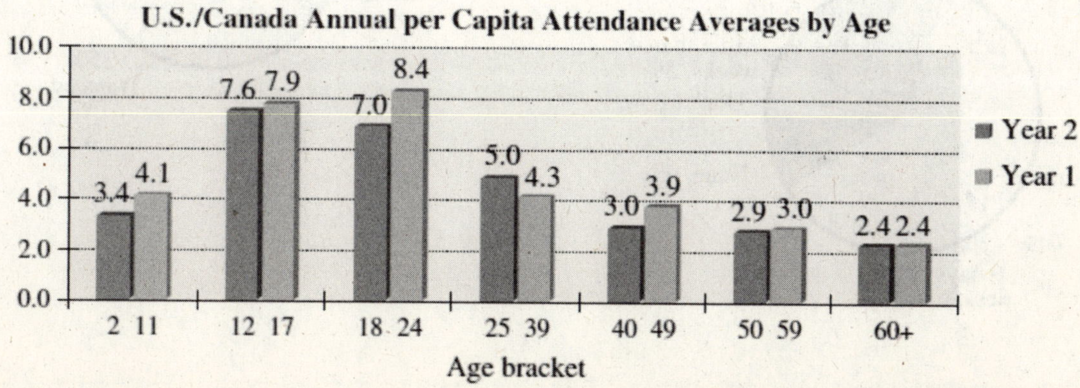

b. Ages 18-24 in Year 1 go to the movies most frequently.
c. Ages 60+ goes to the movies the least.
d. Ages 25-39 had attendance increase from Year 1 to Year 2.

29. a.

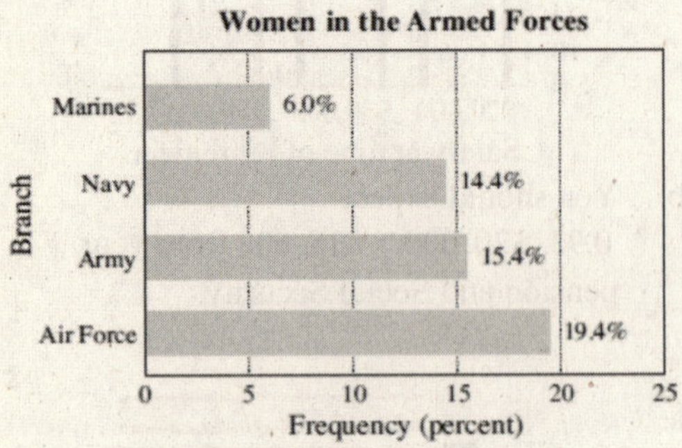

b. The Air Force has the highest percentage of women.
c. The Marines have the lowest percentage of women.
d. No, we need to know the total numbers, not just the percents.

31.

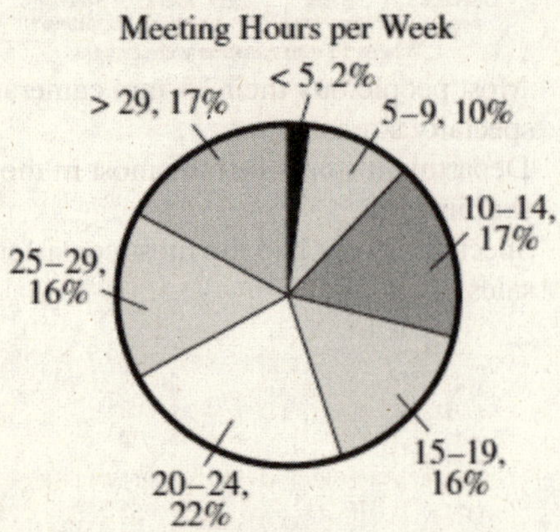

33.

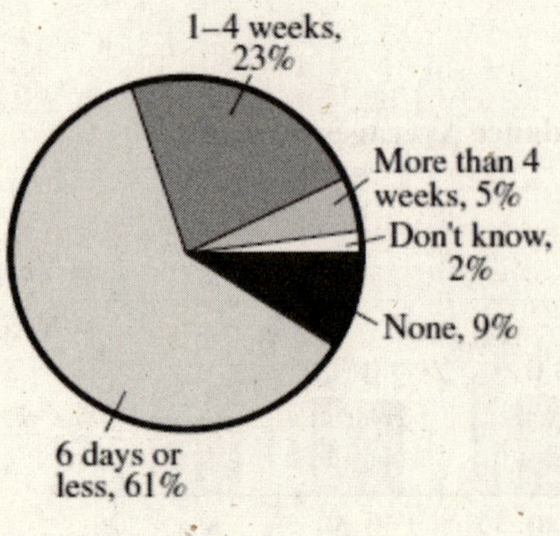

35.

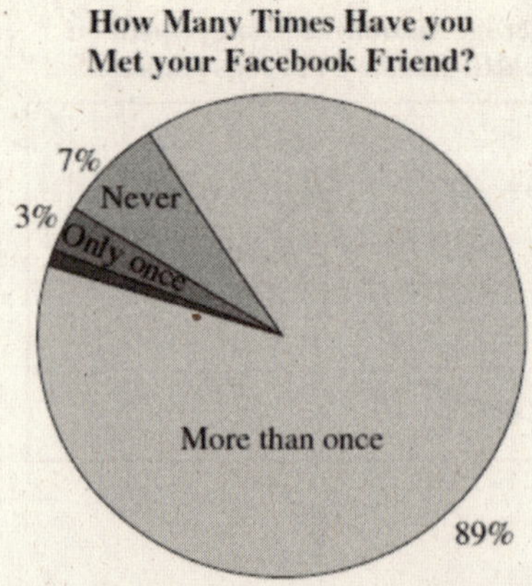

Note: Figures do not add to 100% due to rounding

37. a.

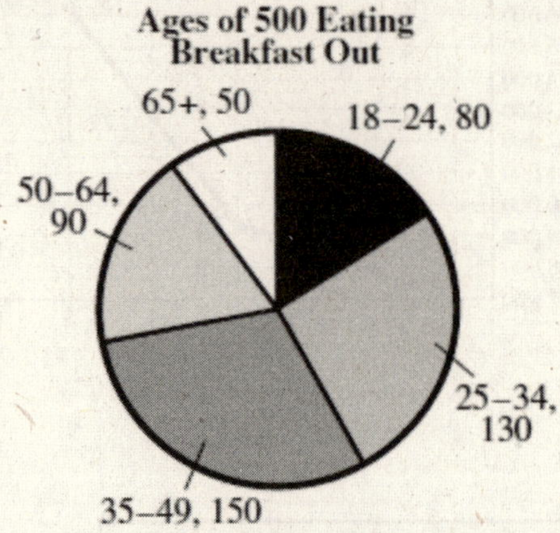

b. You should cater to the 35-49 age group.

39.

41.

Retired Couple's Budget

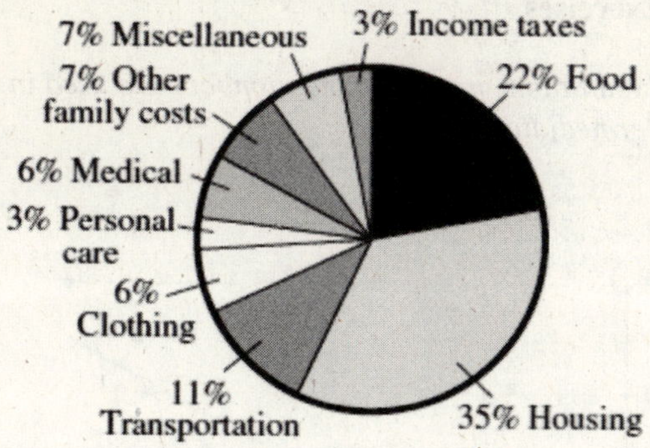

43. a.

Top Six TVs by Brand

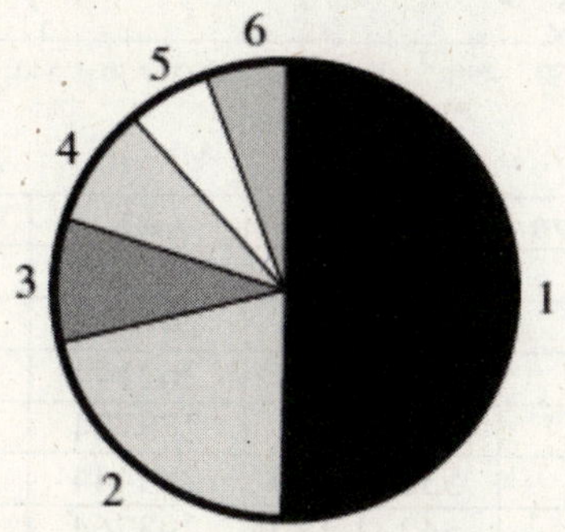

b.

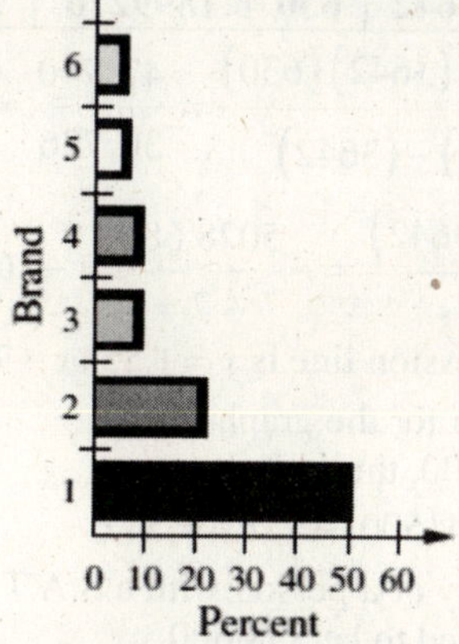

c. The circle graph. The area corresponding to brand 1 overshadows all the rest of the chart.

45. a. Yes. To give a correct visual impression, only the height should be doubled. If both the height and the radius are doubled, the volume is multiplied by 8.

b.

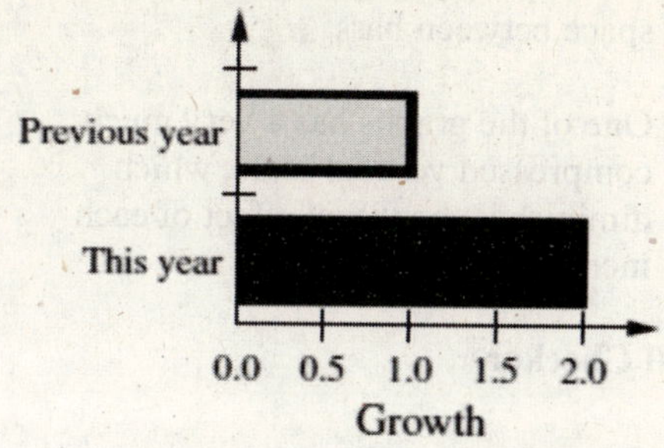

47.

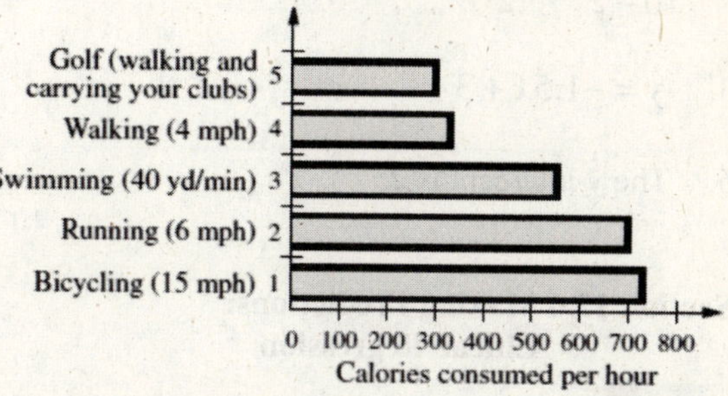

49.

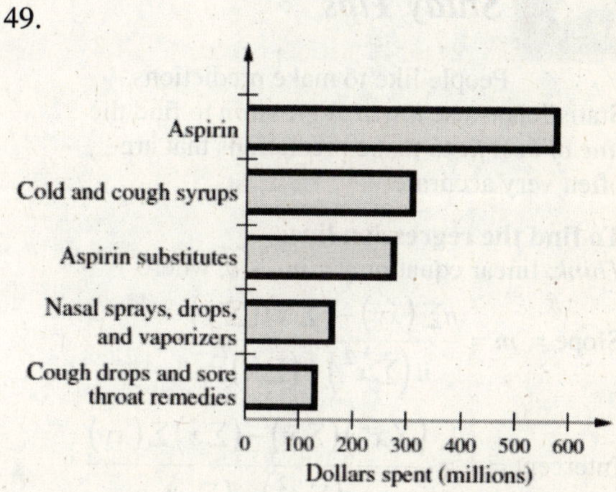

51. a. The sales in year 4 were $16,000,000.
b. The sales started to level off in year 6.

53. a. To make sales look better, use a large scale on the vertical axis, for instance, start the vertical axis at 19. You could use narrow spaced horizontal marks to obtain a steeper looking slope.
b. To make sales look flat, use a very small scale on the vertical axis, for instance start the vertical axis at 0. You could use horizontal marks spaced widely apart to obtain almost zero slope.

55. The area of the bar in a bar graph indicates the amount of the item that is graphed. For this reason, the bars are usually shaded. In a histogram, the height of the bar corresponds to

the frequency in question and there is no space between bars.

57. One of the graphs has a very much compressed vertical scale, which diminishes the visual effect of each increase or decrease.

Skill Checker

1. $\dfrac{3-0}{0-2} = -\dfrac{3}{2} = -1.5$

3. $y = -1.5x + 3$

5. The y-intercept is 2.

Section 12.6 Making Predictions: Linear Regression

Study Tips

People like to make predictions. Statisticians use *linear regression* to find the *line of best fit* to make predictions that are often very accurate.

To find the regression line:
Think: linear equation $y = mx + b$, where

$$\text{Slope} = m = \frac{n\Sigma(xy) - (\Sigma x)(\Sigma y)}{n(\Sigma x^2) - (\Sigma x)^2}$$

$$\text{Intercept} = b = \frac{(\Sigma x^2)(\Sigma y) - (\Sigma x)\Sigma(xy)}{n(\Sigma x^2) - (\Sigma x)^2}$$

Note: The regression line is also called the line of best fit.

Word of Advice
Complete a table of this form to organize your calculations.

Items	*x*	*y*	x^2	*xy*
Totals	Σx	Σy	Σx^2	$\Sigma(xy)$

To make a prediction:
Think: substitute for x in the regression line
Note: The values of x should be close to the original data set, since the regression line could become less reliable as it moves further away from the original data set.

12.6 Exercises

For calculations, non-rounded numbers are used in the intermediate steps.

1. a.

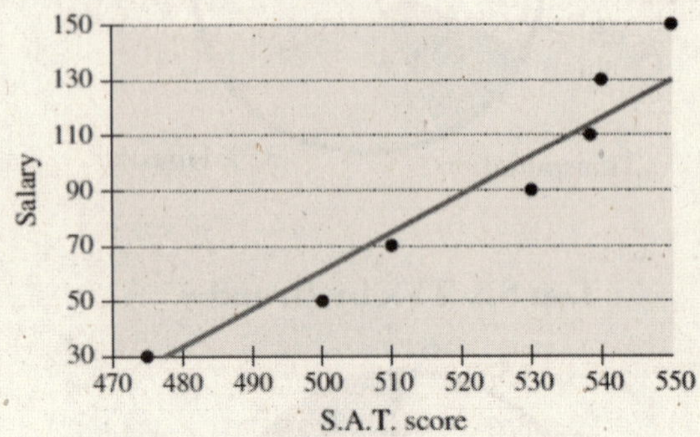

b.

Person	*x*	*y*	x^2	*xy*
1	475	30	225625	14250
2	497	50	247009	24850
3	512	70	262144	35840
4	528	90	278784	47520
5	538	110	289444	59180
6	542	130	293764	70460
7	550	150	302500	82500
Totals	**3642**	**630**	**1899270**	**334600**

$$m = \frac{7(334600) - (3642)(630)}{7(1899270) - (3642)^2} = \frac{47,740}{30,726} \approx 1.554$$

$$b = \frac{630 - 1.554(3642)}{7} = -\frac{5028.686}{7} \approx -718.4$$

The regression line is $y = 1.554x - 718.4$.
See part a for the graph.

c. Let $x = 500$, then
$y = 1.554(500) - 718.4 = 58.6$
The salary of a person with a S.A.T of 500 is predicted to be \$58,600.
Let $x = 540$, then
$y = 1.554(540) - 718.4$
$= 120.76$
The salary of a person with a S.A.T of 540 is predicted to be \$120,760.

3. a.

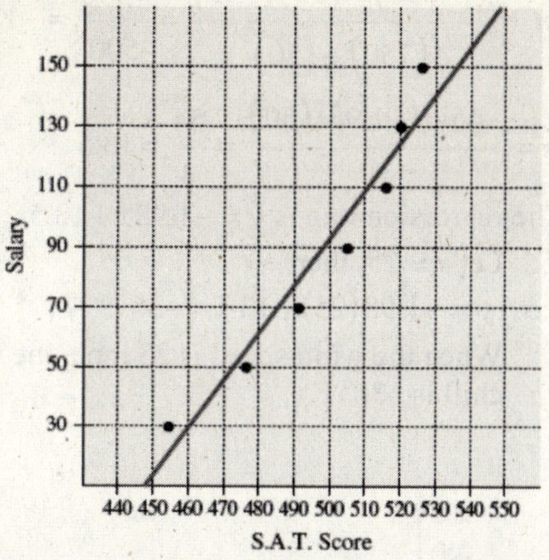

b.

Person	*x*	*y*	x^2	*xy*
1	453	30	205209	13590
2	476	50	226576	23800
3	481	70	241081	34370
4	505	90	255025	45450
5	516	110	266256	56760
6	520	130	270400	67600
7	527	150	277729	79050
Totals	**3488**	**630**	**1742276**	**320620**

$$m=\frac{7(320620)-(3488)(630)}{7(1742276)-(3488)^2}=\frac{46,900}{29,788}\approx 1.574$$

$$b=\frac{630-1.577(3488)}{7}=-\frac{4861.71}{7}\approx -694.5$$

The regression line is $y=1.574x-694.5$.
See part a for the graph.

c. Let $x = 500$, then
$y=1.574(500)-694.5=92.5$
The salary of a person with a S.A.T of 500 is predicted to be \$92,500.
Let $x = 540$, then
$y=1.574(540)-694.5=155.46$
The salary of a person with a S.A.T of 540 is predicted to be \$155,460.

5. a.

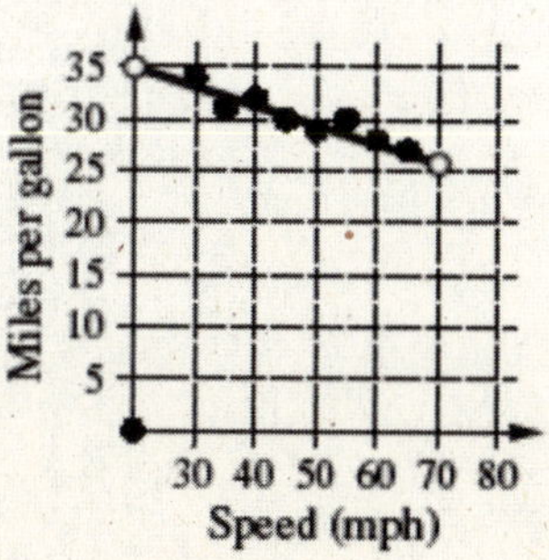

b. If your speed were 70 mph, you would get about 26 mi/gal (may vary).

c.

Car	*x*	*y*	x^2	*xy*
1	30	34	900	1020
2	35	31	1225	1085
3	40	32	1600	1280
4	45	30	2025	1350
5	50	29	2500	1450
6	55	30	3025	1650
7	60	28	3600	1680
8	65	27	4225	1755
Totals	**380**	**241**	**19100**	**11270**

$$m=\frac{8(11270)-(380)(241)}{8(19100)-(380)^2}=-\frac{1420}{8400}\approx -0.169$$

$$b=\frac{241+0.169(380)}{8}=\frac{305.24}{8}\approx 38.155$$

The regression line is $y=-0.169x+38.155$.
Let $x = 70$, then
$y=-0.169(70)+38.155=26.3$
When the speed is 70 mi/hr, you should get 26.3 mi/gal.

7. a.

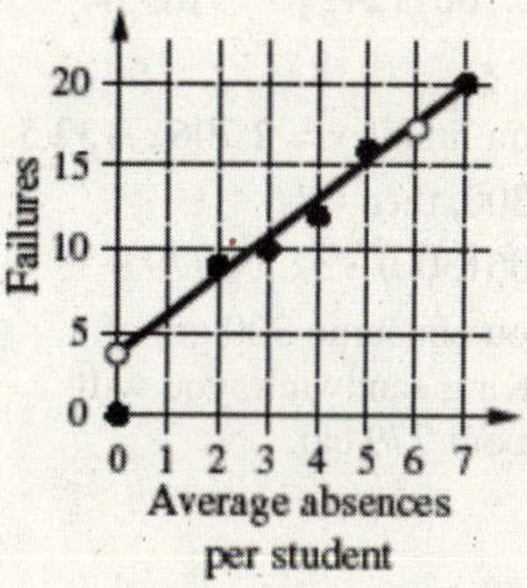

b. For 6 absences per student, about 18 students will fail.

c.

Absences	*x*	*y*	x^2	*xy*
1	5	16	25	80
2	7	20	49	140
3	2	9	4	18
4	4	12	16	48
5	3	10	9	30
Totals	**21**	**67**	**103**	**316**

$$m=\frac{5(316)-(21)(67)}{5(103)-(21)^2}=\frac{173}{74}\approx 2.338$$

$$b=\frac{67-2.338(21)}{5}=\frac{17.91}{5}\approx 3.581$$

The regression line is $y=2.338x+3.581$.

Let $x = 6$, then

$y = 2.338(6) + 3.581 = 17.609$

When the number of ads is 6, about 18 students fail.

9. a.

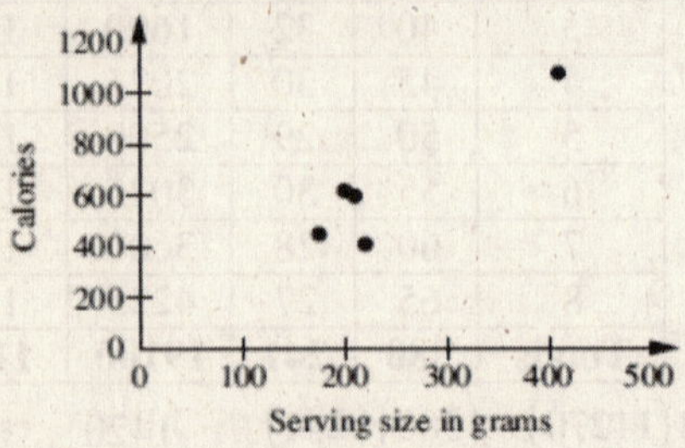

b.

Item	*x*	*y*	x^2	*xy*
1	168	450	28224	75600
2	222	590	49284	130980
3	418	1120	174724	468160
4	209	625	43681	130625
5	231	430	53361	99330
Totals	**1248**	**3215**	**349274**	**904695**

$$m = \frac{5(904695) - (1248)(3215)}{5(349274) - (1248)^2} = \frac{511,155}{188,866} \approx 2.706$$

$$b = \frac{3215 - 2.706(1248)}{5} = -\frac{162.64}{5} \approx -32.5$$

The regression line is $y = 2.706x - 32.5$.

c. Let $x = 300$, then

$y = 2.706(300) - 32.5 = 779.3$

When you consume 300 gm of a Burger King sandwich, you will intake about 779 cal.

11. a.

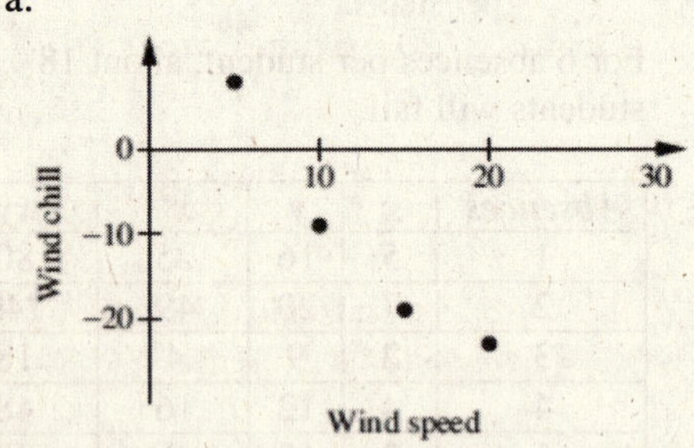

b.

Item	*x*	*y*	x^2	*xy*
1	5	6	25	30
2	10	–9	100	–90
3	15	–18	225	–270
4	20	–24	400	–480
Totals	**50**	**–45**	**750**	**–810**

$$m = \frac{4(-810) - (50)(-45)}{4(750) - (50)^2} = \frac{-990}{500} = -1.98$$

$$b = \frac{-45 - (-1.98)(50)}{4} = \frac{54}{4} = 13.5$$

The regression line is $y = -1.98x + 13.5$.

c. Let $x = 25$, then

$y = -1.98(25) + 13.5 = -36$

When the wind speed is 25 mph, the wind chill is –36° F.

13. a.

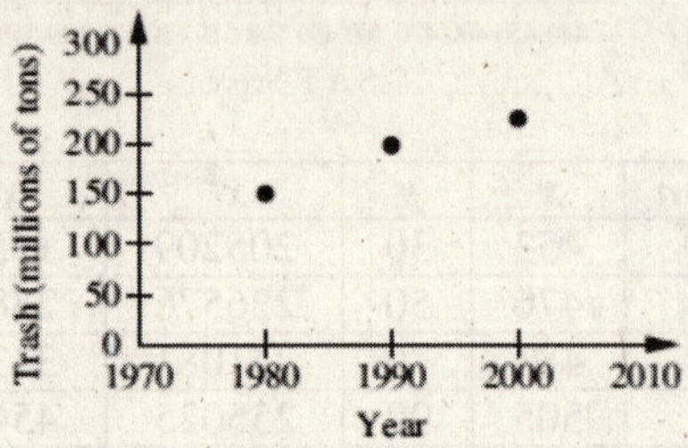

b. Let x be the number of years after 1980.

x	*y*	x^2	*xy*
0	150	0	0
10	200	100	2000
20	220	400	4400
30	**570**	**500**	**6400**

$$m = \frac{3(6400) - (30)(570)}{3(500) - (30)^2} = \frac{2100}{600} = 3.5$$

$$b = \frac{570 - 3.5(30)}{3} = \frac{465}{3} = 155$$

The regression line is $y = 3.5x + 155$.

c. For 2010, let $x = 30$, then

$y = 3.5(30) + 155 = 260$

In 2010, the amount of trash will be 260 million tons.

For 2015, let $x = 35$, then

$y = 3.5(35) + 155 = 277.5$

In 2015, the amount of trash will be 277.5 million tons.

15. a.

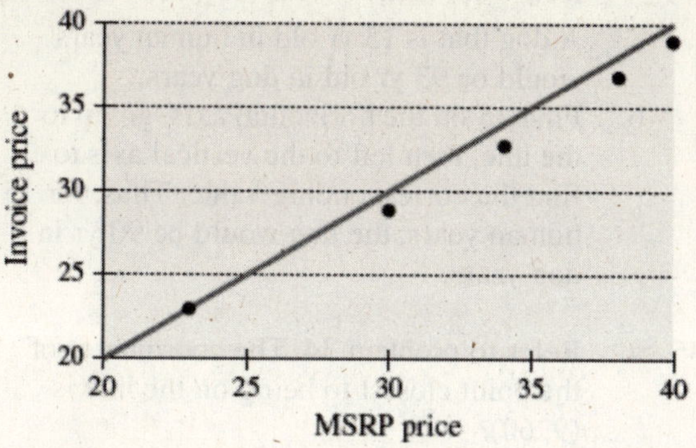

b.

Model	*x*	*y*	x^2	*xy*
1	23	22	529	506
2	30	29	900	870
3	38	36	1444	1368
4	34	33	1156	1122
5	41	39	1681	1599
Totals	**166**	**159**	**5710**	**5465**

$$m = \frac{5(5465)-(166)(159)}{5(5710)-(166)^2} = \frac{931}{994} = 0.94$$

$$b = \frac{159-0.94(166)}{5} = \frac{3.52}{5} \approx 0.7$$

The regression line is $y = 0.94x + 0.7$.

c. For \$30,000 MSRP, let $x = 30$, then
$y = 0.94(30) + 0.7 = 28.9$
For a \$30,000 MSRP, the invoice price will be \$28,900.

17. a.

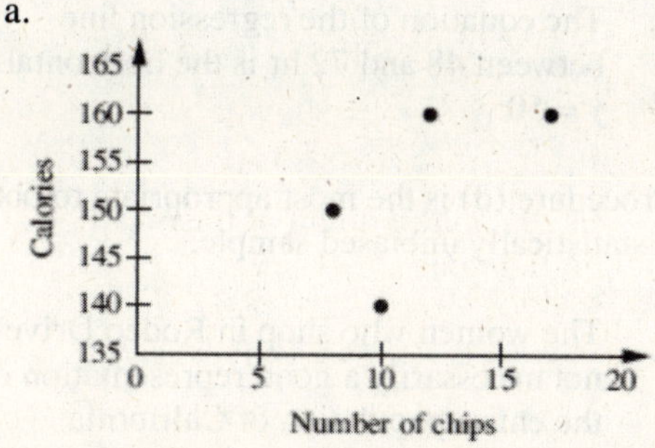

b.

Brand	*x*	*y*	x^2	*xy*
Terra	10	140	100	1400
Taro	10	140	100	1400
Sweet Potato	17	160	289	2720
Glenny's	12	160	144	1920
Sesame	9	150	81	1350
Total	**58**	**750**	**714**	**8790**

$$m = \frac{5(8790)-(58)(750)}{5(714)-(58)^2} = \frac{450}{206} \approx 2.184$$

$$b = \frac{750-2.184(58)}{5} = \frac{623.301}{5} \approx 124.66$$

The regression line is $y = 2.184x + 124.66$.

c. For a serving of 15 chips, let $x = 15$, then
$y = 2.184(15) + 124.66 \approx 157$
For a serving of 15 chips, there will be about 157 cal.

19. $\frac{12}{50} \cdot 3000 = 720$ students would be expected to prefer their hamburgers plain.

21. $\frac{3}{150} \cdot 10,000 = 200$ tires are expected to be defective.

23. $0.727 \cdot 586 \approx 426$ baskets would be expected to be made.

25. a. Punxsutawney Phil: $\frac{10}{15} \cdot 20 \approx 13$ times right

b. Sun Prairie: $\frac{3}{7} \cdot 20 \approx 9$ times right

c. West Orange: $\frac{6}{13} \cdot 20 \approx 9$ times wrong

d. Staten Island: $\frac{1}{7} \cdot 20 \approx 3$ times wrong

e. Lilburn: $\frac{1}{7} \cdot 20 \approx 3$ times wrong

f. Chicago: $\frac{1}{6} \cdot 20 \approx 3$ times right

27. $0.20 \cdot 500 = 100$ liberals and $0.32 \cdot 500 = 160$ conservatives would be expected.

29. a.

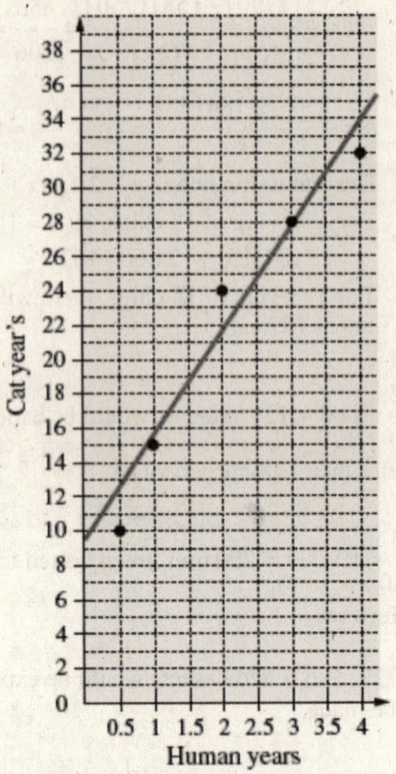

b.

x	y	x^2	xy
0.5	10	0.25	5
1	15	1	15
2	24	4	48
3	28	9	84
4	32	16	128
10.5	**109**	**30.25**	**280**

$$m = \frac{5(280)-(10.5)(109)}{5(30.25)-(10.5)^2} = \frac{255.5}{41} \approx 6.2$$

$$b = \frac{109-6.2(10.5)}{5} = \frac{43.567}{5} \approx 8.7$$

The regression line is $y = 6.2x + 8.7$.

c. For a 2-yr old cat, let $x = 2$, then $y = 6.2(2) + 8.7 = 21.1$

From the formula, a 2-yr old cat would be 21.1 years old.

31. Find 21 on the vertical axis, move to the right to the line; then down to find the corresponding horizontal axis value. The human age 21 corresponds to about 2 yr in human years.

33. a. Use the regression line from problem 32. For dog 15 yr old in human years, let $x = 15$, then $y = 5.4(15) + 12 = 93$.

A dog that is 15 yr old in human years would be 93 yr old in dog years.

b. Find 15 on the horizontal axis, go up to the line, then left to the vertical axis to find the corresponding value. Thus, for 15 human years, the dog would be 90 yr in dog years.

35. a. Refer to problem 34. The coordinates of the point closest to being on the line is (9, 60).

b. Let $x = 9$, then $y = 5.4(9) + 12 = 60.6$

The coordinates are (9, 60.6).

37. a. Since we only have two points, we can find the equation of the line through those points.

$$m = \frac{75-60}{0-36} = -\frac{15}{36} \approx -0.417$$

Using $m = -0.417$ and point (0, 75):

$$y = mx + b$$

$$75 = -0.417(0) + b$$

$$b = 75$$

The regression line is $y = -0.417x + 75$.

b. Let $x = 24$, then

$y = -0.417(24) + 75 = 64.992 \approx 65$ mph

c. From the graph, the wind velocity between 48 and 72 hr after the beginning of the forecast was 50 mph.

d. The equation of the regression line between 48 and 72 hr is the horizontal line $y = 50$.

39. Procedure (d) is the most appropriate to obtain a statistically unbiased sample.

41. a. The women who shop in Rodeo Drive are not necessarily a good representation of the entire population of California.

b. The same can be said of the male population of Berkeley.

c. The same can be said of the people attending an Oakland baseball game.

In general not all elements of the population participate in these activities and, therefore, have zero chance of being selected. The procedures might yield a representative sample, but it is not a random sample because not all elements have an equal chance of selection.

43. The point (0, 19.9) is a y intercept, so the y intercept b of the line is 19.9.

45. The equation of the line is $y = -0.046x + 19.9$.

47. The predicted winning time in the year 2012 using $y = -0.048x + 19.914$ is
$$y = -0.048(8) + 19.914 = 19.53 \text{ sec}.$$

Section 12.7 Scattergrams and Correlation

Study Tips

Sometimes the graph of a set of data will look like a scattered set of points, which explains why we call such a graph a *scattergram.*

Three types of correlation:

Positive: *Negative*: *No correlation*:

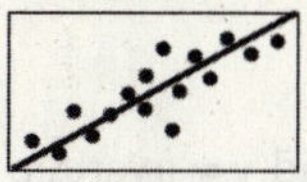
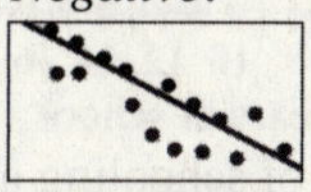
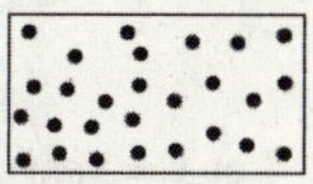

To determine coefficient of correlation r:
Think: measures "goodness" of the linear relationship, use formula

$$r = \frac{n\Sigma(xy) - (\Sigma x)(\Sigma y)}{\sqrt{n(\Sigma x^2) - (\Sigma x)^2} \cdot \sqrt{n\Sigma(y^2) - (\Sigma y)^2}}$$

Note: The value of r is between –1 (perfect negative correlation) and 1 (perfect positive correlation).

Word of Advice
Complete a table of this form to organize your calculations.

x	y	x^2	xy	y^2
Σx	Σy	Σx^2	$\Sigma(xy)$	$\Sigma(y^2)$

To interpret the coefficient of correlation r:
1. Find r for n points of data.
2. Go to line n in Table 12.27.
3. If $|r| >$ number under 95%, then 95% confident that there is a significant linear correlation between the variables.
If $|r| >$ number under 99%, then 99% confident that there is a significant linear correlation between the variables.

12.7 Exercises

1. No. Too many data points lie below the line.

3. Good. Roughly as many points above as below the line.

5. Positive

7. None

9. Negative

11. Positive

13. The value of the linear correlation coefficient r is always (c) between –1 and 1, inclusive.

15. Let $x = 90$, then $y = 10 + 0.9(90) = 91$. The predicted value of Maria's score on the final exam is (c) 91.

17. The least-squares line would predict (a) 1.1 gm of nitrogen oxide per mile driven.

19. A plausible value r would be (c) –0.9.

21. The true statement is (c) There is a positive correlation between the heights of men and women on a date.

23. a.

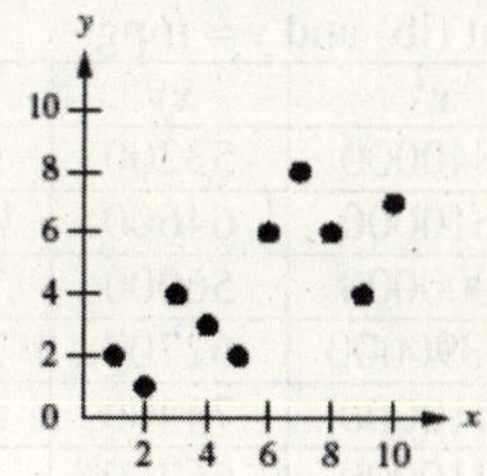

b.

x	y	x^2	xy	y^2
2	1	4	2	1
4	3	16	12	9
7	8	49	56	64
3	4	9	12	16
1	2	1	2	4
5	2	25	10	4
8	6	64	48	36
10	7	100	70	49
6	6	36	36	36
9	4	81	36	16
55	**43**	**385**	**284**	**235**

$$m = \frac{10(284)-(55)(43)}{10(385)-(55)^2} = \frac{475}{825} \approx 0.58$$

$$b = \frac{43-0.58(55)}{10} = \frac{11.333}{10} \approx 1.133$$

The regression line is $y = 0.58x + 1.133$.

c.

$$r = \frac{10(284)-(55)(43)}{\sqrt{10(385)-(55)^2} \cdot \sqrt{10(235)-(43)^2}}$$

$$= \frac{475}{\sqrt{825} \cdot \sqrt{501}} \approx \frac{475}{642.9} \approx 0.7388$$

d. We can be 95% confident that there is a significant positive linear correlation.

25. a.

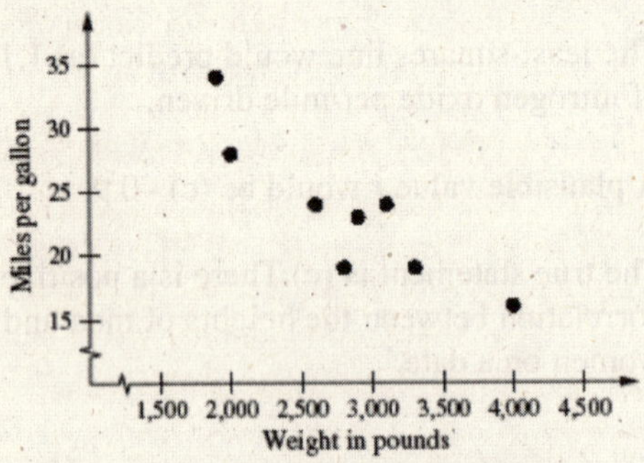

b. Let x = weight (lb) and y = mpg.

x	y	x^2	xy	y^2
2800	19	7840000	53200	361
1900	34	3610000	64600	1156
2000	28	4000000	56000	784
3300	19	10890000	62700	361
3100	24	9610000	74400	576
2900	23	8410000	66700	529
4000	16	1600000	64000	256
2600	24	6760000	62400	576
22600	**187**	**67120000**	**504000**	**4599**

$$m = \frac{8(504,000)-(22,600)(187)}{8(67,120,000)-(22,600)^2}$$

$$= -\frac{194,200}{26,200,000} \approx -0.01$$

$$b = \frac{187-(-0.01)(22,600)}{8} \approx \frac{354.52}{8} \approx 44.315$$

The regression line is $y = -0.01x + 44.315$.

c.

$$r = \frac{8(504000)-(22600)(187)}{\sqrt{8(67120000)-(22600)^2} \cdot \sqrt{8(4599)-(187)^2}}$$

$$= -\frac{194,200}{\sqrt{26,200,000} \cdot \sqrt{1823}}$$

$$\approx -\frac{194,200}{218,546.56} \approx -0.889$$

d. Since $|-0.889| > 0.834$, we can be 99% confident that there is a significant negative linear correlation.

27. a.

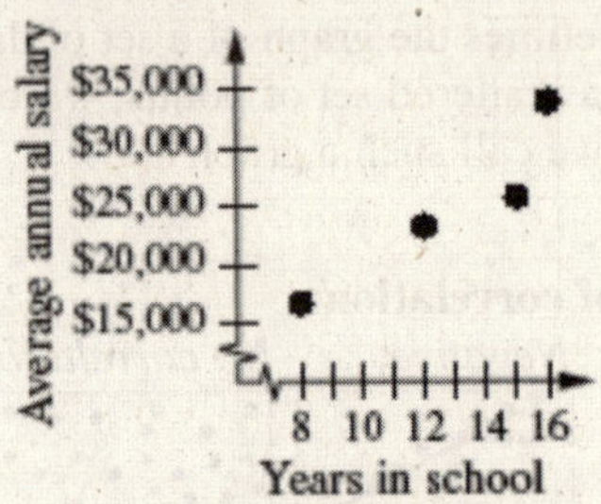

b. Let x = years of schooling and y = average salary in thousands.

x	y	x^2	xy	y^2
8	16.8	64	134.4	282.24
12	23.3	144	279.6	542.89
15	25.8	225	387	665.64
16	33.9	256	542.4	1149.21
51	**99.8**	**689**	**1343.4**	**2639.98**

$$m = \frac{4(1343.4)-(51)(99.8)}{4(689)-(51)^2} = \frac{283.8}{155} \approx 1.83$$

$$b = \frac{99.8-1.83(51)}{4} \approx \frac{6.421}{4} \approx 1.605$$

The regression line is $y = 1.83x + 1.605$ thousands.

c.

$$r = \frac{4(1343.4)-(51)(99.8)}{\sqrt{4(689)-(51)^2} \cdot \sqrt{4(2639.98)-(99.8)^2}}$$

$$= \frac{283.8}{\sqrt{155} \cdot \sqrt{599.88}} \approx \frac{283.8}{304.929} \approx 0.9307$$

d. Since $|0.9307| < 0.950$, we cannot be certain that there is a significant positive linear correlation at the 95% or 99% confidence level.

e. Let $x = 10$, then

$y = 1.83(10) + 1.605 = 19.905$

The predicted salary would be \$19,905.

29\. a.

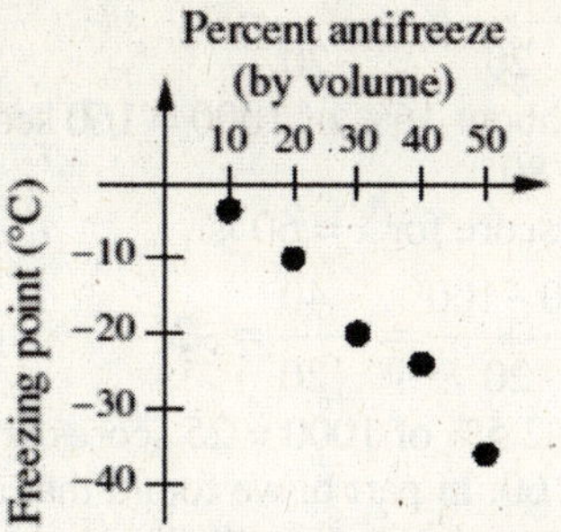

b. Let x = percent of antifreeze and y = freezing point (degrees C).

x	y	x^2	xy	y^2
10	−4	100	−40	16
20	−10	400	−200	100
30	−20	900	−600	400
40	−24	1600	−960	576
50	−36	2500	−1800	1296
150	**−94**	**5500**	**−3600**	**2388**

$$m=\frac{5(-3600)-(150)(-94)}{5(5500)-(150)^2}=-\frac{3900}{5000}=-0.78$$

$$b=\frac{-94-(-0.78)(150)}{5}=\frac{23}{5}=4.6$$

The regression line is $y=-0.78x+4.6$.

c.

$$r=\frac{5(-3600)-(150)(-94)}{\sqrt{5(5500)-(150)^2}\cdot\sqrt{5(2388)-(-94)^2}}$$

$$=-\frac{3900}{\sqrt{5000}\cdot\sqrt{3104}}\approx-\frac{3900}{3939.54}\approx-0.99$$

d. Since $|-0.99|>0.959$, we can be 99% confident that there is a significant negative linear correlation.

e. Let $x = 25$, then

$y=-0.78(25)+4.6=-14.9$

The predicted freezing point would be −14.9°C.

31\. a. There was no outage during the 7th week.

b. The greatest number of outages occurred during the 11th week.

c. A decline in the number of outages seemed to start after the 2nd week.

d. No correlation is shown by the chart.

33\. a. No, correlation does not mean that one variable causes the other.

b. Answers may vary. It is possible that the number of cars with air bags being sold is increasing at a rate that makes them representative of the entire population of cars.

35\. Using a grapher, $r\approx 0.2236$. We cannot be certain that there is a significant linear correlation.

37\. No. inclusion of Ireland makes it appear as though there is no linear correlation.

Chapter 12 Practice Test

Study Tips

There are so many items in this chapter that you should compare your notes with the summary to make sure you understand the terminology and the procedures and formulas that are used to solve the corresponding types of problems. Since this manual gives step-by-step procedures in solving the Exercise Problems, compare your answers with the ones here and be sure you find out how to correct any errors you make.

Once you have reviewed the procedures, take this practice test. Carefully check that you are doing each problem correctly. Here are the solutions to the practice test.

1\. No. Not all members of the student body had the same chance of being chosen (only those in the English Department).

2\.

Scores	*Tally Marks*	*Frequency*
$60<s\le 65$	||	2
$65<s\le 70$	|||	3
$70<s\le 75$	|||	3
$75<s\le 80$	||	2
$80<s\le 85$	|||	3
$85<s\le 90$	卌 ||	7
$90<s\le 95$	|||	3
$95<s\le 100$	||	2

3. a. – b.

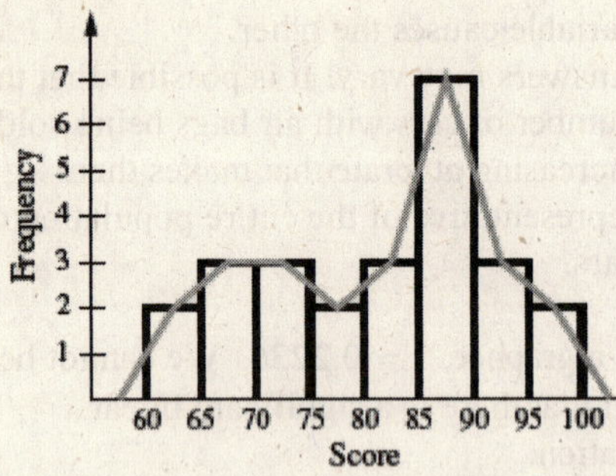

4. a. The mean =

$$\frac{69+70+71+73+78+82+82}{7}=\frac{525}{7}=75°\text{F}.$$

b. The mode is 82°F.

c. Order the numbers: 69, 70, 71, 73, 78, 82, 82.
The median is the 4th term: 73°F

5. a. Range = 82 – 69 = 13°F

b.

x	$x-\bar{x}$	$(x-\bar{x})^2$
78	3	9
82	7	49
82	7	49
71	–4	16
69	–6	36
73	–2	4
70	–5	25
525		**188**
$\bar{x}=\frac{525}{7}=75°\text{F}$		$\frac{188}{6}\approx 31.333$
		$s=\sqrt{31.333}\approx 5.60°\text{F}$

6. We are given $\mu=128$ and $\sigma=8$.

a. Almost all the measurements fall between $128-3(8)=104$ and $128+3(8)=152$.

b. Find the z-score for values of x = 112.

$$z=\frac{112-128}{8}=-\frac{16}{8}=-2$$

The probability associated with 2 standard deviations below the mean is 0.5 – 0.475 = 0.025 = 2.5%.

7. Given $\mu=100$ and $\sigma=20$.

a. The z-score for x = 140 is

$$z=\frac{140-100}{20}=\frac{40}{20}=2$$

Thus, about 2.5% of 1000 = 25 scores are above 140.

b. The z-score for x = 80 is

$$z=\frac{80-100}{20}=-\frac{20}{20}=-1$$

Thus, about 16% of 1000 = 160 scores are below 80.

c. The z-score for x = 60 is

$$z=\frac{60-100}{20}=-\frac{40}{20}=-2$$

About 2.5% of 1000 = 25 scores are below 60. In part b, we found that about 160 scores are below 80. Thus, about 160 – 25 = 135 scores are between 60 and 80.

8. Given $\mu=10$ lb and $\sigma=1$ lb.

a. The z-score for x = 10 is

$$z=\frac{10-10}{1}=\frac{0}{1}=0$$

The probability is that 0.5, or 50%, of the lines has a breaking point of more than 10 lb.

b. The z-score for x = 8 is

$$z=\frac{8-10}{1}=-\frac{2}{1}=-2$$

The probability is that 0.025, or 2.5%, of the lines has a breaking point of less than 8 lb.

9. Given $\mu=50$ and $\sigma=5$.

a. $z=\frac{58-50}{5}=\frac{8}{5}=1.6$

b. $z=\frac{62-50}{5}=\frac{12}{5}=2.4$

10. For the French test, the z-score is $z=\frac{88-76}{18}\approx 0.667$ For the psychology test, the z-score is $z=\frac{90-80}{16}=\frac{10}{16}=0.625$. Thus, the student who scored 88 on the French test has the better score relative to the test.

11. Given $\mu=50$ and $\sigma=5$.

For x = 50: $z=\frac{50-50}{5}=\frac{0}{5}=0.$

For x = 62: $z=\frac{62-50}{5}=\frac{12}{5}=2.4.$

From Table II, we find the probability that a randomly selected score is between the mean and 2.4 standard deviation above the mean to be 0.492.

12. a.

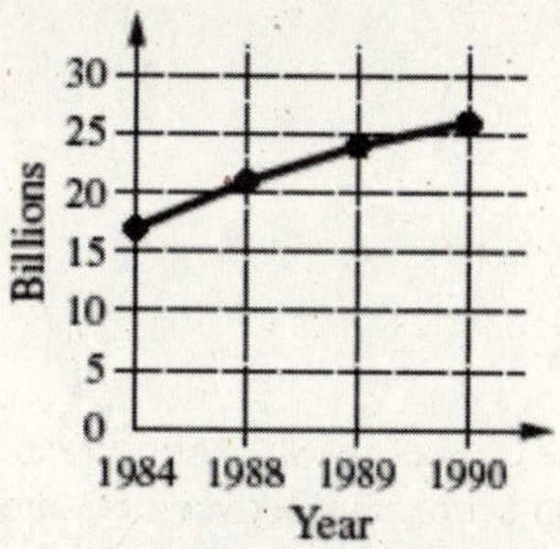

b. In 1986 about \$19 billion was spent on education.

13.

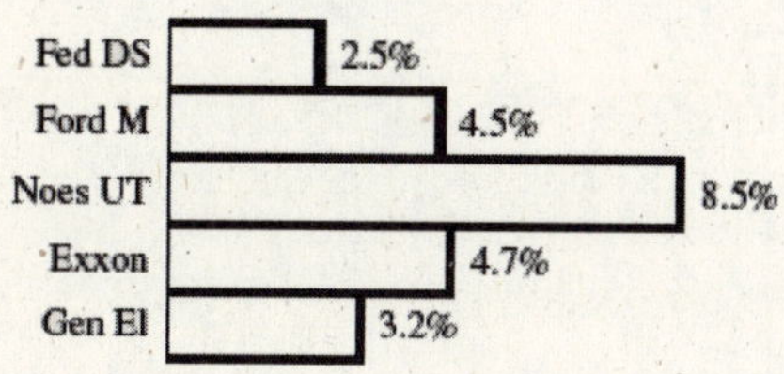

14.

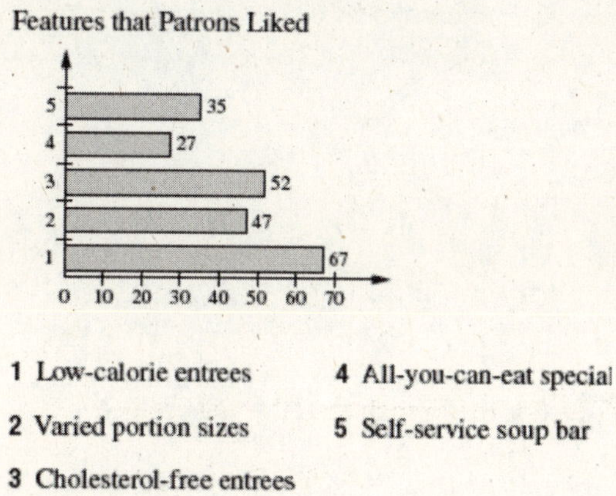

15. The percentage increase is

$$\frac{15-10}{10}=\frac{5}{10}=0.5=50\%.$$

16.

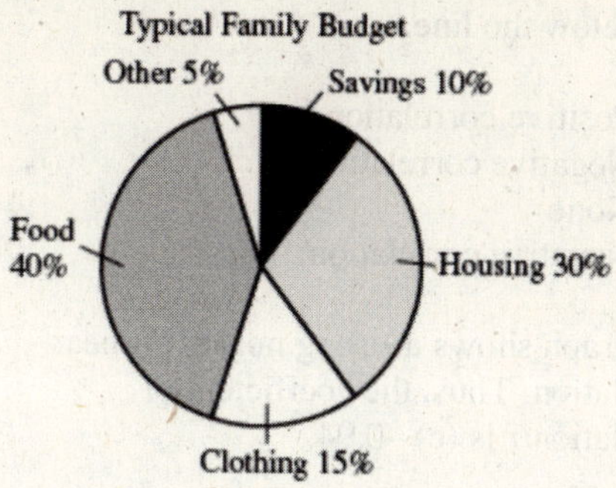

17. For $x = 6$, $y \approx 0$.

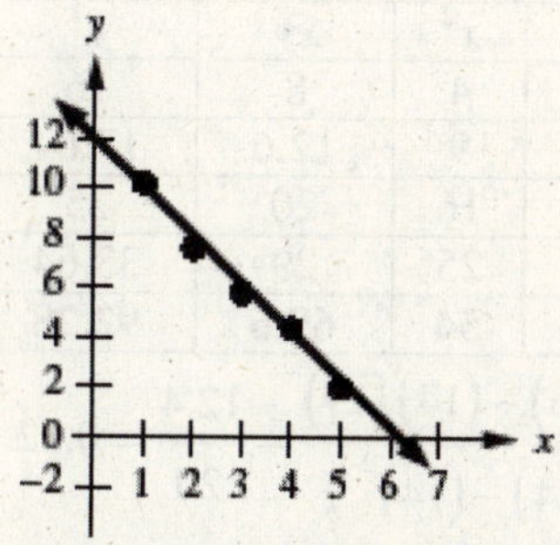

18.

x	y	x^2	xy
1	10.2	1	10.2
2	7.6	4	15.2
3	5.8	9	17.4
4	4.4	16	17.6
5	2.0	25	10
15	**30**	**55**	**70.4**

$$m=\frac{5(70.4)-(15)(30)}{5(55)-(15)^2}=-\frac{98}{50}=-1.96$$

$$b=\frac{30-(-1.96)(15)}{5}=\frac{59.4}{5}=11.88$$

The regression line is $y=-1.96x+11.88$.

c. Let $x = 6$, then

$y=-1.96(6)+11.88=0.12$

19. $\frac{3}{150}\cdot 10,000=200$ tires are expected to be defective.

20. $\frac{4000}{50,000}\cdot 100=8$ students should be expected to have gotten an excellent rating on this test.

21. The best representation for the data is (b). The slope follows the direction of the line and has

about the same number of points above and below the line.

22. a. Positive correlation
    b. Negative correlation
    c. None
    d. Negative correlation

23. The graph shows a strong negative linear correlation. Thus, the coefficient of correlation r is (c) –0.94.

24. Let x = the year and y = the product sales (in billions of dollars).

x	y	x^2	xy	y^2
2	4	4	8	16
3	4.2	9	12.6	17.64
4	5	16	20	25
5	5.8	25	29	33.64
14	**19**	**54**	**69.6**	**92.28**

$$m = \frac{4(69.6)-(14)(19)}{4(54)-(14)^2} = \frac{12.4}{20} = 0.62$$

$$b = \frac{19-0.62(14)}{4} = \frac{10.32}{4} = 2.58$$

The regression line is $y = 0.62x + 2.58$.

25. Refer to table in problem 24.

$$r = \frac{4(69.6)-(14)(19)}{\sqrt{4(54)-(14)^2}\cdot\sqrt{4(92.28)-(19)^2}}$$

$$= \frac{12.4}{\sqrt{20}\cdot\sqrt{8.12}} \approx \frac{12.4}{12.743} \approx 0.97$$

We can be 95% confident that there is a significant correlation between the year x and the sales y.

Chapter 13 Your Money and Your Math

Section 13.1 Interest, Taxes, and Discounts

Study Tips

In order to be financially wise, it is helpful to understand the difference between simple and compound interest. If you are borrowing money, simple interest is best but if you are collecting interest you will want compound interest. The examples in the text give applications to loans, taxes and discounts.

Variables and their meanings:

P Principal, initial amount of money
r Rate, express as a decimal
t Time, express in years
A Final amount, amount after interest has been added to the principal
S Sales tax, result of tax rate times item bought, then added to your purchase
d Discount, similar to interest but subtracted
n Total number of periods that principal is compounded
m Number of times compounded in a year
APY (Effective rate), use formulas
k Number of compoundings per year in APY formula
e **Not** a variable but the number, ≈ 2.71828

To calculate simple interest:
Think: need to know P, r, and t
Formula: $I = Prt$
Note: Based on the information given, this formula can be used to find principal: $P = \frac{I}{rt}$; rate: $r = \frac{I}{Pt}$; and time: $t = \frac{I}{Pr}$

To calculate final amount:
Think: need to know P and I
Formula: $A = P + I$

To calculate future value A_n for compounding n periods:
Think: need to know P, r, and n
Formula: $A_n = P(1+r)^n$
Note: Use this formula when you are compounding annually.

To calculate future value A_n for compound interest:
Think: need to know P, r, n, and m
Formula: $A_n = P\left(1+\frac{r}{m}\right)^n$
Note: Use this formula when you are compounding m times per year, such as monthly (12), weekly (26), quarterly (4), and semiannually (2).

To calculate future value A_n for continuous compounding interest:
Think: need to know P, r, and t
Formula: $A_n = Pe^{rt}$
Note: The text uses n to represent t, time in years. For more accurate results, use the calculator key e instead of the rounded 2.71828.

To calculate APY (effective rate):
Think: need to know r and k
Formula: $\text{APY} = \left(1+\frac{r}{k}\right)^k - 1$
Note: For continuous compounding, you only need to know r, then use the formula $\text{APY} = e^r - 1$.

Caution
Keep in mind that the time t must be in years, so if we are dealing with quarterly interest, then $t = \frac{4}{12}$.

Hint
If you do not have a calculator, you may use Table 13.2 as explained in Example 4. If you can use a calculator, familiarize yourself with its functions, from a y^x key to full calculation menus.

13.1 Exercises

1. a. $I = Prt = 3000(0.08)(1) = \$240$
   b. $I = Prt = 3000(0.04)(1) = \$120$

3. a. $I = Prt = 2000(0.09)(3) = \$540$
   b. $I = Prt = 2000(0.045)(3) = \$270$

5. a. $I = Prt = 4000(0.10)\left(\frac{1}{2}\right) = \$200$
   b. $I = Prt = 4000(0.035)\left(\frac{1}{2}\right) = \$70$

7. a. $I = Prt = 2500(0.10)\left(\frac{1}{4}\right) = \$62.50$

b. $I = Prt = 2500(0.025)\left(\frac{1}{4}\right) = \$15.63$

9. a. $I = Prt = 16{,}000(0.07)\left(\frac{5}{12}\right)$

$= \$466.67$

b. $I = Prt = 16{,}000(0.03)\left(\frac{5}{12}\right) = \$200$

11. a. Sales tax $= 0.06(40.20) = \$2.41$

b. $40.20 + 2.41 = \$42.61$

13. a. FICA tax $= 0.0565(24{,}000) = \$1356$

b. FICA tax $= 0.0765(24{,}000) = \$1836$

15. Maybelle's estimated tax was

$850 + 0.15(25{,}850 - 8500)$

$= 850 + 0.15(17{,}350)$

$= 850 + 2602.50 = \$3452.50$

17. a. Discount $= 0.20(200) = \$40$

b. $200 - 40 = \$160$

c. $160 + 0.05(160) = 160 + 8 = \$168$

19. a. Discount $= 0.25(500) = \$125$

b. $500 - 125 = \$375$

21. a. Discount $= 0.05(28.40) = \$1.42$

b. $28.40 - 1.42 = \$26.98$

NOTE: Problems 23 – 31 can be done using a calculator and the formula: $A_n = P\left(1 + \frac{r}{m}\right)^n$, where P is the principal, r the annual interest rate, m the number of compoundings per year and n the number of periods.

23. a. Amount = (\$100)(1.5938) = \$159.38
Interest = \$159.38 – \$100 = \$59.38
With a calculator, rounded to hundreths:

$A_n = 100(1 + 0.06)^8 = \$159.38$

$I = 159.38 - 100 = \$59.38$

b. Amount = (\$100)(1.3686) = \$136.86
Interest = 136.86 – 100 = \$36.86
With a calculator, rounded to hundreths:

$A_n = 100(1 + 0.04)^8 = \$136.86$

$I = 136.86 - 100 = \$36.86$

25. a. Amount = (\$2580)(2.7731) = \$7154.60
Interest = 7154.60 – 2580 = \$4574.60
With a calculator, rounded to hundreths:

$A_n = 2580(1 + 0.12)^9 = \$7154.54$

$I = 7154.54 - 2580 = \$4574.54$

b. Amount = (\$2580)(1.1951) = \$3083.30
Interest = 3083.30 – 2580 = \$503.30
With a calculator, rounded to hundreths:

$A_n = 2580(1 + 0.02)^9 = \$3083.34$

$I = 3083.34 - 2580 = \$503.34$

27. a. Amount = (\$12,000)(2.1829) = \$26,195
Interest = 26,195 – 12,000 = \$14,195
With a calculator:

$A_n = 12{,}000\left(1 + \frac{0.10}{2}\right)^{16} = \$26{,}195$

$I = 26{,}195 - 12{,}000 = \$14{,}195$

b. Amount = (\$12,000)(1.3728) = \$16,473
Interest = 16,473 – 12,000 = \$4473
With a calculator:

$A_n = 12{,}000\left(1 + \frac{0.04}{2}\right)^{16} = \$16{,}473$

$I = 16{,}473 - 12{,}000 = \$4473$

29. a. Amount = (\$20,000)(1.2682) = \$25,364
Interest = \$25,364 – \$20,000 = \$5364
With a calculator:

$A_n = 20{,}000\left(1 + \frac{0.08}{4}\right)^{12} = \$25{,}365$

$I = 25{,}365 - 20{,}000 = \$5365$

b. Amount = (\$20,000)(1.0615)
= \$21,230
Interest = \$21,230 – \$20,000 = \$1230
With a calculator:

$A_n = 20{,}000\left(1 + \frac{0.02}{4}\right)^{12} = \$21{,}234$

$I = 21{,}234 - 20{,}000 = \$1234$

31. a. Amt = ($40,000)(97.0172)
= $3,880,688
Int = $3,880,688 – $40,000
= $3,840,688
With a calculator:

$$A_n = 40,000\left(1+\frac{0.20}{2}\right)^{48} = \$3,880,689$$

$$I = 3,880,689 - 40,000 = \$3,840,689$$

b. Amt = ($40,000)(2.5871) = $103,484
Int = $103,484 – $40,000 = $63,484
With a calculator:

$$A_n = 40,000\left(1+\frac{0.04}{2}\right)^{48} = \$103,483$$

$$I = 103,483 - 40,000 = \$63,483$$

33. a. Amt = ($1000)(2.8543) = $2854.30
With a calculator, rounded to hundreths:

$$A_n = 1000(1+0.06)^{18} = \$2854.34$$

b. Amount = (1000)(2.8983) = $2898.30
With a calculator, rounded to hundreths:

$$A_n = 1000\left(1+\frac{0.06}{2}\right)^{36} = \$2898.28$$

35. a. Amount = ($3000)(1.3310) = $3993
b. Interest = 3993 – 3000 = $993
With a calculator:

$$A_n = 3000(1+0.10)^3 = \$3993$$

Interest = 3993 – 3000 = $993

37. ($1000)(1.8061 – 1.7908)
= $1000(0.0153) = $15.30
With a calculator, rounded to hundreths:

$$1000\left(1.03^{20} - 1.06^{10}\right) = \$15.26$$

39. $A = Pe^{rt}$, where $P = 2000$, $r = 0.08$ and $t = \frac{1}{2} = 0.5$.

$$A = 2000e^{0.08(0.5)} = 2000e^{0.04}$$

$$= 2000(1.0408108) = \$2081.62$$

41. $A = P\left(1+\frac{r}{m}\right)^n$, where $P = \$10,000$, $r = 0.11$, $m = 2$, $n = 4$

$$A = 10,000\left(1+\frac{0.11}{2}\right)^4 = \$12,388.25$$

$A = Pe^{rt}$, where $P = 10,000$, $r = 0.105$ and $t = 2$

$$A = 10,000e^{0.105(2)} = \$12,336.78$$

The first investment of 11% semiannually yields 12,388.25 – 12,336.78 = $51.47 more.

43. $r = 0.02$, $k = 360$

$$\text{APY} = \left(1+\frac{r}{k}\right)^k - 1 = \left(1+\frac{0.02}{360}\right)^{360} - 1$$

$$= (1.00006)^{360} - 1 \approx 0.02020 = 2.020\%$$

45. $r = 0.06$, $k = 2$

$$\text{APY} = \left(1+\frac{r}{k}\right)^k - 1 = \left(1+\frac{0.06}{2}\right)^2 - 1$$

$$= (1.03)^2 - 1 = 0.0609 = 6.09\%$$

47. $r = 0.08$, $k = 4$

$$\text{APY} = \left(1+\frac{r}{k}\right)^k - 1 = \left(1+\frac{0.08}{4}\right)^4 - 1$$

$$= (1.02)^4 - 1 = 0.0824 = 8.24\%$$

49. $r = 0.09$, $k = 4$

$$\text{APY} = \left(1+\frac{r}{k}\right)^k - 1 = \left(1+\frac{0.09}{4}\right)^4 - 1$$

$$= (1.0225)^4 - 1 \approx 0.0931 = 9.31\%$$

51. $r = 0.18$, $k = 12$

$$\text{APY} = \left(1+\frac{r}{k}\right)^k - 1 = \left(1+\frac{0.18}{12}\right)^{12} - 1$$

$$= (1.015)^{12} - 1 \approx 0.1956 = 19.56\%$$

53. Simple interest means that the interest does not earn additional interest, while compound interest means that the interest earns interest at the same rate as the original principal. For instance, $100 at 10% simple interest would earn $20 in two years, while $100 at compound interest would earn $21 in two years.

55. $r = 0.0398, k = 12$

$$\text{APY} = \left(1+\frac{r}{k}\right)^k - 1 = \left(1+\frac{0.0398}{12}\right)^{12} - 1$$
$$= (1.003317)^{12} - 1 \approx 0.0405 = 4.05\%,$$

instead of 4.06%.

57. $r = 0.0390, k = 52$

$$\text{APY} = \left(1+\frac{r}{k}\right)^k - 1 = \left(1+\frac{0.039}{2}\right)^2 - 1$$
$$= (1.0195)^2 - 1 \approx 0.03938 = 3.94\%,$$

instead of 3.98%.

Section 13.2 Credit Cards and Consumer Credit

Study Tips

If you have a credit card or are planning to get one, this is a "must" section. The examples show applications to finance charges, interest and other fees that you have to pay on credit card accounts.

Terms to know:

Interest rate: percent you pay on the card balance

Annual fee: amount paid for having the card

Grace period: interest-free period

Fixed or variable interest rate

Prime rate: interest available to best customers

Finance charge: amount to pay if you pay for purchases past the grace period

Revolving charge accounts: most charge accounts at department stores

Add-on-interest: interest added to puchase and made part of payment

Note

Revolving charge account finance charges and payments can vary from store to store. Most major credit card companies use the average balance method to determine payments and finance charges.

13.2 Exercises

1. Since the couple's combined gross income must be at least \$40,000, Susan must earn at least \$15,000 per year for them to qualify.

3. $1\frac{1}{2}\%$ of (\$100 − \$10) = (0.015)(\$90) = \$1.35
   New balance = \$90 + \$1.35 + \$50 = \$141.35

5. $1\frac{1}{2}\%$ of (\$134.39 − \$25) = (0.015)(\$109.39) = \$1.64
   New balance = \$109.39 + \$1.64 + \$73.98 = \$185.01

7. $1\frac{1}{2}\%$ of (\$378.93 − \$75) = (0.015)(\$303.93) = \$4.56
   New balance = \$303.93 + \$4.56 + \$248.99 = \$557.48

9. a. $1\frac{1}{2}\%$ of \$85 = (0.015)(\$85) = \$1.28
   b. \$85 + \$1.28 + \$150 = \$236.28
   c. 5% of \$236.28 = (0.05)(\$236.28) = \$11.81

11. a. $1\frac{1}{2}\%$ of \$344 = (0.015)(\$344) = \$5.16
    b. \$344 + \$5.16 + \$60 = \$409.16
    c. 5% of \$409.16 = (0.05)(\$409.16) = \$20.46

13. a. $1\frac{1}{2}\%$ of \$80.45 = (0.015)(\$80.45) = \$1.21
    b. \$80.45 + \$1.21 + \$98.73 = \$180.39
    c. The new balance is under \$200, so the minimum payment is \$10.

15. a. $1\frac{1}{2}\%$ of \$55.90 = (0.015)(\$55.90) = \$0.84
    b. \$55.90 + \$0.84 + \$35.99 = \$92.73
    c. The new balance is under \$200, so the minimum payment is \$10.

17. a. $1\frac{1}{2}\%$ of \$34.76 = (0.015)(\$34.76) = \$0.52
    b. \$34.76 + \$0.52 + \$87.53 = \$122.81
    c. The new balance is under \$200, so the minimum payment is \$10.

19. Finance charge = $1\frac{1}{2}\%$ of \$190 = (0.015)(\$190) = \$2.85

21. Finance charge = $1\frac{1}{2}\%$ of \$90 = (0.015)(\$90) = \$1.35

23. a. Balance = $9000 – $1600 = $7400. Add-on interest is ($7400)(0.09)(3) = $1998.
    b. Total to be paid in 36 months is $7400 + $1998 = $9398. $\frac{9398}{36} = 261.06$, so the monthly payment (to the nearest dollar) is $261.
    c. Use **http://www.onlineloadcalculator.org** Total interest paid is $1071.45. The monthly payment is $235.32.

25. a. Balance = $2400 – $400 = $2000. Add-on interest is ($2000)(0.15)(1.5) = $450.
    b. Total to be paid in 18 months is $2000 +$450 = $2450. Since $\frac{2450}{18} = 136.11$, the monthly payment (to the nearest dollar) is $136.

27. a. Balance = $500 – $100 = $400. Add-on interest is ($400)(0.10)(1.5) = $60.
    b. Total to be paid in 18 months is $400 + $60 = $460. Since $\frac{460}{18} = 25.56$, the monthly payment (to the nearest dollar) is $26.

29. This procedure gives the longest possible time between the purchase date and the date when payment must be made to avoid a finance charge.

31. ($47)(24) = $1128.
    Interest = $1128 – $1000 = $128. Since this is for 2 years, the interest rate is $\frac{\$128}{(\$1000)(2)} = 0.064 = 6.4\%$.

33. ($50)(24) = $1200.
    Interest = $1200 – $1000 = $200. Since this is for 2 years, the interest rate is $\frac{\$200}{(\$1000)(2)} = 0.10 = 10\%$.

Section 13.3 Annual Percentage Rate (APR) and the Rule of 78

Study Tips

Different types of credit accounts will often have different rates. How do you make a decision as to which is the best for you? By using the true annual percentage rate called the APR.

To find the APR of a loan:
Think: determine the actual annual rate of interest based on what you actually pay
1. Multiply the payment and the number of payments to be made.
2. Subtract the amount financed to find the finance charge.
3. Divide the finance charge by the amount financed, then multiply by 100.
4. Use Table 13.3 to find the APR.

Hint
Example 1 tells you how to use an APR table, a portion of which appears as Table 13.3.

To calculate a refund using the Rule of 78:
Think: determine unearned interest, use $u = \frac{r(r+1)}{n(n+1)} \times F$, where r is the rate, n is the number of payments, and F is the finance charge
Note: The refund can also be calculated using the formula $u = \frac{r \cdot \text{PMT} \cdot V}{100 + V}$, where V is the value from the APR table.

To calculate the payoff using the Rule of 78:
Think: subtracting unearned interest from remaining payments times amount, use
Payoff $= r \cdot \text{PMT} - u$
Note: The payoff can also be calculated using the formula Payoff $= \text{PMT} \cdot \left[\frac{1-(1+i)^{-r}}{i}\right]$, where r is the rate and $i = \frac{\text{APR}}{12}$.

13.3 Exercises

1. $\frac{194}{2500} \cdot 100 = 7.76$ = finance charge per $100.
   The nearest table entry for 12 payments is 7.74 corresponding to an APR of 14%.

3. $\frac{264}{1500}\cdot 100 = 17.60$ = finance charge per \$100. The nearest table entry for 24 payments is 17.51 corresponding to 16%.

5. $\frac{210}{1500}\cdot 100 = 14$ = finance charge per \$100. The nearest table entry for 18 payments is 13.99 corresponding to 17%.

7. $\frac{1570}{4500}\cdot 100 = 34.89$ = finance charge per \$100. The nearest table entry for 48 payments is 34.81 corresponding to $15\frac{1}{2}\%$.

9. $\frac{1800}{5000}\cdot 100 = 36$ = finance charge per \$100. The nearest table entry for 48 payments is 36.03 corresponding to 16%.

11. a. Here, $r = 4$ and $n = 12$, so the unearned finance charge is
$$u = \frac{4\cdot 5}{12\cdot 13}\cdot \$15.60 = \frac{5}{39}\cdot \$15.60 = \$2.$$
b. There are 4 payments remaining, so it will take $4\cdot \$25 - \$2 = \$98$ to pay off the loan.

13. a. Here, $r = 6$ and $n = 12$, so the unearned finance charge is
$$u = \frac{6\cdot 7}{12\cdot 13}\cdot \$31.20 = \frac{7}{26}\cdot \$31.20 = \$8.40.$$
b. There are 6 payments remaining, so it will take $6\cdot \$45 - \$8.40 = \$261.60$ to pay off the loan.

15. a. Here, $r = 5$ and $n = 20$, so the unearned finance charge is
$$u = \frac{5\cdot 6}{20\cdot 21}\cdot \$58.50 = \frac{1}{14}\cdot \$58.50 = \$4.18.$$
b. There are 5 payments remaining, so it will take $5\cdot \$10 - \$4.18 = \$45.82$ to pay off the loan.

17. a. $$u = \frac{6\cdot 7}{24\cdot 25}\cdot \$264 = \frac{7}{100}\cdot \$264 = \$18.48$$

b. $$\text{PMT} = \frac{1500 + 264}{24} = \$73.50$$
V = \$4.72 (Table 13.3 - corresponding to 6 and 16% from APR in problem 3)
$$u = \frac{6(73.50)(4.72)}{100 + 4.72} = \$19.88$$

c. $\text{Payoff} = 6(73.50) - 18.48 = \$422.52$

d. $$\text{Payoff} = 73.50\left[\frac{1-\left(1+\frac{0.16}{12}\right)^{-6}}{\frac{0.16}{12}}\right] = \$421.13$$

19. a. $$u = \frac{12\cdot 13}{48\cdot 49}\cdot 1570 = \frac{13}{196}\cdot 1570 = \$104.13$$

b. $$\text{PMT} = \frac{4500 + 1570}{48} = \$126.46$$
V = \$8.59 (Table 13.3 - corresponding to 12 and 15.5% from APR in problem 7)
$$u = \frac{12(126.46)(8.59)}{100 + 8.59} = \$120.04$$

c. $\text{Payoff} = 12(126.46) - 104.13 = \$1413.39$

d. $$\text{Payoff} = 126.46\left[\frac{1-\left(1+\frac{0.155}{12}\right)^{-12}}{\frac{0.155}{12}}\right] = \$1397.43$$

21. a. Finance charge $= 12\cdot 27 - 300 = \$24.$

b. Finance charge per \$100 is $\frac{24}{300}\cdot 100 = \$8.$ The nearest table entry for 12 payments is \$8.03 corresponding to $14\frac{1}{2}\%$.

23. a. Finance charge $= 18\cdot 63 - 1000 = \$134.$

b. Finance charge per \$100 is $\frac{134}{1000}\cdot 100 = \$13.40.$ The nearest table entry for 18 payments is \$13.57 corresponding to $16\frac{1}{2}\%$.

25. a. Here, $r = 4$ and $n = 12$, so the unearned finance charge is
$\frac{4 \cdot 5}{12 \cdot 13} \cdot 31.20 = \frac{5}{39} \cdot 31.20 = \$4.$
b. There are 4 payments remaining, so it will take $4 \cdot 50 - 4 = \$196$ to pay off the loan.

27. a. Finance charge $= 0.10(800 - 80) = 0.10(720) = \$72.$
b. Monthly payment $= \frac{720 + 72}{12} = \$66$
c. Here $r = 4$ and $n = 12$, so the interest refund is
$\frac{4 \cdot 5}{12 \cdot 13} \cdot 72 = \frac{5}{39} \cdot 72 = \$9.23.$
d. There are 4 payments remaining, so it will take $4 \cdot 66 - 9.23 = \$254.77$ to pay off the loan.

29. a. Finance charge $= 0.10(1000 - 200)(1.5) = \$120.$
b. Monthly payment $= \frac{800 + 120}{18} = \$51.11.$
c. Here, $r = 3$ and $n = 18$, so the interest refund is
$\frac{3 \cdot 4}{18 \cdot 19} \cdot 120 = \frac{2}{57} \cdot 120 = \$4.21.$
d. It will take $3 \cdot 51.11 - 4.21 = \$149.12$ to pay off the loan.

For problems 31 – 35, use the links given in the text. Some answers are rounded to the nearest dollar.

31. a. If the interest rate is 7%, the monthly payment \$562.
b. If the interest rate is 6%, the monthly payment \$554.

33. a. When the interest rate is 4%, the total cost to purchase the car is \$22,344.
b. When the interest rate is 5%, the total cost to purchase the car is \$22,637.
c. When the interest rate is 6%, the total cost to purchase the car is \$22,932.92, or \$22,933.
d. We see that with each 1% increase, the differences are \$22,932 – \$22,637 = \$295 and \$22,637 – \$22,344 = \$293. The total cost to purchased when the rate increases by 1% is about \$300.

35. The loan payment would be \$632.99 and the net cost of thé loan is \$10,423.67.

37. $\text{APR} = \frac{2mI}{P(n+1)} = \frac{2 \cdot 12 \cdot 194}{2500 \cdot 13} = 0.143$
$= 14.3\%$
(0.3% more than the answer to problem 1)

39. $\text{APR} = \frac{2mI}{P(n+1)} = \frac{2 \cdot 12 \cdot 264}{1500 \cdot 25} = 0.169$
$= 16.9\%$
(0.9% more than the answer to problem 3)

41. $\text{APR} = \frac{2mI}{P(n+1)} = \frac{2 \cdot 12 \cdot 210}{1500 \cdot 19} = 0.177$
$= 17.7\%$
(0.7% more than the answer to problem 5)

43. $\text{APR} = \frac{2mI}{P(n+1)} = \frac{2 \cdot 12 \cdot 1570}{4500 \cdot 49} = 0.171$
$= 17.1\%$
(1.6% more than the answer to problem 7)

45. $\text{APR} = \frac{2mI}{P(n+1)} = \frac{2 \cdot 12 \cdot 1800}{5000 \cdot 49} = 0.176$
$= 17.6\%$
(1.6% more than the answer to problem 9)

Section 13.4 Buying a House

Study Tips

Buying a home is the single largest credit purchase (and investment) of your life. There are many costs that people overlook or are not aware of when buying a house.

Terms to know:
Points: fees lenders charge to increase profit
Loan origination fee: amount paid to lender
Private Mortgage Insurance (PMI): insurance to protect the lender, required if down payment is less than 20% of the loan
Conventional loans: private lender
FHA loan: Federal Housing Authority is the lender

Note
Keep in mind that an innocent sounding term like "a point" can be worth $800 on an $80,000 loan.

To find down payment and FHA loan amount:
1. Find minimum cash investment: 3% of price.
2. FHA down payment = Acquisition price – FHA
loan amount (Table 13.4 for guidelines).
3. Your down payment: higher value between steps
1 and 2.
4. Loan amount = Acquisition price – step 3 answer.

Note
For a conventional loan, acceptable down payment and loan amount can vary from lender to lender.

To find the monthly payment for a loan:

Think: formula $M = \dfrac{Pi}{1-(1+i)^{-n}}$, where P is the loan amount, i is the monthly rate, and n is the number of months

Word of Advice
The text provides *Guidelines for Purchasing a Home*, which a consumer would be wise to follow. In addition, read the examples, redo the calculations and understand the terminology before you attempt the problems.

13.4 Exercises

1. a. Yes. The maximum they can spend according to the first criterion is $2.5(40,000) = \$100,000$, so they can afford this house.

   b. Yes. $\dfrac{40,000}{52} = \$769.23 > \$750$, so, according to the second criterion, they can afford this house.

   c. Yes. $0.28 \cdot \dfrac{40,000}{12} = \$933.33 > \$750$, so, according to the third criterion, they can afford this house.

3. a. Amount of loan: $= 0.80(77,000) = \$61,600.$

   b. Down payment: $= 77,000 - 61,600 = \$15,400$

   c. For an FHA loan, down payment: $= 0.03(77,000) = \$2310$

5. Using Table 13.5 with 6% interest rate and 20 year payment period, the cost is $7.16 per $1000. The monthly payment is then $7.16(60) = \$429.60$ for principal and interest. Taxes and insurance are $1160 per year or $96.67 per month. The total monthly payment is $429.60 + $96.67 = $526.27.

7. Using Table 13.5 with 9% interest rate and 25 year payment period, the cost is $8.39 per $1000. The monthly payment is then $8.39(90) = \$755.10$ for principal and interest. Taxes and insurance are $2160 per year or $180 per month. The total monthly payment is $755.10 + $180 = $935.10.

9. Using Table 13.5 with 5.5% interest rate and 30 year payment period, the cost is $5.68 per $1000. The monthly payment is then $5.68(173) = \$982.64$ for principal and interest. Taxes and insurance are $3600 per year or $300 per month. The total monthly payment is $982.64 + $300 = $1282.64.

11. a. $0.03(45,000) = \$1350$

    b. $0.9875(45,000) = \$44,437.50$

13. a. $0.03(75,000) = \$2250$

    b. $0.9765(75,000) = \$73,237.50$

15. a. $45,000 + 1200 = \$46,200$

    b. $0.9875(45,000) = \$44,437.50$

    c. $M = \dfrac{44,437.50(0.005)}{1-(1+0.005)^{-180}} = \$374.99$

17. a. $75,000 + 100 = \$75,100$

    b. $75,100 - 0.03(75,000)$
    $= 75,100 - 2250 = \$72,850$

    c. $M = \dfrac{72,850(0.00625)}{1-(1+0.00625)^{-300}} = \$538.36$

19. a. 95% of \$60,000 = (0.95)(\$60,000) = \$57,000, so the down payment is \$60,000 – \$57,000 = \$3000.
    b. Table 13.5 gives \$9.52 as the monthly payment per \$1000 loan at 11% for 30 years. Hence, the mortgage payment is $57(9.52) = \$542.64$. Taxes and insurance amount to \$1500/12 = \$125 per month. The total monthly payment is \$542.64 + \$125 = \$667.64.

21. a. 3% of \$100,000 = (0.03)(\$100,000) = \$3000 is the down payment.
    b. Table 13.5 gives \$6.32 as the monthly payment per \$1000 loan at 6.5% for 30 years. Hence, the mortgage payment is $97(6.32) = \$613.04$. Taxes and insurance amount to \$2400/12 = \$200 per month. The total monthly payment is \$613.04 + \$200 = \$813.04.

23. a. Table 13.5 gives \$8.78 as the monthly payment per \$1000 loan at 10% for 30 years. So, the mortgage payment is $50(8.78) = \$439$.
    b. $30 \cdot 12 = 360$ payments
    c. The total amount paid for principal and interest is $360 \cdot 439 = \$158,040$.
    d. The total interest paid is \$158,040 – \$50,000 = \$108,040.
    e. If 50,000 = 80% of x, then $x = \dfrac{\$50,000}{0.8} = \$62,500$. Thus, the price of the house was \$62,500, which is less than the total interest.

25. a. The down payment is 20% of \$50,000 = \$10,000. The loan fee is 1% of (\$50,000 – \$10,000) = \$400 which makes the closing costs total \$45 + \$150 + \$300 + \$50 + \$220 + \$20 + \$400 = \$1185. Thus, the cash payment is \$10,000 + \$1185 = \$11,185.
    b. Insurance and taxes amount to \$25 + \$50 = \$75 per month. Table 13.5 gives \$11.85 as the monthly payment per \$1000 loan at 14% for 30 years, so the mortgage payment is $40(11.85) = \$474$. The total monthly payment is thus \$474 + \$75 = \$549.
    c. Since the closing costs are \$1185, there would be $1.185(11.85) = \$14.04$ added to the amount found in part b to give a total of \$563.04.

27. a. The down payment is 20% of \$120,000 = \$24,000. The loan amount is \$120,000 – \$24,000 = \$96,000, so the loan fee is 1% of \$96,000 = \$960. The total closing costs are \$45 + \$25 + \$250 + \$960 + \$350 = \$1630 and the down payment plus closing costs is \$24,000 + \$1630 = \$25,630.
    b. Table 13.5 gives \$8.05 as the monthly payment per \$1000 loan at 9% for 30 years. The mortgage payment is $96(8.05) = \$772.80$. Since taxes and insurance come to $\dfrac{1200}{12} = \$100$ per month, the total monthly payment is \$100 + \$772.80 = \$872.80.
    c. To include the closing costs, add $1.63(8.05) = \$13.12$ to the amount in part b to get a total of \$885.92 per month.

29. a. The down payment is 10% of \$84,000 = \$8400. The loan amount is \$84,000 – \$8400 = \$75,600, so the loan fee is 2% of \$75,600 = \$1512. This makes the total closing costs \$1512 + \$235 + \$25 + \$295 = \$2067. The down payment plus closing costs is \$8400 + \$2067 = \$10,467.
    b. Table 13.5 gives \$10.75 as the monthly payment per \$1000 loan at 10% for 15 years. The mortgage payment is $75.6(10.75) = \$812.70$.

31. Answers will vary.

33. Table 13.5 gives \$8.05 as the monthly payment per \$1000 loan at 9% for 30 years. For a \$100,000 loan, the monthly payment is $100(8.05) = \$805$, and in 30 years the total payments would be $360 \cdot 805 = \$289,800$. Thus, the interest paid would be \$289,800 – \$100,000 = \$189,800, which is greater than the mortgage on the house. For a \$50,000 loan, the monthly payment is $50(8.05) = \$402.50$ and in 30 years, the total

payments would be $360(402.50) = \$144,900$. Thus, the interest paid would be $\$144,900 - \$50,000 = \$94,900$, which is more than the price of the house.

Web It Exercises

For mortgage payments, try to visit *Simple Mortgage Calculator* at **http://www.mortgage-calc.com** (Click on *Simple Mortgage Calculator* to determine payment.)
Then for biweekly payments and comparison, then click on *Biweekly Mortgage Calculator*, also found at **http://www.mortgage-calc.com/mortgage/biweekly.html**

1. Answers may vary. There are 13 payments per year when paying your mortgage bi-weekly.

3. Same

Section 13.5 Investing in Stocks, Bonds and Mutual Funds

Study Tips

Long-term investing can be accomplished by investing in stocks, bonds, or mutual funds.

Terms to know if investing in stocks:
Stocks (shares): you become part owner of the company in which you have invested
Dividends: profits
Broker: buys and sells orders from an investor
Commission: fee charged by broker
Types of stock brokers: full service, discount, and deep discount

To find the cost and return on a stock:
1. Total cost of shares = (price per share) × (number of shares) + (commission)
2. Total dividend = (price paid out per share) × (number of shares)
3. Commission on sale = (commission rate) × (sale price)
4. Capital gain = (change in price per share) × 100 – (commission)
5. Total return = (capital gain) + (dividends)
6. Percent of return = $\dfrac{\text{total return}}{\text{total cost of entire transaction}}$

Terms to know if investing in bonds:
Bonds: an IOU, you become a lender to the company in which you have invested
Face value: amount of money being borrowed
Coupon rate or yield: interest rate borrower pays
Coupon payments: interest payments
Maturity dates: deadline for paying back the money

To find the cost and returns for bonds:
1. Periodic amount received = $I = Prt$, where t is the period
2. Total return = $I = Prt$, where t is the total number of years

Terms to know if investing in mutual funds:
Mutual fund: a variety of investments held by a pool of investors
Open-ended: unlimited number of shares can be issued
Net asset value (NAV): price of each share
Making money with mutual funds:
Capital appreciation: an increase in the NAV of the fund
Dividends: deadline for paying back the Money
Capital gain distribution: profits made by selling stocks or bonds that have gone up in price in the fund

To find the net asset value (NAV) of a mutual fund:
Think: use formula $\text{NAV} = \dfrac{A - L}{N}$, where A = total fund assets, L = total fund liabilities, and N = total number of outstanding shares

To find the percent return of a mutual fund:
Think: use formula

$$\text{Return (\%)} = \frac{\text{increase in value + distribution}}{\text{cost of initial investment}}$$

Note: Additional fund performance terminology:

$$\text{Capital appreciation (\%)} = \frac{\text{change in NAV}}{\text{number of shares}}$$

$$\text{Yield (\%)} = \frac{\text{distribution per share}}{\text{NAV (at time of sale)}}$$

Hint
Pay attention to the differences between these three investments. The advantages and disadvantages of stocks, bonds, and mutual funds are given in Section 13.5.

Word of Advice
It would be helpful to create a list with definitions and formulas that apply to each concept. Follow the examples carefully before you begin your work.

13.5 Exercises

1. a. Total cost = cost of stock + commission
$T = 100(\$34.78) + \$50 = \$3528$
b. Dividend = (number of shares) $\cdot \dfrac{\text{dividend}}{\text{share}}$
$D = 100(\$1.06) = \$106$
c. Sell commission = (stock value)(commission rate)
$S = 100(\$36.78)(0.02) = \$73.56$
d. Capital gain = (change in price per share) · (number of shares) – commisions
$G = (\$36.78 - \$35.28) \cdot 100 - \$73.56 = \$76.44$
e. Return = capital gain + dividend
$R = \$76.44 + \$106 = \$182.44$
f. Percent of return = $R / (T + S)$
$$P = \frac{182.44}{3528 + 73.56} \approx 0.0507 = 5.07\%$$

3. a. Total cost = cost of stock + commission
$T = 200(\$28.52) + \$50 = \$5754$
b. Dividend = (number of shares) $\cdot \dfrac{\text{dividend}}{\text{share}}$
$D = 200(\$0.72) = \$144$
c. Sell commission = (stock value)(commission rate)
$S = 200(\$30)(0.02) = \$120$
d. Capital gain = (change in price per share) · (number of shares) commisions
$G = (\$30.00 - \$28.52) \cdot 200 - \$120 = \$126$
e. Return = capital gain + dividend
$R = \$126 + \$144 = \$270$
f. Percent of return = $R / (T + S)$
$$P = \frac{270}{5754 + 120} \approx 0.0460 = 4.60\%$$

5. a. Total cost = cost of stock + commission
$T = 200(\$3.02) + \$604(0.02) = \$616.08$
b. Dividend = (number of shares) $\cdot \dfrac{\text{dividend}}{\text{share}}$
$D = 200(\$0) = \$0$
c. Sell commission $S = \$50$
d. Capital gain = (change in price per share) · (number of shares) – commisions
$G = (\$3.02 - \$3.00) \cdot 200 - \$50 = -\$66.08$
e. Return = capital gain + dividend
$R = -\$66.08 + \$0 = -\$66.08$
f. Percent of return = $R / (T + S)$
$$P = \frac{-66.08}{616.08 + 50} \approx -0.0992 = -9.92\%$$

7. a. Total cost = 1000 (mkt close) + \$25
$T = 1000(\$2.21) + \$25 = \$2235$
b. Return = 1000 (last sale) – total cost
$R = 1000(\$2.60) - \$2235 - \$25 = \$340$
c. Percent of return = $R / (T + 25)$
$$P = \frac{340}{2235 + 25} \approx 0.1504 = 15.04\%$$

9. a. Total cost = 1000 (mkt close) + \$25
$T = 1000(\$3.01) + \$25 = \$3035$
b. Return = 1000 (last sale) – total cost
$R = 1000(\$3.25) - \$3035 - \$25 = \$190$
c. Percent of return = $R / (T + 25)$
$$P = \frac{190}{3035 + 25} \approx 0.0621 \approx 6.21\%$$

11. a. Total cost = 1000 (mkt close) + \$25
$T = 1000(\$18.96) + \$25 = \$18,985$
b. Return = 1000 (last sale) – total cost
$R = 1000(\$19.90) - \$18,985 - \$25 = \$890$
c. Percent of return = $R / (T + 25)$
$$P = \frac{890}{18,985 + 25} \approx 0.0468 = 4.68\%$$

13. a. $I = Prt = \$5000(0.072)(0.5) = \$180$

b. $I = Prt = \$5000(0.072)(5) = \$1800$

15. a. $I = Prt = \$10{,}000(0.0675)(0.5) = \$337.50$

b. $I = Prt = \$10{,}000(0.0675)(2) = \$1350$

17. a. $I = Prt = \$20{,}000(0.0575)(0.5) = \$575$

b. $I = Prt = \$20{,}000(0.0575)(5) = \$5750$

19. a. $A = 200{,}000{,}000 + 100{,}000{,}000 + 250{,}000 = 300{,}250{,}000$

$L = 10{,}000{,}000$

$N = 10{,}000{,}000$

$$NAV = \frac{300{,}250{,}000 - 10{,}000{,}000}{10{,}000{,}000} = \frac{290{,}250{,}000}{10{,}000{,}000} = \$29.025$$

b. $\frac{50{,}000}{29.025} \approx 1722.65 \approx 1722$ shares

21. a. $A = 500{,}000{,}000 + 250{,}000{,}000 + 1{,}500{,}000 = 751{,}500{,}000$

$L = 25{,}000{,}000$

$N = 25{,}000{,}000$

$$NAV = \frac{751{,}500{,}000 - 25{,}000{,}000}{25{,}000{,}000} = \frac{726{,}500{,}000}{25{,}000{,}000} = \$29.06$$

b. $\frac{100{,}000}{29.06} \approx 3441.16 \approx 3441$ shares

23. a. $A = 400{,}000{,}000 + 200{,}000{,}000 + 1{,}000{,}000 = 601{,}000{,}000$

$L = 20{,}000{,}000$

$N = 20{,}000{,}000$

$$NAV = \frac{601{,}000{,}000 - 20{,}000{,}000}{20{,}000{,}000} = \frac{581{,}000{,}000}{20{,}000{,}000} = \$29.05$$

b. $\frac{100{,}000}{29.05} \approx 3442.34 \approx 3442$ shares

25. a. $C = \frac{80 - 70}{100} = \frac{10}{100} = 0.10 = 10\%$

b. $Y = \frac{3}{80} = 0.0375 = 3.75\%$

c. $R = \frac{(80 - 70) \cdot 100 + 3 \cdot 100}{100 \cdot 70} = \frac{1300}{7000} \approx 0.1857 \approx 18.57\%$

27. a. $C = \frac{80 - 75}{300} = \frac{5}{300} \approx 0.0167 \approx 1.67\%$

b. $Y = \frac{1.25}{80} = 0.0156 = 1.56\%$

c. $R = \frac{(80 - 75) \cdot 300 + 1.25 \cdot 300}{300 \cdot 75} = \frac{1875}{22{,}500} \approx 0.0833 \approx 8.33\%$

29. a. $C = \frac{50 - 40}{400} = \frac{10}{400} = 0.025 = 2.5\%$

b. $Y = \frac{4}{50} = 0.08 = 8\%$

c. $R = \frac{(50 - 40) \cdot 400 + 4.00 \cdot 400}{400 \cdot 40} = \frac{5600}{16{,}000} = 0.35 = 35\%$

31. Answers will vary. All are long-term investments. A stock is a share in a company where you become part owner in the company in which you have invested. A bond is lending your money to a company or government. A mutual fund is a pool of many investors who pay a professional manager to purchase a variety of investments.

33. Answers will vary. A dividend is the profits that a company allocates to its owners.

35. $CY = \frac{7.2}{91.242} \cdot 100 \approx 7.89\%$

37. $CY = \frac{6.75}{111.564} \cdot 100 \approx 6.05\%$

39. $CY = \frac{5.75}{108.209} \cdot 100 \approx 5.31\%$

41. $P = \frac{100}{0.07} \approx \$1428.57$

43. $P = \frac{100}{0.05} = \$2000$

45. $P = \frac{100}{0.12} \approx \$833.33$

Chapter 13 Practice Test

Study Tips

Study at the time of day when you learn best, and when you sit down to work, have all your materials together and available. The practice test of the textbook is worked in detail next. Use it as a study tool to master the material before you take the "real" test.

Here are the solutions to the practice test.

1. a. $I = Prt = 800(0.28)(2) = \$448$
   b. $I = Prt = 800(0.28)(0.25) = \$56$
   c. $800 + $448 = $1248

2. a. (0.06)($360) = $21.60
   b. $360 + $21.60 = $381.60

3. a. (0.20)($390) = $78
   b. $390 – $78 = $312

4. a. Interest rate is 4% per period and there are 4 periods. The table gives 1.1699 for these numbers, so the accumulated amount is $100(1.1699) = \$116.99$; the interest is $16.99.
   b. Interest rate is 2% per period and there are 8 periods. The table gives 1.1717 for these numbers, so the accumulated amount is $100(1.1717) = \$117.17$; the interest is $17.17.

5. a. $0.05(185.76) = \$9.29$, so the minimum payment is $10.
   b. Finance charge $= 0.015(185.76 - 10) = \$2.64$.

6. a. $179.64 – $50 = $129.64, so the finance charge is $0.015(129.64) = \$1.94$.
   b. New balance = $129.64 + $1.94 + $23.50 = $155.08.

7. a. $6500 – $1500 = $5000 (unpaid balance) Interest = (0.12)($5000)(4) = $2400
   b. Total to be paid off is $5000 + $2400 = $7400, so the monthly payment is $\frac{7400}{48} = \$154.17$.

8. a. $12(18.10) = \$217.20$ (total amount to be paid)
   $217.20 – $200 = $17.20 (interest charged)
   $\frac{17.20}{200} \cdot 100 = \$8.60$ (interest charge per $100). The closest table entry is $8.59 corresponding to an APR of $15\frac{1}{2}\%$.
   b. Here, $r = 7$ and $n = 12$, so the interest refund is $\frac{7 \cdot 8}{12 \cdot 13} \cdot 17.20 = \$6.17$.
   c. $7(18.10) - 6.17 = \$120.53$ (amount needed to pay off the loan.)

9. a. (0.75)($50,000) = $37,500 (amount of loan)
   b. $50,000 – $37,500 = $12,500 (down payment)
   c. (0.03)($50,000) = $1500 (minimum down payment for FHA loan)
   d. $50,000 – $1500 = $48,500 (maximum FHA loan)

10. The table shows a payment of $12.00 per $1000 loan at 12% for 15 years, so the monthly mortgage payment (for principal and interest) is $37.5(12) = \$450$.

11. a. Total cost = cost of stock + commission
    $T = 100(\$52.50) + \$50 = \$5300$

b. Dividend

$$= (\text{number of shares})\cdot\frac{\text{dividend}}{\text{share}}$$

$$D = 100(\$1.22) = \$122$$

c. Sell comm
= (stock value)(comm rate)

$$S = 100(\$57.75)(0.02) = \$115.50$$

d. Capital gain =
(change in price per share) ·
(number of shares) – commisions

$$G = (57.75 - 52.50)\cdot 100 - 50 - 0.02(5775)$$
$$= \$359.50$$

e. Return = capital gain + dividend

$$R = \$359.50 + \$122 = \$481.50$$

f. Percent of return = $R/(T+S)$

$$P = \frac{481.50}{5300 + 115.50} \approx 0.0889 \approx 8.89\%$$

12. a. $I = Prt = 10{,}000(0.065)(0.5)$
$= \$325$

b. $I = Prt = 10{,}000(0.065)(5)$
$= \$3250$

13. a. $A = 200{,}000{,}000 + 100{,}000{,}000 + 250{,}000 = 300{,}250{,}000$

$$L = 10{,}000{,}000$$
$$N = 20{,}000{,}000$$

$$NAV = \frac{300{,}250{,}000 - 10{,}000{,}000}{20{,}000{,}000}$$

$$= \frac{290{,}250{,}000}{20{,}000{,}000} = \$14.5125 \approx \$14.51$$

b. $\frac{100{,}000}{14.51} \approx 6891.8 \approx 6891$ shares

14. a. $C = \frac{80-70}{100} = \frac{10}{100} = 0.10 = 10\%$

b. $Y = \frac{1.50}{80} = 0.01875 = 1.875\%$

c. $R = \frac{(80-70)\cdot 100 + 1.50\cdot 100}{100\cdot 70}$

$$= \frac{1150}{7000} \approx 0.1643 = 16.43\%$$

15. a. $V = 100(70) = \$7000$

b. $MR = 7000(0.02) = \$140$

c. The annual rate of return would be equivalent to the APY.

$$\text{APY} = \left(1 + \frac{r}{k}\right)^k - 1 = (1 + 0.02)^{12} - 1$$

$$= 1.02^{12} - 1 \approx 0.268 \approx 27\%$$

Chapter 14 Voting and Apportionment

Section 14.1 Voting Systems

Study Tips

In this section we shall study *voting theory,* the mathematical treatment of the process used to express voters' preferences in an election. The different types of voting are *preferential* (asking voters to rank the candidates) and *nonpreferential.*

Preferential Methods: the winner is...

Plurality Method: most first place votes

Plurality with Runoff Method:
1. receives majority of the votes
2. if no majority, eliminate all but *top two,* hold runoff election, winner receives majority of votes

Borda Count Method: candidate with most points
Procedure:
1. voters rank candidates from most to least favorable
2. last place receives no point; each next-to-last-place vote is awarded one point, and so on; determine points for each candidate

Plurality with Elimination (Hare Method):
1. receives majority of the votes
2. if no majority, eliminate the candidate with the *fewest* votes and hold another election; repeat until a candidate receives a majority

Pairwise Comparison Method (Condorcet): candidate with most points
Procedure:
1. voters rank candidates from most to least favorable
2. each candidate is compared with the other candidates
3. winner receives 1 point, if a tie, ½ point

Nonpreferential Method: the winner is...

Approval Voting Method: candidate with most approval votes
Procedure: do not rank but each voter approves or disapproves each candidate

Caution
All the methods use lots of arithmetic to arrive at the answer, so be very careful with the calculations or, if allowed, use a calculator to save time.

Exercise 14.1

1. a. 523,686 + 59,934,814 + 738,475 + 69,456,897 + 603,456 = 131,257,328 votes were cast.

   b. McCain: $\frac{59,934,814}{131,257,328} \approx 0.46 = 46\%$

   Obama: $\frac{69,456,897}{131,257,328} \approx 0.53 = 53\%$

   c. Yes, Obama won 53%.

3. a. 2,911,872 + 2,910,942 + 97,419 + 17,472 = 5,937,705 total votes

   b. 2,911,872 – 2,910,942 = 930

   c. $\frac{930}{5.937.705} \approx 0.000157 \approx 0.016\%$

   d. Yes, the percent different does require a recount as 0.016% is less than 0.5%.

5. a. 6 + 7 + 3 + 4 = 20 votes were cast.
   b. 11 (more than half)
   c. No. Nobody had 11 or more votes.
   d. C(7 votes)

Note: T(120 + 100 = 220) *means that* T *has* 220 *votes.* S(150) *means that* S *has* 150 *votes.*

7. a. T(120 + 100 = 220) and S(150)

   b. If a runoff is computed between T and S only (all others will be eliminated) the new table will look like this:

	Number of Voters			
Place	130	120	100	150
First		T	T	S
Second				
Third	S	S		
Fourth	T		S	T

   The runoff results will be
   S(130 + 150 = 280)
   T(120 + 100 = 220)
   Thus, S is the winner 280 to 220.

9. a. A(13) and C(12) have the most first-place votes.

b. Since the top vote getters were A and C, a runoff is computed between A and C (all others are eliminated). The table now looks like this:

	Number of Voters		
Place	13	12	10
First	A	C	
Second			A
Third			
Fourth	C	A	C

The runoff results will be:
A(13 + 10 = 23)
C(12)
Thus, A is the winner 23 to 12.

11. The vacation destinations are Busch Gardens (B), Disney World (D), Epcot (E), and Sea World (S).

	Number of Voters		
Place	20	15	10
First	D	E	S
Second	B	B	B
Third	E	D	D
Fourth	S	S	E

The number of first place votes is: D(20), E(15), S(10), so none of the destinations received a majority of the vote and B received the least amount of votes and is eliminated.

	Number of Voters		
Place	20	15	10
First	D	E	S
Second			
Third	E	D	D
Fourth	S	S	E

Now D(20), E(15), S(10), so we eliminate S. Here is the new table with D and E only.

	Number of Voters		
Place	20	15	10
First	D	E	
Second			
Third	E	D	D
Fourth			E

Now D(20 + 10 = 30) and E(15), hence Disney World is the winner.

13. The results are: B(5), O(11), G(8) and J(6). Thus, O, the orange juice, is the winner.

15. We assign 0, 1, 2 and 3 points for last, next to last, and so on.

		Number of Voters			
Points	**Place**	5	11	8	6
3	First	B	O	G	J
2	Second	J	J	J	G
1	Third	G	G	O	O
0	Fourth	O	B	B	B

1 orange (O): $0\cdot5+3\cdot11+1\cdot8+1\cdot6$
$=0+33+8+6=47$ points

2 green peppers (G): $1\cdot5+1\cdot11+3\cdot8+2\cdot6$
$=5+11+24+12=52$ points

1 cup cooked broccoli (B):
$3\cdot5+0\cdot11+0\cdot8+0\cdot6$
$=15+0+0+0=15$ points

½ cup fresh orange juice (J):
$2\cdot5+2\cdot11+2\cdot8+3\cdot6$
$=10+22+16+18=66$ points

The Borda winner is J(½ cup fresh orange juice) with 66 points.

17. We have to compare B and G, B and J, B and O, G and J, G and O, J and O
B(5) G(11 + 8 + 6 = 25) G gets 1 point.
B(5) J(11 + 8 + 6 = 25) J gets 1 point.
B(5) O(11 + 8 + 6 = 25) O gets 1 point.
G(8) J(5 + 11 + 6 = 22) J gets 1 point.
G(5 + 8 + 6 = 19) O(11) G gets 1 point.
J(5 + 8 + 6 = 19) O(11) J gets 1 point.
The winner is J(½ cup fresh orange juice) with three points.

19. If Collins drops out, Adams is the winner 6 to 3.

21. a. Latte has 12 votes.
b. Cappuccino has 12 + 5 + 10 + 13 = 40 votes.
c. Mocha has 5 + 10 = 15 votes.
d. Americano has 5 + 13 = 18 votes.
e. Cappuccino is the winner using approval voting.

23. No. You need more than 44 votes for a majority.

25. Beijing (BC) will be selected as it has the most votes.

27.

	Number of Voters								
Pts	3	2	32	3	3	1	8	30	6
5	I	I	BC	M	BG	I	M	S	BG
4	BC	BC	I	BC	BC	S	S	M	S
3	M	BG	BG	BG	I	BC	BG	BG	M
2	BG	M	M	S	S	M	I	I	BC
1	S	S	S	I	M	BG	BC	BC	I

BC($4\cdot 3+4\cdot 2+5\cdot 32+4\cdot 3+4\cdot 3+3\cdot 1+1\cdot 8+1\cdot 30+2\cdot 6=257$)
BG($2\cdot 3+3\cdot 2+3\cdot 32+3\cdot 3+5\cdot 3+1\cdot 1+3\cdot 8+3\cdot 30+5\cdot 6=277$)
I($5\cdot 3+5\cdot 2+4\cdot 32+1\cdot 3+3\cdot 3+5\cdot 1+2\cdot 8+2\cdot 30+1\cdot 6=252$)
M($3\cdot 3+2\cdot 2+2\cdot 32+5\cdot 3+1\cdot 3+2\cdot 1+5\cdot 8+4\cdot 30+3\cdot 6=275$
S($1\cdot 3+1\cdot 2+1\cdot 32+2\cdot 3+2\cdot 3+4\cdot 1+4\cdot 8+5\cdot 30+4\cdot 6=259$)
Berlin with 277 points will be selected.

29. Since there are five cities, there will be $\frac{5\cdot 4}{2}=10$ comparisons. They are

1) BC(3 + 2 + 32 + 3 + 1 = 41) vs. BG(3 + 8 + 30 + 6 = 47) BG gets 1 point.
2) BC(32 + 3 + 3 + 6 = 44) vs. I(3 + 2 + 1 + 8 + 8 = 22) BC gets 1 point.
3) BC(3 + 2 + 32 + 3 + 1 = 41) vs. M(3 + 8 + 30 + 6 = 47) M gets 1 point.
4) BC(3 + 2 + 32 + 3 + 3 = 43) vs. S(1 + 8 + 30 + 6 = 45) S gets 1 point.
5) BG(3 + 3 + 8 + 30 + 6 = 50) vs. I(3 + 2 + 32 + 1 = 38) BG gets 1 point.
6) BG(2 + 32 + 3 + 6 = 43) vs. M(3 + 3 + 1 + 8 + 30 = 45) M gets 1 point.
7) BG(3 + 2 + 32 + 3 + 3 + 6 = 49) vs. S(1 + 8 + 30 = 39) BG gets 1 point.
8) I(3 + 2 + 32 + 3 + 1 = 41) vs. M(3 + 8 + 30 + 6 = 47) M gets 1 point.
9) I(3 + 2 + 32 + 3 + 1 = 41) vs. S(3 + 8 + 30 + 6 = 47) S gets 1 point.
10) M(3 + 2 + 32 + 3 + 8 = 48) vs. S(3 + 1 + 30 + 6 = 40) M gets 1 point.

BC(1 pt), BG(3pts), I(0 pt), M(4 pts), and S(2 pts), so Manchester (M) is selected.

31. I(3 + 2 + 1 + 32 = 38)
S(30 + 1 + 8 + 6 = 45)
BC(32 + 3 + 2 + 3 + 3 = 43)
M(3 + 8 + 30 = 41)
BG(3 + 6 = 9)
S(Sydney) is selected with 45 votes.

33. Rio de Janeiro won with 66 votes in the third round.

35. A majority is more than $\frac{94}{2}=47$ votes.
No city received more than 47 votes in the second round.

37. Madrid with the most votes would be selected to host the 2016 Olympics.

39. A(10), B(13), C(8 + 7 = 15), D(12).
Thus, C(Coala Cola) wins with 15.

41. We assign 0, 1, 2 and 3 points for last, next to last, and so on.

	Number of Voters				
Points	10	13	8	7	12
3	A	B	C	C	D
2	B	A	D	B	A
1	C	D	B	A	B
0	D	C	A	D	C

A($3\cdot 10+2\cdot 13+0\cdot 8+1\cdot 7+2\cdot 12=87$ pts)
B($2\cdot 10+3\cdot 13+1\cdot 8+2\cdot 7+1\cdot 12=93$ pts)
C($1\cdot 10+0\cdot 13+3\cdot 8+3\cdot 7+0\cdot 12=55$ pts)
D($0\cdot 10+1\cdot 13+2\cdot 8+0\cdot 7+3\cdot 12=65$ pts)
The preferred Cola is B(Best Cola) with 93 points.

Section 14.2 Voting Objections

Study Tips

What is a fair method of voting? For many years, mathematicians and sociologists have been looking for a voting procedure that would satisfy a prescribed set of *fairness criteria*.

Majority Criterion
Think: majority of 1st place votes, should win
Never violates this criterion: plurality, plurality with elimination and pairwise comparison
Potential to violate this criterion: Borda

Head-to-Head (Condorcet) Criterion
Think: if favored when compared head-to-head with other candidates, should win
Never violates this criterion: pairwise comparison
Potential to violate this criterion: plurality, Borda, plurality with elimination

Monotonicity Criterion
Think: if winner gains support without losing, should also be winner of 2nd election
Potential to violate this criterion: plurality with elimination

Irrelevant Alternatives Criterion
Think: if candidate wins, and in 2nd election one or more of losing candidates are removed, the winner of the 1st election should win the 2nd election
Potential to violate this criterion: plurality, Borda, plurality with elimination, and pairwise comparison

Caution
Is is impossible to find a voting method that satisfies all four criteria at the same time. This is stated in *Arrow's Impossibility Theorem*.

Word of Advice
Follow the methods of solution presented here when you do your assignment. Keep your work organized!

14.2 Exercises

1.

		Number of Voters		
Points	**Place**	25	5	20
2	First	W	S	P
1	Second	P	W	S
0	Third	S	P	W

a. Using plurality W(25) is the winner.
b. Using Borda
$W(2\cdot 25+1\cdot 5=50+5=55)$,
$S(2\cdot 5+1\cdot 20=10+20=30)$
$P(1\cdot 25+2\cdot 20=25+40=65)$
Thus, P is the winner with 65 votes.
c. Yes. W won under plurality but P wins under Borda.
d. We have to compare W and S, W and P, and S and P.
W(25) S(5 + 20 = 25)
Each gets ½ point.
W(25 + 5 = 30) P(20) W gets 1 point.
S(5) P(25 + 20 = 45) P gets 1 point.
W is the winner with 1½ points.

3.

		Number of Voters			
Points	**Place**	10	20	30	10
3	First	C	C	L	L
2	Second	A	M	C	C
1	Third	M	A	M	A
0	Fourth	L	L	A	M

a. A (20 + 20 + 10 = 50)
C (30 + 60 + 60 + 20 = 170)
L (90 + 30 = 120)
M (40 + 10 + 30 = 80)
Under the Borda count C(Cappucino) is the winner.
b. No. L(Latte) is the plurality winner with 30 + 10 = 40 votes but C is the winner under Borda.

5. a. We have to compare S and U, S and H, S and B, U and H, U and B, and H and B.
Here are the comparisons:
S(50 + 25 + 15 = 90) U(10)
S gets 1 point.
S(50 + 10 = 60) H(25 + 15 = 40)
S gets 1 point.
S(50 + 25 = 75) B(15 + 10 = 25)
S gets 1 point.
U(50 +10 = 60) H(25 + 15 = 40)
U gets 1 point.

U(50 + 25 + 10 = 85) B(15)
U gets 1 point.
H(50 + 25 = 75) B(15 + 10 = 25)
H gets 1 point.
S(Staples) is the commercial preferred to U, H, and B.

b. Yes. S wins under plurality too.

7. Yes. The results of the initial election (D, A and H) are as follows: D(20), A(18) and H(12). When H is eliminated, the results are: D(20 + 12 = 32) and A(18), so D is still the winner. This means that the irrelevant alternatives criterion is satisfied.

9. a. Using the plurality methods the results are as follows: A(27), B(24) and C(2), so A is the winner.

b. Yes, A does.

c. We have to compare A and B, A and C, and B and C.

A(27) B(24 + 2 = 26) A gets 1 point.
A(27) C(24 + 2 = 26) A gets 1 point.
B(24) C(27 + 2 = 29) C gets 1 point.
A is the winner with 2 points.

d. Yes, A beats every other candidate.

e.

	Number of Voters			
Points	**Place**	27	24	2
2	First	A	B	C
1	Second	C	C	B
0	Third	B	A	A

$A(2 \cdot 27 = 54)$
$B(2 \cdot 24 + 1 \cdot 2 = 50)$
$C(1 \cdot 27 + 1 \cdot 24 + 2 \cdot 2 = 55)$
The winner is C.

f. The majority criterion. A wins using plurality but C wins using Borda.

g. B drops out, so we have

	Number of Voters			
Points	**Place**	27	24	2
1	First	A	C	C
0	Second	C	A	A

A(27) C(24 + 2 = 26)
No. A is the new winner 27 to 26 over C.

h. Majority Criterion (A has a majority but C wins under Borda)
Condorcet Criterion (A wins head-to head but C wins under Borda)
Irrelevant Alternatives (C was the winner (part e) under Borda but B's dropping out causes C to lose to A (part g)).

11. a. The results for the first election are:
A(14) B(4) C(11) D(8)
Thus, B is eliminated.

	Number of Voters				
Place	14	4	10	1	8
First	A		C	C	D
Second		D		D	C
Third	C	C	D		
Fourth	D	A	A	A	A

The results for the second election are:
A(14) C(10 + 1 = 11)
D(4 + 8 =12)
Thus, C is eliminated.

	Number of Voters				
Place	14	4	10	1	8
First	A				D
Second		D		D	
Third			D		
Fourth	D	A	A	A	A

The results of the third election are:
A(14) D(4 + 10 + 1 + 8 = 23)
Thus, D has a majority and is the winner.

b. We have to compare A and B, A and C, A and D, B and C, B and D, C and D. Here are the comparisons:
A(14) B(23) B gets 1 point.
A(14) C(23) C gets 1 point.
A(14) D(23) D gets 1 point.
B(18) C(19) C gets 1 point.
B(28) D(9) B gets 1 point.
C(25) D(12) C gets 1 point.
Thus, C is the pairwise comparison winner with 3 points.

c. Yes. C beats A, B and D.

d. The Condorcet criteria. C wins head to head (part c) but D wins plurality with elimination (part a).

13. a. B wins. Here are the results:
A(20) B(19) C(5).
No majority, C is eliminated.

	Number of Voters		
Place	20	19	5
First	A	B	
Second	B		B
Third		A	A

A(20) B(19 + 5 = 24)
B has a majority and is the winner.

b. A drops out, we have

Number of Voters			
Place	20	19	5
First		B	C
Second	B	C	B
Third	C		

B(20 + 19 = 39) C(5), B is the winner.
Yes. After A drops out B(39) and C(5), so B still wins.

c. None. B wins even after A drops out.

15. a. For three candidates $\frac{3\cdot 2}{2}=3$ comparisons need to be made.

b. For four candidates $\frac{4\cdot 3}{2}=6$ comparisons need to be made.

c. For five candidates $\frac{5\cdot 4}{2}=10$ comparisons need to be made.

d. For n candidates $\frac{n(n-1)}{2}$ comparisons need to be made.

17. Answers will vary. If a candidate receives a majority of first-place votes, then that candidate should be the winner (see page 834).

19. Answers will vary. In general, if a candidate holds a majority of first-place votes, this candidate always wins every pairwise (head-to-head) comparison (see page 836).

21. Answers will vary. Elimination might add to the original plurality winner but add even more to the second place plurality entry, enough to overtake the original difference. Thus, plurality with elimination method will always satisfy the majority criterion, and may not satisfy the monotonicity criterion (see page 839).

23. Answers will vary. The winner of each pair-wise comparison would also be the winner of a head-to-head comparison (see page 838).

25. Answers will vary.

Section 14.3 Apportionment Methods

Study Tips

The U.S. House of Representatives has to divide 435 seats among 50 states. The challenge is to do that fairly! In most apportionment problems, the items to be divided are identical and indivisible, but the problem is that each participant may be entitled to a different percentage of the items to be apportioned.

To find the standard divisor and standard quota:
Think: standard divisor relates to total population; standard quota to population in the group
Note: You will always find the standard divisor (SD) first, then use the SD to find the standard quota (SQ).

$$\text{SD}=\frac{\text{total population}}{\text{total number to be apportioned}}$$

$$\text{SQ}=\frac{\text{population in the group}}{\text{SD}}$$

To use the Hamilton method of apportionment:
1. Find $\text{SD}=\frac{\text{total population}}{\text{total seats to be apportioned}}$
2. Find $\text{SQ}=\frac{\text{state population}}{\text{SD}}$
3. Round SQ *down* to nearest integer; each state gets that many seats but must get at least one
4. Additional seats one at a time to *largest fractional part* of the standard quotas

To use the Jefferson method of apportionment:
1. Find modified divisor MD, so modified quotient MQ is rounded *down* to the nearest integer; sum equals number of items to be apportioned
2. Apportionment for each group corresponds to rounded down MQs found in step 1.

To use the Webster method of apportionment:
1. Find MD, the modified divisor.
2. Find MQ, the modified quota for each group:

$$\text{MQ}=\frac{\text{total population}}{\text{MD}}$$

3. Round MQ (*up* for 0.5 or more, *down* for less than 0.5)
4. Apportionment for each group corresponds to values found in step 3; sum must equal total number of items to be apportioned.

To use the Adam method of apportionment:
1. Find modified divisor MD, so modified

quotient MQ is rounded *up* to the nearest integer; sum equals number of items to be apportioned

2. Apportionment for each group corresponds to rounded up MQs found in step 1.

Hint

Finding the "right" MD: start with calculating the SD, and then change it up or down as necessary. If you make the MD *larger*, the resulting MQ *will be smaller*; make the MD *smaller*, the resulting MQ *will be larger*.

Caution

Each of these methods (the Hamilton method, the Jefferson method, the Webster method, and the Adams method) requires intense calculation, so you must be very careful when doing the arithmetic.

14.3 Exercises

1. a. $\frac{2,001,102,854}{10} = \$200,110,285$

b. $\frac{2,001,102,854}{216,221} = \$9255$

c. $\frac{2,001,102,854}{136,363} = \$14,675$

3. University of South Florida:

$\frac{18,176}{136,363} \cdot 2,001,102,854 = \$266,729,578$

University of West Florida:

$\frac{4556}{136,363} \cdot 2,001,102,854 = \$66,858,492$

5. a. $SD = \frac{136,363}{200} = 681.815$

b. Florida Atlantic:

$SQ = \frac{10,725}{681.815} = 15.730$

University of Central Florida:

$SQ = \frac{18,312}{681.815} = 26.858$

7. $SD = \frac{136,363}{200} = 681.815$

	FTE	*Quota*	*Initial*	*Extra*	*Final*
UF	29,646	43.4810	43		43
FSU	21,195	31.0861	31		31
FAMU	8064	11.8273	11	1	12
USF	18,176	26.6583	26	1	27
FAU	10,725	15.7301	15	1	16
UWF	4556	6.6822	6	1	7
UCF	18,312	26.8577	26	1	27
FIU	17,434	25.5700	25		25
UNF	6697	9.8223	9	1	10
FGC	1558	2.2851	2		2
Total	**136,363**		**194**	**6**	**200**

The number of positions for each university are on the final column. Note that one additional seat was allocated to the universities with the *largest* fractional part of the standard quota. No, the number of positions for each university is not the same.

9. a. Total population = 5637 + 5683 + 7655 + 3715 + 2061 = 24,751

$SD = \frac{\text{total population}}{\text{total seats(positions, money) to be apportioned}}$ and $SQ = \frac{\text{state population}}{SD}$, so that

$SQ = \frac{\text{state population} \cdot \text{total seats to be apportioned}}{\text{total population}}$

CAL: $\frac{5637 \cdot 100}{24,751} = 22.7748$

FLA: $\frac{5683 \cdot 100}{24,751} = 22.9607$

NY: $\frac{7655 \cdot 100}{24,751} = 30.9280$ TX: $\frac{3715 \cdot 100}{24,751} = 15.0095$ NJ: $\frac{2061 \cdot 100}{24,751} = 8.3269$

b.

State	*Cases*	*Quota*	*Initial*	*Extra*	*Final*
CAL	5637	22.7748	22	1	23
FL	5683	22.9607	22	1	23
NY	7655	30.9280	30	1	31
TEX	3715	15.0095	15		15
NJ	2061	8.3269	8		8
Total	**24,751**		**97**	**3**	**100**

The final column shows the number of millions for each state. One additional seat was allocated to the state with the *largest* fractional part of the standard quota.

11. a. Note: SD = 5.35; use MD = 5.25 to find MQ = $\frac{\text{participants}}{\text{MD}}$.

Exercise: $\frac{150}{5.25} = 28.5714$ Sports: $\frac{90}{5.25} = 17.1429$

Charity W: $\frac{85}{5.25} = 16.1905$ Home R: $\frac{130}{5.25} = 24.7619$ Computer H: $\frac{80}{5.25} = 15.2381$

b. Note that the modified quotas are *rounded down* in the final column.

Activity	*Participants*	*Quota* 5.35	*Modified* 5.25	*Final*
Exercise	150	28.0374	28.5714	28
Sports	90	16.8224	17.1429	17
Charity	85	15.8879	16.1905	16
Home repair	130	24.2991	24.7619	24
Computer hobbies	80	14.9533	15.2381	15
Total	**535**			**100**

According to the table, Exercise will get \$28 million, Sports \$17 million, Charity \$16 million, Home Repair \$24 million and Computer Hobbies \$15 million.

13. a. Use MD = 1.08 to find MQ = $\frac{\text{population in group}}{\text{MD}}$.

Dogs: $\frac{190}{1.08} = 175.9259$ Cats: $\frac{110}{1.08} = 101.8519$

Birds: $\frac{10}{1.08} = 9.2593$ Horses: $\frac{230}{1.08} = 212.9630$

b. Note that the modified quotas are *rounded* in the usual manner (*up* for 0.5 or more, *down* for less than 0.5).

Pet	*Cost*	*Quota* 1.08	*Modified* 1.08	*Final*
Dogs	190	175.9259	175.9259	176
Cats	110	101.8519	101.8519	102
Birds	10	9.2593	9.2593	9
Horses	230	212.9630	212.9630	213
Total	**540**			**500**

According to the table for every \$500 spent on pets, \$176, \$102, \$9 and \$213 are spent on dogs, cats, birds and horses respectively.

15. a. Note: $SD = \frac{1500}{75} = 20$. Use MD = 20.5 to find $MQ = \frac{\text{population in group}}{\text{MD}}$.

Lake Park: $\frac{600}{20.5} = 29.2683$ E.G. Simmons: $\frac{470}{20.5} = 22.9268$

Lettuce Lake: $\frac{240}{20.5} = 11.7073$ Lithia Springs: $\frac{160}{20.5} = 7.8049$

Eureka Springs: $\frac{30}{20.5} = 1.4634$

b. Note that the modified quotas are *rounded up.* The final column shows the number of rangers assigned to each park.

Park	*Acreage*	*Quota* 20	*Modified* 20.5	*Final*
Lake Park	600	30	29.2683	30
E.G. Simmons	470	23.5	22.9268	23
Lettuce Lake	240	12	11.7073	12
Lithia Springs	160	8	7.8049	8
Eureka Springs	30	1.5	1.4634	2
Total	**1500**			**75**

17. a. $SD = \frac{\text{total population}}{\text{total money to be apportioned}} = \frac{33{,}000+21{,}000+17{,}000+11{,}000+11{,}000}{100} = 930$

b. $SQ = \frac{\text{population in group}}{\text{SD}}$

Dade: $\frac{33{,}000}{930} = 35.4839$ Broward: $\frac{21{,}000}{930} = 22.5806$

Hillsborough: $\frac{17{,}000}{930} = 18.2796$ Orange: $\frac{11{,}000}{930} = 11.8280$

Pinellas: $\frac{11{,}000}{930} = 11.8280$

c. The initial appropriation summed to 97. One additional seat was allocated to the county with the *largest* fractional part of the standard quota.

	Dade	*Broward*	*Hills*	*Orange*	*Pinellas*
SD = 930	33000	21000	17000	11000	11000
Standard Quota	35.48	22.58	18.28	11.83	11.83
Hamilton's Method	35	22 + 1 = 23	18	11 + 1 = 12	11 + 1 = 12

The bottom row shows each county's apportionment (in millions of dollars) using Hamilton's method.

19. a. Use MD = 948.60 to find $MQ = \frac{\text{population in group}}{\text{MD}}$.

Dade: $\frac{33{,}000}{948.60} = 34.7881$ Broward: $\frac{21{,}000}{948.60} = 22.1379$ Hillsborough: $\frac{17{,}000}{948.60} = 17.9211$

Orange: $\frac{11{,}000}{948.60} = 11.5960$ Pinellas: $\frac{11{,}000}{948.60} = 11.5960$

b. Results are then *rounded up* as shown.

	Dade	*Broward*	*Hills*	*Orange*	*Pinellas*
MD = 948.60	33000	21000	17000	11000	11000
Modified Quota	34.79	22.14	17.92	11.60	11.60
Adam's Method	35	23	18	12	12

The allocations are 35, 23, 18, 12 and 12 million dollars respectively as shown.

21. a. $SD = \dfrac{\text{total population}}{\text{total specialists to be apportioned}} = \dfrac{100+72+62+56+50}{200} = 1.7$

b. $SQ = \dfrac{\text{population in group}}{SD}$

SF Gen: $\dfrac{100}{1.7} = 58.8235$ JH: $\dfrac{72}{1.7} = 42.3529$

Mass: $\dfrac{62}{1.7} = 36.4706$ UCSF: $\dfrac{56}{1.7} = 32.9412$ Sloan: $\dfrac{50}{1.7} = 29.4118$

c. The initial appropriation summed to 197. One additional seat was allocated to the county with the *largest* fractional part of the standard quota.

	SF Gen	*JH*	*Mass*	*UCSF*	*Sloan*
SD = 1.7	100	72	62	56	50
Standard Quota	58.82	42.35	36.47	32.94	29.41
Hamilton's Method	58 + 1 = 59	42	36 + 1 = 37	32 + 1 = 33	29

The bottom row shows each hospital's apportionment of the 200 AIDS specialists using Hamilton's method.

23. a. Use MD = 1.7238 to find $MQ = \dfrac{\text{population in group}}{MD}$.

SF Gen: $\dfrac{100}{1.7238} = 58.0114$ JH: $\dfrac{72}{1.7238} = 41.7682$

Mass: $\dfrac{62}{1.7238} = 35.9670$ UCSF: $\dfrac{56}{1.7238} = 32.4864$ Sloan: $\dfrac{50}{1.7238} = 29.0057$

b. Results are then *rounded up* as shown.

	SF Gen	*JH*	*Mass*	*UCSF*	*Sloan*
MD = 1.7238	100	72	62	56	50
Modified Quota	58.011	41.768	35.967	32.4586	29.006
Adams's Method	59	42	36	33	30

The bottom row shows each hospital's apportionment of the 200 AIDS specialists using Adam's method.

25. a. Sport A gets $\dfrac{2000}{20{,}000} \cdot 200{,}000 = \$20{,}000$ Sport B gets $\dfrac{4000}{20{,}000} \cdot 200{,}000 = \$40{,}000$

Sport C gets $\dfrac{6000}{20{,}000} \cdot 200{,}000 = \$60{,}000$ Sport D gets $\dfrac{8000}{20{,}000} \cdot 200{,}000 = \$80{,}000$

b. $SD = \dfrac{\text{total population}}{\text{total money to be apportioned}} = \dfrac{2000+4000+6000+8000}{200{,}000} = 0.1$

$SQ = \dfrac{\text{population in group}}{SD}$

Sport A: $\dfrac{2000}{0.1} = 20{,}000$ Sport B: $\dfrac{4000}{0.1} = 40{,}000$

Sport C: $\dfrac{6000}{0.1} = 60{,}000$ Sport D: $\dfrac{8000}{0.1} = 80{,}000$

Since the sum of these quotas is equal to 200,000, the divisors would not need to be modified. The apportionment would be the same for Hamilton's, Jefferson's, Adam's, and Webster's methods.

c. The answers are the same in parts a and b.

27. Answers will vary.

Section 14.4 Apportionment Objections

Study Tips

Just as there were several objections, or flaws, associated with the voting methods, there are several objections, or flaws, associated with apportionment methods, especially the Hamilton method.

Alabama Paradox:
Think: an *increase* in the number of items to be apportioned results in a *loss* of items for the group
Note: Objection is to the number of items to be apportioned.

Population Paradox:
Think: faster *increase* in population, yet *loses* items
Note: Objection is to the population in the group.

New-states Paradox:
Think: *addition* of a new group *changes* the apportionment of another group
Note: Objection is to the addition of new groups that require apportionment.

Hint
For these paradoxes, note that it is always an increase of some sort, that will cause what could considered a negative effect.

To apply the Quota Rule:
Think: only two choices for apportionment, either the upper quota or the lower quota

Caution
Is is impossible to find an apportionment method that satisfies the quota rule and avoids paradoxes. This is stated in *Balinski and Young's Impossibility Theorem.*

14.4 Exercises

1. a. For 30 seats: SD $= \dfrac{18{,}000}{30} = 600$

For 31 seats: SD $= \dfrac{18{,}000}{31} = 580.6452$

	SQ (30)	*Final*	SQ (31)	*Final*
α	5.37	5 + 1 = 6	5.59	5
β	8.33	8	8.6	8 + 1 = 9
δ	16.3	16	16.8	16 + 1 = 17

Yes, the Alabama paradox occurs. α loses one seat (from 6 to 5) when seats are increased from 30 to 31.

b. For 600 seats: SD $= \dfrac{18{,}000}{60} = 300$

For 61 seats: SD $= \dfrac{18{,}000}{61} = 295.0820$

	SQ (60)	*Final*	SQ (61)	*Final*
α	10.73	10 + 1 = 11	10.91	10 + 1 = 11
β	16.67	16 + 1 = 17	16.94	16 + 1 = 17
δ	32.6	32	33.14	33

No, the Alabama paradox does not occur.

3. For 71 tickets: SD $= \dfrac{1020}{71} = 14.366$

For 72 tickets: SD $= \dfrac{1020}{72} = 14.167$

	SQ (71)	*Final*	SQ (72)	*Final*
B	28.19	28	28.59	28 + 1 = 29
K	21.3	21	21.6	21 + 1 = 22
W	14.2	14	14.4	14
S	7.31	7 + 1 = 8	7.41	7

Yes, the Alabama paradox does occur. Six Flags Magic Mountain goes from 8 to 7 tickets.

5. a. Most recent year:

$$SD = \frac{740}{11} = 67.2727$$

	SQ	*Final*
DE	1.64	1
NE	3.72	3 + 1 = 4
KS	5.65	5 + 1 = 6

Delaware will receive \$1 million, Nebraska \$4 million, and Kansas \$6 million.

b. Projected: $SD = \frac{850}{11} = 77.2727$

	SQ	*Final*
DE	1.58	1 + 1 = 2
NE	3.88	3 + 1 = 4
KS	5.54	5

Delaware will receive \$2 million, Nebraska \$4 million, and Kansas \$5 million.

c. Delaware:

$$\frac{122-110}{110} \approx 0.1091 = 10.91\%$$

Kansas: $\frac{428-380}{380} \approx 0.1263 = 12.63\%$

Kansas has the higher percent increase of enrollees.

d. Yes, the population paradox occurred. Kansas population increased by 12.63% but contributions decreased from \$6 to \$5 million, while Delaware, with a smaller 10.91% increase, got an extra million.

7. a. Now:

$$SD = \frac{439,000}{13} = 33,769.23077$$

	SQ	*Final*
A	2.64	2
B	3.70	3 + 1 = 4
C	6.66	6 + 1 = 7

County A will receive 2 seats, county B 4 seats, and county C 7 seats.

b. 10 Years:

$$SD = \frac{489,000}{13} = 37,615.38$$

	SQ	*Final*
A	2.58	2 + 1 = 3
B	3.85	3 + 1 = 4
C	6.57	6

County A will receive 3 seats, county B 4 seats, and county C 6 seats.

c. Yes, the population paradox occurred. Even though C's growth rate was

$$\frac{247,000-225,000}{225,000} \approx 0.0978 = 9.78\%$$

and A's was

$$\frac{97,000-89,000}{89,000} \approx 0.0899 = 8.99\%, \text{ C}$$

lost a seat and A gained one.

d. County C will be unhappy as they had a greater increase yet lost a seat.

9. a. $SD = \frac{558}{41} = 13.6098$

	Number	SQ	*Final*
P	402	29.54	29 + 1 = 30
S	156	11.46	11
Total	**558**		**41**

Division P will receive 30 managers and S will receive 11.

b. $SD = \frac{672}{49} = 13.714$

	Number	SQ	*Final*
P	402	29.31	29
S	156	11.38	11 + 1 = 12
A	114	8.31	8
Total	**672**		**49**

Yes, the new-states paradox occurred. The new apportionment is 29 for division P, 12 for S and 8 for A, so P lost one manager and S gained one.

11. a. $SD = \frac{4990}{100} = 49.9$

	Number	SQ	*Final*
A	4470	89.58	89 + 1 = 90
B	520	10.42	10
Total	**4990**		**100**

State A will receive 90 seats and B 10 seats.

b. $SD = \frac{5253}{106} = 49.5567$

	Number	SQ	*Final*
A	4470	90.20	90
B	520	10.49	10 + 1 = 11
C	263	5.31	5
Total	**5253**		**106**

Yes, the new-states paradox does occur. The new apportionment is 90 for A and 11 for B, so B gained one seat.

13. Answers will vary.

Chapter 14 Practice Test

Study Tips

What does this statement mean?

You don't know it until you can tell it to the wall.

Verbalizing concepts helps you fully understand them. This is so true, try it sometime! Consider your past performance on previous tests. What went well for you? What did not go well? As you study, determine what habits you would like to keep and what changes you could do. You may want to study alone and then study in a group setting to reinforce what you've learned.

Here are the solutions to the practice test.

1. a. No. It takes over $\frac{10+8+7+5}{2}=15$, or 16 votes, for a majority.
   b. D with 7 + 5 = 12 votes
   c. A with 10 votes
   d. B

2. D and A have the most first place votes so they have a runoff.
   8 + 7 + 5 = 20 people prefer D over A
   10 people prefer A over D
   D wins the runoff 20 to 10.

3. Give 3 points for first place, 2 for second 1 for third, none for last.

	Number of Voters			
Points	10	8	7	5
3	A	C	D	D
2	C	D	C	B
1	B	B	B	C
0	D	A	A	A

Here are the results:
$A(3\cdot 10=30\text{ pts})$
$B(1\cdot 10+1\cdot 8+1\cdot 7+2\cdot 5=35\text{ pts})$
$C(2\cdot 10+3\cdot 8+2\cdot 7+1\cdot 5=63\text{ pts})$
$D(2\cdot 8+3\cdot 7+3\cdot 5=52\text{ pts})$
C wins with 63 points.

4. Eliminate B first.

Number of Voters			
10	8	7	5
A	C	D	D
C	D	C	
			C
D	A	A	A

Second election. A: 10, C: 8, D: 7 + 5 = 12
Eliminate C.

Number of Voters			
10	8	7	5
A		D	D
	D		
D	A	A	A

Third election: A(10) D(8 + 7 + 5 = 20)
D is the winner.

5. a. A(5 + 9 = 14) B(3) A gets 1 point.
   b. A(5) C(3 + 9 = 12) C gets 1 point.
   c. B(5 + 3 = 8) C(9) C gets 1 point.
   d. A(1 point) C(2 points), C is the winner.

6. a. Adams has 5 votes, Barnes 3 votes, and Collins 2 votes, so Adams wins.
   b. Barnes has 3 votes and Collins 2 votes, so Barnes wins.

7. a. B with 36 votes wins using the plurality method. Yes, since 36 is more than $\frac{36+16+12}{2}=32$.
   b. Give 2 points for first place, 1 for second none for last.

	Number of Voters		
Points	36	16	12
2	B	A	A
1	A	B	C
0	C	C	B

Here are the results:
$A(1\cdot 36+2\cdot 16+2\cdot 12=92\text{ pts})$
$B(2\cdot 36+1\cdot 16=88\text{ pts})$
$C(1\cdot 12=12\text{ pts})$
A is the winner. No, this method does not satisfy the majority criterion as B has the majority but A wins the Borda count.

c. Eliminate C.

Number of Voters		
36	16	12
B	A	A
A	B	
		B

B(36) A(16 + 12 = 28)
B is the winner. Yes, the majority criterion is satisfied.

d. We have to compare A and B, A and C, and B and C.
A(16 + 12 = 28) B(36) B gets 1point.
A(36 + 16 + 12 = 64) C(0)
A gets 1point.
B(36 + 16 = 52) C(12) B gets 1 point.
A(1 point) B(2 points), B is the winner.
Yes, the majority criterion is satisfied.

8. a. 74 voters prefer T over V.
75 voters prefer P over V.
99 voters prefer V over C.
75 voters prefer P over T.
144 voters prefer T over C.
119 voters prefer P over C.
P is the winner. Yes, the head-to-head criterion is satisfied.

b. C with 45 votes wins using the plurality method. No, it does not satisfy the head to-head criterion.

c. Give 3 points for first place, 2 for second 1 for third, none for last.

	Number of Voters				
Points	15	25	29	30	45
3	V	V	T	P	C
2	T	T	V	V	P
1	P	C	P	T	T
0	C	P	C	C	V

Here are the results:
$V(3\cdot15+3\cdot25+2\cdot29+2\cdot30=238\text{ pts})$
$T(2\cdot15+2\cdot25+3\cdot29+1\cdot30+1\cdot45=242\text{ pts})$
$P(1\cdot15+1\cdot29+3\cdot30+2\cdot45=224\text{ pts})$
$C(1\cdot25+3\cdot45=160\text{ pts})$
T is the winner using the Borda method. No, it does not satisfy the head-to-head criterion.

d. Eliminate P.

Number of Voters				
15	25	29	30	45
V	V	T		C
T	T	V	V	
	C		T	T
C		C	C	V

Second election.
V(15 + 25 + 30 = 70)
T(29)
C(45)
Eliminate T.

Number of Voters				
15	25	29	30	45
V	V			C
		V	V	
	C			
C		C	C	V

Third election.
V(15 + 25 + 29 + 30 = 99) C(45)
V wins using the plurality with elimination method.
No, it does not satisfy the head-to-head criterion.

e. We have to compare V and T, V and P, V and C, T and P, T and C, and P and C.
1) V(15 + 25 + 30 = 70)
T(29 + 45 = 74) T gets 1 point.
2) V(15 + 25 + 29 = 69)
P(30 + 45 = 75) P gets 1 point.
3) V(15 + 25 + 29 + 30 = 99)
C(45) V gets 1 point.
4) T(15 + 25 + 29 = 69)
P(30 + 45 = 75) P gets 1 point.
5) T(15 + 25 + 29 + 30 = 99)
C(45) T gets 1 point.
6) P(15 + 29 + 30 = 74)
C(25 + 45 = 70) P gets 1 point.
T(2 points) V(1 point) P(3 points)
P wins using pairwise comparison. Yes, it does satisfy the head-to-head criterion.

9. a. Yes. C with 56 votes has the majority (more than $\frac{56+24+20}{2}=50$) and wins under plurality.

b. Yes. 56 voters prefer C over A and over B. C has the majority and wins head-to-head against all other candidates.

10. a. Yes. C will still win the second election under plurality.
    b. Yes. C will still win the second election under plurality when either A or B drops out. Note that C starts with a majority, not simply plurality, so cannot lose the majority.

11. Arrow's Impossibility Theorem states that there is no voting method that will always simultaneously satisfy each of the four fairness criteria.

12. $\text{SD} = \dfrac{40,500}{162,000,000} = 0.00025$

Standard Quotas (in millions) for:

A: $\dfrac{15,000}{0.00025} = 60$ B: $\dfrac{8500}{0.00025} = 34$ C: $\dfrac{6500}{0.00025} = 26$

D: $\dfrac{6000}{0.00025} = 24$ E: $\dfrac{4500}{0.00025} = 18$

Note that the total is \$162 million.

13. $\text{SD} = \dfrac{180}{100} = 1.8$

Department	*Faculty*	*SQ*	*RD*	*Actual Number*
Mathematics	30	30/1.8 ≈ 16.67	16	16 + 1 = **17**
English	60	60/1.8 ≈ 33.33	33	33
Language	24	24/1.8 ≈ 13.33	13	13
Art	16	16/1.8 ≈ 8.89	8	8 + 1 = **9**
Chemistry	50	50/1.8 ≈ 27.78	27	27 + 1 = **28**
Total	**180**	**100**	**97**	**100**

14. Use 1.95 as the modified divisor then round down.

Department	*Faculty*	*MQ*	*Actual*
Mathematics	30	30/1.95 ≈ 15.38	15
English	**60**	60/1.95 ≈ 30.77	30
Language	24	24/1.95 ≈ 12.31	12
Art	16	16/1.95 ≈ 8.21	8
Chemistry	50	50/1.95 ≈ 25.64	25
Total	**180**	**90**	**90**

15. MD = 49,800;

$\text{MQ} = \dfrac{34,730}{49,800} \approx 0.6974, \text{ or } 1$

16. Use 2.05 as the modified divisor, then round up.

Department	*Faculty*	*MQ*	*Actual*
Mathematics	30	30/2.05 ≈ 14.63	15
English	60	60/2.05 ≈ 29.27	30
Language	24	24/2.05 ≈ 11.71	12
Art	16	16/2.05 ≈ 7.80	8
Chemistry	50	50/2.05 ≈ 24.39	25
Total	**180**	**90**	**90**

17.

State	Population	SQ	RD	Extra	Final
A	50,4000	$\frac{50,400}{1960.78} \approx 25.70$	25	1	26
B	46,600	$\frac{46,600}{1960.78} \approx 23.77$	23	1	24
C	3000	$\frac{3000}{1960.78} \approx 1.53$	1	0	1
Total	**100,000**		**49**		**51**

The Alabama paradox occurs because even though the number of seats was increased from 50 to 51, state C lost one seat, from 2 to 1.

18. a.

Positions	Number	New SQ	Positions
10	484	9.6338	**9**
19	990	19.7054	**20**
71	3550	70.6608	**71**

b. No. Even though Math's growth rate was $\frac{484-476}{476} \approx 0.01681 = 1.681\%$ and English's was $\frac{990-975}{975} \approx 0.01538 = 1.538\%$, Math lost a seat and English gained one.

c. Yes, this apportionment is an example of the population paradox. Math lost one position and English gained one even though math was growing at a faster rate, 8/476 = 1.681%, than English 15/975 = 1.538%.

19. a. $SD = \frac{1116}{41} = 27.2195$

	Number	SQ	Final
A	804	29.54	29 + 1 = 30
B	312	11.46	11
Total	**1116**		**41**

State A will receive 30 seats and B will receive 11.

b. $SD = \frac{1344}{49} = 27.4286$

	Number	SQ	Final
A	804	29.31	29
B	312	11.37	**11 + 1 = 12**
C	228	8.31	8
Total	**1344**		**49**

Yes, the new-states paradox occurred. The addition of the new state C caused A to lose 1 seat (from 30 to 29) and B to gain 1 (from 11 to 12).

20. There is no apportionment method that satisfies the quota rule and avoids all paradoxes.

Chapter 15 Graph Theory

Section 15.1 Introduction to Graph Theory

Study Tips

Graph theory is used to reach cost-effective solutions by individuals and business.

Terms to know:

Graph: collection of points consisting of vertices and edges
Vertex: endpoint of any edge
Edge: arcs or line segments
Degree of a vertex: number of edges that have the vertex as a common endpoint
Subgraph: graph that consists of some of the vertices and edges of another graph
Path: sequence of edges from one vertex to another, can be traced without lifting your pencil
Simple path: path with no repeated edges
Circuit: path that begins and ends at the same vertex
Connected graph: every pair of vertices can be joined by a path
Disconnected graph: not connected, you would have to lift your pencil
Complete graph: every pair of vertices is joined by exactly one edge, K_n

Word of Advice

Concepts such as *degree*, *path* and *circuit* will be used repeatedly throughout Chapter 15, so be sure to understand them now.

15.1 Exercises

1. The degree of a vertex is found by counting the number of edges that meet it. So, the degrees of the vertices *A*, *B*, *C*, *D* and *E* are 1, 1, 1, 0 and 3, respectively.

3. The degree of a vertex is found by counting the number of edges that meet it. So, the degrees of vertices *A*, *B*, *C* and *D* are 5, 3, 4 and 6, respectively.

5. Since *C* is not one of the vertices listed, this does not determine a subgraph.

7. Since both vertices of edge *g* are included, this defines a subgraph.

9. By tracing along the edges listed (in order) you can see that [*a*, *b*, *c*] is a path from *A* to *D*.

11. By tracing along the edges listed (in order) you can see that [*j*, *d*, *k*, *a*] is a circuit based at *A*.

13. By tracing along the edges listed (in order) you can see that [*h*, *g*] is a path from *A* to *G*.

15. Getting from edge *h* to edge *f* requires a "jump," so [*h*, *f*, *g*] is not a path.

17. The graph consists of "one piece," so it is connected.

19. The graph consists of "two pieces," so it is disconnected.

21. The graph is complete. Here's an easy way to see this: if you compare this graph with K_4 in figure 15.9 you can see that they are "the same" graph drawn in two different ways.

23. There is a pair of vertices that is not joined by any edge, so this graph is not complete.

25. This is K_3 so it is complete (see Figure 15.9).

27. Since no pair of vertices is joined by more than one edge, this graph is simple.

29. Since no pair of vertices is joined by more than one edge, this graph is simple.

31. Answers may vary.

33. Represent each computer as a vertex and the connecting wires as edges as shown in the following diagram.

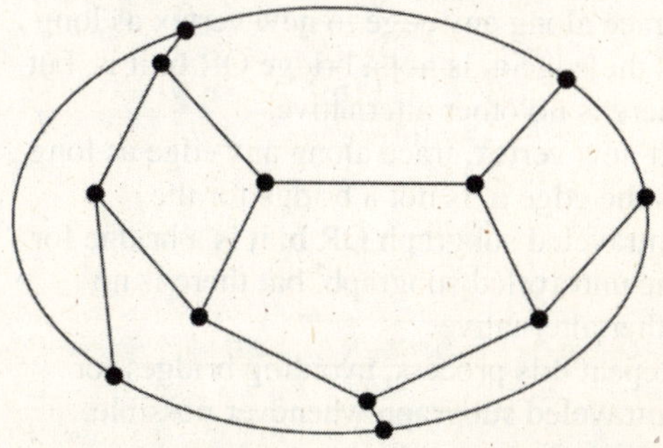

Section 15.2 Euler Paths and Euler Circuits

Study Tips

The *traversability* rules are the keys to deciding whether a graph has an *Euler path* or *circuit.*

Terms to know:
Euler path: uses every edge of a graph once and only once
Euler circuit: Euler path that begins and ends at the same vertex
Traversable: Euler paths and circuits
Bridge: an edge in a graph, when removed, creates a disconnected graph
Untraveled subgraph: created by removing the edge that has been traced on a graph, each time a new edge is traced, a new untraveled subgraph is created

Hint
You might want to review the definition of traversability that was given in Section 8.6.

To determine if a graph has an Euler path or circuit:
Think: traversability rules

A graph is traversable if…
1. connected, *no* vertices of odd degree
2. connected, *exactly two* vertices of odd degree

A graph is not traversable if…
1. *more than two* vertices of odd degree

To find an Euler circuit:
Think: Fleury's Algorithm, only connected and traversable graphs
Procedure summary:
1. Pick any vertex as starting point.
2. Trace along any edge to new vertex as long as the edge **a.** is not a bridge OR **b.** it is, but there is no other alternative.
3. At new vertex, trace along any edge as long as the edge **a.** is not a bridge for the untraveled subgraph OR **b.** it is a bridge for the untraveled subgraph, but there is no other alternative
4. Repeat this process, avoiding bridges for untraveled subgraph whenever possible, until ending up at the starting point.

Word of Advice
If the graph you are working with has an Euler circuit, use *Fleury's algorithm* to find one. Using this algorithm makes the job of finding Euler circuits easy: don't cross a bridge unless you have no other choice. It might be helpful to work through Example 7 before trying the exercises.

Exercise 15.2

1. It is easy to see that [*a, b, c, d, e*] is a path from *D* to *B*. Since each edge is traversed once, we actually have an Euler path from *D* to *B*.

3. It is easy to see that [*a, b, c, d, e*] is a circuit based at *A*. Since each edge is traversed once, we actually have an Euler circuit based at *A*.

5. Every vertex in this graph has even degree. According to the traversability rules there is an Euler circuit based at any of the vertices.

7. Every vertex in this graph has even degree. According to the traversability rules there is an Euler circuit based at any of the vertices.

9. Every vertex in this graph has even degree. According to the traversability rules there is an Euler circuit based at any of the vertices.

11. Every vertex in this graph has even degree. According to the traversability rules there is an Euler circuit based at any of the vertices. There are many correct answers (for example, [*a, b, c, d, e, f, g, i, j, h*]).

13. Every vertex in this graph has even degree. According to the traversability rules there is an Euler circuit based at any of the vertices. There are many correct answers (for example, [*a, b, e, f, h, g, d, c*]).

15. Every vertex in this graph has even degree. According to the traversability rules there is an Euler circuit based at any of the vertices. There are many correct answers (for example, [*a, b, c, d, e, f, g, h, i*]).

17. Every vertex in this graph has even degree. According to the traversability rules there is an Euler circuit based at any of the vertices. There are many correct answers (for example, [*a, b, c, d, e, f, g, h, i, j, k, l*]).

19. We represent the streets as labeled edges and the intersections as vertices. The result is shown.

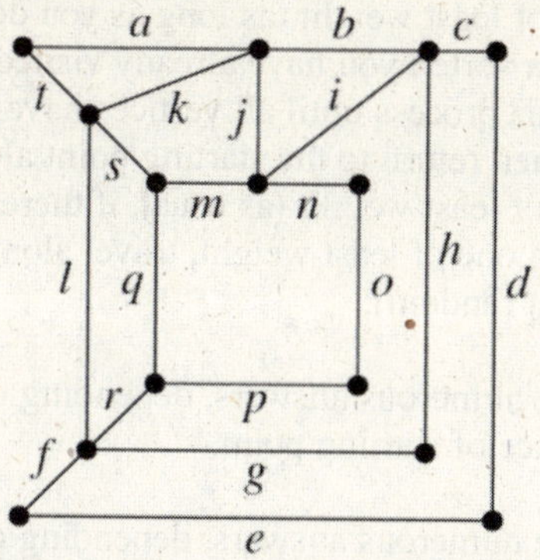

There are many correct solutions that start at either vertex of odd degree. For example, a possible Euler path is: [*s, t, a, b, c, d, e, f, g, h, i, j, k, l, r, p, o, n, m, q*].

21. We represent the streets as labeled edges and the intersections as vertices. The result is shown.

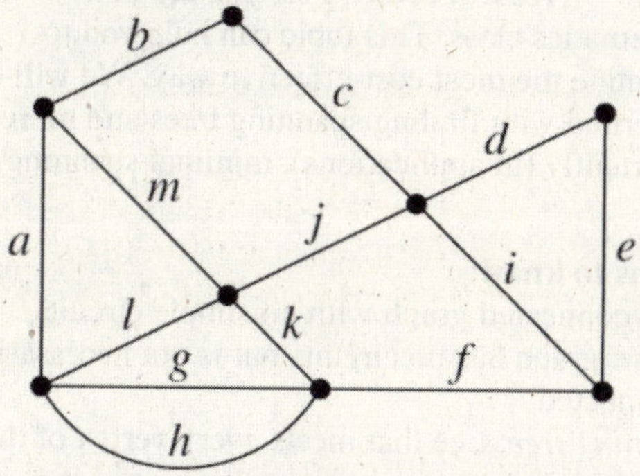

There are many correct solutions that start at either vertex of odd degree. For example, [*a, l, m, b, c, j, k, g, h, f, i ,d, e*] is a possible Euler path.

23. Answers may vary.
   a. A graph is traversable if it has an Euler path.
   b. An Euler circuit is circuit that traverses each edge exactly one time.
   c. An Euler path is a path that traverses each edge exactly one time.
   d. If you remove an edge from a graph and get a disconnected graph, then that edge is a bridge.

25. There are 3 regions, 5 edges and 4 vertices. Thus, $X(G) = 3 - 5 + 4 = 2$.

27. There are 9 regions, 12 edges and 5 vertices. Thus, $X(G) = 9 - 12 + 5 = 2$.

29. There are 6 regions, 8 edges and 4 vertices. Thus, $X(G) = 6 - 8 + 4 = 2$.
Possible redraw:

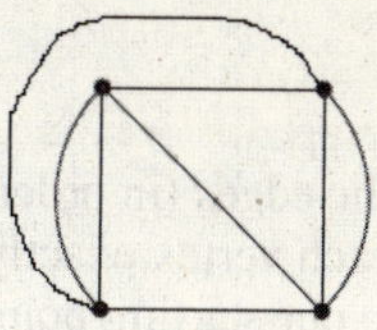

Section 15.3 Hamilton Paths and Hamilton Circuits

Study Tips

Finding *Hamilton* circuits is a difficult process (there is no known algorithm for even finding them). The *Nearest- Neighbor Method* allows us to find *nearly minimal* Hamilton circuits in *weighted graphs*, which has many real-world applications.

Terms to know:
Hamilton path: path that meets each vertex once and only once
Hamilton circuit: path that meets every vertex once and then returns to the starting point
Weighted graph: each edge has a number called a weight associated to it
Length of a path: the sum of the weights of each of its edges
Traveling salesman problem (TSP): find the most efficient circuit to visit each city once and only once
Nearly minimal Hamilton circuit: circuit found using the nearest-neighbor method

To find the length of a path or circuit:
Think: determine the path, add the weights

To apply the nearest-neighbor method:
Think: process to find most efficient route
Procedure summary:
1. Begin at indicated starting point or choose one at random
2. Travel along edge with smallest weight.
3. From new vertex, repeat, do not return to previous vertex.
4. From each new vertex, travel to an unvisited vertex along edge of east weight.
5. Return to the starting point along the edge with smallest weight.

Word of Advice
Applications of Nearest-Neighbor Method include efficient solutions to the *Traveling Salesman*

Problem. Before trying the exercises, it might be helpful to carefully work through Example 7.

15.3 Exercises

1. By tracing along the edges (in order) we see that we meet each vertex exactly once before returning to our starting point. In other words, we complete a Hamilton circuit.

3. When tracing the sequence of edges [*a, b, c, d, e*] we land on vertex *C* twice: it's neither a Hamilton circuit nor a Hamilton path.

5. Add the weights. The length of the path is $8+3+2+4=17$.

7. Add the weights. The length of the path is $5+5+7+4=21$.

9. Recall that the graph K_4 has $\frac{(4-1)!}{2}=3$ distinct Hamilton circuits based at vertex *A*. Their lengths are $5+7+5+5=22$, $5+3+5+4=17$, and $5+3+7+4=19$.

11. The nearest neighbor method determines the circuit with length $2+2+4+6=14$.

13. There is more than one answer. One possibility is the circuit with length $2+1+1+5+2=11$.

15. There is more than one answer. One possibility is the circuit with length $2+1+1+1+9+2=16$.

17. Using the Nearest-Neighbor Method we get $8+12+19+11=50$ miles.

19. The most efficient delivery route (applying the Nearest-Neighbor Method) has length $15+13+17+25=70$ miles.

21. Answers may vary. How can one travel between a number of cities, visiting each destination once before returning to the starting point, in the most cost-efficient manner?

23. Answers may vary. From the starting point, choose to travel along the edge of least weight -- if there is more than one, choose any of them at random. When you have arrived at a new vertex choose to travel along the edge of least weight (as long as you don't return to a vertex you have already visited). Repeat this process until all vertices have been visited, then return to the starting point along the edge of least weight (as usual, if there is more than one of least weight, travel along any of them at random).

25. There are numerous answers, depending on your choice of starting point.

27. There are numerous answers, depending on your choice of starting point.

Section 15.4 Trees

Study Tips

Trees? Forest? Yes, you are in a mathematics class. This topic can help you to determine the most cost-effective way. We will be concerned with finding spanning trees and more importantly (in applications), minimal spanning trees.

Terms to know:
Tree: connected graph with no simple circuits
Forest: graph has no circuits but is not necessarily connected
Spanning tree: tree that meets every vertex of the graph
Minimal spanning tree: spanning tree with the least weight

To find a minimal spanning tree:
Think: Kruskal's algorithm
Procedure summary:
1. Start with edge of least weight
2. From edges not selected, choose the edge of minimal weight.
3. From remaining edges, choose any edge of minimal weight as long as the subgraph does not form a circuit.
4. Repeat step 3 until subgraph contains all vertices of the original graph.

Hint
You can use *Theorem 15.1 Unique Path Property* to show that certain graphs are not trees: if there is more than one simple path joining a pair of vertices, the graph cannot be a tree!

Word of Advice
When faced with a weighted graph use *Kruskal's algorithm* to find minimal spanning trees. You will always be able to find a spanning tree since *Theorem 15.2 Existence of Spanning Trees* tells us that a spanning tree always exists.

15.4 Exercises

1. This graph is a connected graph with no simple circuits, so it is a tree.

3. This graph is not a tree. Here's an easy way to see this: Since there is more than one simple path from any pair of vertices, Theorem 15.1 is not satisfied – the graph can't be a tree!

5. There are many correct answers. One possibility is {*a*, *b*, *c*}.

7. There are many correct answers. One possibility is {*a*, *b*, *c*, *d*}.

9. There are many correct answers. One possibility is {*a*, *b*, *c*, *d*}.

11. The graph below indicates a minimal spanning tree (there is more than one).

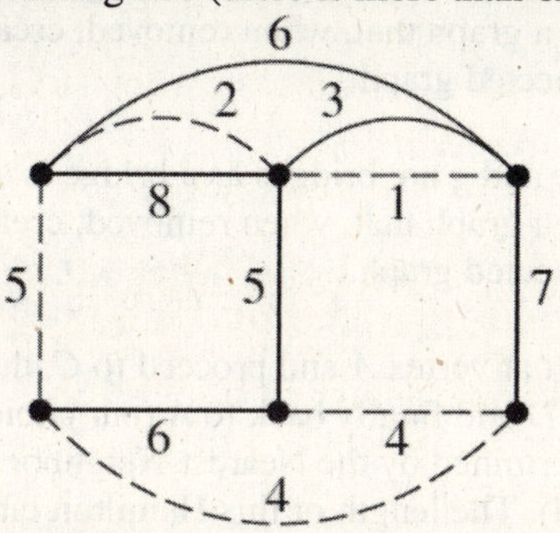

This spanning tree was found using Kruskal's algorithm. Its weight is $1 + 2 + 4 + 4 + 5 = 16$.

13. Applying Kruskal's algorithm provides the solution indicated.

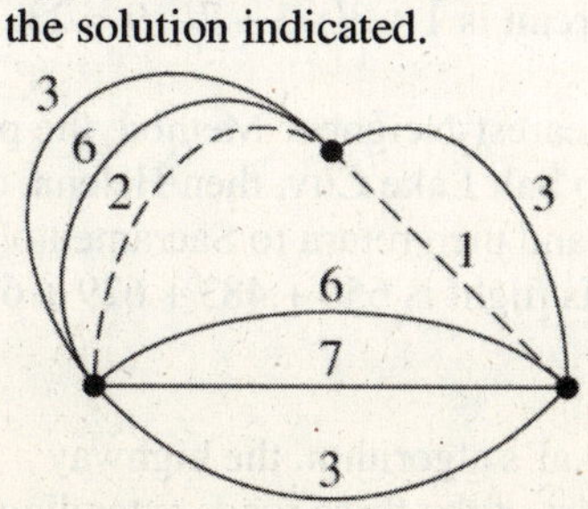

Its weight is $1 + 2 = 3$.

15. Applying Kruskal's algorithm provides the solution indicated.

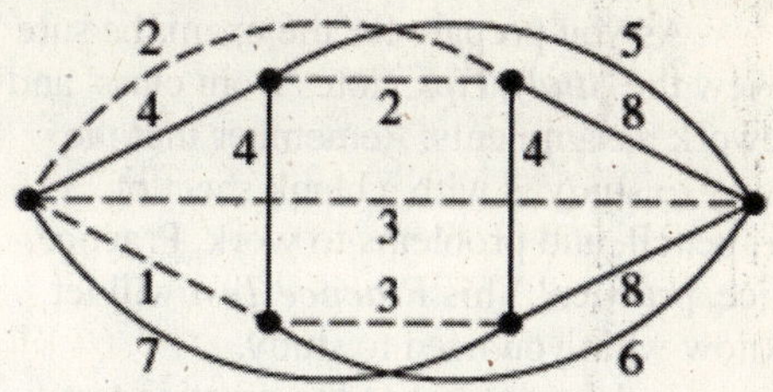

Its weight is $1 + 2 + 2 + 3 + 3 = 11$.

17. Applying Kruskal's algorithm, we find that the most economical way to link all cities is a network of length $3.8 + 4.9 + 5.4 + 7.8 = 21.9$ representing a cost of \$21.9 million.

19. Applying Kruskal's algorithm, we find that the most economical way to create an irrigation system has cost $126 + 132 + 135 + 141 =$ \$534.

21. Answers may vary. A spanning tree is a subgraph that is connected, has no simple circuits and contains all of the vertices of the original graph.

23. Answers may vary. For each pair of vertices in a tree, there is exactly one simple path between them.

25. Neither Euler circuits nor paths would be appropriate here because, both require that every edge be used. For this problem, we are not concerned about edges and, in fact, if we can leave out an edge and still connect all the cities, it will be cheaper and thus, more desirable. The tree shown is a possible answer! Is it the cheapest way? Can you find a better way?

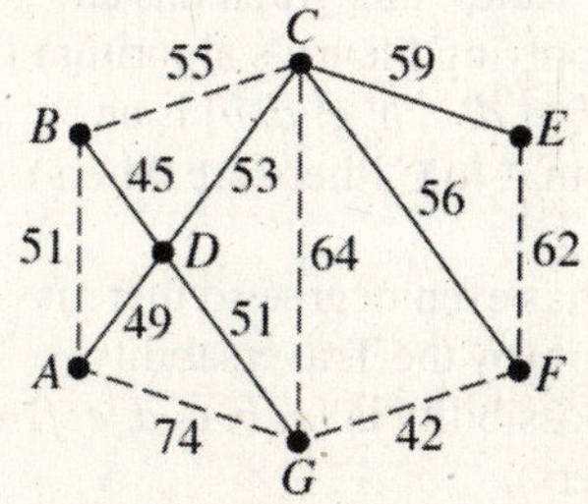

Chapter 15 Practice Test

Study Tips

As you prepare for the exam, be sure to review the ***Study Tips***, notes from class, and homework assignments. Remember that the best way to study is with a blank sheet of paper, pencil, and problems to work. Practice, practice, practice! This *Practice Test* will let you know what you need to study.

Here are the solutions to the practice test.

1. To find the degree of a vertex you count the number of edges that meet it. So, the degrees of vertices *A*, *B*, *C*, *D*, *E*, *F* and *G* are 4, 4, 3, 4, 1, 2 and 0, respectively.

2. As before, count the edges: the degrees of vertices *A*, *B*, *C*, *D* and *E* are 5, 2, 4, 2 and 1, respectively.

3. If you trace along the edges *a*, *b*, *c* and *d* (in order) you find that we have a path from *A* to *B*.

4. If you trace along the edges *g*, *a*, *e* and *d* (in order) you find that we have a circuit based at *B*.

5. If you trace along the edges *b*, *c* and *f* (in order) you get a path from *B* to *E*.

6. Tracing along the edges *g* and then *d* gives us a path from *A* to *D*. BUT, there is no way to get from vertex *D* to edge *b* without lifting your pencil so this sequence of edges does not represent a path (or circuit).

7. Vertices *A* and *C* have odd degree and the others have even degree. According to the Traversability Rules, this graph has an Euler path. Applying Fleury's algorithm we find that [*f, e, d, c, h, g, a, b*] is an Euler path from *A* to *C* (there are others).

8. Every vertex has even degree so there is an Euler circuit (by the Traversability Rules). One possibility is [*a, b, c, d, e, f*] (there are others).

9. This graph has four vertices of odd degree (*A*, *D*, *E*, and *F*) so there is no Euler circuit (or path) by the Traversability Rules.

10. Vertices *A* and *D* have odd degree, while all other vertices have even degree. According to the Traversability Rules, this graph has an Euler path. Applying Fleury's algorithm we find that [*a, b, g, f, e, c, d*] is an Euler path from *A* to *D* (there are others).

11. By representing the streets that the mail carrier must cover by edges and each intersection as a vertex, we see that there are more than two vertices with odd degree. Thus, it is impossible to find an Euler a path or circuit; the mail carrier will have to retrace his steps at some point.

12. By representing each of the roads by an edge and each intersection as a vertex, we see that there are two vertices with odd degree (the intersection of Fair Oaks Blvd. at San Juan Ave. and the intersection of Sunset Ave. at San Juan Ave). Thus, according to the Traversability Rules there is an Euler path in the graph. It is possible for the employee to drive along each road once and only once – he must start at one of the intersections above and end at the other.

13. Edges *c* and *e* are bridges as a bridge is an edge in a graph that, when removed, creates a disconnected graph.

14. Edges *e* and *g* are bridges as a bridge is an edge in a graph that, when removed, creates a disconnected graph.

15. We start at vertex *A* and proceed to *C*, then *B*, then to *D* and finally back to *A* (our choices are determined by the Nearest-Neighbor Method). The length of this Hamilton circuit is $2 + 3 + 6 + 5 = 16$.

16. In order we visit the vertices *A*, *B*, *E*, *C*, *D* and then back to *A* (as determined by the Nearest-Neighbor Method). The length of this Hamilton circuit is $1 + 3 + 3 + 7 + 9 = 23$.

17. Using the Nearest-Neighbor Method, the pilot should fly to Salt Lake City, then Helena, on to Olympia and then return to Sacramento. The length of this flight is $650 + 483 + 629 + 693 = 2455$ miles.

18. Using Kruskal's algorithm, the highway should consist of the three roads extending

from Guadalajara. The total cost will be $8 + 12 + 13 = \$33$ million.

19. The three edges extending from vertex A determine the minimal spanning tree.

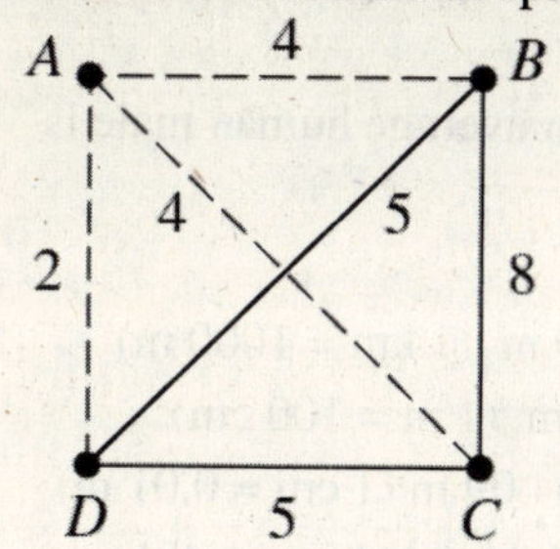

20. The edges with weights 1, 2, 3 and 4 determine the minimal spanning tree.

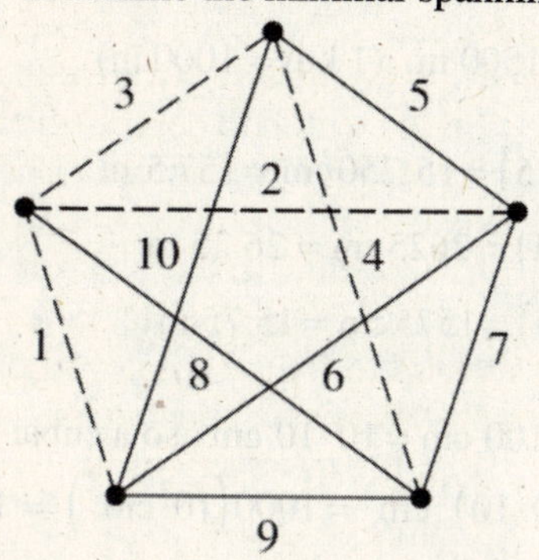

Appendix The Metric System

Section A.1 Metric Units of Measurement

Study Tips

The metric system is a decimal system of weights and measures. Multiples and subdivisions of the basic unit are given in powers of 10. This makes conversion within the metric system very easy.

To convert from one metric measurement to another:
Think: small unit to large, move decimal to the left
Large unit to small, move decimal to the right
Note: Changing from one metric unit to another involves only a shift in the decimal point.
Kilo Hecto Deka Unit Deci Centi Milli

Hint
Going from Hecto to Deci, three units right, involves a shift of the decimal point three units right. Thus, 382 hectoliters is equivalent to 382,000 deciliters. This idea works not only for liters but other metric measurements.

Word of Advice
Memorize the multiples, subdivisions and prefixes of the basic units in the metric system (See Table A.1). To remember prefixes, remember the phrase "**K**ing **H**enry **D**ied **M**onday **D**rinking **C**hocolate **M**ilk."

Metric units of measure for…
Length: meter (m), a little longer than a yard
Volume: liter (l), a little more than a quart
Weight: gram (g), about weight of paper clip
Time: second (sec)

To convert temperatures from Celsius to Fahrenheit or Fahrenheit to Celsius:
Think: use formulas

$$C = \frac{5(F-32)}{9} \text{ and } F = \frac{9}{5}C + 32$$

A.1 Exercises

1. a. 1000 (kilo means thousand)
   b. 0.001 (milli means one-thousandth)
   c. 100 (centi means one-hundredth)
   d. 1000 (kilo means thousand)

3. The height of a professional basketball play is (c) 200 cm.

5. The diamter of an aspirin tablet is (a) 1 cm.

7. The weight of an average human male is (a) 70 kg.

9. a. 8 km = 8000 m (1 km = 1000 m)
   b. 4 m = 400 cm (1 m = 100 cm)
   c. 3409 cm = 34.09 m (1 cm = 0.01 m)
   d. 49.4 mm = 4.94 cm (1 mm = 0.1 cm)

11. The bed is 2.10 m. (1 cm = 0.01 m)

13. The race is 1500 m. (1 km = 1000 m)

15. a. $300(52.5) = 15{,}750\text{cm} = 157.5$ m
    b. $50(52.5) = 2625\text{cm} = 26.25$ m
    c. $30(52.5) = 1575\text{cm} = 15.75$ m

17. A meter is $100 \text{ cm} = 10 \cdot 10 \text{ cm}$, so a cubic meter = $(10 \cdot 10)^3 \text{ cm}^3 = 1000(10^3 \text{cm}^3)$ = 1000 liters. Note: $10^3 \text{ cm}^3 = 1000 \text{ cm}^3$ = 1 liter.

19. 1000 ml = 1 liter, so 1 liter of sea water weights 1 $1 \cdot 3.5$ g/l = 3.5 g.

21. The volume is $50 \cdot 20 \cdot 10 = 10{,}000 \text{ cm}^3$ = 10 liters.

23. Since 5/20 = 1/4, 1/4 of a liter or 250 ml are needed.

25. a. 14 kg = 14,000 g (1 kg = 1000 g)
    b. 4.8 kg = 4800 g (1 kg = 1000 g)
    c. 2.8 g = 0.0028 kg (1 g = 0.001 kg)
    d. 3.9 g = 3900 mg (1 g = 1000 mg)

27. 1 liter = 1000 cm^3, so 1 liter weighs 1000 g or 1 kg.

29. The water needed to fill a 1-liter bottle weights (b) 1 kg. (See problem 27.)

31. $59°\text{F} = \frac{5}{9}(59-32) = \frac{5}{9} \cdot 27 = 15°\text{C}$

33. $86°\text{F} = \frac{5}{9}(86-32) = \frac{5}{9}\cdot 54 = 30°\text{C}$

35. $-22°\text{F} = \frac{5}{9}(-22-32) = \frac{5}{9}\cdot(-54) = -30°\text{C}$

37. $10°\text{C} = \frac{9}{5}\cdot 10 + 32 = 18 + 32 = 50°\text{F}$

39. $-10°\text{C} = \frac{9}{5}\cdot(-10) + 32 = -18 + 32 = 14°\text{F}$

41. $500°\text{F} = \frac{5}{9}(500-32) = \frac{5}{9}\cdot 468 = 260°\text{C}$

43. $41°\text{C} = \frac{9}{5}\cdot 41 + 32 = 73.8 + 32 = 105.8°\text{F}$

45. $98.6°\text{F} = \frac{5}{9}(98.6-32) = \frac{5}{9}\cdot 66.6 = 37°\text{C}$

47. a. $41°\text{F} = \frac{5}{9}(41-32) = \frac{5}{9}\cdot 9 = 5°\text{C}$

b. $212°\text{F} = \frac{5}{9}(212-32) = \frac{5}{9}\cdot 180 = 100°\text{C}$

49. $-78°\text{C} = \frac{9}{5}\cdot(-78) + 32$

$= -140.4 + 32 = -108.4°\text{F}$

51. $2.75(31) + 71.48 = 156.73$, so the female with a 31 cm long humerus was about 157 cm tall.

53. (i) A person ≈ (d) 72 kg
(ii) A book ≈ (c) 1 kg
(iii) A small automobile ≈ (e) 1000 kg
(iv) A common pin ≈ (a) 80 mg
(v) An orange ≈ (b) 68 g

Section A.2 Convert If You Must

Study Tips

Most people think that changing to the metric system means that you will have to be converting units from one system to the other. But you buy sodas and wine by the liter and you do not convert it to quarts! Conversions should be done only to give you a good idea of the relative size of the units involved.

To convert measurement from metric to U.S. and from U.S. to metric:
Think: use unit and metric equivalents
Note: Table A.5 will give you the conversion factors you need.

Hint
Keep in mind that the inch is defined to be 2.54 cm exactly, that is, 1 in. = 2.54 cm. All the other items in table are approximate values, so refer to the examples to know how to do computations with approximate numbers.

A.2 Exercises

1. 8 in. $= 8(2.54) = 20.32$ cm

3. 12 cm $= 12(0.394) = 4.73$ in.

5. 51 yd $= 51(0.914) = 46.61$ m

7. 3.7 m $= 3.7(1.09) = 4.03$ yd

9. 4 mi $= 4(1.61) = 6.44$ km

11. 3.7 km $= 3.7(0.621) = 2.30$ mi

13. 6 lb $= 6(0.454) = 2.72$ kg

15. 5 kg $= 5(2.20) = 11.00$ lb

17. 5 qt $= 5(0.946) = 4.73$ liters

19. 8.1 liters $= 8.1(1.06) = 8.59$ qt

21. 75 cm $= 75(0.394) \div 12 = 2.46$ ft

23. $1\ \text{in}^3 = (2.54)^3 = 16.39\ \text{cm}^3$

25. 2 yd $= 2(36)(2.54) = 182.88$ cm

27. 78 cm $= 78(0.394) \div 36 = 0.85$ yd

29. $40 \text{ mph} = 40(1.61) = 64.4 \approx 64 \text{ km/hr}$

31. $63 \text{ yd} = 63(0.914) = 57.6 \text{ m}$

33. $8848 \text{ m} = 8848(1.09)(3) = 28{,}900 \text{ ft}$

35. $1234 \text{ lb} = 1234(0.454) = 560 \text{ kg}$

37. $30 \text{ mi} = 30(1.61) = 48.3 \text{ km}$

39. $52 \text{ kg} = 52(2.20) = 114 \text{ lb}$

41. $2(52.3 + 96.84) = 298.28 \approx 298 \text{ m}$
$= 298(1.09) = 325 \text{ yd}$

43. 14 tons =
$14 \cdot 2000 \div 2200 = 12.7$ metric tons

45. $\dfrac{50(1.61)}{4(0.946)} = 21.3 \text{ km/liter}$

47. $1900 \text{ m} = 1.9 \text{ km} = 1.9(0.621) = 1.18 \text{ mi}$

49. $316 \text{ ft} = 3792 \text{ in.}$, so $316 \text{ ft } 5\frac{3}{4} \text{ in.} =$
3797.75 in., or
$3797.75(2.54) = 9646.285 \text{ cm} \approx 96.5 \text{ m}$

51. $39 \text{ in.} = 39(2.54) \approx 99.1$
$23 \text{ in.} = 23(2.54) \approx 58.4$
$33 \text{ in.} = 33(2.54) \approx 83.8$
To three significant digits, the measurements would be
99.1 cm –58.4 cm –83.8 cm.

53. $15 \text{ mi} = 15(1.61) \approx 24 \text{ km}$

55. $4 \text{ pt} = 4(0.5) = 2 \text{ qt}$ and $4.5 \text{ pt} = 4.5(0.5) = 2.25 \text{ qt}$, since 1 qt is about a liter, you should drink between 2 and 2.25 liters per day.

57. 1234 lb =
$1234(0.454)(1000) \approx 560{,}000 \text{ g}$